Mathematics Study Resources

Volume 19

Series Editors

Kolja Knauer, Departament de Matemàtiques Informàtic, Universitat de Barcelona, Barcelona, Spain

Elijah Liflyand, Department of Mathematics, Bar-Ilan University, Ramat-Gan, Israel

This series comprises direct translations of successful foreign language titles, especially from the German language. Powered by advances in automated translation, these books draw on global teaching excellence to provide students and lecturers with diverse materials for teaching and study.

Dirk W. Hoffmann

Limits of Mathematics

A Journey Through the Key Areas of Mathematical Logic

Dirk W. Hoffmann
Fakultät für Informatik und Wirtschaftsinformatik
Hochschule Karlsruhe - Technik und Wirtschaft
Karlsruhe, Germany

ISSN 2731-3824　　　　　　ISSN 2731-3832　(electronic)
Mathematics Study Resources
ISBN 978-3-662-70998-6　　ISBN 978-3-662-70999-3　(eBook)
https://doi.org/10.1007/978-3-662-70999-3

Translation from the German language edition: "Grenzen der Mathematik. Eine Reise durch die Kerngebiete der mathematischen Logik" by Dirk W. Hoffmann, © Springer 2018. Published by Springer Berlin Heidelberg. All Rights Reserved.

© The Editor(s) (if applicable) and The Author(s), under exclusive license to Springer-Verlag GmbH, DE, part of Springer Nature 2025

This work is subject to copyright. All rights are solely and exclusively licensed by the Publisher, whether the whole or part of the material is concerned, specifically the rights of translation, reprinting, reuse of illustrations, recitation, broadcasting, reproduction on microfilms or in any other physical way, and transmission or information storage and retrieval, electronic adaptation, computer software, or by similar or dissimilar methodology now known or hereafter developed.
The use of general descriptive names, registered names, trademarks, service marks, etc. in this publication does not imply, even in the absence of a specific statement, that such names are exempt from the relevant protective laws and regulations and therefore free for general use.
The publisher, the authors and the editors are safe to assume that the advice and information in this book are believed to be true and accurate at the date of publication. Neither the publisher nor the authors or the editors give a warranty, expressed or implied, with respect to the material contained herein or for any errors or omissions that may have been made. The publisher remains neutral with regard to jurisdictional claims in published maps and institutional affiliations.

This Springer imprint is published by the registered company Springer-Verlag GmbH, DE, part of Springer Nature.
The registered company address is: Heidelberger Platz 3, 14197 Berlin, Germany

If disposing of this product, please recycle the paper.

Preface

Perceiving the impossible is an intellectual achievement that sets humans apart. For instance, Einstein's theory of relativity and Heisenberg's uncertainty principle impose physical boundaries we will never surpass. As disturbing as they are, these discoveries possess an irresistible allure; throughout human history, the impossible has captivated more people than the possible. However, such phenomena are not limited to physics. Twentieth-century mathematical logic has shown that similar negative results impact nearly all branches of mathematics. Today, we know that the notion of truth and the notion of provability do not coincide, even for seemingly straightforward theories such as number theory. It is impossible to encapsulate mathematics within a formal system in which all true statements are provable, and all false statements are not.

I invite you on a journey through the realms of mathematical logic. My goal when writing this book was to present the concepts, methods, and findings of this fascinating branch of mathematics as clearly as possible without sacrificing depth. Wherever appropriate, I have motivated definitions and theorems with examples and placed them in their factual and historical context through numerous side notes. Additionally, I have chosen to present theorems that play only a marginal role in a sketchy manner or to indicate where an elaborated proof can be found. In this sense, the book cannot replace the formally precise literature in mathematical logic in all aspects – and it does not intend to. Above all, I aim to convey the fascination that this branch of mathematics undoubtedly exudes. It is up to you, dear readers, to judge how much I have succeeded in this endeavor.

I want to take the opportunity to thank my publisher, as this text would not exist without permitting me to prepare English translations of two of my logic books initially published in German. One of these books is the one you are reading right now. It is the revised translation of the third edition of my German language book *Grenzen der Mathematik* [103]. The other one is *Gödel's Incompleteness Theorems – A Guided Tour Through Kurt Gödel's Historic Proof* [104], the revised translation of the second edition of my German book *Die Gödel'schen Unvollständigkeitssätze* [101]. It details Gödel's paramount article from 1931, which sent shockwaves through large parts of mathematics that still reverberate today. Gödel's two incompleteness theorems are also central to this book, as their negative propositions permeate almost all areas of mathematical logic. However, I kept their presentation concise due to the wide range of covered topics. If you become as mesmerized by Gödel's incompleteness theorems as I became many years ago, you might find my other book an enjoyable follow-up read.

For me, as a native German speaker, the translation was a considerable linguistic challenge with an open outcome. At the beginning of the project, I was unsure whether I could complete it to a reasonable standard. After several months of hard work, however, I finished the manuscripts with a quality I am personally satisfied with, and I am pleased to present the result to an international readership. Undoubtedly, this book is not perfect. I am therefore grateful to any reader pointing out errors or suggesting improvement.

Karlsruhe, 1 January 2025   Dirk W. Hoffmann

Symbols

 Definition

 Theorem, Lemma, Corollary

 Easy exercise

 Moderate exercise

 Hard exercise

Solutions to the exercises

Download the solutions to the exercises on the website for this book:
www.dirkwhoffmann.de/LM

Contents

1 Historic Notes — **1**
  1.1 Truth and Provability 1
  1.2 The Path to Modern Mathematics 6
      1.2.1 Riddles of the Continuum 6
      1.2.2 Chaising the Infinite 12
      1.2.3 The Power of Symbols 24
      1.2.4 The Dawn of a New Century 29
      1.2.5 Foundational Crisis 33
      1.2.6 Axiomatic Set Theory 38
      1.2.7 Hilbert's Program 40
      1.2.8 The Limits of Computability 48
      1.2.9 Risen from Ruins 55
  1.3 Exercises 61

2 Formal Systems — **65**
  2.1 Concepts 65
  2.2 Decision Procedures 76
  2.3 Propositional Logic 80
      2.3.1 Syntax and Semantics 80
      2.3.2 Propositional-Logic Calculus 86
  2.4 First-Order Predicate Logic 96
      2.4.1 Syntax and Semantics 97
      2.4.2 Predicate-Logic Calculus 103
  2.5 Predicate Logic with Equality 106
  2.6 Higher-Order Predicate Logic 110
      2.6.1 Syntax and Semantics 110
      2.6.2 Henkin Semantics 114
  2.7 Exercises 117

3 Foundations of Mathematics — **127**
  3.1 Peano Arithmetic 128
      3.1.1 Syntax 128
      3.1.2 Semantics 129
      3.1.3 Axioms and Inference Rules 133
  3.2 Axiomatic Set Theory 141
      3.2.1 Zermelo-Fraenkel Set Theory 142

		3.2.1.1	ZF Axioms 143

            3.2.1.1   ZF Axioms 143
            3.2.1.2   Axiom of Choice 151
            3.2.1.3   Set Theory as the Foundation of Mathematics 157
            3.2.1.4   Embedding of the Natural Numbers 165
        3.2.2   Ordinal Numbers 167
            3.2.2.1   Definition and Properties 167
            3.2.2.2   Towards Infinity 172
            3.2.2.3   Order Types and Well-Orders 178
            3.2.2.4   Transfinite Induction 181
        3.2.3   Cardinal Numbers 184
   3.3   Exercises 186

4 Proof Theory   191
   4.1   Gödel's Incompleteness Theorems 191
   4.2   The First Incompleteness Theorem 192
        4.2.1   Arithmetization of Syntax 195
        4.2.2   Primitiv-Recursive Functions 200
        4.2.3   Arithmetic Representability 202
        4.2.4   Gödel's Diagonal Argument 209
        4.2.5   Rosser's Trick 214
        4.2.6   The Diagonalization Lemma 218
        4.2.7   Tarski's Truth Predicate 222
        4.2.8   Berry Paradox 227
   4.3   The Second Incompleteness Theorem 231
        4.3.1   Hilbert–Bernays-Löb Provability Conditions 233
        4.3.2   Löb's Theorem 237
   4.4   Common Misconceptions 240
   4.5   Goodstein's Theorem 245
   4.6   Exercises 253

5 Computability Theory   257
   5.1   Models of Computation 258
        5.1.1   Turing Machines 258
            5.1.1.1   Extensions to the Basic Model 261
            5.1.1.2   Alternative Representations 263
            5.1.1.3   The Universal Turing Machine 266
        5.1.2   Register Machines 271
   5.2   Church's Thesis 273
   5.3   Limits of Computability 280
        5.3.1   Halting Problem 280
        5.3.2   Rice's Theorem 282
   5.4   Impact on Mathematics 284
        5.4.1   Undecidability of PL1 286
        5.4.2   Incompleteness of Arithmetic 292

	5.4.3	Hilbert's Tenth Problem 300
		5.4.3.1 Diophantine Representability 303
		5.4.3.2 Encoding of Register Machines 305
5.5	Exercises 316	

6 Algorithmic Information Theory 325
6.1 Algorithmic Complexity 326
6.2 Chaitin's Constant 333
6.3 Incompleteness of Formal Systems 342
6.4 Exercises 345

7 Model Theory 349
7.1 Meta-Results on Predicate Logic 350
    7.1.1 Model Existence Theorem 353
    7.1.2 Compactness Theorem 355
    7.1.3 Löwenheim-Skolem Theorem 358
7.2 Nonstandard Models of PA 361
    7.2.1 Denumerable Nonstandard Models 361
    7.2.2 Uncountable Nonstandard Models 364
7.3 Skolem's Paradox 370
7.4 Boolean Models 376
    7.4.1 Boolean Algebras 378
    7.4.2 Boolean Sets 383
    7.4.3 Boolean Semantics 384
7.5 Forcing 387
7.6 Exercises 393

Bibliography 399

Image Credits 409

Name Index 411

Timeline 413

Index 415

1 Historic Notes

"Mathematics takes us still further from what is human, into the region of absolute necessity, to which not only the world, but every possible world, must conform."

Bertrand Russell [178]

1.1 Truth and Provability

Few phenomena captivate scientists as much as the mysteries of nature. Driven by curiosity, we constantly search for structure in a world with more questions than answers. However, summarizing the developments of the last few centuries, we can proudly look back on a remarkable track record. Over and over again, scientists have succeeded in reducing complex relationships to simpler, less complex ones, thereby providing adequate explanations. In doing so, science has not only changed our daily lives on a massive scale but has also created the foundation of our modern view of the world.

Until the beginning of the twentieth century, no one seriously doubted that nature follows elementary rules. After all, it corresponds to our intuition and experience that a cause precedes an effect, and nothing in the world happens without a reason. This principle of sufficient reason (Latin: *principium rationis sufficientis*, French: *principe de la raison suffisante*) is the unspoken assumption of all natural sciences. Without it, the scientific method would be a blunt sword.

The principle of sufficient reason personifies itself in Gottfried Wilhelm Leibniz, probably the world's last universal scholar. The principle is a cornerstone of Leibniz's philosophy in the form of a metatheoretical tenet. It states that the world we live in is the most perfect possible, a world of complete harmony, in which not only every physical process has a cause but also every metaphysical truth has a reason. Accordingly, complex statements are never factual on their own. Instead, they are always justified by other statements whose truth has already been established.

"And that of sufficient reason, in virtue of which we hold that there can be no fact real or existing, no statement true, unless there be a sufficient reason, why it should be so and not otherwise, although these reasons usually cannot be known by us." [124]

Gottfried Wilhelm Leibniz
(1646 – 1716)

Figure 1.1: Gottfried Wilhelm Leibniz ranks among the most famous and universally talented scholars of the late 17th and early 18th centuries. A sheer endless list of publications and letters forms an unparalleled legacy, covering areas from philosophy, mathematics, and the natural sciences to history and law.

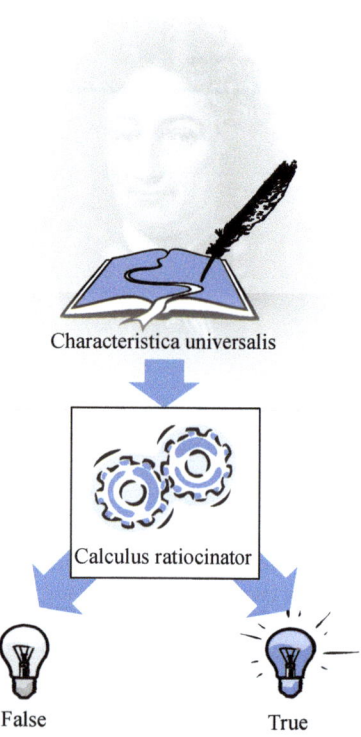

Figure 1.2: Throughout his life, Leibniz was captivated by the invention of a *characteristica universalis*, a universal language capable of representing the objects, concepts, and relationships of the material world on a symbolic level. He believed in the possibility of establishing a set of rules powerful enough to systematically determine the truth value of a symbolic statement in the form of a mechanical procedure: the *calculus ratiocinator*.

In his visionary thinking, Leibniz was far ahead of his time. He dreamed of a *characteristica universalis*, a universal language capable of capturing all facets of human knowledge. Leibniz had no spoken language in mind. Instead, he conceived a symbolic notation in which individual alphabetic characters represent objects or concepts of the natural world, and their relationships become visible on the symbolic level. Due to the formal nature of his envisioned language, Leibniz believed that the truth content of a statement could be calculated systematically by applying a fixed set of rules, the *calculus ratiocinator* (Figure 1.2).

Leibniz was well aware of the scale of his project and never attempted to achieve his ambitious goal alone. Nevertheless, he had concrete ideas for its realization. The first step was to create an encyclopedia that would unite all the knowledge available to humanity up to this point. The second step was to define a formal language powerful enough to describe all the concepts and relationships of the created knowledge base. The final step was to transfer the logical rules of inference to the symbolic level, thereby establishing the formal reasoning apparatus that would allow true statements to be generated and verified mechanically. Leibniz believed he could complete the project with a group of selected scientists in about five years. However, he was never given a chance during his lifetime, so the characteristica universalis remained in the realm of dreams. As perhaps the most extraordinary visionary of his time, Gottfried Wilhelm Leibniz died on November 14, 1716, aged 70, and with him his ambitious project.

It would take another 200 years before his dream was at least partially realized. In the nineteenth century, advancements in symbolic logic paved the way for the development of formal systems that, in many respects, closely resemble Leibniz's concept of a characteristica universalis. Propositional and predicate logic emerged as artificial languages capable of encoding mathematical statements in symbolic form and deriving logical conclusions through the application of well-defined rules of inference. Both logics are the subject of Chapter 2. In Chapter 3, we will utilize the predicate-logic apparatus to construct axiomatic set theory. The resulting formal system will be powerful enough to describe all areas of classical mathematics and is widely regarded as the formal foundation for all modern mathematics.

As mathematics became increasingly formalized, new questions emerged that focused not on the theorems derivable within a formal system but on the properties and limitations of the systems themselves. This branch of research flourished at the beginning of the twentieth century and is known today as *metamathematics*. The insights gained since then have been both profound and unsettling.

1.1 Truth and Provability

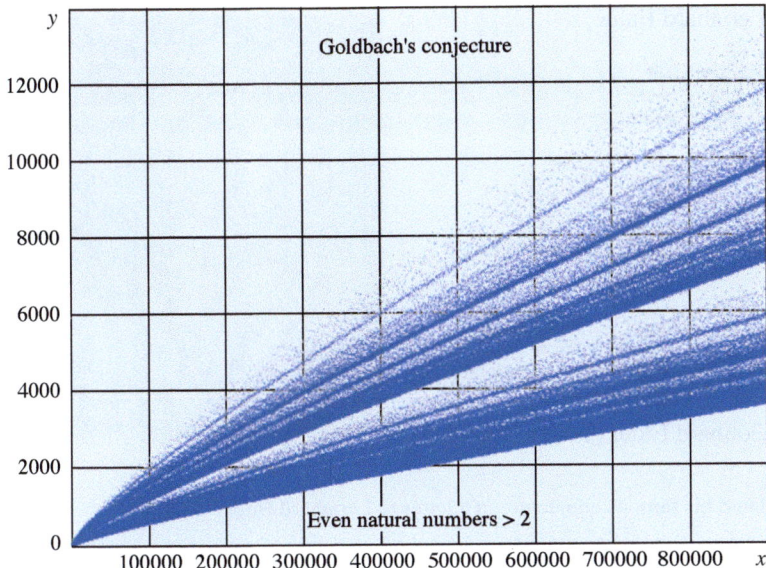

Figure 1.3: Goldbach's conjecture states that all even numbers $n > 2$ can be expressed as the sum of two prime numbers. The diagram nearby plots the even natural numbers on the x-axis and the number of possible decompositions on the y-axis. Goldbach believed that the x-axis remains free of data points. Even though there is strong support for the truth of the conjecture, a formal proof is yet to be found.

Well into the twentieth century, hardly any mathematician seriously doubted that, if searched for long enough, either a proof or a refutation could be found for every mathematical statement. It was the unwritten dogma of mathematics that *truth* and *provability* formed a harmonious relationship. Today, we know that the concepts of truth and provability do not fully align, even for simple theories such as number theory. It is impossible to encapsulate mathematics within a formal system in which all true mathematical statements, and only those, can be proven.

The discoveries of the twentieth century have altered our mathematical perspective for good. By uncovering fundamental limitations that are impossible to overcome, these findings hold a significance in mathematics comparable to the theory of relativity in physics. Today, we understand that a calculus ratiocinator cannot exist. Leibniz's vision of mechanizable mathematics, as alluring as it may be, will remain an unattainable dream forever.

The ideas leading to this result form the core of this book, and we will explore them in detail in the chapters to come. This much in advance: These ideas are so fundamental that there is no escape; mathematics resists its complete formalization.

At this juncture, let us look at two examples to understand the implications for ordinary mathematics.

From a letter from Christian Goldbach to Leonhard Euler

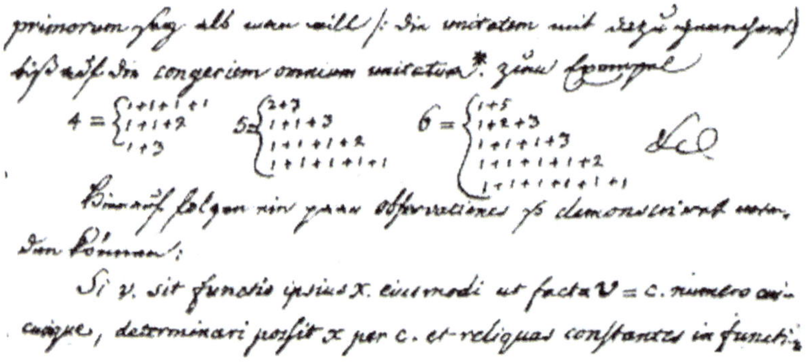

Leonhard Euler (1707 – 1783)

Figure 1.4: In 1742, Christian Goldbach formulated his famous conjecture in a letter to Leonhard Euler.

 Conjecture 1.1 (Goldbach)

Every even natural number $n > 2$ can be written as the sum of two prime numbers.

Goldbach's conjecture is one of the oldest and most famous problems in number theory (Figure 1.3). It is named after the German mathematician Christian Goldbach, who proposed in a letter to Leonhard Euler in 1742 that every natural number greater than two can be written as the sum of *three* prime numbers[1] (Figure 1.4). Conjecture 1.1 is also known as *Goldbach's strong conjecture*, as it implies the validity of the original variant.

The second example also originates from number theory and is just as prominent:

 Conjecture 1.2 (Twin prime conjecture)

There are infinitely many numbers n such that both n and $n+2$ are prime numbers.

Table 1.1 provides an overview of the first 35 twin primes.

---

[1] In Goldbach's definition, 1 is considered a prime number. Otherwise, the case $n = 4$ would already refute his conjecture.

1.1 Truth and Provability

"Cubum autem in duos cubos, aut quadratoquadratum in duos quadratoquadratos, et generaliter nullam in infinitum ultra quadratum potestatem in duos eiusdem nominis fas est dividere. Cuius rei demonstrationem mirabilem sane detexi. Hanc marginis exiguitas non caperet."

Pierre de Fermat
(1607 – 1665)

Figure 1.5: Pierre de Fermat wrote his famous Latin quote onto the margin of his *Arithmetica* edition in 1637 (see Section 1.2.1). Translated into English, it reads: *"It is impossible to separate a cube into two cubes, or a fourth power into two fourth powers, or in general, any power higher than the second, into two like powers. I have discovered a truly marvelous proof of this, which this margin is too narrow to contain."* For over 300 years, mathematicians searched to no avail for Fermat's *"truly marvelous proof."* Meanwhile, they consider it a certainty that no short proof of his conjecture exists.

Each of the two conjectures makes a statement about the natural numbers and must be substantively true or false. Nevertheless, all efforts to reveal their nature of truth have been in vain. Even more troubling, we do not know whether it is possible to prove the conjectures using the classical tools of number theory. The intensity with which both have so far evaded a solution may nourish the suspicion of unprovability but does not provide certainty, of course.

Another famous conjecture in number theory also resisted all attempts at proof for over three hundred years. In 1637, the French mathematician Pierre de Fermat asserted that the equation

$$a^n + b^n = c^n$$

has no solutions in the positive integers for $n > 2$ (Figure 1.5). It was not until 1995 that the Briton Andrew Wiles proved the Taniyama-Shimura conjecture, which bears *Fermat's Last Theorem* as a corollary. It remains to be seen whether proof will be found for Goldbach's conjecture or the conjecture about the existence of an infinite number of twin primes. Even though there is growing evidence, uncertainty prevails.

The discovery of incompleteness is the greatest achievement of mathematical logic in the twentieth century and undoubtedly one of the most astounding mathematical results of all time. Chapter 4 will explore this subject in detail and carefully derive the key results.

Chapter 5 will move one step further and supplement the notion of provability with the concept of *computability*. This key term is crucial for

(3, 5)	(5, 7)
(11, 13)	(17, 19)
(29, 31)	(41, 43)
(59, 61)	(71, 73)
(101, 103)	(107, 109)
(137, 139)	(149, 151)
(179, 181)	(191, 193)
(197, 199)	(227, 229)
(239, 241)	(269, 271)
(281, 283)	(311, 313)
(347, 349)	(419, 421)
(431, 433)	(461, 463)
(521, 523)	(569, 571)
(599, 601)	(617, 619)
(641, 643)	(659, 661)
(809, 811)	(821, 823)
(827, 829)	(857, 859)
(881, 883)	

Table 1.1: The twin primes between 0 and 1000

us in two respects. On the one hand, it will take us down an alternative path to the limits of provability, much shorter than the one we chose in Chapter 4. On the other hand, it plays a central role in computer science, where the limits of computability have a tangible effect. Today, we know it is impossible to formulate an algorithm that correctly decides for every computer program whether or not it fulfills a given functional property. Seemingly simple tasks such as checking for infinite loops are algorithmically impossible, which is why even the most advanced compilers can hardly do more than syntactic source-code checking. Again, we face fundamental limitations we cannot overcome.

1.2 The Path to Modern Mathematics

Before devoting ourselves to the technical details of the outlined ideas, let us revisit the turbulent history of mathematical logic. Only then will it become possible to adequately contextualize and understand the results in all their dimensions. So, let's not waste any time!

1.2.1 Riddles of the Continuum

We start our journey through the history of mathematics in Greece in the third century, where a thirteen-volume work was written that later laid the foundations for modern algebra. We are talking about the *Arithmetica*, a collection of over one hundred algebraic puzzles and their solutions (Figure 1.6). Only volumes 1 to 3 and 8 to 10 are still available in the original today. Arabic translations have been found for volumes 4 to 7, but the remaining three disappeared without a trace. The Arithmetica is the work of Diophantos of Alexandria, about whom we know little. Only a riddle verse from the time after his death gives us tentative clues about the course of his life. In an English translation, it reads as follows [151]:

Figure 1.6: The *Arithmetica* consists of thirteen volumes in which Diophantus of Alexandria compiled more than one hundred algebraic puzzles with their solutions. The general solvability of Diophantine equations is part of Hilbert's tenth problem, discussed in detail in Section 5.4.3.

"Here lies Diophantus,' the wonder behold. Through art algebraic, the stone tells how old: 'God gave him his boyhood one-sixth of his life, One twelfth more as youth while whiskers grew rife; And then yet one-seventh ere marriage begun; In five years there came a bouncing new son. Alas, the dear child of master and sage After attaining half the

measure of his father's life chill fate took him. After consoling his fate by the science of numbers for four years, he ended his life.'"

Denoting Diophantos' age by x, the riddle verse can be translated into the following equation:

$$x = \frac{x}{6} + \frac{x}{12} + \frac{x}{7} + 5 + \frac{x}{2} + 4$$

Multiplying by 84 eliminates all fractions:

$$84x = 14x + 7x + 12x + 420 + 42x + 336$$

We can now determine his age as the solution to the equation

$$9x - 756 = 0 \tag{1.1}$$

which is 84. However, whether Diophantus really reached that age and had to cope with the pain of seeing his son die, we will likely never know.

Equation (1.1) is a trivial example of what is today called a *Diophantine equation*. Such an equation has the general form

$$p(x_1, x_2, \ldots, x_n) = 0, \tag{1.2}$$

where p denotes a multivariable polynomial with integer coefficients. The solution to a Diophantine equation is the set of integer roots of p.

When discussing equations in the subsequent sections, we will adjust the symbol set slightly whenever appropriate and write, e.g., x for x_1 and y for x_2. Equation

$$x_1^3 + x_2^3 + x_1 + x_2 - 380 = 0$$

then reads as follows:

$$x^3 + y^3 + x + y - 380 = 0 \tag{1.3}$$

This equation has a geometric interpretation that solves a problem from the fourth book of the Arithmetica. As shown in Figure 1.7, x and y can be interpreted as the side lengths of two cubes whose combined volume is 370 and the sum of their side lengths is 10. With $x = 7, y = 3$ and $x = 3, y = 7$, the equation has exactly two solutions in the natural numbers.

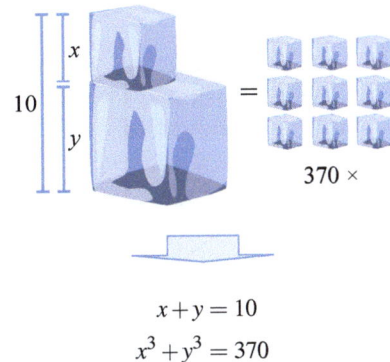

$x + y = 10$
$x^3 + y^3 = 370$

Figure 1.7: In the fourth volume of the Arithmetica, Diophantos asked to determine the side lengths x and y of two cubes such that the sum of the side lengths equals 10 and the sum of the cube volumes equals 370.

■ Pythagorean triple

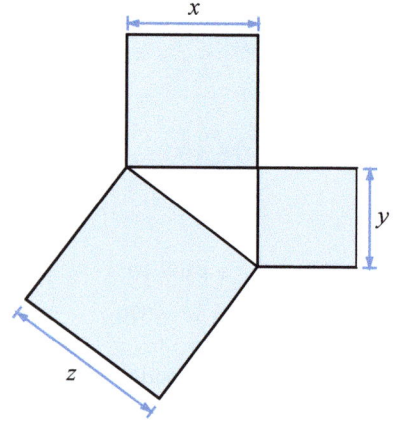

Pythagorean triples can be calculated using the formulas

$$x = m(u^2 - v^2)$$
$$y = m(2uv)$$
$$z = m(u^2 + v^2)$$

where m, u, v are positive natural numbers with $u > v$.

■ Examples

(m,u,v)			(x,y,z)		
1	2	1	3	4	5
1	3	1	8	6	10
1	3	2	5	12	13
1	4	1	15	8	17
1	4	2	12	16	20
1	4	3	7	24	25
2	2	1	6	8	10
2	3	1	16	12	20
2	3	2	10	24	26
2	4	1	30	16	34
2	4	2	24	32	40
2	4	3	14	48	50
...					

Figure 1.8: Pythagorean triples

The following Diophantine equation, for instance, has an infinite number of solutions:

$$x^2 + y^2 - z^2 = 0 \quad (1.4)$$

It refers to a problem from the second book of the Arithmetica, which is to divide a square into two squares with no change in area. The solutions to this equation are the so-called *Pythagorean triples*. According to the Pythagorean theorem, they comprise all triples of natural numbers that occur as the side lengths of right-angled triangles.

Generalizing equation (1.4) leads to

$$x^n + y^n - z^n = 0, \quad (1.5)$$

the legendary equation that gave rise to Pierre de Fermat's famous conjecture. Today, we know that it has no solutions in the positive integers for $n > 2$.

Notice that (1.5) is not an ordinary Diophantine equation since the variable n appears as an exponent. It belongs to the larger group of *exponential Diophantine equations*, which we will encounter again in Section 5.4.3. There, we will discuss in detail whether the solvability of Diophantine equations can be determined by a systematic procedure. This much in advance: We will receive an astounding answer.

The fact that we exclusively seek solutions of Diophantine equations in integers today has no historical precedent. Diophantus asked the reader of the Arithmetica to solve the Pythagorean equation (1.4) for the case $z^2 = 16$. Under this constraint, the equation has no solution in the integers, and with $\frac{12}{5}$ and $\frac{16}{5}$ precisely one solution in the set of *rational numbers*.

Just like natural numbers, which allow us to count entities, rational numbers have a practical meaning, too: they arise naturally whenever two geometric lengths, p and q, are related to each other and, in this sense, are the basic algebraic building blocks of geometry.

At this point, we emphasize that fractional notation is only one of several possible representations. For example, we can write any rational number $\frac{p}{q}$ as a periodic decimal fraction:

$$\tfrac{1}{8} = 0{,}125 = 0{,}125\overline{0}$$
$$\tfrac{1}{3} = 0{,}3333\ldots = 0{,}\overline{3}$$
$$\tfrac{1}{1} = 1{,}\overline{0} = 0{,}\overline{9}$$

1.2 The Path to Modern Mathematics

Likewise, every periodic decimal fraction can be systematically converted into its fractional representation. E.g., to represent the number

$$x = 0{,}0\overline{238095} \qquad (1.6)$$

in the form $\frac{p}{q}$, we employ a simple trick. First, we multiply both sides by 1,000,000:

$$1{,}000{,}000\,x = 23{,}809.5\overline{238095}$$

Now, we subtract (1.6) from this equation, which makes the periodic part disappear:

$$999{,}999\,x = 23{,}809.5$$

This gives us the following representation for x:

$$x = \frac{23{,}809.5}{999{,}999} = \frac{238{,}095}{9{,}999{,}990} = \frac{1}{42}$$

Unlike natural numbers, rational numbers are *dense* on the number line. This means that we can approximate any point with arbitrary precision using a sequence of rational numbers. The approximation is always possible because for any two numbers $x, y \in \mathbb{Q}$, the arithmetic mean $\frac{x+y}{2}$ is also a rational number (Figure 1.9).

Nevertheless, the set of rational numbers has gaps. The Pythagoreans already knew that no rational number equals the length of the diagonal of a square with side length 1 (Figure 1.10). But what is this mysterious diagonal length all about? We know when multiplied by itself, the result is 2, legitimating us to denote the value by $\sqrt{2}$. The value of $\sqrt{2}$ is approximately

$$\sqrt{2} \approx 1{,}4142135623730950488016887242096980785 6$$

and the accuracy can be increased arbitrarily by adding more digits. However, we can never write down the value precisely, as $\sqrt{2}$ does not have a fractional representation. Its decimal fraction representation is non-periodic; it consists of an infinite number of irregularly occurring digits.

By closing the gaps between rational numbers, we arrive at the set of *real numbers* $\mathbb{R}$, familiar from classical analysis. Since it spans the entire number line without interruptions, the set of real numbers is called the *continuum*.

A closer look at the number $\sqrt{2}$ reveals another interesting aspect. It is a real-valued solution of the algebraic equation

$$x^2 - 2 = 0.$$

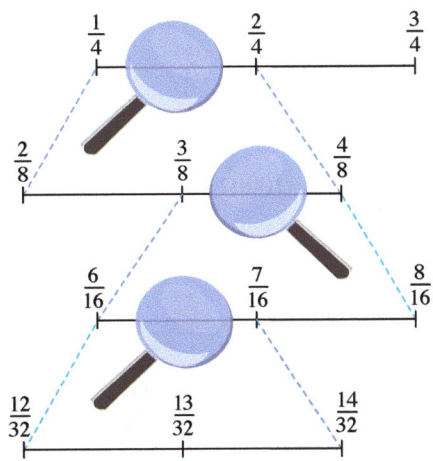

Figure 1.9: Because the arithmetic mean of any two rational numbers $x, y \in \mathbb{Q}$ is also rational, we can approximate any point with arbitrary precision.

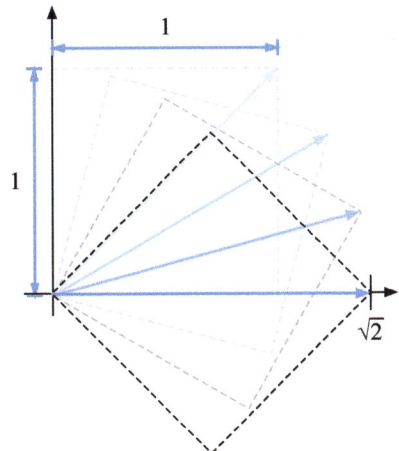

Figure 1.10: The rational numbers do not cover the number line seamlessly. For example, the diagonal length of a square with side length 1 cannot be expressed by a rational number.

$$L = \sum_{k=1}^{\infty} 10^{-k!}$$

$$= 0{,}1100010\ldots010\ldots010\ldots$$

with arrows pointing to digits at positions $1!$, $2!$, $3!$, $4!$, $5!$.

Joseph Liouville
(1809 – 1882)

Figure 1.11: In 1844, the French mathematician Joseph Liouville proved the existence of transcendental numbers.

This observation lets us step directly into the realm of *algebraic numbers*. A complex number x is said to be algebraic if it solves an equation of the form

$$a_n x^n + a_{n-1} x^{n-1} + \ldots + a_1 x + a_0 = 0 \qquad (1.7)$$

where all coefficients a_i originate from the set of integers.

It is easy to verify that every rational number $\frac{p}{q}$ is algebraic since it is the solution of the following equation:

$$q \cdot x - p = 0$$

With $\sqrt{2}$, we have already become acquainted with an irrational number that is also algebraic, which raises a **natural question**: Is every real number algebraic? One thing we know for sure: If there were a real number that was not the solution to an algebraic equation, it would not be easy to grasp, as we could neither write it down as a decimal fraction nor indirectly characterize it as the root of an algebraic expression.

One of the first to firmly believe in the existence of such *transcendental numbers* was Leonhard Euler. In particular, he assumed that the number $a^{\sqrt{b}}$ is not algebraic for all rational numbers $a \neq 1$ and all natural numbers b that are not square numbers. Nevertheless, he failed to prove his conjecture during his lifetime.

It was not until 1844 that Euler's conjecture became a certainty. This year, the French mathematician Joseph Liouville first succeeded in proving the existence of transcendental numbers beyond doubt [19]. Liouville carried out the proof constructively and devised a specific number that an algebraic equation could not describe (Figure 1.11). It is the famous number

$$L := \sum_{k=1}^{\infty} 10^{-k!},$$

now referred to as *Liouville number* after its discoverer. From 1844, transcendence was no longer a mere possibility but a mathematical reality.

However, Liouville's discovery was not to remain in isolation. In 1873, the French mathematician Charles Hermite proved the transcendence of Euler's famous number e, the basis of the natural logarithm. In 1882, the German mathematician Ferdinand von Lindemann made another fascinating discovery. He proved that the equation

$$\beta_1 e^{\alpha_1} + \ldots + \beta_n e^{\alpha_n} = 0,$$

1.2 The Path to Modern Mathematics

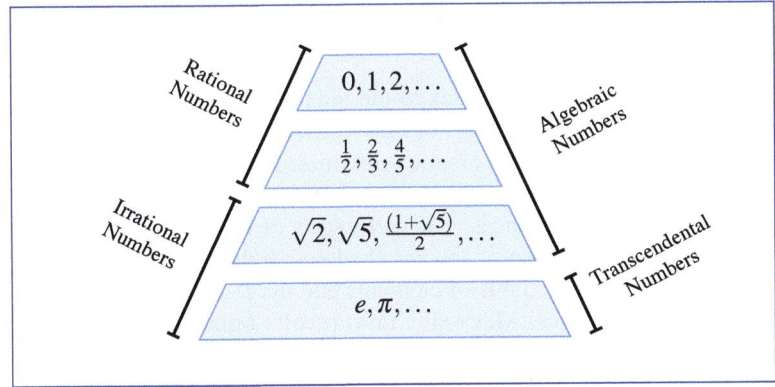

Figure 1.12: Real numbers divide themselves into rational numbers and irrational numbers. Every rational number is also algebraic, but not vice versa. For example, the number $\sqrt{2}$ can be represented as a solution to an algebraic equation but not as a fraction. Since 1844, we have been aware of the existence of numbers that are not solutions to an algebraic equation. They form the set of transcendental numbers and include, besides many others, Euler's number e and the number π.

in which $\alpha_1, \ldots, \alpha_n$ and $\beta_1, \ldots, \beta_n$ are algebraic numbers, only has the trivial solution

$$\beta_1 = \ldots = \beta_n = 0$$

if all α_i are pairwise distinct. This finding is the content of the famous Lindemann-Weierstrass theorem.

From this theorem and the known relationship

$$e^{i\pi} = -1,$$

it was possible to conclude that π must also be transcendental. At this point, two of the most important mathematical constants, Euler's number e and the number π, were identified as transcendental (Figure 1.12).

Lindemann's result also answered one of the most famous questions in geometry, called *squaring the circle*. The term refers to the challenge of constructing a square with the same area for a given circle under the constraint that the construction is carried out with a compass and straightedge in a finite number of steps (Figure 1.13). Since every length constructible this way can be characterized as the solution of an algebraic equation, the transcendence of $\sqrt{\pi}$, which is derivable from the transcendence of π, implies the unsolvability of the squaring problem [211]. Today, squaring the circle is a widely used idiom to describe an unsolvable problem.

The discovery of transcendental numbers quickly raised the question of how many of those elusive numbers actually exist. Is transcendence a scarce property of some distinguished numbers, or should it pervade the continuum silently and for a long time unnoticed like the hypothetical dark matter of our universe? Knowledge about infinity will provide us with an astonishing answer to this question. We will return to it in a moment.

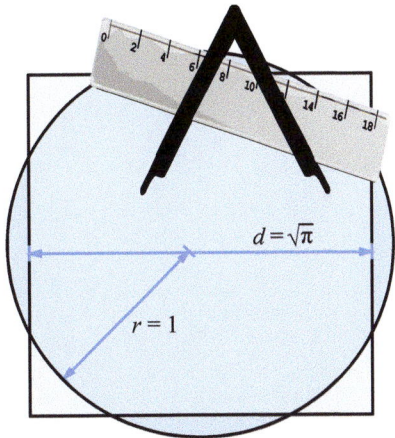

Figure 1.13: The problem of squaring the circle describes the challenge of constructing a square with the area of a given circle with a finite number of specific geometric operations. The transcendence of π suffices to derive the transcendence of $\sqrt{\pi}$, from which the squaring of the circle can be proved impossible.

1.2.2 Chaising the Infinite

Modern mathematics roots back to the nineteenth century, a century of sheer unlimited progress that not only provided new answers to social and political questions but also revolutionized economics and science. New discoveries created a spirit of optimism in all areas of engineering and the natural sciences. Mendel's laws of heredity and Darwin's discoveries on the origin of species shed new light on nature. The disclosure of the periodic table of elements laid the foundation for modern chemistry. In physics, Maxwell's laws revolutionized our conception of the world, and with the development of the first vaccine by Louis Pasteur, humanity took the horror out of deadly diseases. At the turn of the century, the scientific community believed itself to be on the verge of omniscience, and for many, it was only a matter of time before the world's last riddle would be solved. All-encompassing theories seemed within reach.

In the nineteenth century, mathematics also progressed rapidly. Cauchy and Weierstrass put the infinitesimal calculus on solid ground, and the infinitely large and the infinitely small were freed from their mystical aura. Riemann and Gauss gave geometry a new face through rigorous analytical methods, while Dedekind and Kronecker made significant contributions to number theory. It was a century of specialization in which interest in epistemological questions gradually faded.

Despite all the progress made, there was one thing the most precise of all sciences had yet to achieve: the creation of a unified foundation on which mathematics as a whole could be built. The fact that set theory provides us with such a foundation today is not a matter of course, and, as so often, it was mere chance that brought about the turning point.

It is a twist of history that a question in calculus gave rise to set theory, triggered by the French mathematician Jean Baptiste Fourier, who conjectured in 1822 that any function is representable in the form of a trigonometric series[2]. Fourier's conjecture was largely proven for continuous functions, and more and more mathematicians began to apply the results to the discontinuous case. The German mathematician Georg Cantor was one of them (Figure 1.14). Cantor pursued the plan of gradually weakening the continuity assumption until it would be eliminated entirely. His work soon bore fruit. First, he succeeded in showing that Fourier's conjecture applies to functions with a finite number of discontinuity points. Inspired by his initial achievement, he transferred his re-

---

[2] Today, we know that Fourier's original assumption was wrong, but its pioneering nature secures it a prominent spot in the history of science.

1.2 The Path to Modern Mathematics

 The German mathematician Georg Cantor was born in St. Petersburg on March 3, 1845. From 1862 to 1867, he attended universities in Zurich, Göttingen, and Berlin, where he was fortunate to study under renowned figures such as Karl Weierstrass, Ernst Eduard Kummer, and Leopold Kronecker. In 1867, he received his doctorate from the University of Berlin. Afterward, he moved to Halle, where he worked as a private lecturer, associate professor, and finally, full professor.

Cantor ranks among the most influential late nineteenth and early twentieth-century mathematicians. He is recognized as the founder of set theory, and with the concept of *cardinality*, he laid the foundation for dealing with the infinite. The notion of *countability* also traces back to Cantor, as does the *diagonalization method*, which we will employ several times in this book.

Cantor never hesitated to break new ground. Nevertheless, the high degree of misunderstanding, distrust, and hostility he encountered along his lonely path left scars on his psyche. It is a tragic aspect of his life that his teacher, Leopold Kronecker, was particularly rebellious against his new ideas and tried to fight Cantor with blind rage [226]. Kronecker, who saw him as a spoiler of youth, skillfully used his influence to prevent Cantor from transferring to the prestigious University of Berlin. As a result, Halle would be Cantor's first and last stop in his academic career.

Aged 39, Cantor fell ill with manic depression – a condition that would accompany him until the end of his life. Shortly after his seventieth birthday, he was admitted to the University Hospital in Halle after another outbreak of the disease. It was there that Georg Cantor died on January 6, 1918, at the age of 72.

sults to functions with an infinite number of such points. Cantor did not master the general case but succeeded in proving the assumption when the distribution of the points of discontinuity satisfied specific properties. By grouping the discontinuities into sets, initially called *manifolds*, he was able to show that their distribution properties could be associated with the structural properties of the constructed sets. Back then, Cantor's manifolds were still perceived by many mathematicians as strange obscurities that did not fit in with the conventional concepts of the time. There was still a long way to go before set theory could establish itself as the accepted foundation of mathematics.

Cantor had created a mathematical apparatus so general that he could examine both finite and infinite sets in a unified way. His crucial idea was to classify a set M based on its *cardinality*, denoted by:

$$|M|$$

The cardinality of a finite set corresponds to the number of its elements. For example, the following relationships apply:

$$M_1 = \emptyset \quad \Rightarrow \quad |M_1| = 0$$
$$M_2 = \{\square, \lozenge, \circ\} \quad \Rightarrow \quad |M_2| = 3$$
$$M_3 = \{2, 3, 5\} \quad \Rightarrow \quad |M_3| = 3$$

The sets M_2 and M_3 are *equipotent*, as they contain the same number of elements. In this and only in this case, we can establish a one-to-one

Georg Cantor
(1845 – 1918)

Figure 1.14: Georg Cantor is the founder of modern set theory. With his extensive investigation of the infinite, he gave mathematics its contemporary face.

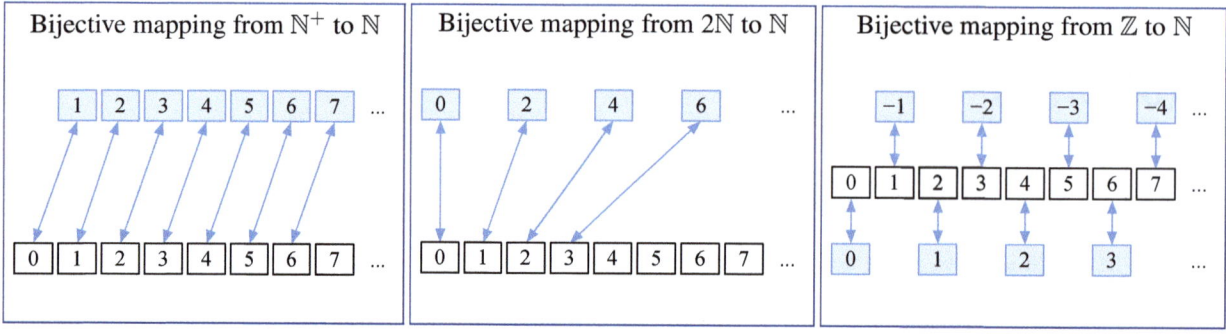

Figure 1.15: The existence of a bijective mapping between the natural numbers, the positive numbers, the even numbers, and the integers proves these sets to be equipotent.

mapping between the elements of both sets. For our example, such a mapping may look as follows:

$$\square \mapsto 2, \quad \diamondsuit \mapsto 3, \quad \circ \mapsto 5$$

If the number of elements does not match, such an assignment is impossible. Hence, we can link the concept of cardinality to the existence of certain mappings:

 Definition 1.1 (Cardinality)

> We call two sets M_1 and M_2 *equipotent*, denoted as
>
> $$|M_1| = |M_2|$$
>
> if there exists a bijective mapping $f : M_1 \to M_2$. We write
>
> $$|M_1| \leq |M_2|$$
>
> if there exists an injective mapping $f : M_1 \to M_2$.

By definition, two infinite sets are equipotent if their elements can be mapped one-to-one to each other. At first glance, this approach seems unnatural and unnecessarily cumbersome. At a second glance, however, it becomes clear that the definition does not rely on explicitly counting the elements. This property is the key to comparing the cardinality of two sets, even if they contain infinitely many elements.

Surprising consequences arise. First, let us consider the set of natural numbers $\mathbb{N}$ and the set of positive integers $\mathbb{N}^+$. Although the set $\mathbb{N}^+$ is

1.2 The Path to Modern Mathematics

a proper subset of $\mathbb{N}$, it maps bijectively to the natural numbers via the following function (Figure 1.15 left):

$$f : x \mapsto (x-1)$$

Similarly, we can create a mapping between $2\mathbb{N}$, the set of even non-negative numbers, and $\mathbb{N}$ (Figure 1.15 center):

$$f : x \mapsto \frac{x}{2}$$

Likewise, as shown in Figure 1.15 (right), we can map the integers bijectively to the set of natural numbers. The following mapping is one of – you guessed it – infinitely many:

$$f : x \mapsto \begin{cases} -2x-1 & \text{if } x < 0 \\ 2x & \text{if } x \geq 0 \end{cases}$$

Thus, the set of the natural numbers and the set of the integers are equipotent. But not only that. The set of all pairs of natural numbers also maps bijectively to $\mathbb{N}$. Figure 1.16 illustrates the construction of a suitable mapping. First, we arrange all elements of $\mathbb{N}^2$ in a matrix extending infinitely to the right and bottom. Now, we assign a unique number $\pi_{\mathbb{N}}(x,y) \in \mathbb{N}$ for each element (x,y) by starting at $(0,0)$, the top left, and then moving diagonally through the matrix.

The resulting mapping is the *Cantor pairing function*

$$\pi_{\mathbb{N}}(x,y) := y + \sum_{i=0}^{x+y} i = y + \frac{(x+y)(x+y+1)}{2}.$$

The existence of a bijective mapping between $\mathbb{N}$ and $\mathbb{N}^2$ proves both sets equipotent.

The Cantor pairing function is powerful enough to identify further sets as equipotent. Its recursive application enables us to assign a unique element in $\mathbb{N}$ not only to each pair $(x,y) \in \mathbb{N}^2$, but also to each triple $(x,y,z) \in \mathbb{N}^3$. The function $\pi_{\mathbb{N}}^3$ with

$$\pi_{\mathbb{N}}^3(x,y,z) := \pi_{\mathbb{N}}(\pi_{\mathbb{N}}(x,y),z)$$

serves this purpose. Continuing this way brings forth a bijective function from $\mathbb{N}^n$ to $\mathbb{N}$ for all $n \in \mathbb{N}$:

$$\pi_{\mathbb{N}}^1(x_1) := x_1 \tag{1.8}$$

$$\pi_{\mathbb{N}}^{n+1}(x_1,\ldots,x_n,x_{n+1}) := \pi_{\mathbb{N}}(\pi_{\mathbb{N}}^n(x_1,\ldots,x_n),x_{n+1}) \tag{1.9}$$

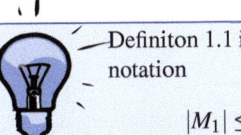

Definition 1.1 introduced the notation

$$|M_1| \leq |M_2|$$

to express that the set M_1 can be mapped injectively into the set M_2. Several times in this book, we will exploit that the relation '$\leq$' is *total* or *complete*. Both terms indicate that two arbitrary sets M_1 and M_2 are comparable. That is, the following applies:

$$|M_1| \leq |M_2| \text{ or } |M_2| \leq |M_1|$$

While this property may seem intuitively self-evident, its proof is by no means trivial.

The selection of the symbol '$\leq$' is no coincidence either, as the relation fulfills all properties of an ordering. Reflexivity and transitivity, i.e., the relationships

$$|M_1| \leq |M_1|$$
$$|M_1| \leq |M_2|, |M_2| \leq |M_3| \Rightarrow |M_1| \leq |M_3|,$$

are easy to see; both follow with little effort from the properties of injective functions. More challenging is the property of antisymmetry, namely:

$$|M_1| \leq |M_2|, |M_2| \leq |M_1| \Rightarrow |M_1| = |M_2|.$$

This relationship is guaranteed by the famous Cantor-Schröder-Bernstein theorem, discussed in detail on page 21.

Figure 1.16: The depicted *pairing function* assigns

each tuple $(x,y) \in \mathbb{N}^2$
a number $\pi_{\mathbb{N}}(x,y) \in \mathbb{N}$.

The mapping is bijective and proves $\mathbb{N}^2$ and $\mathbb{N}$ as equipotent.

In the same way, we can assign each reduced fraction of the form $\frac{x}{y} \in \mathbb{Q}$ an individual matrix cell by interpreting x as the column number and y as the row number. Skipping all cells with an unreduced fraction or a fraction with denominator 0 results in a bijective mapping between the rational numbers $\mathbb{Q}$ and the natural numbers $\mathbb{N}$. Consequently, these sets are also equipotent.

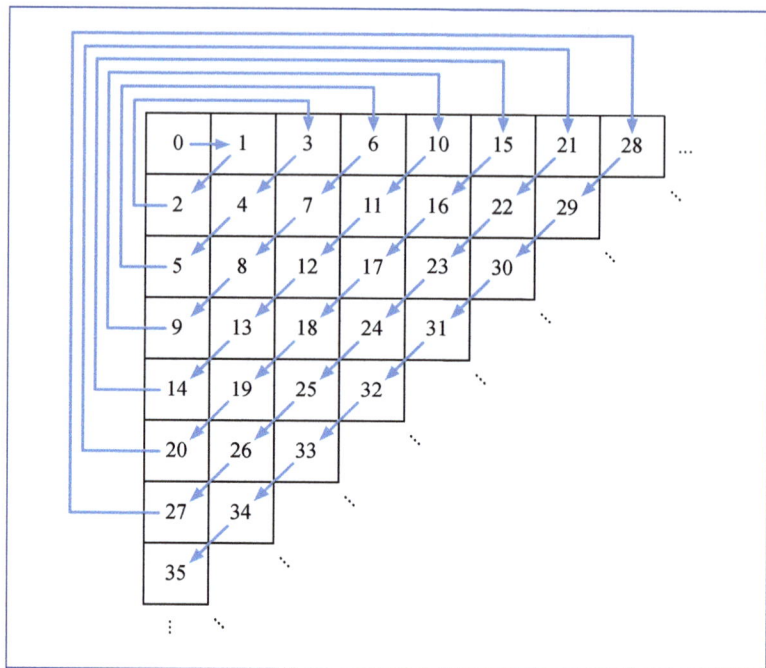

This proves the n-dimensional number space $\mathbb{N}^n$ to be equipotent to the set $\mathbb{N}$ itself, regardless of the dimension $n \in \mathbb{N}$.

In a famous paper published in 1874, Cantor proved the existence of a bijective mapping between the algebraic and the natural numbers [20]. For this purpose, he commenced by assigning each algebraic equation of the form (1.7) a *height*, which he calculated as follows:

$$N := n - 1 + |a_n| + \ldots + |a_3| + |a_2| + |a_1| + |a_0|$$

Only a finite number of algebraic equations exist for each value of N, each possessing a maximum of N solutions. This enables us to enumerate the algebraic numbers one after the other, leading to a unique mapping to the natural numbers.

Cantor's more significant discovery, however, was a different one. He demonstrated in the same paper that the continuum resists a similar mapping. He proved that the number of real numbers exceeds that of the natural numbers to such an extent that it is impossible to establish a one-to-one correspondence between the two sets. Thus, Cantor showed that the set of natural numbers and the set of real numbers represent different infinities. Conceptually, we express the difference as follows:

1.2 The Path to Modern Mathematics

 Definition 1.2 (Countable, Denumerable, Uncountable)

A set M is called

- *countable* if $|M| \leq |\mathbb{N}|$,
- *denumerable* or *countably infinite* if $|M| = |\mathbb{N}|$,
- *uncountable* if $|M| \not\leq |\mathbb{N}|$.

In colloquial terms, a set is countable if it is either finite or of the same size as the set of natural numbers. Or, more concisely, if its size does not exceed the size of the natural numbers.

Cantor's First Uncountability Proof

To demonstrate the uncountability of the real numbers, Cantor conducted a classical proof by contradiction. He started with the assumption that the real numbers can be arranged in an infinite list:

$$\omega_1, \omega_2, \omega_3, \ldots \qquad (1.10)$$

Each element ω_i denotes a real number, and for each real number x there is, by assumption, an index i with $\omega_i = x$. Cantor successfully demonstrated that every non-empty interval (α_1, β_1) must contain at least one real number ν that does not appear in the list. He deduced the contradiction by extending the start interval (α_1, β_1) to a sequence of intervals of the following form:

$$(\alpha_1, \beta_1), (\alpha_2, \beta_2), (\alpha_3, \beta_3), \ldots$$

The interval $(\alpha_{i+1}, \beta_{i+1})$ is determined by traversing the list of real numbers from left to right until two numbers inside the interval (α_i, β_i) are found. The smaller of the two forms the left boundary, and the larger the right boundary of the new interval (Figure 1.17). Afterward, Cantor distinguishes two cases:

- Case 1: The number of nested intervals is finite (Figure 1.18 above). Then, there would be a last interval (α_ν, β_ν), and among the numbers $\frac{\alpha_\nu + \beta_\nu}{2}$ and $\frac{2\alpha_\nu + \beta_\nu}{3}$, at least one would not appear in (1.10).

- Case 2: The number of nested intervals is infinite. Then, the sequences $\alpha_1, \alpha_2, \ldots$ and $\beta_1, \beta_2, \ldots$ approach a limit as they are

---

Cantor's work from 1874 bears the inconspicuous title *"Über eine Eigenschaft des Inbegriffs aller reellen algebraischen Zahlen"* [20] (*"On a Property of the Collection of All Real Algebraic Numbers"* [25]). Historians debate why Cantor chose a title that does not give the reader any indication of his main result, the uncountability of the real numbers.

On the one hand, there is good reason to assume that Cantor did not foresee his work's revolutionary nature at the time of publication and was only interested in an alternative proof of Liouville's theorem. A letter from Cantor to Richard Dedekind from December 2, 1873, hints in this direction [149].

On the other hand, Cantor biographer Joseph Dauben suspected a political motive. He believed the primary purpose of the title was to avoid drawing Kronecker's attention, Cantor's archenemy at that time [42].

"Had Cantor been more direct with a title like 'The set of real numbers is non-denumerably infinite' or 'A new and independent proof of the existence of transcendental numbers', he could have counted on a strongly negative reaction from Kronecker. After all, when Lindemann later established the transcendence of π in 1882, Kronecker asked what value the result could possibly have, since irrational numbers did not exist."

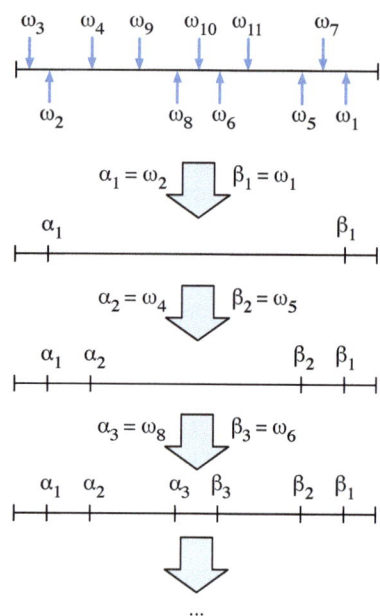

Figure 1.17: The beginning of Cantor's first proof of the continuum's uncountability. Starting with an enumeration of the real numbers, Cantor constructed a sequence of intervals.

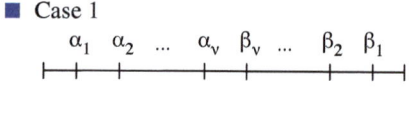

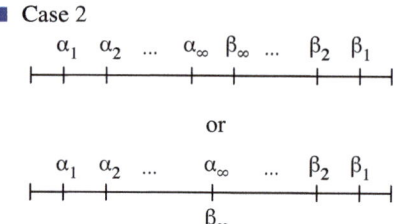

Figure 1.18: Regardless of how the interval sequence unfolds, all possibilities lead to a contradiction.

bounded and strictly monotonic. Let us denote these limits, as Cantor did, by α_∞ and β_∞, respectively. If $\alpha_\infty < \beta_\infty$, the number $\frac{\alpha_\infty + \beta_\infty}{2}$ would not appear in (1.10) (Figure 1.18 middle). However, $\alpha_\infty = \beta_\infty$, also leads to a contradiction (Figure 1.18 bottom). On the one hand, the limit must lie in every interval. On the other hand, the construction ensures that every ω_i lies outside any interval beyond a certain iteration step. Thus, the limit cannot appear in (1.10).

There is no escape! The contradictions unmistakenly demonstrate that a bijective mapping between the real numbers and the natural numbers cannot exist.

Cantor's results had far-reaching consequences for the set of transcendental numbers. Since the algebraic numbers are countable while the real numbers are not, a bijective mapping between the transcendental and the natural numbers cannot exist. Like the continuum, the set of transcendental numbers contains uncountably many elements.

Whilst only a handful of transcendental numbers were known in the middle of the nineteenth century, Cantor's groundbreaking discoveries revealed that transcendence is anything but an exotic property of selected numbers. Apart from a tiny subset, all elements of the continuum are transcendental!

Cantor had provided a truly elegant proof, even though the assertion itself was nothing new; a few years earlier, Liouville had achieved a similar result. The historically significant part of Cantor's work hides in one of his intermediate results: the uncountability of the continuum.

Cantor's Second Uncountability Proof

Three years later, Cantor proved his statement again – this time in an astonishingly simple way. At its core, the proof utilized a diagonalization argument, an equally powerful and intuitive method for identifying uncountable sets. Cantor had the following train of thought: If the two sets $\mathbb{N}$ and $\mathbb{R}$ were equipotent, a bijective mapping $f : \mathbb{N} \to \mathbb{R}$ would exist that uniquely maps each element $y \in \mathbb{N}$ to an element $f(y) \in \mathbb{R}$. If we list the decimal parts of $f(1), f(2), f(3), \ldots$ from top to bottom, we obtain a matrix as sketched in Figure 1.19. Formally, the element in column x and row y corresponds to the x-th decimal digit of the fractional representation of $f(y)$. Of course, we can only draw a tiny section of the resulting matrix, as the function f is defined for an infinite number

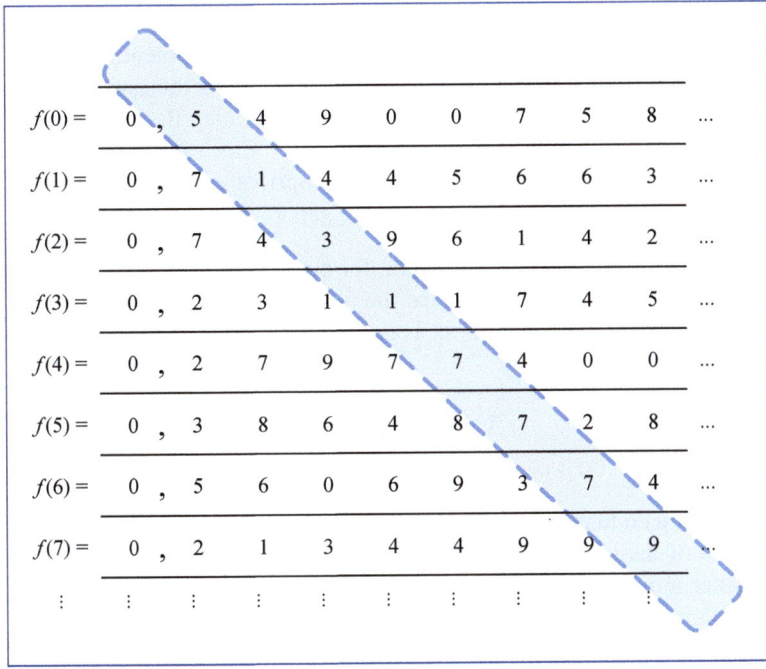

Figure 1.19: The argument of diagonalization. If there were a bijective mapping between the natural and the real numbers, each real number's (infinitely long) series of digits would appear somewhere in the depicted matrix. However, it is easy to construct a missing number regardless of the actual assignment, simply by moving along the main diagonal from top left to bottom right and increasing or decreasing the digits by one. The assembled sequence cannot occur anywhere in the matrix, as the i-th digit of the constructed number differs in row i by design. The argument shows that a bijective mapping from $\mathbb{R}$ to $\mathbb{N}$ cannot exist. In short, the set of real numbers must be uncountable.

of values $y \in \mathbb{N}$, and the decimal fraction of a real number $f(y)$ extends infinitely.

By diagonalization, we can demonstrate that the matrix is incomplete. Regardless of the specific choice of f, the matrix misses certain real numbers, contradicting the assumption of bijectivity. To construct such a number, we move along the main diagonal from the top left to the bottom right and increase or decrease every digit by one. The resulting sequence of digits forms the fractional representation of a real number. If f were a bijective mapping from $\mathbb{N}$ to $\mathbb{R}$, this number would have to appear in some line. However, the chosen construction scheme ensures that the number in the i-th row differs from the newly created diagonal number in the i-th digit. Assuming the existence of a bijective assignment between $\mathbb{N}$ and $\mathbb{R}$ immediately leads to a contradiction, making any attempt to enumerate the real numbers doomed to fail.

Nevertheless, some properties we have worked out for the natural numbers also apply to the real numbers. For instance, we can also map a tuple $(x,y) \in \mathbb{R}^2$ bijectively into $\mathbb{R}$. Figure 1.20 outlines the basic idea. The two real numbers $x \in \mathbb{R}$ and $y \in \mathbb{R}$ are merged into a single number $\pi_\mathbb{R}(x,y) \in \mathbb{R}$ by combining the digits before and after the decimal point in a zipper-like manner.

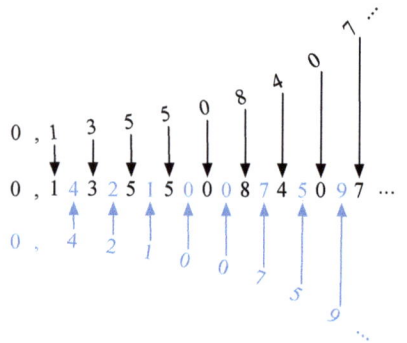

Figure 1.20: The zipper method merges two real numbers into one. That way, a bijective mapping from $\mathbb{R}^2$ to $\mathbb{R}$ arises, proving both sets as equipotent.

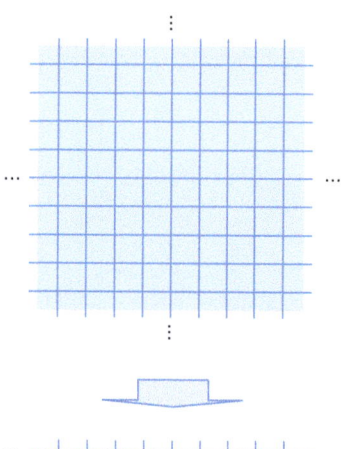

Figure 1.21: The two-dimensional plane and the one-dimensional straight line contain the same "amount" of points. Each point of one geometric object uniquely maps to a point of the other.

Once again, Cantor's approach to tackling the infinite has revealed an astonishing property of certain number sets. The equipotency of $\mathbb{R}$ and $\mathbb{R}^2$ signifies that a straight line in the plane contains the same number of points as the plane itself. As a result, we can map the points of the plane to the points of a straight line without allocating a single spot twice. Conversely, every straight line comprises enough points to fill the entire plane without leaving a single gap.

Applying $\pi_\mathbb{R}$ recursively, we obtain, for every natural number $n \in \mathbb{N}$, a function $\pi_\mathbb{R}^n$ that bijectively maps the n-dimensional number space $\mathbb{R}^n$ to $\mathbb{R}$. Formally, the function $\pi_\mathbb{R}^n$ is defined analogous to equations (1.8) and (1.9):

$$\pi_\mathbb{R}^1(x_1) := x_1$$
$$\pi_\mathbb{R}^{n+1}(x_1,\ldots,x_n,x_{n+1}) := \pi_\mathbb{R}(\pi_\mathbb{R}^n(x_1,\ldots,x_n),x_{n+1})$$

We have seen that the set of real numbers has a greater cardinality than the set of natural numbers. This result raises the question of whether another infinity exists that, in turn, has a greater cardinality than the set of real numbers. The following theorem by Cantor affirms this question:

> **Theorem 1.1** (Cantor's Theorem)
>
> For any given set M, the power set $\mathcal{P}(M)$ has a greater cardinality than M itself.

We can prove this statement with a similar diagonal argument that we used to show the uncountability of the real numbers. Again, we assume the existence of a bijective mapping $f : M \to \mathcal{P}(M)$ and subsequently derive a contradiction.

Let f be a bijection between M and $\mathcal{P}(M)$. For each element $x \in M$, either x is contained in the image $f(x)$ ($x \in f(x)$) or it is not ($x \notin f(x)$). Now, let T be the set containing all elements with the latter property:

$$T := \{x \in M \mid x \notin f(x)\}$$

Since f is bijective, and therefore surjective, there must be a pre-image x_T with $f(x_T) = T$. As with all elements of M, the element x_T satisfies either $x_T \in T$ or $x_T \notin T$. However, both cases lead to a contradiction:

$$x_T \in T \Rightarrow x_T \notin f(x_T) \Rightarrow x_T \notin T$$
$$x_T \notin T \Rightarrow x_T \in f(x_T) \Rightarrow x_T \in T$$

1.2 The Path to Modern Mathematics

We have thus shown that no bijection $f: M \to \mathcal{P}(M)$ can exist.

Cantor's theorem has significant consequences. Firstly, it shows that there is no *maximum infinity*; that is, we have no chance of constructing a universal set with a cardinality that no other set exceeds. Infinity once again manages to evade any limit. Secondly, the proposition establishes a hierarchical order on the different infinities:

$$|\mathbb{N}| < |\mathcal{P}(\mathbb{N})| < |\mathcal{P}(\mathcal{P}(\mathbb{N}))| < |\mathcal{P}(\mathcal{P}(\mathcal{P}(\mathbb{N})))| < \cdots$$

Cantor employed the Hebrew letter Aleph ($\aleph$) to describe the cardinality of an infinite set. The smallest infinity is denoted by the *cardinal number* $\aleph_0$, representing the cardinality of the natural numbers. There can be no smaller infinity than $|\mathbb{N}|$ since all infinite subsets of $\mathbb{N}$ can be mapped bijectively to $\mathbb{N}$. The cardinal number $\aleph_1$ represents the next larger infinity, and so on. If a set M has the cardinality $\aleph_n$, we denote the cardinality of the power set $\mathcal{P}(M)$ by $2^{\aleph_n}$.

At this juncture, we shall introduce the famous *Cantor-Schröder-Bernstein theorem* (*CSB theorem*), a powerful tool for proving many sets as equipotent. To fully understand the meaning of the theorem, let's take another look at Definition 1.1. It introduced the notation $|M_1| \leq |M_2|$, stating the existence of an injective mapping from the set M_1 to the set M_2. In colloquial terms, such a mapping allows us to embed the elements of M_1 in M_2 without occupying a single element of M_2 twice. Since every bijective mapping is also injective, $|M_1| = |M_2|$ implies $|M_1| \leq |M_2|$ and $|M_2| \leq |M_1|$. Now, the Cantor-Schröder-Bernstein theorem states that the opposite direction is also true:

Theorem 1.2 (Cantor-Schröder-Bernstein Theorem)

For any two sets M_1 and M_2, the following holds:

If $|M_1| \leq |M_2|$ and $|M_2| \leq |M_1|$, then $|M_1| = |M_2|$.

A detailed proof of this theorem is presented in [51] or [224].

Let us bring the Cantor-Schröder-Bernstein theorem to life with two examples. In the first example, we prove the open interval $(-1;1)$ and the closed interval $[-1;1]$ as equipotent. On the one hand, the set $(-1;1)$ can be trivially embedded in the set $[-1;1]$ via the identical mapping $f: x \mapsto x$. On the other hand, $g: x \mapsto \frac{x}{2}$ is an injective mapping of the closed interval into the open interval. At this point, we have already crossed the finish line. From the CSB theorem, it follows that the two

■ Injective functions f and g

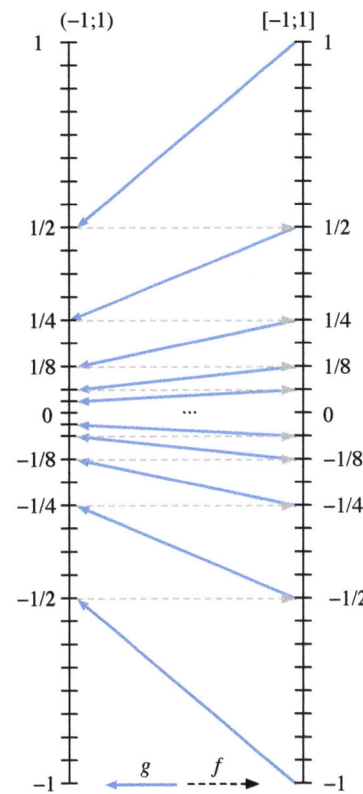

■ Bijective function $h: (-1;1) \to [-1;1]$

$$h: x \mapsto \begin{cases} \pm 1 & \text{if } x = \pm\frac{1}{2} \\ \pm\frac{1}{2} & \text{if } x = \pm\frac{1}{4} \\ \pm\frac{1}{4} & \text{if } x = \pm\frac{1}{8} \\ \pm\frac{1}{8} & \text{if } x = \pm\frac{1}{16} \\ \cdots & \\ x & \text{otherwise} \end{cases}$$

Figure 1.22: If a set maps injectively into another set and vice versa, a bijection between the two sets can be established by cleverly combining both embeddings. This is the contents of the famous Cantor-Schröder-Bernstein theorem, demonstrated here with the intervals $(-1;1)$ and $[-1;1]$.

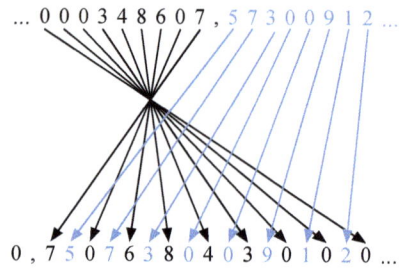

Figure 1.23: By rearranging the digits, all real numbers can be mapped injectively into the interval $[0;1]$.

intervals are equipotent. Once again, the result puts our intuition to the test, as the closed interval $[-1;1]$ appears to contain two more elements than its open counterpart $(-1;1)$. However, Figure 1.22 assures that we are still on solid ground. It reveals what a bijective mapping between the two intervals may look like in concrete terms.

Similarly, we can prove the existence of a bijective mapping between the set of real numbers and the unit interval $[0;1]$. Embedding $[0;1]$ in $\mathbb{R}$ is trivial. Conversely, the assignment

$$\sum_{i=-\infty}^{\infty} b_i 10^i \mapsto b_0 10^{-1} + \sum_{i=1}^{\infty} \left(b_{-i} 10^{-2i} + b_i 10^{-2i-1} \right)$$

maps every real number into the interval $[0,1]$ without allocating an element of the image set twice (Figure 1.23). Thus, we have once again fulfilled the requirements of the CSB theorem, proving the equipotency of $[0;1]$ and $\mathbb{R}$.

Having at hand a bijective mapping between the real numbers and the unit interval $[0;1]$ leads to a considerable simplification, which we will utilize several times in the subsequent chapters. Instead of treating the real numbers as a whole, we can restrict ourselves to those with the decimal part 0 without the loss of generality.

At this point, we are ready to establish a close connection between the real numbers and the power set of the natural numbers. In binary, a real number x from the unit interval $[0;1]$ can be represented as follows:

$$x = \sum_{i=1}^{\infty} b_i 2^{-i}$$

The coefficients b_i form an infinitely long sequence of zeros and ones. Thus, we can uniquely assign x to a subset of the natural numbers by including the number $n \in \mathbb{N}$ in the subset precisely when the n-th decimal place of x equals 1:

$$\sum_{i=1}^{\infty} b_i 2^{-i} \mapsto \{n \in \mathbb{N} \mid b_n = 1\} \tag{1.11}$$

Conversely, we can injectively embed any subset of $\mathbb{N}$ into the interval $[0;1]$ as follows:

$$\{n_1, n_2, \ldots\} \mapsto \sum_i 2^{-2n_i - 1} \tag{1.12}$$

The CSB theorem assures us the result we were looking for. It reveals the equipotency of the set of real numbers and the power set of the natural numbers:

$$|\mathbb{R}| = |\mathcal{P}(\mathbb{N})| = 2^{\aleph_0} \tag{1.13}$$

1.2 The Path to Modern Mathematics

Cantor intensively studied the question of whether other infinities were hiding between $\mathbb{N}$ and $\mathbb{R}$. From early on, he suspected this was not the case.

Accordingly, the real numbers would be in second place ($\aleph_1$) in the infinitely long list of infinities. This assertion is the content of the famous *continuum hypothesis*, symbolically expressed as:

$$|\mathbb{R}| \stackrel{?}{=} \aleph_1 \qquad (1.14)$$

Due to (1.13), we can rewrite equation (1.14) in the form

$$2^{\aleph_0} \stackrel{?}{=} \aleph_1$$

and generalize it to:

$$2^{\aleph_n} \stackrel{?}{=} \aleph_{n+1} \qquad (1.15)$$

This assumption is called the *generalized continuum hypothesis*, claiming that the power set operation does not skip any infinities while traversing from one infinity to the next.

The continuum hypothesis haunted Cantor for the rest of his life. Sometimes, he believed himself to have proven the hypothesis; other times, he thought to have found a refutation. But as time passed, flaws in his proofs kept cropping up, undermining his perceived successes. No matter how hard he tried, he failed to solve this grand riddle of the continuum. Tragically, Cantor had no chance to know how doomed he was to failure.

The fact that Cantor's notion of set was rejected by some of his contemporaries and fiercely opposed by others can only be understood in a historical context. Cantor shaped his ideas at a time when the debate about the nature of infinity was in full swing. Two concepts were at the center of the discourse: *Potential infinity* and *actual infinity*.

A closer look at the natural numbers

$$0, 1, 2, 3, \ldots$$

reveals the difference between these terms. Except for 0, each element in this infinitely long list emerges from its predecessor by applying the *successor operation*. This process allows us to continuously generate new numbers with no upper limit on the iteration count. We say the number of iterations is *potentially infinite*. This kind of infinity does not bear any risks. Even if the number of iterations is open-ended, we will reach every natural number after a finite number of iterations, thus never having to apply the successor operation infinitely often.

Assignment (1.11) injectively embeds the real numbers from the unit interval $[0;1]$ into the set $\mathcal{P}(\mathbb{N})$ by utilizing the binary representation of a real number. This approach involves a technical difficulty often overlooked at first glance, namely the existence of real numbers with multiple binary representations. For instance, the fraction $\frac{1}{2}$ has two representations, 0.1 and $0.0\overline{1}$. Consequently, function (1.11) maps the number $\frac{1}{2}$ to both $\{1\}$ and $\{2,3,4,\ldots\}$, contradicting the property of being a function. Luckily, this problem is easy to solve. The ambiguity disappears if we agree to assign the representation with the lowest number of ones.

Care must also be taken when embedding the power set $\mathcal{P}(\mathbb{N})$ into the unit interval $[0;1]$. For instance, if we used the mapping

$$\{n_1, n_2, \ldots\} \mapsto \sum_i 2^{-n_i - 1},$$

the embedding would no longer be injective, as both the sets $\{0\}$ and $\{1,2,3,\ldots\}$ would map to $\frac{1}{2}$. This is the reason why n_i is multiplied by 2 in equation (1.12). Only through this trick does the mapping become injective, as different subsets of the natural numbers now map to different real numbers.

In contrast, by considering the numbers generated by the successor operation as a totality, we have already transitioned from the potential infinite to the *actual infinite*. This totality is none other than the set of natural numbers itself, with infinitely many elements mentally caged inside a closed entity. Whether we are allowed to regard the natural numbers as a self-contained totality or merely accept the potentially infinite as the sole foundation of mathematics has been the subject of controversial debates. Criticism of actual infinity can be traced back even to the works of Aristotle [177].

"[...] I protest above all the use of an infinite quantity as a completed one, which in mathematics is never allowed." [41]

"[...] so protestiere ich gegen den Gebrauch einer unendlichen Größe als einer Vollendeten, welches in der Mathematik niemals erlaubt ist." [69]

The putative contradictions that arise almost naturally when dealing with the infinite have fueled the criticism. For instance, we have already learned about the possibility of relating the integers and the natural numbers on-to-one, although the relation $\mathbb{N} \subset \mathbb{Z}$ suggests the opposite. Similarly, it is easy to see that every infinite subset of $\mathbb{N}$ has the same cardinality as the set of the natural numbers itself. Some scientists, such as the renowned astronomer Galileo Galilei, considered this as a confirmation that size comparisons between infinite sets are inadmissible and only meaningful for finite collections [67, 119]. Other scholars, such as the famous mathematician Carl Friedrich Gauss, rejected the idea of treating infinite sets as self-contained quantities from the outset (Figure 1.24).

Carl Friedrich Gauß
(1777 – 1855)

Figure 1.24: The German mathematician Carl Friedrich Gauss was one of the most brilliant mathematicians of the late eighteenth and early nineteenth centuries. He made ground-breaking achievements in various fields of mathematics, astronomy, and physics, elevating Göttingen to one of the world's most renowned centers for mathematics. A memorial coin issued a year after his death honors the brilliant mathematician with the title *"Mathematicorum Principi"*, *"The Prince of Mathematicians"*.

For Cantor, the alleged paradoxes were nothing more than properties. He saw that the apparent contradictions merely stemmed from the unfounded assumption that infinite manifolds must obey the same principles as finite sets. Cantor found an advocate in Richard Dedekind. Like him, Dedekind considered in what others called paradoxes a defining property of infinite sets: A set has an infinite number of elements if and only if there is a proper subset with the same cardinality.

Although the high level of incomprehension, mistrust, and hostility left scars on Cantor's psyche, he stayed on course. Undeterred, he headed towards a new mathematics that would put the actual infinite into a primary position, freeing it once and for all from its role as an extra. Cantor had yet to learn that he built his train of thought on shaky ground.

1.2.3 The Power of Symbols

The German mathematician Gottlob Frege, three years younger than Cantor, was another advocate of the actual infinite (Figure 1.25). From early on, Frege foresaw that the different approaches to tackling the infinite would develop into a controversial debate with the potential of

1.2 The Path to Modern Mathematics

"[Cantors] refutation of the objection to the infinite seems to me on the whole well-done and to the point. The objections arise when the infinite is assigned properties it does not have, either because the properties of the finite are carried over to the infinite as a matter of course or because a property belonging only to the absolute infinite is carried over to all of the infinite. To point emphatically to the differences within the infinite is one of the merits of this work." [59]

"For surely the infinite cannot ultimately be denied in arithmetic, and on the other hand it is incompatible with that epistemological outlook. Here, it seems, is the field where a decisive battle will be fought." [59]

Gottlob Frege (1848 – 1925)

Figure 1.25: The German mathematician Gottlob Frege founded a new school of thought called *logicism*. Like Cantor, he considered the actual infinite a conceptual milestone on the path to a new mathematics.

dividing the mathematics community for a long time. Nevertheless, he firmly believed that the actual infinite would eventually become accepted. Like Cantor, he saw mathematics dominated by a *powerful academic-positivist skepticism*[3] capable of delaying progress but not putting it on hold forever.

In 1879, Gottlob Frege published his most influential work, the *Begriffsschrift*. In hindsight, this booklet of about a hundred pages marks a milestone in the history of mathematical logic and is one of the most significant single publications in this field. In this work, Frege created what is now known as *symbolic logic*. He succeeded in developing an artificial language that was expressive enough to formalize all of ordinary mathematics, together with a logical reasoning apparatus. At the time of publication, however, the significance of Frege's work was widely underestimated. Most of his contemporaries reacted either indifferently or even dismissive of his ideas. Even Cantor considered the *Begriffsschrift* as largely irrelevant.

But what exactly made Frege's work so special? With propositional logic, George Boole had already created the basic framework for expressing logical relationships between elementary statements using symbolic operators a few years earlier [11, 12]. The new approach, however, went far beyond. Frege realized that logical directives were

---

[3] *"mächtigen akademisch-positivistischen Skepsis"* [59]

Friedrich Ludwig Gottlob Frege was born on November 8, 1848 in Wismar, Mecklenburg. In 1869, he enrolled at the University of Jena, where he found an influential teacher and lifelong supporter in Ernst Abbe, the director of Carl Zeiss. Probably, Abbe's suggestion led Frege to transfer to the renowned mathematics faculty at Göttingen University after four semesters, where he received his doctorate in geometry in 1873. Back in Jena, he submitted his habilitation thesis in 1874. After teaching a few years as a private lecturer, he was appointed associate professor in 1879 and finally full professor in 1896. Frege is one of the founders of mathematical logic and analytical philosophy. In his famous *Begriffsschrift* from 1879 [64], he developed an axiomatic logic that went far beyond George Boole's propositional logic calculus. Many contemporary mathematicians consider Frege's conceptual framework the origin of modern predicate logic.

For most of his life, Frege viewed mathematics as a branch of logic, thus being a firm advocate of *logicism*. According to Frege, all truths must be reducible to a set of axioms which, in his words, were *neither capable of nor in need of proof* (*"eines Beweises weder fähig noch bedürftig"* [58]). By doing so, he positioned himself against other mathematicians of his time, who considered logic an isolated branch of mathematics.

After the devastating discovery of Russell's antinomy, Frege largely withdrew from academia and was not to publish any other relevant work. Mentally trapped in the ruins of his conceptual framework, Frege died as a bitter man on 26 July 1925, aged 76.

not only capable of describing the relationships between elementary statements but also of formalizing the elementary statements themselves. Frege thus abolished a fundamental restriction of Boolean logic, which strictly distinguished between the level of elementary statements (Boolean variables, *primary propositions*) and the level of logical relations (propositional expressions, *secondary propositions*).

Let's look at an example. In Frege's logic, the statement

"All humans are mortal"

is represented in form of an implication:

"For all x: If x is human, then x is mortal."

Similarly, the statement

"Some people are noble"

can be expressed in form of a conjunction:

"For some x: x is human and x is noble."

Frege's structural framework became the cornerstone of modern predicate logic. In contemporary notation, the two statements above read as follows:

$$\forall x \, (\text{Human}(x) \rightarrow \text{Mortal}(x)) \qquad (1.16)$$

$$\exists x \, (\text{Human}(x) \wedge \text{Noble}(x)) \qquad (1.17)$$

1.2 The Path to Modern Mathematics

Or, even shorter:

$$\forall x \, (H(x) \to M(x)) \tag{1.18}$$
$$\exists x \, (H(x) \wedge N(x)) \tag{1.19}$$

The *universal quantifier* '$\forall$' and the *existential quantifier* '$\exists$' make quantitative statements about the elements of a *domain*. $\forall x$ is read as "For all x, ..." and $\exists x$ as "There exists an x such that ...". The symbols '$\to$' and '$\wedge$' are the commonly used symbols for the logical implication and conjunction.

While the conceptual core of the *Begriffsschrift* hardly differs from that of modern predicate logic, its appearance couldn't be more distinct. The reason is Frege's two-dimensional approach to layout formulas (Figure 1.26). His sophisticated notation not only posed new challenges for the printers' guild but also made the book extremely cumbersome to read. To get a more vivid impression, Figure 1.27 shows how the formulas (1.18) and (1.19) look like in Frege's notation.

With his *Begriffsschrift*, Frege had successfully raised logical thinking to a symbolic level. However, his ambitions went much further. In contrast to many of his contemporaries, such as Cantor and Boole, he did not consider logic as a part of mathematics, but rather mathematics as a part of logic. He passionately sought to derive all mathematical terms and concepts from elementary logical concepts, thus providing a solid foundation for all branches of mathematics. With his ambitious project, Frege founded a new philosophical movement called *logicism*.

Frege reached a significant milestone in 1884 with the publication of *The Foundations of Arithmetic* [66] (*Die Grundlagen der Arithmetik* [63]). In this work, he attempted to define the concept of numbers formally and explained how he planned to realize his logicism program. Unlike the *Begriffsschrift*, his new publication was purely written in colloquial language.

Over the next twenty years, he focused on formalizing his ideas, culminating in the *Basic Laws of Arithmetic* [62] (*Grundgesetze der Arithmetik* [60, 61]), a two-volume book that we may regard as Frege's second seminal work alongside the *Begriffsschrift* (Figure 1.28). Frege established a connection between the concept of a number and the concept of a set to put arithmetic on logical grounds. For example, when considering the set of all days of the week or the set of all world wonders, we count seven elements in each case. Even if we knew nothing about the number 7, we could conclude that all mentioned sets contained *equally*

■ Negation

■ Implication

■ Conjunction

■ Disjunction

■ Universal quantification

■ Existential quantification

Figure 1.26: Comparison between the notation of the *Begriffsschrift* and modern predicate logic

■ "All humans are mortal"

■ "Some humans are noble"

Figure 1.27: Compound expressions in Frege's notation

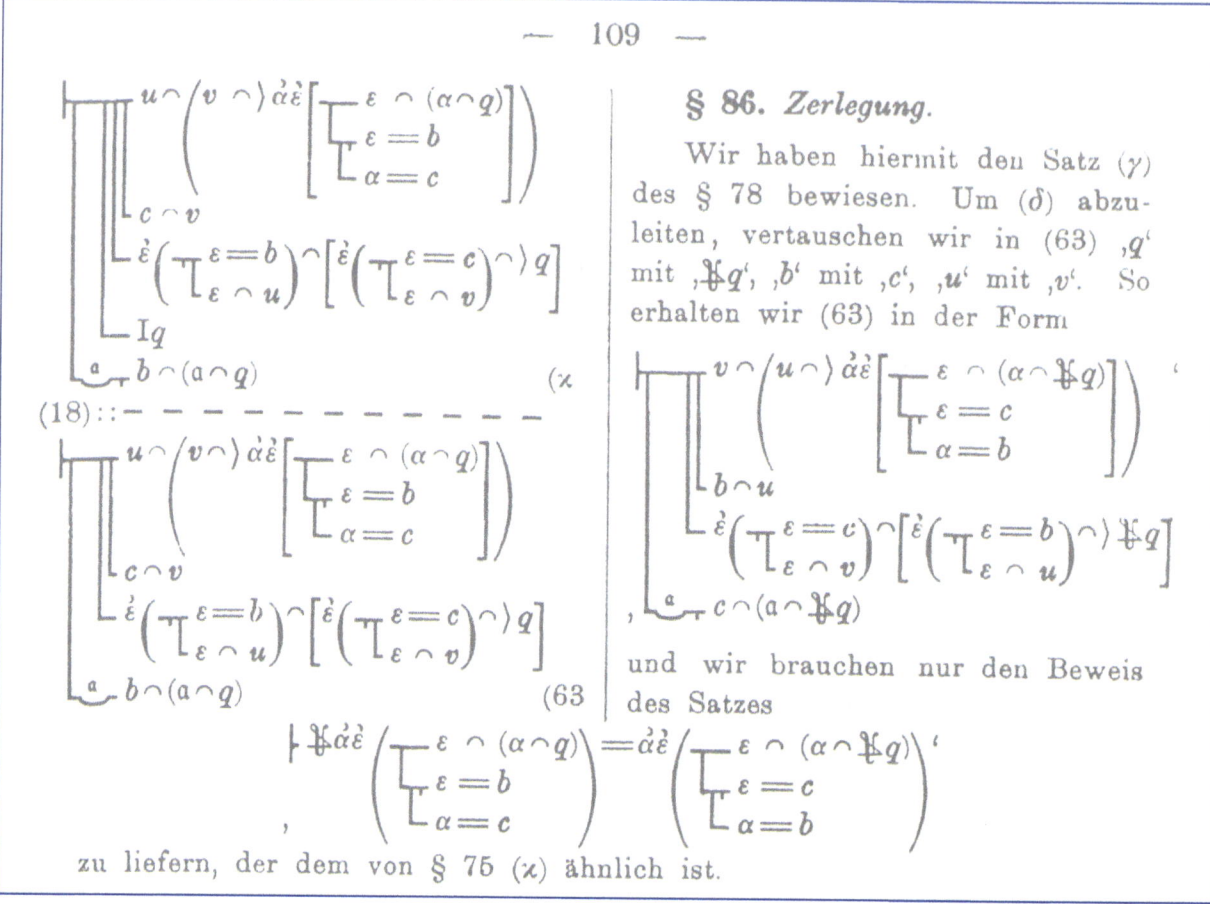

Figure 1.28: Excerpt from the 1st volume of the *Basic Laws of Arithmetic*. Gottlob Frege wrote the two volumes as part of his logicist program. It was the first comprehensive effort to give mathematics a logical foundation.

many elements. The insight that we can talk about the size of sets without necessarily naming the number of their elements makes it possible to reduce the concept of a number to a property of sets. This is precisely what Frege had in mind. He designed his logical framework to identify the number 7 with the set of all sets that map bijectively to one of the example sets above.

Like Cantor, Frege sincerely believed in the soundness of his approach. Yet the clouds that would soon darken the bright sky of this new kind of mathematics were out of sight.

1.2 The Path to Modern Mathematics

David Hilbert was born on January 23, 1862, in Königsberg as the eldest child of an East Prussian family of lawyers. His hometown's Albertus University (Albertina) offered him ideal conditions to foster his talents. He completed his mathematics studies with a doctorate degree in 1884, followed by his habilitation in 1886. After a few years as a private lecturer, he was appointed professor in 1892.

In 1895, Hilbert accepted an offer from the mathematics department at the University of Göttingen. Renowned figures such as Gauss, Dirichlet, and Riemann once brought fame to the picturesque university town. Towards the end of the nineteenth century, however, this fame slowly faded due to a lack of adequate successors. Hilbert's call should mark a turning point. It marked a fresh start, leading Göttingen's mathematics to new glory.

Hilbert was an extraordinarily talented and unusually versatile mathematician. He changed his research focus several times throughout his career and made his mark in mathematical logic, geometry, number theory, analysis, and theoretical physics.

Hilbert influenced the mathematics of the early twentieth century more than anyone else. In 1900, he gave his famous speech at the International Congress of Mathematicians in Paris, in which he presented 23 unsolved problems that would engage mathematicians for decades to come. Even today, some of Hilbert's problems are still unsolved.

Hilbert died on February 14, 1943, at the age of 81. An unobtrusive gravestone in Göttingen's city cemetery quietly and modestly commemorates one of the greatest visionaries of his time. Carved in the gravestone are his famous words: *"Wir müssen wissen. Wir werden wissen."* (*"We must know. We will know"*).

1.2.4 The Dawn of a New Century

On New Year's Eve, December 31, 1899, people greeted the new century in euphoria. The scientific discoveries of the nineteenth century had created a sense of omnipotence in many people's minds, with the mathematical community being no exception. Given this background, it is no surprise that the 2nd International Congress of Mathematicians in Paris saw a lecture solely aimed at looking forward to the coming century. The speech took place on the morning of August 8, 1900, and began with the following words:

> *"Who of us would not be glad to lift the veil behind which the future lies hidden; to cast a glance at the next advances of our science and at the secrets of its development during future centuries? What particular goals will there be toward which the leading mathematical spirits of coming generations will strive? What new methods and new facts in the wide and rich field of mathematical thought will the new centuries disclose?"* [98]

The podium speaker was the 38-year-old David Hilbert (Figure 1.29). Despite his young age, the German mathematician already had a wide reputation. Through numerous successes in various fields, he rose early to the Olympus of the most prominent mathematicians of all time.

David Hilbert (1862 – 1943)

Figure 1.29: David Hilbert was one of his time's most famous and influential mathematicians. His seminal speech at the International Congress of Mathematicians in Paris guided the direction of mathematics for decades to come.

Hilbert strongly advocated the axiomatic method and brought it to the forefront of interest by the end of the nineteenth century. For him, it was the only adequate response to the decades-long debate about the nature of mathematics.

Compared to Frege, Hilbert considered it futile to explain the concept of numbers by other concepts, as those were hardly more obvious than the concept of the natural numbers themselves. He also did not share the view of the prominent number theorist Leopold Kronecker that *"the good Lord made the integers; all else is the work of man."*[4] Such a view would have rendered all attempts to define the concept of numbers as superfluous.

Hillbert's way out of the dilemma was a formalistic one. Instead of explaining the essential mathematical elements by their nature, he limited himself to naming the logical relationships between the objects under consideration. In 1899, he achieved a breakthrough with the reformulation of *Euclidean geometry*. From a total of 20 axioms, divided into 5 axiom groups, all theorems of Euclidean geometry could be derived without giving the symbols any specific interpretation [91]. This work paved a promising path on which many mathematicians followed Hilbert for many years to come. Subsequently, large areas of mathematics were axiomatized and thus opened up to precise observation. In this modern sense, mathematics became a symbolic game solely determined by its rules and not the meaning of its figures. Hilbert's formalistic method brought a degree of clarity and accuracy that mathematicians have striven for all along, free from interpretations of any kind.

In his Paris speech, Hilbert addressed 23 problems that were crucial to mathematics and yet to be solved. Hilbert presented the first 10 in person; the remaining 13 are contained only in the written version. Hilbert was aware of the seminal role the infinite would play in the future of mathematics, letting him put the continuum hypothesis at the top of his list:

> *"The investigations of Cantor on such assemblages of points suggest a very plausible theorem, which nevertheless, in spite of the most strenuous efforts, no one has succeeded in proving. This is the theorem: Every system of infinitely many real numbers, i.e., every assemblage of numbers (or points), is either equivalent to the assemblage of natural integers, 1, 2, 3,... or to the assemblage of all*

---

[4] *"Die ganzen Zahlen hat der liebe Gott gemacht, alles andere ist Menschenwerk."* [218]

1.2 The Path to Modern Mathematics

> *real numbers and therefore to the continuum, that is, to the points of a line; as regards equivalence there are, therefore, only two assemblages of numbers, the countable assemblage and the continuum."* [98]

In second place, Hilbert called for a consistency proof for the arithmetical axioms.

> *"But above all I wish to designate the following as the most important among the numerous questions which can be asked with regard to the axioms: To prove that they are not contradictory, that is, that a finite number of logical steps based upon them can never lead to contradictory results."* [98]

Hilbert refers to the so-called *Peano axioms*, named after the Italian mathematician Giuseppe Peano. They appeared first in 1889 in the *Arithmetices Principia*, which, in hindsight, is one of the most significant publications in mathematical history [152]. The booklet was written in Latin and later translated into English under *The Principles of Arithmetic* [153].

When contemporary mathematicians refer to the Peano axioms, they typically mean axioms 1, 6, 7, 8, and 9 in Figure 1.30. These five axioms encapsulate the fundamental properties that uniquely define the structure of the natural numbers. Peano's decision to let the natural numbers start with 1 and not with 0 has little significance, as we have shown in Section 1.2.2 how to map the sets $\{0, 1, 2, \ldots\}$ and $\{1, 2, 3, \ldots\}$ bijectively onto each other. In Section 3.1, we will adopt the axioms again in a slightly revised form and translate them into modern predicate logic.

The consistency of the arithmetic axioms is significant for entire mathematics, as almost all of its branches build upon the theory of numbers. As long as we cannot guarantee consistency, we cannot rule out that the equations $1 + 1 = 2$ and $1 + 1 \neq 2$ are both derivable from the axioms. The consequences would be disastrous.

Hilbert strongly believed that proving the consistency of axiomatic systems was possible, and his initial successes backed him up. As part of his reformulation of geometry, he constructed a specific number range with the property that every provable relationship between geometric objects corresponded to a provable relationship between numbers and vice versa. Consequently, any contradiction arising from the geometric

Giuseppe Peano (1858 – 1932)

Figure 1.30: Excerpt from the English translation of Peano's original work from 1889, showing the first formal axiomatization of the natural numbers. Peano used a mirrored 'C' to denote logical implication. Later, it was written as '⊃' and equates to the modern implication operator '→'.

Explanations

The sign N means *number (positive integer)*.
The sign 1 means *unity*.
The sign $a + 1$ means *the successor of a*, or *a plus* 1.
The sign = means *is equal to*. We consider this sign as new, although it has the form of a sign of logic.

Axioms

1. $1 \, \varepsilon \, N$.
2. $a \, \varepsilon \, N \, .\supset. \, a = a$.
3. $a, b \, \varepsilon \, N \, .\supset: a = b \, .=. \, b = a$.
4. $a, b, c \, \varepsilon \, N \, .\supset\therefore \, a = b \, . \, b = c \, :\supset. \, a = c$.
5. $a = b \, . \, b \, \varepsilon \, N \, :\supset. \, a \, \varepsilon \, N$.
6. $a \, \varepsilon \, N \, .\supset. \, a + 1 \, \varepsilon \, N$.
7. $a, b \, \varepsilon \, N \, .\supset: a = b \, .=. \, a + 1 = b + 1$.
8. $a \, \varepsilon \, N \, .\supset. \, a + 1 -= 1$.
9. $k \, \varepsilon \, K \therefore 1 \, \varepsilon \, k \therefore x \, \varepsilon \, N \, . \, x \, \varepsilon \, k \, :\supset_x. \, x + 1 \, \varepsilon \, k ::\supset. \, N \supset k$.

axioms would become visible as a contradiction in arithmetic. In other words, if we trust arithmetic, the consistency of the geometric axioms follows.

What Hilbert accomplished is called a *relative proof of consistency* because he reduced the consistency of geometry to the consistency of arithmetic. For arithmetic itself, however, Hilbert demanded an *absolute proof* that did not rely on the consistency of another system. Much like the chicken-and-egg problem, any relative proof merely shifts the question of consistency to another system. At the time of his speech, Hilbert was in no doubt that an absolute consistency proof of arithmetic did exist. For him, it was merely a matter of time before it would be found.

In the tenth place, Hilbert demanded a method for solving Diophantine equations (Figure 1.31).

"Given a Diophantine equation with any number of unknown quantities and with rational integral numerical co-

efficients: To devise a process according to which it can be determined by a finite number of operations whether the equation is solvable in rational integers." [98]

As outlined in Section 1.2.1, a Diophantine equation has the form

$$p(x_1, x_2, \ldots, x_n) = 0,$$

where p is a multivariable polynomial with integer coefficients. The solution of a Diophantine equation is the set of the integer roots of p. What Hilbert called a *method* back then is what we call an *algorithm* today. At the time of his speech, however, the computer was still a long way off, and there was only a vague idea of what exactly was meant by a procedure in Hilbert's sense. More than 30 years would pass before Alan Turing and Alonzo Church captured the concept of computability with mathematical precision. Chapter 5 will discuss computability theory in detail and show why any attempt to construct such a procedure is doomed to fail.

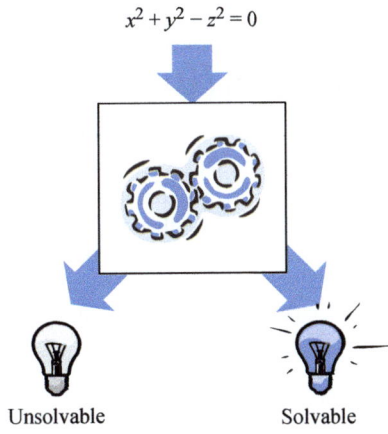

Figure 1.31: In his legendary Paris speech, Hilbert requested the development of a procedure for deciding the solvability of a given Diophantine equation.

1.2.5 Foundational Crisis

The new century had barely begun when Gottlob Frege received a letter from the British mathematician and philosopher Bertrand Russell in June 1902. What Frege was to read would not only shake his research to the core but also plunge mathematics into its greatest crisis. The letter reached Frege on the verge of completing the second volume of his Fundamental Laws of Arithmetic. He had devoted many years to this work, now finding it in ruins in one fell swoop. As it was too late for substantial changes, the second volume closes with the following epilogue:

> *"Hardly anything more unfortunate can befall a scientific writer than to have one of the foundations of his edifice shaken after the work is finished. This was the position I was placed in by a letter of Mr. Bertrand Russell, just when the printing of this volume was nearing its completion."*
>
> Gottlob Frege [65]

What in the world could have had the potential to shake Frege's work so fundamentally that he saw his life's work in jeopardy? The answer hides in *Proposition V*, his fifth fundamental law. This law is sufficiently

Bertrand Arthur William Russell was born as the third child of a liberal aristocratic family on May 18, 1872, and was the grandson of two-time British Prime Minister Lord John Russell. At the age of 2, his mother and sister died of diphtheria. When he also lost his father in 1876, his grandparents fought for custody. Two years later, his grandfather died, leaving his grandmother to raise him alone. From early on, Russel's unique talent for mathematics and philosophy became apparent. Initially, he was educated privately and later at the renowned Trinity College in Cambridge.

Between 1890 and 1894, he studied mathematics and became familiar with his teacher and later friend Alfred North Whitehead. After graduation, he seized the opportunity to conduct research in Cambridge without teaching duties until 1901. In 1908, he became a member of the Royal Society. The First World War brought about a drastic change in his life. In 1916, he was fined for repeated pacifist activities and dismissed from his position at Trinity College [82]. In a subsequent trial, he was sentenced to two years in prison. After that, he published several seminal articles on philosophical and social issues and was awarded the Nobel Prize for Literature in 1950. His literary achievements made him world-famous, and many people today associate his name exclusively with his philosophical writings. Many people do not know that the famous philosopher Bertrand Russell was also one of the twentieth century's greatest mathematicians.

strong to derive the *general comprehension schema*, which reads as follows in its contemporary form:

$$\exists y \, \forall x \, ((x \in y) \leftrightarrow \varphi(x))$$

Herein, φ is an arbitrary formula not containing the variable y. In plain English, the general axiom of comprehension reads like this: There exists a set y that includes precisely those elements x to which the property φ applies. For example, if φ formalizes the property of being a prime number, the axiom of comprehension entitles us to speak of the set of all prime numbers.

It is a central element of Frege's logic that the formula φ is not subject to any restrictions. Russell recognized the danger of this liberty and made the following choice:

$$\varphi(x) := (x \notin x)$$

The formula describes a seemingly harmless property: It applies to all sets x that do not contain themselves as an element. φ is true for most sets. For instance, the set of all humans is not itself a human, and the set of all prime numbers is not a prime number itself. In contrast, φ is false for the set of all sets. This set contains itself as an element.

For the chosen formula φ, the comprehension schema guarantees the existence of a set y with the following property:

$$\forall x \, ((x \in y) \leftrightarrow (x \notin x))$$

In colloquial terms, y is the set of all sets that do not contain themselves. We can eliminate the all-quantifier by so-called *universal instantiation*,

1.2 The Path to Modern Mathematics

allowing us to replace x with any element. If we choose y for x, we get the contradiction that the set y contains itself exactly when it does not:

$$(y \in y) \leftrightarrow (y \notin y)$$

The general axiom of comprehension was a cornerstone of Frege's logical framework and Cantor's set theory, and its collapse left the new mathematics on shaky ground.

Russell's antinomy reveals that Frege and Cantor approached the actual infinite with too little caution. As harmless as the general comprehension schema may seem, it allows us to construct sets that must not be regarded as a totality. If we consider the set of all sets that do not contain themselves as actually existent, the observed contradiction is inevitable.

The circular reasoning of Russell's antinomy resembles that of the *Barber paradox*, a paradox Russell employed himself to make his antinomy accessible to a broader audience.

The widespread recognition of Russell's antinomy often belies the fact that the concept of sets had been the source of obscurities before. As early as 1897, Cantor noticed that the set of all cardinal numbers couldn't include its own cardinal number. Two years later, he hit upon the *Burali-Forti paradox*, which we will return to in Section 3.2.2. It is named after the Italian mathematician Cesare Burali-Forti, who was the first to notice that the conception of the set of all ordinal numbers gave rise to contradictions.

Cantor never published his discovery; we only know he was aware of the antinomies from correspondence with David Hilbert and Richard Dedekind. Back then, all three seemed to view the antinomies merely as curiosities arising from a too informal, thus inadmissible, use of specific terms and concepts.

Russell's antinomy was different, though. For one thing, it was so fundamental that it affected all areas of mathematics that relied on the concept of sets in any form. For another, Russell showed not only the contradictory character of the set of all sets not containing themselves but also the constructibility of this set with standard logical means. Unlike the antinomies found in ordinal or cardinal number theory, which many viewed as curiosities at the edge of an otherwise intact mathematical core, Russell's antinomy could not be ignored. Russell discovered a tectonic fault of epic proportion at the very center of mathematics.

Frege perceived the discovery of the antinomy as a severe blow that burst his life's work like a soap bubble. When his wife Margarete died

BARBER PARADOX

"You can define the barber as 'one who shaves all those, and those only, who do not shave themselves'. The question is, does the barber shave himself?" [179]

Case 1: The barber shaves himself. It follows that...

Case 2: The barber does not shave himself. It follows that...

Figure 1.32: The barber in question shaves precisely those men who do not shave themselves. Asking whether the barber shaves himself leads to the same circular reasoning that underlies Russell's antinomy.

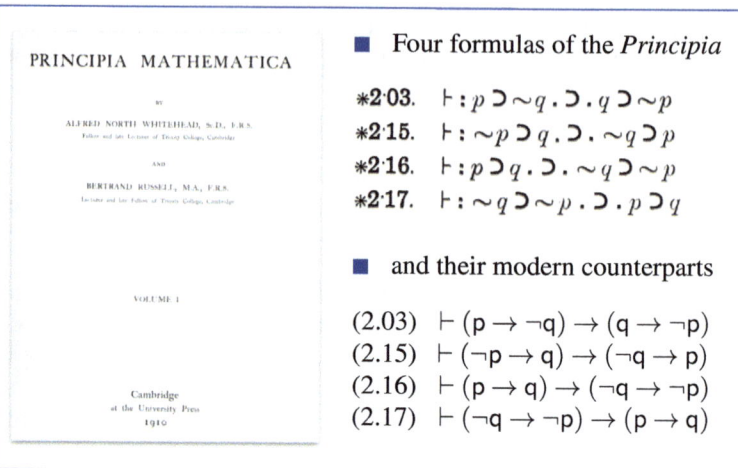

Figure 1.33: *Principia Mathematica*. This monumental work by Russell and Whitehead is a challenging read for us today, not only because its notation differs significantly from what we are accustomed to but also due to some poor choices. For instance, the dot in the *Principia* has a double meaning. Depending on its position, it either symbolizes a logical connective or separates two subexpressions. The illustration shows four formulas from the original edition of the *Principia Mathematica* and their translation into today's standard notation.

two years later, he fell into a deep depression from which he never recovered for the rest of his life.

Bertrand Russell disagreed with Frege's pessimism. He realized that the discovered antinomies arise from the construction of sets being *too large* to make sense as a self-contained whole. By cleverly modifying the underlying axioms, Russell saw a chance to gain sufficient control over the concept of a set, and together with the British mathematician Alfred North Whitehead, he gave the realization of Frege's dream another try. It was the second major attempt to supplement mathematics with a secure footing.

Ten years of intensive work made the result tangible: The *Principia Mathematica*, published between 1910 and 1913, had been completed (Figure 1.33). Russell and Whitehead created a monumental work that far surpassed Frege's account in scope and depth. In over 1800 pages, spread over three volumes, the authors systematically derived large parts of mathematical knowledge from a small set of axioms. Undoubtedly, the *Principia Mathematica* is one of world history's most remarkable writings.

Russell and Whitehead's monumental work highlights the advantages and disadvantages of fully formalized mathematics. On the one hand, the *Principia* shows that almost all areas of ordinary mathematics can be captured with a precision unattainable in any other science. All proofs are meticulously elaborated, formally deriving the theorem from a fixed set of axioms by applying a sequence of inference rules. On the other hand, the high degree of precision takes its toll in an enormous increase

1.2 The Path to Modern Mathematics

---

360   PROLEGOMENA TO CARDINAL ARITHMETIC   [PART II

∗54·42.   ⊢ :: α ϵ 2 . ⊃ :. β ⊂ α . ⱻ! β . β ≠ α . ≡ . β ϵ ι''α

   Dem.

⊢ . ∗54·4 .   ⊃ ⊢ :: α = ι'x ∪ ι'y . ⊃ :.

                β ⊂ α . ⱻ! β . ≡ : β = Λ . v . β = ι'x . v . β = ι'y . v . β = α : ⱻ! β :

[∗24·53·56.∗51·161]   ≡ : β = ι'x . v . β = ι'y . v . β = α   (1)

⊢ . ∗54·25 . Transp . ∗52·22 . ⊃ ⊢ : x ≠ y . ⊃ . ι'x ∪ ι'y ≠ ι'x . ι'x ∪ ι'y ≠ ι'y :

[∗13·12]   ⊃ ⊢ : α = ι'x ∪ ι'y . x ≠ y . ⊃ . α ≠ ι'x . α ≠ ι'y   (2)

⊢ . (1) . (2) . ⊃ ⊢ :: α = ι'x ∪ ι'y . x ≠ y . ⊃ :.

                β ⊂ α . ⱻ! β . β ≠ α . ≡ : β = ι'x . v . β = ι'y :

[∗51·235]   ≡ : (ⱻz) . z ϵ α . β = ι'z :

[∗37·6]   ≡ : β ϵ ι''α   (3)

⊢ . (3) . ∗11·11·35 . ∗54·101 . ⊃ ⊢ . Prop

∗54·43.   ⊢ :. α, β ϵ 1 . ⊃ : α ∩ β = Λ . ≡ . α ∪ β ϵ 2

   Dem.

⊢ . ∗54·26 . ⊃ ⊢ :. α = ι'x . β = ι'y . ⊃ : α ∪ β ϵ 2 . ≡ . x ≠ y .

[∗51·231]   ≡ . ι'x ∩ ι'y = Λ .

[∗13·12]   ≡ . α ∩ β = Λ   (1)

⊢ . (1) . ∗11·11·35 . ⊃

   ⊢ :. (ⱻx, y) . α = ι'x . β = ι'y . ⊃ : α ∪ β ϵ 2 . ≡ . α ∩ β = Λ   (2)

⊢ . (2) . ∗11·54 . ∗52·1 . ⊃ ⊢ . Prop

   From this proposition it will follow, when arithmetical addition has been defined, that $1 + 1 = 2$.

Figure 1.34: Formal proof of the arithmetic relationship $1 + 1 = 2$ in the logical framework of the *Principia Mathematica*

in complexity. Figure 1.34 shows one of the most famous passages of the *Principia*, depicting the final step of the proof of the arithmetic relationship $1 + 1 = 2$.

In light of the historical events, it is clear why a large part of the *Principia* is devoted to *type theory*, a variant of set theory incapable of replicating Frege's contradictions. To circumvent the antinomies, Russell opted for a hierarchical arrangement of sets. At the lowest level are the sets of type 1, which only contain elements of the individual domain.

At the next level are the type 2 sets, which are allowed to contain elements of the individual domain, as well as sets of type 1. Type 3 sets may also contain type 2 sets, and so on. Since a type-n-set must never contain an element of type n, no set that conforms to the typing rules of the *Principia* can ever contain itself. By introducing a hierarchy of sets, Russell and Whitehead successfully avoided the kind of self-reference that plunged mathematics into its most profound crisis a few years earlier.

However, Russell's type theory failed to stand the test of time for two main reasons. First, it restricted the construction of sets to such an extent that many presumably harmless sets could no longer be formed. Second, the clunky hierarchy made reasoning in the logic of the Principia much more laborious than in Frege's logic.

1.2.6 Axiomatic Set Theory

To a large extent, modern set theory has been shaped by the formal axiomatic approaches of Ernst Zermelo (Figure 1.35) and Abraham Fraenkel [55, 229]. Zermelo laid the cornerstone in 1907 by introducing seven axioms, initially expressed in colloquial terms [229] (Figure 1.36). It wasn't until 1929 that Thoralf Skolem formalized these axioms using predicate logic [190]. While Zermelo's set theory significantly improved Russell and Whitehead's type theory, it remained too restrictive. For example, it could not accommodate the formation of the (well-behaved) set of all sets with n elements, meaning it could not fully represent the consistent core of Frege's logic.

In the following years, mathematics walked a fine line. On the one hand, it became necessary to extend Zermelo's axioms so that all benign sets could be guaranteed to exist. On the other hand, it was paramount to avoid antinomies, necessitating a cautious and conservative extension of the axioms. In the course of events, Zermelo's set theory was supplemented by Abraham Fraenkel in 1922 with the *axiom of replacement* and by Zermelo in 1930 with the *axiom of foundation* [55, 230] (Figure 1.36). In his 1930 version, Zermelo temporarily dropped the axiom of infinity and considered the axiom of choice no longer belonging to general set theory.

Today, *Zermelo-Fraenkel set theory*, or ZF for short, is made up of the following nine axioms:

Ernst Friedrich Ferdinand Zermelo (1871 – 1953) [107]

Figure 1.35: With seven colloquially formulated axioms, the German mathematician Ernst Zermelo laid the foundations of axiomatic set theory.

1.2 The Path to Modern Mathematics

- **Axiom I (Axiom of Extensionality)**
  "If every element of a set M is also an element of N and vice versa, if, therefore, both $M \subseteq N$ and $N \subseteq M$, then always $M = N$; or, more briefly: Every set is determined by its elements."

- **Axiom II (Axiom of Elementary Sets)**
  "There exists a (fictitious) set, the null set, 0, that contains no element at all. If a is any object of the domain, there exists a set $\{a\}$ containing a and only a as element; if a and b are any two objects of the domain, there always exists a set $\{a,b\}$ containing as elements a and b but no object x distinct from both."

- **Axiom III (Axiom of Separation)**
  "Whenever the propositional function $\Phi(x)$ is definite for all elements of a set M, M possesses a subset M_Φ containing as elements precisely those elements x of M for which $\Phi(x)$ is true."

- **Axiom IV (Axiom of the Power Set)**
  "To every set T there corresponds another set $\mathfrak{U}T$, the power set of T, that contains as elements precisely all subsets of T."

- **Axiom V (Axiom of the Union)**
  "To every set T there corresponds a set $\mathfrak{S}T$, the union of T, that contains as elements precisely all elements of the elements of T."

- **Axiom VI (Axiom of Choice)**
  "If T is a set whose elements all are sets that are different from 0 and mutually disjoint, its union $\mathfrak{S}T$ includes at least one subset S_1 having one and only one element in common with each element of T."

- **Axiom VII (Axiom of Infinity)**
  "There exists in the domain at least one set Z that contains the null set as an element and is so constituted that to each of its elements a there corresponds a further element of the form $\{a\}$, in other words, that with each of its elements a it also contains the corresponding set $\{a\}$ as an element."

- **Axiom of Extensionality (B)**
  "Every set is determined by its elements, provided that it has any elements at all."

- **Axiom of Separation (A)**
  "Every propositional function $\mathfrak{f}(x)$ separates from every set m a subset $m_{\mathfrak{f}}$ containing all those elements x for which $\mathfrak{f}(x)$ is true. Or: To each part of a set there in turn corresponds a set containing all elements of this part."

- **Axiom of Pairing (P)**
  "If a and b are any two elements, then there is a set that contains both of them as its elements."

- **Axiom of the Power Set (U)**
  "To every set m there corresponds a set $\mathfrak{U}m$ that contains as elements all subsets of m, including the null set and m itself. Here, an arbitrarily chosen 'Urelement' u_0 takes the place of the 'null set'."

- **Axiom of the Union (V)**
  "To every set m there corresponds a set $\mathfrak{S}m$ that contains the elements of its elements."

- **Axiom of Replacement (E)**
  "If the elements x of a set m are replaced in a unique way by arbitrary elements x' of the domain, then the domain contains also a set m' that has as its elements all these elements x'."

- **Axiom of Foundation (F)**
  "Every (decreasing) chain of elements, in which each term is an element of the preceding one, terminates with finite index at an urelement. Or, what amounts to the same thing: Every partial domain T contains at least one element t_0 that has no element t in T."

Figure 1.36: Left: Zermelo set theory as formulated by Ernst Zermelo in 1908 [229]. Right: Zermelo-Fraenkel set theory as formulated by Ernst Zermelo in 1930 [230].

- Axiom of extensionality (Zermelo, 1908)
- Axiom of the empty set (Zermelo, 1908)
- Axiom of pairing (Zermelo, 1908)
- Axiom of union (Zermelo, 1908)
- Axiom of separation (Zermelo, 1908)
- Axiom of infinity (Zermelo, 1908)
- Axiom of power set (Zermelo, 1908)
- Axiom of replacement (Fraenkel, 1922)
- Axiom of foundation (Zermelo, 1930)

When the *axiom of choice* is added to the axioms, we speak of ZFC set theory (*Zermelo-Fraenkel with Choice*).

Section 3.2 will deal in detail with Zermelo-Fraenkel set theory. It will thoroughly discuss the meaning of the individual axioms and show how to formally express the colloquial formulations in predicate logic.

Zermelo-Fraenkel set theory has become the backbone of modern mathematics. On the one hand, it has created a logical foundation that is expressive enough to formalize all notions and concepts of ordinary mathematics. On the other hand, to this day, no one has succeeded in deriving a contradiction within ZF or ZFC. But does this mean that Zermelo-Fraenkel set theory is free of contradictions, or is this framework at risk of collapsing as Frege's logic did at the turn of the century? The honest answer is that we do not know, and Section 4.3 will reveal that we will never know.

1.2.7 Hilbert's Program

Over time, the increasing interest in various formal systems led to the development of a *meta-mathematics* that did not deal with the derivation of propositions *within* the system but with assertions *about* the system. In particular, three questions stood at the center of interest:

- **Completeness**

    A formal system is called *complete* if every true statement expressible in the logic under consideration is provable within the system.

In other words, for every true statement φ, there must be a finite sequence of derivation steps that deduces φ from the axioms. Note that a complete formal system is not required to reveal how a suitable sequence can be found. Completeness merely guarantees its existence.

■ **Consistency**

A formal system is called *consistent* if no statement φ is simultaneously provable with its negation, written as $\neg\varphi$. A formal system lacking this property couldn't be more useless as it would allow us to prove any statement.

■ **Decidability**

A formal system is called *decidable* if there is a systematic procedure capable of determining whether a present formula is provable within the calculus or not. Decidability paves the way to mechanized mathematics. For instance, if number theory were complete and decidable, it would be possible to construct a proof for every true number-theoretical statement by mechanical means. The dream of many mathematicians would come true.

Hilbert believed in the existence of a complete, consistent, and decidable axiomatization of mathematics. His efforts to establish such a system materialized in the 1920s. In particular, he sought to replicate ordinary mathematics in a formal system comprising all ordinary methods of proof. This system would appear as a formal variant of classical mathematics when viewed from the inside. However, when viewed from the outside, it would present itself as a formal collection of axioms and inference rules. If Hilbert could ensure that no contradictions arose from the axioms, the correctness of all the proof methods embodied within the system would follow.

Little would be gained, of course, if the proof of consistency utilized the same controversial proof methods that were meant to be backed up. Therefore, Hilbert sought to prove the consistency of mathematics exclusively by *finite means*. Roughly speaking, this term encompassed all proof methods whose correctness was beyond question. I.e., it excluded all methods stressing the concept of infinity and all non-constructive methods, such as the indirect proof (*reductio ad absurdum*), which builds upon the law of excluded middle (*tertium non datur*). The exclusion of these methods was a tribute to the philosophical movement of *intuitionism*, which gained increasing popularity at the beginning of the twentieth century and called for strictly constructive mathematics.

Intuitionism, as opposed to logicism and formalism, was the third philosophical movement in twentieth-century mathematics. It was established in 1907 by the Dutch mathematician Luitzen Egbertus Jan Brouwer and found prominent advocates in Arend Heyting, Stephen Kleene, and Michael Dummett.

In Brouwer's view, mathematics builds upon intuitively evident concepts that do not require a definition, such as natural numbers or the continuous passage of time. In other words, he only considered mentally constructible objects as existing. Brouwer identified the truth of a statement with its provability. For example, he considered a statement of the form $\varphi \vee \psi$ as true only if a proof for φ or a proof for ψ could be constructed. Consequently, the statement $\varphi \vee \neg\varphi$ is not universally valid in the intuitionistic sense; it is only true if it is possible to come up with a proof for φ or $\neg\varphi$. As a result, established basic assumptions, such as the *law of excluded middle (tertium non datur)* and the resulting proof by contradiction, lose their legitimacy.

In [93], Hilbert commented on the intuitionist movement as follows:

"*Taking the principle of excluded middle from the mathematician would be the same, say, as proscribing the telescope to the astronomer or to the boxer the use of his fists.*" [95]

Today, intuitionism has mainly historical relevance. Towards the end of the twentieth century, the intuitionistic line of thought was almost entirely displaced from school mathematics, accompanied by a fading interest in philosophical questions.

If Hilbert were to succeed, it would be clear once and for all that modern mathematics is built on safe grounds. We would no longer need to fear a tectonic quake, such as the one that struck Frege's logic years before. Incidentally, Hilbert would also have been victorious in another arena. The program's success would be a finishing blow to intuitionism, which Hilbert tried to combat throughout his life.

Initially, the program proceeded as planned. Two of Hilbert's students, Wilhelm Ackermann and John von Neumann, found a finite consistency proof for a weakened variant of Peano arithmetic. Back then, it only seemed a matter of time before solving the technical problems and extending the proof to cover all parts of Peano arithmetic. However, the hope of a quick success did not materialize; instead, one failure followed another. Peano arithmetic seemed to be shielded by an invisible barrier, strong enough to repel all attempts of proof.

In 1929, Hilbert's hopes were nourished by the work of the young mathematician Kurt Gödel (Figure 1.37), who proved in his doctoral thesis the completeness of a system known as *engerer Funktionenkalkül*, which, in large parts, is what we call first-order predicate logic today [71]. Gödel demonstrated the existence of a formal system capable of deriving every universally valid first-order predicate-logic formula from the axioms in a finite number of steps. His proof suggested that the logical reasoning apparatus was strong enough to realize Hilbert's program. In those days, the program appeared to be on track, and it seemed only a matter of time before Hilbert's vision became reality.

In 1930, however, developments took an unexpected twist. On September 8, in his hometown, Königsberg, Hilbert reaffirmed his deep conviction about the solvability of every scientific problem. An excerpt from his speech was broadcast in a radio address (Figure 1.38).

At the time of his speech, Hilbert was still unaware of the events that had occurred elsewhere in Königsberg the day before and were to change mathematics forever. The events took place at the 2nd Conference on Epistemology of the Exact Sciences, a three-day conference held by the Berlin Society for Empirical Philosophy from September 5 to 7, 1930. The first day began with several lectures about the three major philosophical currents in mathematics. Logicism was represented by Rudolf Carnap, intuitionism by Arend Heyting, a student of Brouwer, and formalism by John von Neumann. Among the speakers on the second day was Kurt Gödel, who gave a brief twenty-minute lecture on the proof of completeness he had developed in his dissertation.

Figure 1.37: Kurt Gödel is widely regarded as the greatest logician of all time. His groundbreaking discoveries have forced us to rethink our understanding of the mathematical method from the ground up.

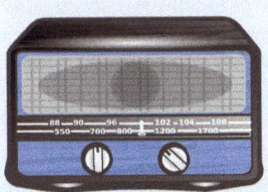

"The instrument that mediates between theory and practice, between thought and observation, is mathematics; it builds the connecting bridge and makes it stronger and stronger. Thus it happens that our entire present-day culture, insofar as it rests on intellectual insight into and harnessing of nature, is founded on mathematics. Already, Galileo said: Only he can understand nature who has learned the language and signs by which it speaks to us; but this language is mathematics and its signs are mathematical figures. Kant declared, 'I maintain that in each particular natural science there is only as much true science as there is mathematics.' In fact, we do not master a theory in natural science until we have extracted its mathematical kernel and laid it completely bare. Without mathematics today's astronomy and physics would be impossible; in their theoretical parts, these sciences unfold directly into mathematics. These, like numerous other applications, give mathematics whatever authority it enjoys with the general public.

Nevertheless, all mathematicians have refused to let applications serve as the standard of value for mathematics. Gauss spoke of the magical attraction that made number theory the favorite science for the first mathematicians, not to mention its inexhaustible richness, in which it so far surpasses all other parts of mathematics. Kronecker compared number theorists with the Lotus Eaters, who, once they had sampled that delicacy, could never do without it. With astonishing sharpness, the great mathematician Poincaré once attacked Tolstoy, who had suggested that pursuing 'science for science's sake' is foolish. The achievements of industry, for example, would never have seen the light of day had the practical-minded existed alone and had not these advances been pursued by disinterested fools. The glory of the human spirit, so said the famous Königsberg mathematician Jacobi, is the single purpose of all science. We must not believe those, who today with philosophical bearing and a tone of superiority prophesy the downfall of culture and accept the ignorabimus. For us there is no ignorabimus, and in my opinion even none whatever in natural science. In place of the foolish ignorabimus let stand our slogan: We must know, We will know."

Figure 1.38: Excerpt from David Hilbert's radio address from 1930 [193]

The bombshell dropped on the third day when Gödel spoke up during the final panel discussion. He pointed out that the consistency of a formal system, such as the *Principia Mathematica*, could not guarantee that all derived theorems were substantively true statements. Even if the Principia were proven consistent, it could not be ruled out that a statement about the natural numbers was derivable within the system that turned out to be substantively false when viewed outside the system. Then came the crucial sentence:

> "One can – assuming the consistency of classical mathematics – even give examples of sentences (namely those of Goldbach's or Fermat's type) that are indeed correct in content, but unprovable in the formal system of classical mathematics."
>
> Kurt Gödel, 1930 [27]

This is the earliest public formulation of Gödel's first incompleteness theorem. The bombshell had exploded, yet no one seemed to feel its

Kurt Gödel was born on April 28, 1906, in the Austro-Hungarian city of Brno. His hometown became part of the newly founded Czechoslovak Republic in 1918, which he always perceived as an exile. Aged 17, he took Austrian citizenship and moved to Vienna a year later to study theoretical physics. However, the legendary lecture on number theory by Philipp Furtwängler soon directed Gödel's interest toward the fundamentals of mathematics.

Gödel was attracted to Vienna's social and cultural life. He became part of the *Vienna Circle*, an academic assembly dedicated to the philosophical questions of science headed by Moritz Schlick. During that time, Gödel discovered the two incompleteness theorems that have fundamentally changed our mathematical view of the world.

Although conditions in Vienna gradually worsened after Hitler seized power, Gödel seemed unaware of the danger. It was not until 1940 that he took one of the last opportunities to flee to the USA. He joined the Institute for Advanced Study in Princeton, where he had previously spent several months as a visiting scholar. Due to his unusual personality and tendency towards hypochondria, Gödel was not without controversy, and it wasn't until 1953 that the IAS offered him a position as a professor. There, he found a loyal advocate in Albert Einstein, who soon became a lifelong friend. Gödel's mental condition had been unstable since his childhood and was to worsen steadily with increasing age. Marked by severe hypochondria, paranoia, and depression, Kurt Gödel died on January 14, 1978, as a result of self-induced malnutrition.

seismic waves. We do not know whether his reserved demeanor or the boldness of his statement prevented anyone in the room from commenting on Gödel's contribution. It is likely that hardly anyone in the audience fully understood what the young mathematician was talking about.

The Hungarian mathematician John von Neumann (Figure 1.40) was the only one to ask Gödel for a brief chat after the panel discussion. Like Gödel, von Neumann was an exceptional mathematician and legendary for his rapid comprehension, even during his lifetime. He seemed to be the only person in the room who fully grasped the implications of incompleteness for modern mathematics. A crucial aspect of Gödel's discoveries was their general nature, as they apply to any axiomatic system expressive enough to formalize number theory. As a result, any attempt to fix the flaw in the *Principia Mathematica* or a similar system was doomed to failure. In fact, Gödel revealed a general limit to the entire formal method. He discovered that no formal system will ever capture mathematics as a whole.

Euphorized by Gödel's findings, von Neumann began examining the consequences of the first incompleteness theorem in the coming weeks. In a letter of November 20, 1930, he informed Gödel about a striking discovery:

Dear Mr. Gödel!

I have recently been dealing with logic again, using the methods that you have so successfully used to reveal unde-

1.2 The Path to Modern Mathematics

cidable properties. In doing so, I have achieved a result that seems remarkable to me. I was able to show that the consistency of mathematics is unprovable.

John von Neumann [78]

Von Neumann had come across the second incompleteness theorem, but the letter arrived too late. Gödel discovered the theorem independently a few weeks earlier and had already submitted it for publication along with his first incompleteness theorem (Figure 1.39). His article bears the inconspicuous name *On formally undecidable propositions of Principia Mathematica and related systems I*. The publication deals primarily with deriving the first incompleteness theorem and only sketches the proof of the second. Gödel intended to work out the details in a subsequent publication, yet this never happened. His first article was so well received that he no longer felt the necessity to publish the announced second part. It was only later that a formal proof of the second incompleteness theorem was worked out by David Hilbert and Paul Bernays [97].

What is the significance of the second incompleteness theorem for mathematics? Gödel and von Neumann showed that the consistency of a formal system with a certain expressiveness cannot be proved with the system's own means. From this, it follows that the consistency of ordinary mathematics is unprovable by ordinary mathematical reasoning. But this was precisely the plan Hilbert had pursued so vehemently for years: proving the consistency of ordinary mathematics with finite means. Gödel's second theorem struck Hilbert's program a heavy blow, from which it would never recover.

Unlike von Neumann, Gödel did not regard Hilbert's program as a failure because even if the consistency of ordinary mathematics is unprovable by ordinary mathematics's means, this does not rule out the existence of any system in which a consistency proof could be carried out. In Gödel's words,

> *"it should be expressly noted that Theorem XI (and the corresponding results about M and A) in no way contradicts Hilbert's formalistic standpoint. For the latter presupposes only the existence of a consistency proof carried out by finitary methods, and it is conceivable that there might be finitary proofs which cannot be represented in P (or in M or A)."*

Kurt Gödel, 1931 [133]

John von Neumann
(1903 – 1957)

Figure 1.40: John von Neumann ranks among the leading mathematicians of the twentieth century. He was born under the name Neumann János Lajos in the Austro-Hungarian city of Budapest. He later called himself Johann von Neumann and finally took the name John von Neumann after he emigrated to the USA. His scientific legacy includes numerous contributions to various fields of mathematics. Today, his name is primarily associated with the *von Neumann architecture*, the primary organizing principle of modern computer systems.

> **Über formal unentscheidbare Sätze der Principia Mathematica und verwandter Systeme I**[1]).
>
> Von **Kurt Gödel** in Wien.
>
> 1.
>
> Die Entwicklung der Mathematik in der Richtung zu größerer Exaktheit hat bekanntlich dazu geführt, daß weite Gebiete von ihr formalisiert wurden, in der Art, daß das Beweisen nach einigen wenigen mechanischen Regeln vollzogen werden kann. Die umfassendsten derzeit aufgestellten formalen Systeme sind das System der Principia Mathematica (PM)[2]) einerseits, das Zermelo-Fraenkelsche (von J. v. Neumann weiter ausgebildete) Axiomensystem der Mengenlehre[3]) andererseits. Diese beiden Systeme sind so weit, daß alle heute in der Mathematik angewendeten Beweismethoden in ihnen formalisiert, d. h. auf einige wenige Axiome und Schlußregeln zurückgeführt sind. Es liegt daher die Vermutung nahe, daß diese Axiome und Schlußregeln dazu ausreichen, alle mathematischen Fragen, die sich in den betreffenden Systemen überhaupt formal ausdrücken lassen, auch zu entscheiden. Im folgenden wird gezeigt, daß dies nicht der Fall ist, sondern daß es in den beiden angeführten Systemen sogar relativ einfache Probleme aus der Theorie der ge-

Figure 1.39: In 1931, Kurt Gödel published two incompleteness theorems that would fundamentally change our understanding of mathematics [73]. Gödel's theorems manifest that the concepts of provability and truth cannot be aligned. They impose limits on the mathematical method that we will never be able to overcome.

In this quote, *P*, *M*, and *A* refer to a formalized variant of Peano arithmetic, set theory, and all of classical mathematics.

But what could such a system look like? First, it needed to include novel proof techniques not found in ordinary mathematics. Second, these techniques had to count as *finite means*; they had to be correct for apparent reasons. Even if Gödel's incompleteness theorems do not entirely rule out their existence, no one ever had a definite idea about their construction. Today, hardly anybody believes in the existence of such a system.

Gödel's incompleteness theorems undoubtedly belong to the paramount discoveries of twentieth-century mathematics. Chapter 4 will thoroughly examine the theorems and explain their tremendous impact on our understanding of the mathematical method.

It is reported that Hilbert initially reacted to the incompleteness theorems with anger [227]. Nevertheless, he did not ignore reality for long

1.2 The Path to Modern Mathematics

Figure 1.41: The ENIAC (Electronic Numerical Integrator And Computer) was the first fully functional calculating machine that meets virtually all definitions of a modern computer. The giant apparatus was built at the Moore School of Electrical Engineering at the University of Pennsylvania under the direction of J. Presper Eckert and John W. Mauchly. It was impressive in its sheer size. The ENIAC consisted of 30 units arranged in a U-shape across the entire space. The whole construction weighed almost 30 tons and consumed 150 kW of electricity.

and soon accepted Gödel's results as undeniable facts. For John von Neumann, however, logic would never be the same again. After lecturing on the incompleteness theorems for some time, he soon shifted his interests. He worked as a scientific advisor on the construction of the ENIAC, which we can look back on as the world's first universal computer (Figure 1.41). In 1946, von Neumann published a groundbreaking concept for the organization of microcomputers. The *von Neumann architecture* is the basis of most computer systems to this day [147].

Bertrand Russell also withdrew almost entirely from logic over the following years. In hindsight, it is difficult to fully grasp the immense intellectual effort required to write *Principia Mathematica*. What is certain, however, is that the decade devoted to this monumental work left a profound and lasting impact on Russell's remarkable mind (Figure 1.42).

Other mathematicians reacted with ignorance to the publication of the incompleteness theorems, mainly because of the following reason: To come up with a substantively true and, at the same time, unprovable statement, Gödel constructed a complex formula that essentially claimed its own unprovability. Many of his critics considered such self-reference a necessity for obtaining unprovable statements. Consequently, they viewed Gödel's formulas as strange curiosities at the edge of an intact mathematical core – and did not bother about them anymore.

> *"[...] I always found myself hoping that perhaps Principia Mathematica would be finished some day. Moreover the difficulties appeared to me in the nature of a challenge, which it would be pusillanimous not to meet and overcome. So I persisted, and in the end the work was finished, but my intellect never quite recovered from the strain. I have been ever since definitely less capable of dealing with difficult abstractions than I was before. This is part, though by no means the whole, of the reason for the change in the nature of my work."* [180]

Figure 1.42: At an advanced age, Bertrand Russell wrote his three-volume autobiography, published between 1967 and 1969.

However, is incompleteness really just a whim of logic that plays virtually no role in ordinary mathematics? This hope was shattered in 1977 when the two mathematicians Jeff Paris and Leo Harrington devised a variant of *Ramsey's theorem* that could be formulated but not proved within Peano arithmetic. Ramsey's theorem is a problem from combinatorics named after the British mathematician Frank Plumpton Ramsey [167]. On the surface, it looks radically different from the cleverly constructed formulas Gödel used to prove the incompleteness theorems. It lacks self-reference, yet it is an unprovable formula in Gödel's sense. Today, we know that Paris and Harrington's statement is equivalent to the consistency of Peano's arithmetic. Thus, its unprovability inevitably follows from Gödel's second incompleteness theorem.

Section 4.5 will discuss *Goodstein's theorem*, an equally innocent-looking proposition that the English logician Reuben Louis Goodstein proved in 1944 with set-theoretical means [80]. Goodstein's theorem is a substantively true statement from number theory that is expressible yet unprovable within Peano arithmetic. Laurie Kirby and Jeff Paris discovered this remarkable result in 1982 [116].

Both examples show that the incompleteness discovered by Gödel is anything but a curious property of pathologically constructed statements. On the contrary, it is an omnipresent phenomenon that permeates almost all areas of mathematics.

1.2.8 The Limits of Computability

Gödel's work exposed a fundamental limit of formal mathematics; it made it unmistakably clear that a consistent and, at the same time, complete formal system cannot exist for the theory of natural numbers. Nevertheless, mathematicians persisted in hoping that at least the question of decidability could be answered positively. The incompleteness theorems do not rule out the existence of a systematic procedure capable of determining whether or not a given statement is derivable from the axioms.

To solve the decision problem mathematically, it became essential to concretize the vague concept of a *systematic procedure*. Today, our algorithmic thinking is well-trained through the daily use of computers. In the 1930s, however, the computer was still on the drawing board, and mathematicians only had vague ideas of what it meant to perform a calculation.

1.2 The Path to Modern Mathematics

> ON COMPUTABLE NUMBERS, WITH AN APPLICATION TO
> THE ENTSCHEIDUNGSPROBLEM
>
> *By* A. M. TURING.
>
> [Received 28 May, 1936.—Read 12 November, 1936.]
>
> The "computable" numbers may be described briefly as the real numbers whose expressions as a decimal are calculable by finite means. Although the subject of this paper is ostensibly the computable *numbers*, it is almost equally easy to define and investigate computable functions of an integral variable or a real or computable variable, computable predicates, and so forth. The fundamental problems involved are, however, the same in each case, and I have chosen the computable numbers for explicit treatment as involving the least cumbrous technique. I hope shortly to give an account of the relations of the computable numbers, functions, and so forth to one another. This will include a development of the theory of functions of a real variable expressed in terms of computable numbers. According to my definition, a number is computable

Figure 1.43: In 1936, Alan Turing solved the riddles around Hilbert's decision problem once and for all [212]. With the *Turing machine*, he created an abstract machine model that quickly became the cornerstone of modern computability theory.

This hurdle was taken in 1936 when the British mathematician Alan Turing published his seminal work *On computable numbers, with an Application to the Entscheidungsproblem* (Figure 1.43). To formally grasp the concept of computability, Turing constructed an abstract machine model that resembles the basic operating principles of modern computers. In his original paper, he motivated the conception of his machine, which we now call a *Turing machine*, with the following words:

> "Computing is normally done by writing certain symbols on paper. We may suppose this paper is divided into squares like a child's arithmetic book."

The quote exemplifies Turing's open-minded thinking, visible throughout his work. He begins his exploration of computability with a familiar tool from his childhood: a blank piece of squared paper. Immediately

- Replace symbol

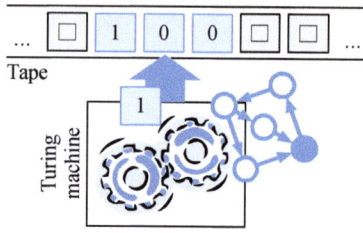

- Move head

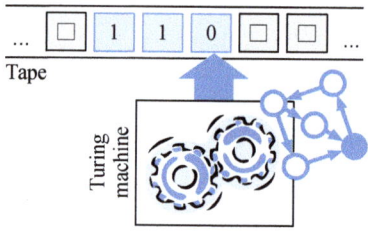

- Change state

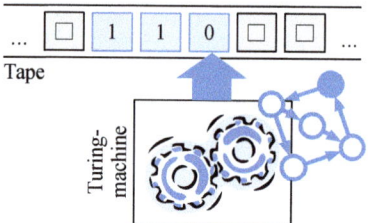

Figure 1.44: Turing defines a few primitive elementary operations from which complex calculations arise. In each processing step, a Turing machine may replace the currently viewed symbol with another and move the observed square, followed by a potential change of the inner state (state of mind).

after, Turing makes his first abstraction. He stressed that the paper's two-dimensional shape is irrelevant from a theoretical point of view. All the calculations doable by hand on paper are also feasible on a one-dimensional tape, albeit less elegant:

> "[...] I think that it is agreed that the two-dimensional character of paper is no essential of computation. I assume then that the computation is carried out on one-dimensional paper, i.e. on tape divided into squares."

Turing continues with further assumptions. First, he postulates that each tape cell may host a symbol stemming from a finite set. He further assumes that the human brain is in one of a finite number of states at any given time:

> "We may suppose that there is a **bound B** to the number of symbols or squares which the computer can observe at one moment. [...] We will also suppose that the number of states of mind which will be taken into account is finite."

Afterward, Turing defines a set of elementary operations as the basic building blocks of complex calculations. They allow the machine to replace the symbol in the currently observed cell and to shift attention to one of the surrounding cells:

> "The simple operations must therefore include: (a) Changes of the symbol on one of the observed squares. (b) Changes of one of the squares observed to another square within L squares of one of the previously observed squares."

A possible change of state accompanies both actions:

> "It may be that some of these changes necessarily involve a change of state of mind. The most general single operation must therefore be taken to be one of the following: (A) A possible change (a) of symbol together with a possible change of state of mind. (B) A possible change (b) of observed squares, together with a possible change of state of mind."

1.2 The Path to Modern Mathematics

> "The machine is to have the four m-configurations 'b', 'c', 'f', 'e' and is capable of printing '0' and '1'. The behaviour of the machine is described in the following table in which 'R' means 'the machine moves so that it scans the square immediately on the right of the one it was scanning previously'. Similarly for 'L'. 'E' means 'the scanned symbol is erased' and 'P' stands for 'prints'." [212]

Configuration		Behaviour	
m-config.	symbol	operations	final m-config.
b	None	P0, R	c
c	None	R	e
e	None	P1, R	f
f	None	R	b

Alan Mathison Turing (1912 – 1954)

Figure 1.45: The first example of a Turing machine. In his original article, Turing referred to it simply as *computing machine*. Alonzo Church coined the term Turing machine in 1937 [89].

Figure 1.44 summarizes the aforementioned elementary operations.

Turing devised his machine to characterize the set of *computable numbers*. To better understand his objective, let us examine the Turing machine depicted in Figure 1.45, the first example machine from Turing's original paper.

The machine has four states: b, c, e, and f. It starts in state b (begin) on an empty tape. The instruction table determines its exact behavior. After launching, it performs the action $P0, R$. $P0$ denotes *"Print 0"* and writes 0 onto the tape. R denotes *"Right"* and instructs the machine to move the read-write head one cell to the right. Afterward, the machine switches from state b to c. In this state, the tape remains unchanged; the machine merely moves the read-write head one cell to the right and enters state e. Now the machine writes 1 onto the tape, moves the read-write head again one cell to the right, and enters state f. After another move to the right, it reenters the initial state b (Figure 1.46). By continuously repeating this cycle, the machine produces the following output:

The generated digits represent the decimal digits of a real number; in

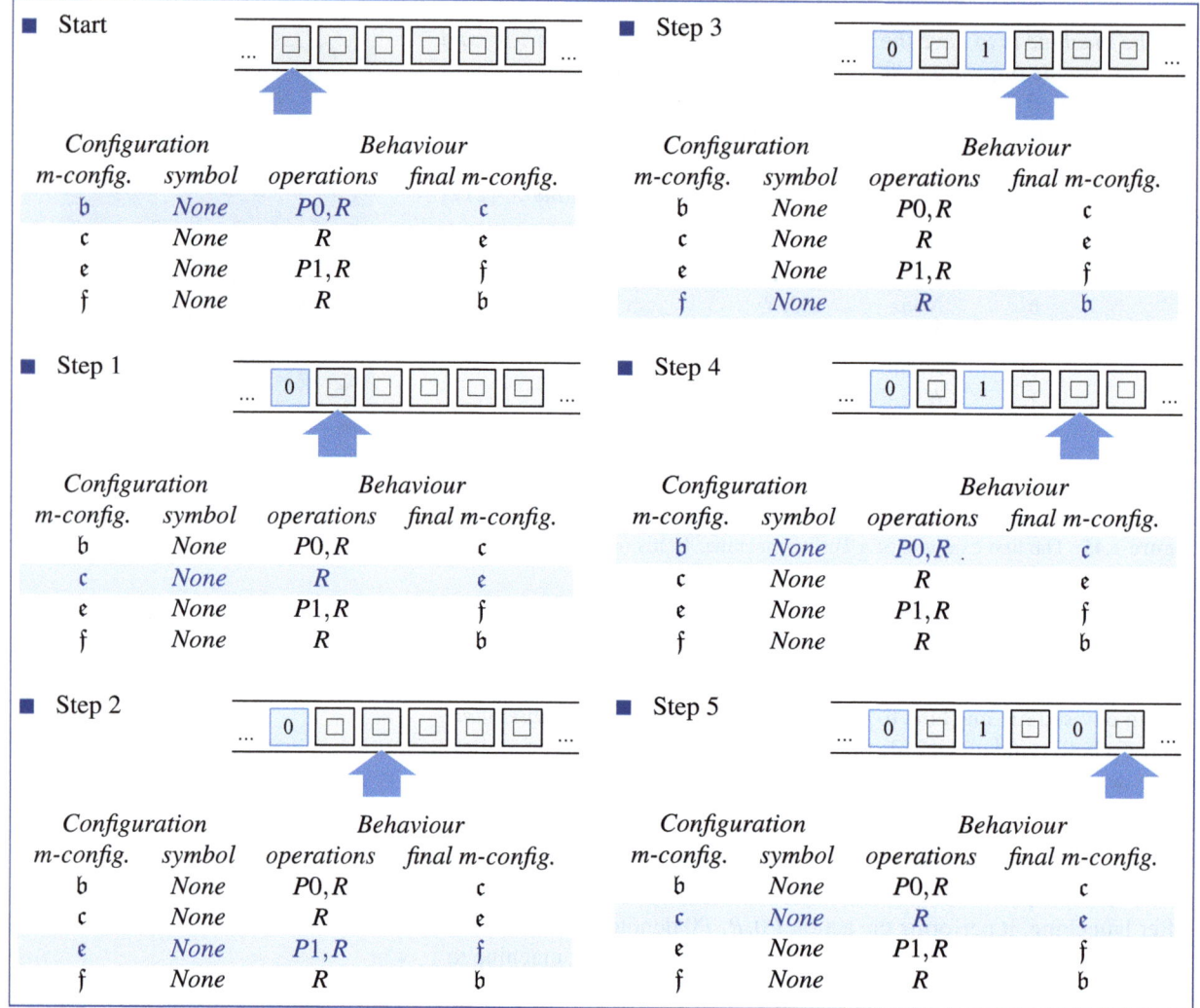

Figure 1.46: The first machine from Turing's original article in action

the depicted case, the decimal digits of the number

$$0,0101010101010101\ldots$$

The blank spaces between the digits are irrelevant to the numerical value. Turing's machine definition explicitly allows for empty cells or cells filled with symbols other than digits. Nonetheless, it is no coincidence that the examined machine leaves every other field empty. Turing

1.2 The Path to Modern Mathematics

Alan Mathison Turing was born in London Paddington on June 23, 1912. Together with his older brother, Alan grew up in England with family friends, while his mother and father, a civil servant of the British Empire, spent most of their time in Chatrapur, India. Turing's extraordinary mathematical talent was already apparent in his early youth, as was his inability to get along with social norms and state authorities.

Turing began his education at an all-day school in St. Michael's and moved to the well-known Sherborne boarding school in Dorset at 14. After graduation, he enrolled as a mathematics student at King's College, Cambridge. It was only Turing's second choice; he was denied the more prestigious Trinity College due to poor grades in non-scientific subjects.

Just one year after graduation, he achieved his scientific breakthrough. In 1936, he published *On computable numbers, with an application to the Entscheidungsproblem*, which is, in hindsight, one of the historically most pivotal publications in mathematical logic.

Afterward, the Second World War dictated the course of events. Turing went to Bletchley Park, where he worked secretly with other scientists to break the German Wehrmacht's encryption code. During this time, he created the *Turing bomb*, a computing machine capable of decoding enemy radio communication in just a few hours. Its construction was of particular significance to Turing. Although the machine differed in crucial aspects from his theoretically conceived *computing machine*, it brought many elements of his idea to life.

After the Second World War, Turing reverted his attention to theoretical topics. In 1950, he proposed the *Turing test*, originally called the *imitation game*, a technique for transferring the concept of intelligence to machines.

In 1952, Turing's career came to an abrupt end. When the police investigated his house after a burglary, he admitted to having a homosexual relationship. The prudish England of the 1950s reacted mercilessly and found Turing guilty of sexual perversion in criminal proceedings. The forced therapy turned him into a broken man. Two years later, shortly before his 42nd birthday, he was found dead next to the remains of a poisoned apple.

often fills blank spaces with auxiliary symbols that control the program flow but are irrelevant to the calculated numerical value.

With the developed conceptual framework, Turing walked a tightrope. On the one hand, the Turing machine fulfills the requirements of a formal model in all respects, allowing for a mathematically precise analysis of the concept of computability. On the other hand, it exhibits an inner simplicity and clarity that opens up a surprisingly intuitive approach to this complex subject. In contrast to purely mathematical models, such as the *lambda calculus* [34, 35] developed around the same time by Alonzo Church or the theory of *recursive functions* [53], the Turing machine seems almost close enough to touch.

The vast popularity of the Turing machine often lets us forget that the machine model was merely a means to an end. Turing aimed to solve the decision problem of mathematical logic, which required capturing the concept of computability with mathematical precision. Once he had created the necessary conceptual framework, he was able to prove two main results:

- It is impossible to devise a procedure that correctly decides, for any given Turing machine, whether or not it will write a 0 onto the tape. If such a procedure existed, it would be possible to construct a Turing machine that receives another Turing machine in encoded form and correctly calculates the answer. However, this assumption about such a machine's existence leads to a contradiction, as can be demonstrated by the same diagonalization technique used by Cantor to prove the uncountability of the continuum.

■ A Turing machine can be translated into a formula of first-order predicate logic that is universally valid if and only if the translated machine eventually outputs 0. Gödel's completeness theorem states that every universally valid formula of first-order predicate logic is provable. Thus, if Hilbert's decision problem had a solution, that is, if there were a procedure to determine the provability or unprovability of every predicate-logic formula correctly, we could decide for every Turing machine whether it outputs 0 at some point. However, as stated above, this is impossible.

With his groundbreaking result, Turing drew a line under the long-standing hunt for a decision procedure for first-order predicate logic. Leibniz's dream of mechanized mathematics was over; it was a hunt for the mathematical perpetual motion machine with no chance of success.

Turing's theory of computability has become an invaluable tool in computer science. For one thing, it lets us grasp the central concept of an *algorithm* with mathematical precision. For another, it reveals the existence of problems unsolvable by systematic procedures. However, computability theory is valuable in other respects, too. It establishes an elegant pathway to formal proof theory, allowing us to derive many of its central results more quickly than before. For example, in Section 5.4.2, we will work out how to exploit the arithmetization of Turing machines to develop an equally elegant and concise proof of Gödel's first incompleteness theorem.

In the subsequent years, computability theory evolved into a powerful tool, allowing mathematicians to identify numerous problems as formally undecidable. Hope was growing to solve problems that had persistently withstood all earlier attacks. Besides others, Hilbert's tenth problem once again attracted attention. In 1944, the mathematician Emil Leon Post commented on this problem with the following words (Figure 1.47):

> "One of the problems posed by Hilbert in his Paris address of 1900 is the problem of determining for an arbitrary Diophantine equation with rational integral coefficients whether it has, or has not, a solution in rational

---

Turing's first main result is, in essence, what we today call the *unsolvability of the halting problem*. There is a simple reason why the term *halting problem* does not appear even once in his original work. The definition of his machine model differs from the modern one in essential aspects. Turing imagined his *computable machines* to calculate real numbers, thus letting them write an infinite sequence of digits onto the tape and never stop under normal conditions.

It was not until 1958 that Martin Davis brought the machine model into its current form and mentioned the halting problem for the first time [45]. This problem addresses the question of the existence of a systematic procedure that always correctly decides whether a given Turing machine terminates for a particular input. We can answer this question negatively with the same means that Turing used to obtain his first main result, which largely legitimizes the frequently voiced claim that Turing proved the undecidability of the halting problem, even though the term was not coined until four years after his death. In Section 5.1.1, we will look closely at the structure and operation of Turing machines in their modern form.

1.2 The Path to Modern Mathematics

integers. [...]. *The above problem of Hilbert begs for an unsolvability proof."* [161]

Post's student Martin Davis was among the first to take on the challenge. In 1953, he achieved a remarkable intermediate result [44], which he later extended, together with Hilary Putnam and Julia Robinson, to an almost complete proof of the undecidability of Hilbert's tenth problem [48]. In 1961, the authors proved that a decision procedure for *exponential* Diophantine equations cannot exist. In such equations, variables may also occur as exponents.

Yuri Matiyasevich closed the gap in 1970 (Figure 1.48) [132]. The young Russian mathematician demonstrated that exponential Diophantine equations can be reduced to ordinary Diophantine equations. Consequently, a decision procedure for ordinary Diophantine equations can solve the exponential case, too. From now on, it was clear that the decision procedure sought by Hilbert could not exist. The riddle of Hilbert's tenth problem was solved, albeit in a different sense than initially anticipated.

In 1984, James Jones and Yuri Matiyasevich published a new proof demonstrating the unsolvability of Hilbert's tenth problem in a remarkably simple manner [109]. Their key idea was to translate a given *register machine* into a Diophantine equation, solvable in the integers precisely when the machine terminates. Had Hilbert's envisioned decision procedure existed, it would have solved the halting problem for register machines and, consequently, also for Turing machines. In retrospect, Turing's groundbreaking 1936 proof not only dismantled Hilbert's decision problem but also provided a plausible explanation for the unsolvability of Hilbert's tenth problem.

1.2.9 Risen from Ruins

Gödel and Turing's works were a head-on attack on the foundations of mathematics. By the end of the 1930s, Hilbert's program lay in ruins, as did the vision of mechanized mathematics. Still, the insights of the twentieth century should inspire a specific area: set theory. Gödel intensified his study of set-theoretical problems at the end of the 1930s and quickly reaped his labor's first fruits in relative consistency proofs. In such a proof, the consistency of some system B is not shown directly; it is merely proved that the consistency of another system, A, carries over to B.

Emil Leon Post
(1897 – 1954)

Figure 1.47: Emil Post is primarily associated with the *Post correspondence problem* today, one of the most famous undecidable problems ever discovered.

Yuri Matiyasevich
(geb. 1947) [131]

Figure 1.48: In 1970, the Russian mathematician Yuri Matiyasevich closed the final gap in the proof of the unsolvability of Hilbert's tenth problem.

- **Construction schema**

$$0 := \emptyset$$
$$n+1 := n \cup \{n\}$$

- **Examples**

$$1 = \{0\}$$
$$= \{\emptyset\}$$
$$2 = \{0,1\}$$
$$= \{\emptyset, \{\emptyset\}\}$$
$$3 = \{0,1,2\}$$
$$= \{\emptyset, \{\emptyset\}, \{\emptyset, \{\emptyset\}\}\}$$

Figure 1.49: Set representation of natural numbers

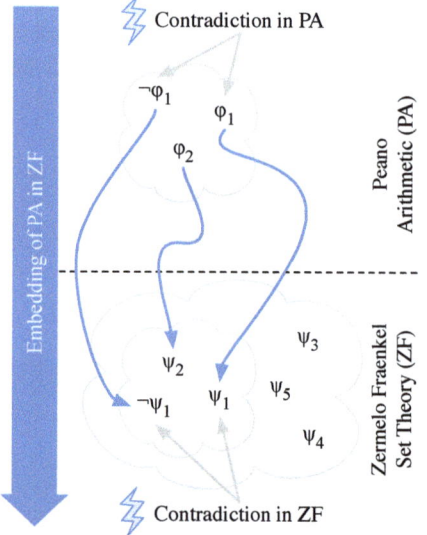

Figure 1.50: Relative proof of consistency. Each formula φ_i in Peano Arithmetic (PA) is mapped to a corresponding formula ψ_i in Zermelo-Fraenkel set theory (ZF) such that if φ_i is provable in PA, then its counterpart ψ_i is provable in ZF. As a result, any contradiction within PA would manifest in ZF, meaning that the consistency of ZF implies the consistency of PA.

Taking Zermelo-Fraenkel set theory (ZF) and Peano arithmetic (PA) as examples, let us outline the process of conducting a relative consistency proof. Anticipating Section 3.2.2, we note that every natural number is representable as a set (Figure 1.49) [143, 146]. This allows us to treat PA as a subset of ZF, enabling the translation of any number-theoretic statement from PA into an equivalent set-theoretic statement within ZF. For example, an arithmetic statement of the form

"For all numbers x, it holds that ..."

translates to:

""For all sets x, if x represents a number, it holds that ..."

We have crossed the finish line once the translation guarantees that every statement provable in PA becomes a statement provable in ZF. Then, every contradiction derivable within PA would create a contradiction in ZF (Figure 1.50). In other words, the consistency of ZF implies the consistency of PA.

With a similar construction, Gödel established the relative consistency between ZF and ZFC [75]. His approach built upon John von Neumann's idea of arranging sets hierarchically [143]. Analogous to the von Neumann hierarchy V, Gödel imposed a similar hierarchy L on sets generated by the repeated application of specific construction rules. Gödel called these sets *constructible*. Since every constructible set is also a set in the broader hierarchy, the relationship $L \subseteq V$ applies.

Gödel was interested in the consequences of the assumption that every set is constructible. To this end, he enriched the ZF axioms with the *axiom of constructibility*, $(V=L)$ for short. Through clever construction, he embedded the theory ZF+$(V=L)$ in ZF such that every statement provable in ZF+$(V=L)$ translates into a statement provable in ZF. As every contradiction derivable in ZF+$(V=L)$ leads to a contradiction in ZF, the consistency of ZF implies the consistency of ZF+$(V=L)$. This is an exciting result on its own, but the crucial step is yet to come: The axiom of choice is a theorem in ZF+$(V=L)$. Accordingly, the axiom of choice is compatible with the axioms of ZF+$(V=L)$ and, thus, even more so with the axioms of ZF. In other words, if ZF set theory is free of contradictions, we can safely add the axiom of choice as a further axiom. In the same way, Gödel showed that adding the continuum hypothesis does not create any contradiction either.

Had Gödel's latest coup accomplished what Cantor had been striving for to the end of his life? Had he finally unveiled the last great riddle the continuum kept hidden for so long? Even though Gödel's achievement

1.2 The Path to Modern Mathematics

Paul Joseph Cohen was born on April 2, 1934, in Long Branch, New Jersey. From an early age, Cohen was considered a mathematical prodigy. After attending Brooklyn College in New York for two years, he transferred to the University of Chicago, where he received his master's degree in 1954, followed by a doctorate four years later. His professional career continued at the Massachusetts Institute of Technology, with some time spent at the Institute for Advanced Study in Princeton between 1959 and 1961. Cohen joined Stanford University in 1961 and was appointed professor in 1964.

Cohen made California his new home, which felt less stressful than his previous stations on the East Coast. At Stanford, he found the ideal environment to intensify his work on the fundamental problems of set theory. 1963 was the year of his breakthrough when he confirmed the long-held conjecture that neither the axiom of choice nor the continuum hypothesis can be proved nor refuted within Zermelo-Fraenkel set theory.

In 1966, he was honored with the Fields Medal for his groundbreaking discovery. The medal, awarded every four years, is the most prestigious in mathematics. It is of similar importance as the Nobel Prize in scientific disciplines.

In the 1970s, Cohen intensified his work on number-theoretical problems. He shifted his interest to the *Riemann conjecture*, one of modern mathematics's most significant unsolved problems. Cohen passionately worked on this conjecture until the end of his life but was denied the chance of finding an answer. Paul J. Cohen died on March 23, 2007, of a rare lung disease.

is precious, it was only a partial success, as the compatibility of the continuum hypothesis with the ZF axioms does not imply its truth. Gödel already suspected that negating the continuum hypothesis would not create a contradiction either. Should his conviction become a certainty, the continuum hypothesis would thus be undecidable within Zermelo-Fraenkel set theory – neither provable nor refutable. In the years to come, Gödel thought to have crossed the finish line several times, but subtle proof errors kept him from fully resolving the issue.

It was not until 1963 that Gödel's conjecture became a certainty when the Americal mathematician Paul Cohen proved that both the negation of the axiom of choice and the negation of the continuum hypothesis can be consistently added to the ZF axioms [37–39] (Figure 1.51). Combined with Gödel's results, it followed that neither the axiom of choice nor the continuum hypothesis can be proved nor disproved within Zermelo-Fraenkel set theory.

Cohen's work is a landmark on the path to modern set theory. To prove the independence of both axioms, he introduced a new method called *forcing*, which provides the necessary means for deriving *models* with certain controllable properties. Unlike Gödel, who utilized the constructibility axiom ($V=L$) to construct an *inner model* of set theory, Cohen decided to expand the model. Cohen managed to enforce specific properties in the larger model, such as the falsity of the continuum hypothesis. In this way, Cohen proved that the negation of the continuum hypothesis is compatible with ZF and can thus be safely added to the existing axioms. One of the outstanding properties of Cohen's forc-

> ### THE INDEPENDENCE OF THE CONTINUUM HYPOTHESIS
>
> By Paul J. Cohen*
>
> DEPARTMENT OF MATHEMATICS, STANFORD UNIVERSITY
>
> *Communicated by Kurt Gödel, September 30, 1963*
>
> This is the first of two notes in which we outline a proof of the fact that the Continuum Hypothesis cannot be derived from the other axioms of set theory, including the Axiom of Choice. Since Gödel[3] has shown that the Continuum Hypothesis is consistent with these axioms, the independence of the hypothesis is thus established. We shall work with the usual axioms for Zermelo-Fraenkel set theory,[2] and by Z-F we shall denote these axioms without the Axiom of Choice, (but with the Axiom of Regularity). By a model for Z-F we shall always mean a collection of actual sets with the usual ϵ-relation satisfying Z-F. We use the standard definitions[3] for the set of integers ω, ordinal, and cardinal numbers.
>
> THEOREM 1. *There are models for Z-F in which the following occur:*
>
> (1) *There is a set a, $a \subseteq \omega$ such that a is not constructible in the sense of reference 3, yet the Axiom of Choice and the Generalized Continuum Hypothesis both hold.*
>
> (2) *The continuum (i.e., $\mathcal{P}(\omega)$ where $\mathcal{P}$ means power set) has no well-ordering.*
>
> (3) *The Axiom of Choice holds, but $\aleph_1 \neq 2^{\aleph_0}$.*
>
> (4) *The Axiom of Choice for countable pairs of elements in $\mathcal{P}(\mathcal{P}(\omega))$ fails.*
>
> Only part 3 will be discussed in this paper. In parts 1 and 3 the universe is well-

Figure 1.51: In 1963, Paul Cohen proved that adding the negation of the continuum hypothesis to the axioms of Zermelo-Fraenkel set theory does not lead to contradictions. Combined with earlier results, his proof assured that the continuum hypothesis could neither be proved nor disproved within ZF set theory.

ing method is its generality. In subsequent years, mathematicians successfully applied the framework to other properties, making Cohen's method one of the standard instruments for deriving new independence results over time.

The forcing method is closely related to the theory of *Boolean models*, introduced in the 1960s by Dana Scott, Robert Solovay, and Petr Vopěnka, with the rationale of making Cohen's approach more intuitive [7, 185]. Section 7.4 will dissect the structure of a Boolean model in a little more detail.

For Gödel, however, the mystery of the continuum hypothesis was still unsolved. Long before Cohen published his proof, he had explicitly pointed out that the undecidability proof, which still was to be found

1.2 The Path to Modern Mathematics

back then, would not clarify the continuum hypothesis. What Gödel expressed at that time was testimony to his Platonic view on mathematics. For him, sets were truly existing entities in a world of thought, implying that the continuum hypothesis must be true or false in an absolute sense. Accordingly, undecidability merely means that the underlying axioms are insufficient to determine the truth value of a proposition. To finally settle the riddle of the continuum hypothesis, Gödel considered it inevitable to add further axioms to set theory.

In a publication from 1947, Gödel expresses his view as follows:

> "There might exist axioms so abundant in their verifiable consequences, shedding so much light upon a whole discipline, and furnishing such powerful methods for solving given problems [...] that quite irrespective of their intrinsic necessity they would have to be assumed at least in the same sense as any well established physical theory." [76]

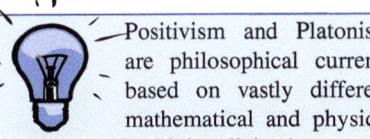

Positivism and Platonism are philosophical currents based on vastly different mathematical and physical worldviews. Positivism links the concept of existence to the observable. Accordingly, a particular question is only meaningful if it is decidable by objective experiments. Metaphysical views or theories are considered meaningless. The question of whether, for instance, natural numbers exist independently as part of an independent world of thought or whether they are an exclusive product of human imagination is neither affirmed nor denied by a positivist; instead, it is rejected as devoid of meaning.

Platonism, on the other hand, acknowledges mathematical concepts and relationships as actual entities of thought that exist in an objective sense. Therefore, a statement's truth or falsity is a property that exists independently of proof or refutation. Consequently, in physics and mathematics, Platonists see the scientist as a discoverer rather than a creator.

His words underline the dramatic changes the axiomatic method has undergone throughout history. While the Greeks once regarded the geometric axioms as objective, apparent truths that did not require any justification, the axioms of modern set theory are far from being alike. Their legitimacy does not stem from their plausibility but rather from their consequences. Modern mathematics follows the procedure of modern physics in this respect. For instance, quantum mechanics stretches human intuition to the limit. Yet, we accept it as a legitimate theory because its conclusions align with the fundamental phenomena of nature.

But which axioms are the most appropriate to let Zermelo-Fraenkel set theory adequately describe the concept of a set while simultaneously providing answers to previously unresolved problems? Gödel suspected the answer in the area of *infinity axioms*. Roughly speaking, an axiom of this kind postulates the presence of numbers so large that their existence can neither be proved nor disproved within the existing framework.

Using the weakest axiom of infinity, we want to provide an insight into how adding such an axiom alters the expressive power of the resulting system. The axiom in question is the *axiom of infinity*, which we have already discussed above. It is part of ZF and requires the existence of a set with infinitely many elements. The theory resulting from removing the axiom of infinity is called $ZF^{-\omega}$. ZF has a surprising property: it can prove the consistency of $ZF^{-\omega}$. However, $ZF^{-\omega}$ itself is strong enough to formalize number theory and, according to Gödel's second incompleteness theorem, cannot prove its own consistency. Consequently, by adding the infinity axiom, we can prove theorems undecidable in the old theory.

The study of infinity axioms is an active research area. Numerous new axioms have been found that expand Zermelo-Fraenkel set theory into ever-stronger theories. However, we do not know whether the numbers postulated by the addition of large infinity axioms exist; the existence of these numbers is neither provable nor disprovable without the axiom. But should we refer to a large infinite number as existing at all? Shall we grant these numbers an absolute existence in a Platonic sense, or do we have to resort to a positivist view that does not distinguish between existence and provability? As you can see, modern set theory resurrects a philosophical debate that almost wholly fell victim to the increasing formalization of the twentieth century. And so we come full circle.

1.3 Exercises

Exercise 1.1

The axioms listed below are part of Frege's famous *Begriffsschrift*:

- Propositional logic

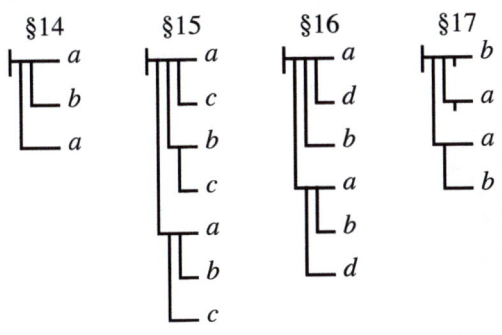

- Predicate logic

Translate the formulas into contemporary notation.

Exercise 1.2

We have worked out in Section 1.2.2 how Cantor proved the countability of algebraic numbers in his 1874 publication. For this purpose, he assigned each equation of the form

$$a_n x^n + a_{n-1} x^{n-1} + \ldots + a_1 x + a_0$$

a *height N*, which was defined as follows:

$$N = n - 1 + |a_n| + \ldots + |a_3| + |a_2| + |a_1| + |a_0|$$

Was Cantor constrained to choose the definition in this form, or could he have replaced it with one of the simplified definitions below?

a) $N = n$

b) $N = |a_n| + \ldots + |a_3| + |a_2| + |a_1| + |a_0|$

c) $N = n + |a_n| + \ldots + |a_3| + |a_2| + |a_1| + |a_0|$

Exercise 1.3

In Section 1.2.1, you have learned how to turn the periodic decimal number $0.0\overline{238095}$ into the fraction $\frac{1}{42}$.

a) Utilize this method to prove the relationship $1 = 0.\overline{9}$.

b) Did the trick used in the transformation make you feel uncomfortable? If so, you already have a good sense of the dangers of dealing with the actual infinite. Try to prove the relationship $1 = 0.\overline{9}$ without interpreting an infinite sequence of decimal digits as a closed entity.

Exercise 1.4

Figure 1.20 shows how to merge two real numbers into a single number using the zipper method. The construction results in a bijective mapping between $\mathbb{R}^2$ and $\mathbb{R}$, thereby proving the equality of both sets.

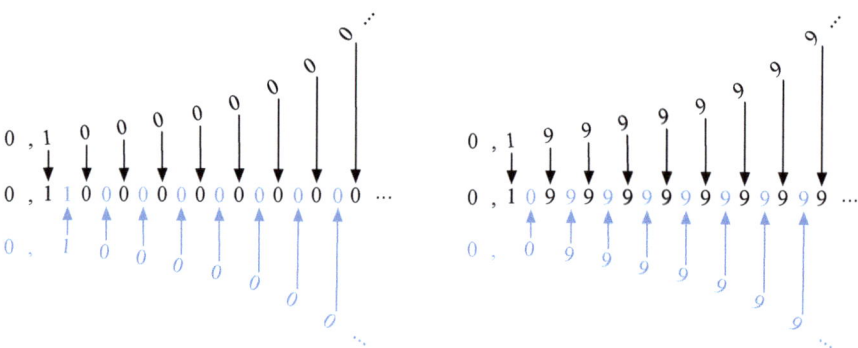

In fact, we have been cheating here a little, as the representation of a real number as a decimal fraction is sometimes ambiguous. For instance, $0.11 = 0.10\overline{9}$.

Consequently, the constructed mapping between $\mathbb{R}^2$ and $\mathbb{R}$ is not injective, thus, a fortiori, not bijective. Solve this problem without abandoning the basic idea of zipper construction.

Exercise 1.5

With *Goldbach's conjecture*, you have become acquainted with one of the most prominent yet unsolved problems in number theory. Its strong form reads like this:

"*Every even natural number $n > 2$ can be written as the sum of two prime numbers.*"

Show that the strong Goldbach conjecture implies the following variant:

"*Every odd natural number $n > 5$ can be written as the sum of three prime numbers.*"

1.3 Exercises

Exercise 1.6

This exercise deals again with Goldbach's conjecture.

a) Suppose the conjecture is wrong. In that case, could it be disproved using ordinary arithmetic?

b) Assume that Goldbach's conjecture is unprovable with the methods of ordinary mathematics. In this case, may we conclude that the conjecture is true or false?

c) Can the result be carried over to the conjecture about infinitely many twin primes?

d) Is Fermat's conjecture a mathematical theorem of Goldbach type?

Exercise 1.7

The *Erdős-Straus conjecture* states that the equation

$$\frac{4}{n} = \frac{1}{a} + \frac{1}{b} + \frac{1}{c}$$

has a solution in the natural numbers for every natural number $n > 1$.

Could we solve the conjecture with a decision procedure for Diophantine equations?

Exercise 1.8

Supplement the statements below.

The set…	empty	finite	countable	uncountable
$\{M \in \mathcal{P}(\mathbb{N}) \mid \mathbb{N} \subseteq M\}$ is	○	○	○	○
$\{M \in \mathcal{P}(\mathbb{N}) \mid |M| = |\mathbb{N}|\}$ is	○	○	○	○
$\{M \in \mathcal{P}(\mathbb{N}) \mid |M| < |\mathbb{N}|\}$ is	○	○	○	○
$\{M \in \mathcal{P}(\mathbb{N}) \mid |M| > |\mathbb{N}|\}$ is	○	○	○	○

2 Formal Systems

> *"When we are engaged in investigating the foundations of a science, we must set up a system of axioms which contains an exact and complete description of the relations subsisting between the elementary ideas of that science. The axioms so set up are at the same time the definitions of those elementary ideas; and no statement within the realm of the science whose foundation we are testing is held to be correct unless it can be derived from those axioms by means of a finite number of logical steps."*
>
> David Hilbert [98]

2.1 Concepts

In Chapter 1, we have identified the axiomatic method as the foundation of modern mathematical reasoning and showed how it has changed the face of mathematics over time. In its modern sense, conducting a proof is conceived as deriving propositions from a set of assumptions, the so-called *axioms*, by applying well-defined *inference rules*. It was only through the precise deductive nature of this approach that mathematics could develop into the exact science that we know today.

A *formal system*, or *logic calculus*, aims to mechanize the axiomatic method. To clarify this concept, we commence with investigating a specific formal system called E. We will introduce the system through a series of steps, successively defining E's syntax, axioms, inference rules, and semantics.

Syntax

Syntax defines the rules governing the structure of the expressions (formulas) the formal system deals with. At this stage, a formula is merely a sequence of uninterpreted symbols whose structure complies with formally defined construction rules.

The formulas of the example system E consist of the symbols '0', 's', '=', '>', '(', ')', and '¬'. The set of these symbols is the *alphabet* of E. Naturally, only some sequences of such characters constitute a formula, and we consider only those sequences well-formed whose structure complies with certain construction rules. The set of all formulas is called the *language* of E.

For the language of the example system E, we agree on the following construction rules:

- 0 is a term.
- If σ is a term, then $\mathsf{s}(\sigma)$ is also a term.
- If σ and τ are terms, then the following expressions are formulas:
  $$(\sigma = \tau), (\sigma > \tau), \neg(\sigma = \tau), \neg(\sigma > \tau)$$

Terms are the basic building blocks of the formal system's artificial language and symbolically represent the objects we can talk about in E. Repeatedly applying the first two formation rules allows us to generate the following terms:

$$0, \mathsf{s}(0), \mathsf{s}(\mathsf{s}(0)), \mathsf{s}(\mathsf{s}(\mathsf{s}(0))), \mathsf{s}(\mathsf{s}(\mathsf{s}(\mathsf{s}(0)))), \ldots$$

The third rule defines how to combine terms into formulas. Among others, the following formulas belong to the language of E:

$$(0 = 0), (0 > 0), \neg(0 = 0), (\mathsf{s}(\mathsf{s}(0)) = 0), \neg(0 = \mathsf{s}(0)), \ldots$$

Axioms and Inference Rules

Table 2.1 summarizes the axioms and the inference rules. Six inference rules allow us to derive new formulas from a single axiom. For each derivable formula φ, we write $\vdash \varphi$ and call φ a *theorem* of E. The expression $\vdash \varphi$ is thus the symbolic notation for the statement: "φ is provable in E."

The variables in the inference rules are placeholders representing arbitrary terms. For example, the following *formula instances* arise from the formula schema $(\mathsf{s}(\sigma) > \tau)$:

- **Substitution 1:** $\mathfrak{S} := [\sigma \leftarrow \mathsf{s}(\mathsf{s}(0)), \tau \leftarrow 0]$
  $(\mathsf{s}(\sigma) > \tau)\mathfrak{S} = (\mathsf{s}(\mathsf{s}(\mathsf{s}(0))) > 0)$

From now on, we will constantly deal with two different language levels. On one level resides the formal system's artificial language, and on the other, the ordinary language of mathematics. The latter is also known as the *meta-level* because it allows us to make statements *about* a formal system. At this level, we move outside the system, allowing us to employ the entire mathematical toolkit to analyze its axioms and inference rules. The *object level*, housing the system's formal language, does not offer this freedom. Precise rules govern syntax and semantics, imposing a corset we cannot escape.

Blending the object and meta-level is a recurring source of difficulties in understanding mathematical logic and the origin of many putative paradoxes.

To visually separate both levels, we display all formulas of the formula language in a sans-serif font (e.g., $\mathsf{s}(0) = \mathsf{s}(0)$). Especially for formal systems with a rich language, the font selection helps to distinguish ordinary mathematical statements from statements formulated in the artificial language of the system under consideration.

Sometimes, formulas appear in a mixed notation in this book, with Greek letters replacing individual formula components (e.g., $\mathsf{s}(\mathsf{s}(0)) = \sigma$). Such an expression is a *formula schema* and not a proper formula. However, every schema becomes a well-formed formula by substituting the placeholders with suitable expressions. For instance, replacing σ with $\mathsf{s}(0)$ in $\mathsf{s}(\mathsf{s}(0)) = \sigma$ results in the proper formula $\mathsf{s}(\mathsf{s}(0)) = \mathsf{s}(0)$.

2.1 Concepts

- **Substitution 2:** $\mathfrak{S} := [\sigma \leftarrow s(0), \tau \leftarrow s(s(0))]$

  $(s(\sigma) > \tau)\mathfrak{S} = (s(s(0)) > s(s(0)))$

With the groundwork laid, we can breathe life into the calculus and derive new theorems by systematically applying inference rules.

- **Example 1:** Derivation of $(s(s(s(s(0)))) > s(s(0)))$

  1. $\vdash (0 = 0)$ (A1)
  2. $\vdash (s(0) = s(0))$ (S1, 1)
  3. $\vdash (s(s(0)) = s(s(0)))$ (S1, 2)
  4. $\vdash (s(s(s(0))) > s(s(0)))$ (S2, 3)
  5. $\vdash (s(s(s(s(0)))) > s(s(0)))$ (S3, 4)

- **Example 2:** Derivation of $\neg(s(s(0)) = s(s(s(0))))$

  1. $\vdash (0 = 0)$ (A1)
  2. $\vdash (s(0) = s(0))$ (S1, 1)
  3. $\vdash (s(s(0)) = s(s(0)))$ (S1, 2)
  4. $\vdash (s(s(s(0))) > s(s(0)))$ (S2, 3)
  5. $\vdash \neg(s(s(0)) = s(s(s(0))))$ (S5, 4)

Both examples illustrate the symbolic nature of formal proofs. Due to the precise formulation of the axioms and inference rules, it becomes possible to derive theorems at the syntactic level without assigning meaning to individual formula components; conducting a proof is reduced to the mechanical manipulation of symbol strings.

The way is paved to capture the concept of a *proof* with mathematical precision.

Axioms (System E)	
$(0 = 0)$	(A1)

Inference Rules (System E)	
$\dfrac{(\sigma = \tau)}{(s(\sigma) = s(\tau))}$	(S1)
$\dfrac{(\sigma = \tau)}{(s(\sigma) > \tau)}$	(S2)
$\dfrac{(\sigma > \tau)}{(s(\sigma) > \tau)}$	(S3)
$\dfrac{(\sigma > \tau)}{\neg(\sigma = \tau)}$	(S4)
$\dfrac{(\sigma > \tau)}{\neg(\tau = \sigma)}$	(S5)
$\dfrac{(\sigma > \tau)}{\neg(\tau > \sigma)}$	(S6)

Table 2.1: Axioms and inference rules of the example calculus E. All inference rules follow a uniform format. The premise is noted above the middle bar and describes the formulas to which the inference rule applies. The statement below the middle line is the conclusion derived from the premise.

Definition 2.1 (Proof)

A formal proof is a chain of formulas $\varphi_1, \varphi_2, \ldots, \varphi_n$, which is constructed according to the following construction rules:

- φ_i is an axiom, or
- φ_i is derived from the preceding members of the proof chain by the application of an inference rule.

The last formula in this chain is the proven *theorem*.

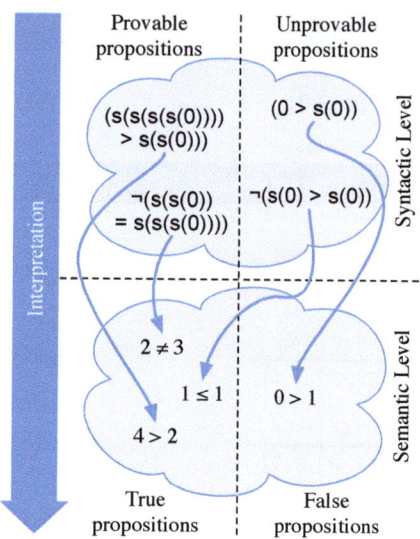

Figure 2.1: An interpretation assigns meaning to the formulas of a formal system. In this example, terms are interpreted as natural numbers, the symbol '=' as equality, and the symbol '>' as the greater-than relation on natural numbers. The symbol '¬' has the same meaning in all formal systems and represents logical negation (negation).

Semantics

After defining the language and providing axioms and inference rules for deriving new theorems, it is time to add a semantic level to our formal system. Semantics determines how the individual formula components are to be interpreted, thus giving the formulas a substantive meaning. We can only speak of true and false formulas once we have chosen a specific interpretation (Figure 2.1). Remember that the truth value of a formula generally varies between interpretations. Depending on the chosen interpretation, a formula may resemble a true or false statement.

Of particular interest are those interpretations that render all theorems of a formal system true. Such an interpretation is called a *model*. In Chapter 7, we will explore model theory, which deals in detail with the structure, properties, and systematic construction of these very special interpretations.

Before defining the semantics for our example calculus E, let's agree on the following notation:

$$\overline{n} := \underbrace{\mathsf{s}(\mathsf{s}(\ldots \mathsf{s}(0)\ldots))}_{n \text{ times}} \tag{2.1}$$

As the examples below demonstrate, the new notation allows us to write many formulas of E in a very concise way:

$$(\overline{4} > \overline{2}) \text{ symbolizes } (\mathsf{s}(\mathsf{s}(\mathsf{s}(\mathsf{s}(0)))) > \mathsf{s}(\mathsf{s}(0)))$$
$$\neg(\overline{2} = \overline{3}) \text{ symbolizes } \neg(\mathsf{s}(\mathsf{s}(0)) = \mathsf{s}(\mathsf{s}(\mathsf{s}(0))))$$
$$(\overline{0} > \overline{1}) \text{ symbolizes } (0 > \mathsf{s}(0))$$

Using the agreed notation, we define the semantics of the formal system E as follows:

$$\overline{n} \text{ represents the natural number } n \in \mathbb{N}$$
$$(\overline{n} = \overline{m}) \text{ corresponds to the statement } n = m$$
$$(\overline{n} > \overline{m}) \text{ corresponds to the statement } n > m$$
$$\neg(\overline{n} = \overline{m}) \text{ corresponds to the statement } n \neq m$$
$$\neg(\overline{n} > \overline{m}) \text{ corresponds to the statement } n \leq m$$

We have already introduced the *provability relation* '⊢', which provides a notation for expressing a formula's derivability. Similarly, we will use the *model relation* '⊨' to express a statement's truth.

2.1 Concepts

The following relationships are easy to identify:

$$\models (\overline{4} > \overline{2})$$
$$\models \neg(\overline{2} = \overline{3})$$
$$\not\models (\overline{0} > \overline{1})$$
$$\not\models \neg(\overline{1} > \overline{0})$$

In general, we define the model relation '$\models$' as follows:

$$\models (\overline{n} = \overline{m}) \;:\Leftrightarrow\; n = m \qquad (2.2)$$
$$\models (\overline{n} > \overline{m}) \;:\Leftrightarrow\; n > m \qquad (2.3)$$
$$\models \neg \varphi \;:\Leftrightarrow\; \not\models \varphi \qquad (2.4)$$

The definition ensures that $\models \varphi$ and $\models \neg\varphi$ cannot apply simultaneously to any formula φ. This property reflects the fixed semantics of the negation operator '$\neg$' and holds for every model relation. However, even more applies to our example. The relationship (2.4) ensures that for every formula φ, either $\models \varphi$ or $\models \neg\varphi$ applies. This property holds because we have deliberately linked the truth value of a formula to a specific interpretation. In this *standard interpretation*, each formula φ corresponds to an arithmetic statement being either true or false, implying $\models \varphi$ or $\models \neg\varphi$, respectively. This will no longer hold for propositional and predicate logic, where arbitrary interpretations will be considered. In these logics, $\models \varphi$ expresses that φ is universally valid; that is, it is true in every possible interpretation. If a formula φ is true for some interpretations and false for others, the same applies to $\neg\varphi$. In this case, none of the formulas φ or $\neg\varphi$ is universally valid, thus neither $\models \varphi$ nor $\models \neg\varphi$ applies. To sum up:

Theorem 2.1

- $\models \varphi$ and $\models \neg\varphi$ never hold simultaneously.
- In some logics, $\not\models \varphi$ does not imply $\models \neg\varphi$.

The provability relation '$\vdash$' and the model relation '$\models$' allow us to define fundamental properties that will be omnipresent through all remaining chapters. Remember these terms well!

Definition 2.2

A formal system is

- *consistent* if $\vdash \varphi$ implies $\not\vdash \neg\varphi$,
- *negation complete* if $\not\vdash \neg\varphi$ implies $\vdash \varphi$,
- *correct* if $\vdash \varphi$ implies $\models \varphi$,
- *complete* if $\models \varphi$ implies $\vdash \varphi$.

Accordingly, a formal system is consistent if it is impossible to derive a formula φ together with its negation $\neg\varphi$. It is negation-complete if at least one of the two alternatives φ or $\neg\varphi$ is derivable. Formal systems being both consistent and negation-complete are of particular interest. In these formal systems, and only these, either φ or $\neg\varphi$ is a theorem for each formula φ.

Consistency and negation completeness are syntactic properties, as their definitions make no use of the meaning of the individual symbols. In contrast, correctness and completeness are semantic properties; they establish a relationship between the notions of provability and truth. A formal system is correct if all its theorems are substantively true, and it is complete if every true statement is a theorem, that is if every true formula is derivable from the axioms by applying a finite number of inference steps. There arises a natural desire to define formal systems that are simultaneously correct and complete, as in these systems, the difference between provability and truthness disappears.

The syntactic concepts of consistency and negation completeness and the semantic concepts of correctness and completeness relate in two interesting ways:

Theorem 2.2

- If a formal system is complete and for all formulas, either $\models \varphi$ or $\models \neg\varphi$ holds, it is also negation-complete.
- Every correct formal system is consistent.

The first statement follows from the fact that for every formula φ, either φ itself or its negation $\neg\varphi$ is a true statement (either $\models \varphi$ or $\models \neg\varphi$ applies according to the prerequisite). Consequently, at least one of

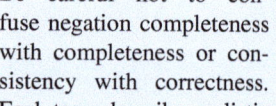

Be careful not to confuse negation completeness with completeness or consistency with correctness. Each term describes a distinct property of formal systems.
In German literature, there is an equally sharp distinction between these terms. A negation-complete calculus is referred to as *negationsvollständig*, and a complete calculus as *vollständig*. A consistent calculus is called *widerspruchsfrei*, and a correct calculus *korrekt*.
In both English and German, negation completeness is sometimes referred to as *syntactic completeness* and completeness as *semantic completeness*. If there is no risk of confusion, the prefixes *syntactic* and *semantic* are occasionally dropped, and the formal system under investigation is simply labeled as complete. Thus, when speaking of completeness, always pay attention to whether it refers to the syntactic or the semantic level. In both cases, its meaning is entirely different.

2.1 Concepts

the two formulas can be derived in a complete formal system, making it negation-complete. The second statement follows from the fact that only true statements are provable in a correct formal system. Since only one among the two formulas φ and $\neg\varphi$ can be true, only one can be derived, making the formal system consistent.

The time has come to revisit our example system E and to examine which of the properties introduced in Definition 2.2 are satisfied and which are not. A closer analysis of the axioms and inference rules allows us to draw the following conclusions:

- **E is correct and consistent.**

  The correctness follows from the special meaning assigned to the formulas of E. Under the given interpretation, the sole axiom (A1) becomes a true statement about the natural numbers, and the inference rules ensure that the truth of the premise carries over to the conclusion. Thus, by Theorem 2.2, the calculus is consistent.

- **E is neither negation complete nor complete.**

  The formal system would only be negation-complete if at least one of the formulas φ and $\neg\varphi$ were derivable for each formula φ. However, neither $(0 > 0)$ nor $\neg(0 > 0)$ is a theorem of E. Thus, the formal system is negation incomplete and also incomplete because of $\models \neg(0 > 0)$.

At this juncture, we want to enrich the formal system E with additional inference rules, summarized in Table 2.2. Our center of interest is how each of the proposed extensions affects the formal system's properties mentioned above.

We begin by discussing the formal system E_2, which differs from E by adding the inference rule (S7) (Table 2.2 left). At first glance, the new rule sacrifices correctness since it allows us to derive the conclusion $(\bar{n} > \bar{n})$ from the premise $\neg(\bar{n} > \bar{n})$. When interpreted over the natural numbers, the first formula represents a true statement, but the second is false.

Let us try to demonstrate the contradiction within E_2. To derive a false statement using (S7), we must first prove the formula $\neg(\bar{n} > \bar{n})$ for any natural number n. Afterward, we could apply (S7) to derive the theorem $(\bar{n} > \bar{n})$. Voilà: We found a formula φ with the property that both φ and $\neg\varphi$ are theorems, exposing the formal system E_2 as contradictory.

Axioms of E_2		Axioms of E_3		Axioms of E_4	
$(0 = 0)$	(A1)	$(0 = 0)$	(A1)	$(0 = 0)$	(A1)
Inference rules of E_2		**Inference rules of E_3**		**Inference rules of E_4**	
$\dfrac{(\sigma = \tau)}{(s(\sigma) = s(\tau))}$	(S1)	$\dfrac{(\sigma = \tau)}{(s(\sigma) = s(\tau))}$	(S1)	$\dfrac{(\sigma = \tau)}{(s(\sigma) = s(\tau))}$	(S1)
$\dfrac{(\sigma = \tau)}{(s(\sigma) > \tau)}$	(S2)	$\dfrac{(\sigma = \tau)}{(s(\sigma) > \tau)}$	(S2)	$\dfrac{(\sigma = \tau)}{(s(\sigma) > \tau)}$	(S2)
$\dfrac{(\sigma > \tau)}{(s(\sigma) > \tau)}$	(S3)	$\dfrac{(\sigma > \tau)}{(s(\sigma) > \tau)}$	(S3)	$\dfrac{(\sigma > \tau)}{(s(\sigma) > \tau)}$	(S3)
$\dfrac{(\sigma > \tau)}{\neg(\sigma = \tau)}$	(S4)	$\dfrac{(\sigma > \tau)}{\neg(\sigma = \tau)}$	(S4)	$\dfrac{(\sigma > \tau)}{\neg(\sigma = \tau)}$	(S4)
$\dfrac{(\sigma > \tau)}{\neg(\tau = \sigma)}$	(S5)	$\dfrac{(\sigma > \tau)}{\neg(\tau = \sigma)}$	(S5)	$\dfrac{(\sigma > \tau)}{\neg(\tau = \sigma)}$	(S5)
$\dfrac{(\sigma > \tau)}{\neg(\tau > \sigma)}$	(S6)	$\dfrac{(\sigma > \tau)}{\neg(\tau > \sigma)}$	(S6)	$\dfrac{(\sigma > \tau)}{\neg(\tau > \sigma)}$	(S6)
$\dfrac{\neg(\sigma > \tau)}{(\tau > \sigma)}$	(S7)	$\dfrac{\neg(\sigma > \tau)}{(\tau > \sigma)}$	(S7)		
		$\dfrac{(\sigma = \tau)}{\neg(\sigma > \tau)}$	(S8)	$\dfrac{(\sigma = \tau)}{\neg(\sigma > \tau)}$	(S8)

Table 2.2: Axioms and inference rules of the formal systems E_2, E_3, and E_4.

However, a closer look at the inference rules reveals that we cannot derive the formula $\neg(\bar{n} > \bar{n})$ within E_2. The ongoing lack of negation completeness prevents us from proving a false statement. As a result, E_2 remains correct despite the addition of the semantically invalid inference rule (S7). According to Theorem 2.2, it also remains consistent.

Next, consider the calculus E_3, which arises from E_2 by adding another inference rule (Table 2.2 center). The new rule (S8) makes E_3 complete; that is, every substantively true statement expressible in the limited language of our example system is now derivable. According to Theorem 2.2, E_3 is a fortiori negation complete.

2.1 Concepts

At the same time, adding (S8) has strengthened E_3 to the point where we can derive the contradiction mentioned above within the formal system. The following proof demonstrates how the complementary formula pair $\neg(0 > 0)$ and $(0 > 0)$ can be inferred:

1. $\vdash (0 = 0)$ (A1)
2. $\vdash \neg(0 > 0)$ (S8, 1)
3. $\vdash (0 > 0)$ (S7, 2)

E_3 is contradictory and thus all the more incorrect, according to Theorem 2.2.

Removing the problematic inference rule (S7) takes us straight to E_4 (Table 2.2 right). Although this system has fewer rules, it remains complete. In addition, all contradictions have disappeared. E_4 offers what we have been looking for: a complete and correct formal system capable of deriving every true statement from the axioms by applying a finite number of inference steps.

Next, we consider the calculus E_5, whose axioms and inference rules are summarized in Table 2.3. It differs from the correct and complete calculus E_4 in that all premises and conclusions appear negated. As a result, E_5 behaves complementary to E_4; that is, a statement φ is derivable in E_5 precisely when its negation is derivable in E_4. E_5 is the perfect liar, as each of its theorems is a substantively false statement, and each false statement is a theorem. Hence, E_5 is neither correct nor complete but possesses the syntactic properties of consistency and negation completeness. This system is witness to the fact that the implications formulated in Theorem 2.2 only hold in one direction.

 Theorem 2.3

- Not every negation-complete calculus is complete.
- Not every consistent calculus is correct.

Axioms of E_5	
$\neg(0 = 0)$	(A1')

Inference rules of E_5	
$\dfrac{\neg(\sigma = \tau)}{\neg(s(\sigma) = s(\tau))}$	(S1')
$\dfrac{\neg(\sigma = \tau)}{\neg(s(\sigma) > \tau)}$	(S2')
$\dfrac{\neg(\sigma > \tau)}{\neg(s(\sigma) > \tau)}$	(S3')
$\dfrac{\neg(\sigma > \tau)}{(\sigma = \tau)}$	(S4')
$\dfrac{\neg(\sigma > \tau)}{(\tau = \sigma)}$	(S5')
$\dfrac{\neg(\sigma > \tau)}{(\tau > \sigma)}$	(S6')
$\dfrac{\neg(\sigma = \tau)}{(\sigma > \tau)}$	(S8')

Table 2.3: Axioms and inference rules of the formal system E_5.

All formal systems share the idea of deriving theorems from a set of axioms by applying well-defined inference rules. Their differences usually become apparent only through closer examination. Some systems, such as the ones discussed so far, are built up from a few axioms and obtain their expressiveness through an extensive repertoire of inference rules. Others are rich in axioms and get by with just a few inference rules. It is not uncommon for such systems to provide infinitely many axioms generated from one or more *axiom schemata*.

Axioms of E_6	
$(0 = 0)$	(A1)
$(\sigma = \tau) \to (s(\sigma) = s(\tau))$	(A2)
$(\sigma = \tau) \to (s(\sigma) > \tau)$	(A3)
$(\sigma > \tau) \to (s(\sigma) > \tau)$	(A4)
$(\sigma > \tau) \to \neg(\sigma = \tau)$	(A5)
$(\sigma > \tau) \to \neg(\tau = \sigma)$	(A6)
$(\sigma > \tau) \to \neg(\tau > \sigma)$	(A7)
$(\sigma = \tau) \to \neg(\tau > \sigma)$	(A8)

Inference rules of E_6	
$\dfrac{\sigma,\ \sigma \to \tau}{\tau}$	(MP)

Table 2.4: Axioms and inference rules of the formal system E_6.

Table 2.4 portrays an example. In addition to the familiar axiom (A1), the depicted formal system E_6 features seven axiom schemata, allowing the retrieval of an infinite number of axioms. In contrast, there is only one inference rule called the *modus ponens* (MP). A comparison of E_6 with the complete and correct calculus E_4 reveals a structural but no substantive difference between the two. The theorems in both formal systems are the same.

■ **Example 1:** Derivation of $(s(s(s(s(0)))) > s(s(0)))$

1. $\vdash (0 = 0)$ (A1)
2. $\vdash (0 = 0) \to (s(0) = s(0))$ (A2)
3. $\vdash (s(0) = s(0))$ (MP, 1,2)
4. $\vdash (s(0) = s(0)) \to (s(s(0)) = s(s(0)))$ (A2)
5. $\vdash (s(s(0)) = s(s(0)))$ (MP, 3,4)
6. $\vdash (s(s(0)) = s(s(0))) \to (s(s(s(0))) > s(s(0)))$ (A3)
7. $\vdash (s(s(s(0))) > s(s(0)))$ (MP, 5,6)
8. $\vdash (s(s(s(0))) > s(s(0))) \to (s(s(s(s(0)))) > s(s(0)))$ (A4)
9. $\vdash (s(s(s(s(0)))) > s(s(0)))$ (MP, 7,8)

■ **Example 2:** Derivation of $\neg(s(s(0)) = s(s(s(0))))$

1. $\vdash (0 = 0)$ (A1)
2. $\vdash (0 = 0) \to (s(0) = s(0))$ (A2)
3. $\vdash (s(0) = s(0))$ (MP, 1,2)
4. $\vdash (s(0) = s(0)) \to (s(s(0)) = s(s(0)))$ (A2)
5. $\vdash (s(s(0)) = s(s(0)))$ (MP, 3,4)
6. $\vdash (s(s(0)) = s(s(0))) \to (s(s(s(0))) > s(s(0)))$ (A3)
7. $\vdash (s(s(s(0))) > s(s(0)))$ (MP, 5,6)
8. $\vdash (s(s(s(0))) > s(s(0))) \to \neg(s(s(0)) = s(s(s(0))))$ (A6)
9. $\vdash \neg(s(s(0)) = s(s(s(0))))$ (MP, 7,8)

Most formal systems discussed later in this book will exhibit this structure, thus encoding the relationships between the objects of the conceptional domain in the axioms rather than the inference rules. An advantage of this approach is its generality. A formal system dealing with number theory can utilize the same reasoning apparatus as a formal system dealing with set theory or another area of mathematics. Section 2.3 and Section 2.4 will elaborate on the inference apparatus in detail. They will present the propositional and the predicate-logic calculus, which have become the standard frameworks of formal logical reasoning.

2.1 Concepts

At this point, we enhance our formal systems with *assumptions*, another crucial building block. In ordinary mathematics, assumptions are made in various forms and do not have to be substantively true. To model statements of the form

$$\text{"From } M \text{ it follows that ..."},$$

we allow us to augment a proof by a fixed set of assumptions. In such an extended calculus, a proof is a chain of formulas $\varphi_1, \varphi_2, \ldots, \varphi_n$, formed according to the following rules:

- φ_i is an axiom, or
- φ_i is an assumption, or
- φ_i is derived from the preceding members of the proof chain by the application of an inference rule.

If M denotes the set of assumptions, we write $M \vdash \varphi$ if the formula φ is derivable using the described construction rules. Using this notation, we can interpret the expression $\vdash \varphi$ introduced earlier as a shorthand for $\emptyset \vdash \varphi$.

The following relationships are almost obvious:

Theorem 2.4

- $\{\varphi\} \cup M \vdash \varphi$
- $M \vdash \varphi, M \subset N$ implies $N \vdash \varphi$
- $M \vdash \varphi_1, \ldots, M \vdash \varphi_n, \{\varphi_1, \ldots, \varphi_n\} \vdash \varphi$ implies $M \vdash \varphi$
- $M \vdash \varphi \Leftrightarrow$ For some finite subset $M' \subseteq M: M' \vdash \varphi$

The first three statements are direct consequences of the construction rules mentioned above. Only the last statement deserves our attention. It holds since every proof is a finite chain of formulas. Consequently, we can easily construct M' from an existing proof by collecting all the assumptions used. The finite number of proof steps ensures that M' is also finite.

2.2 Decision Procedures

> *"The decision problem is solved if one knows a procedure that allows a decision about the validity or satisfiability of a given logical expression through a finite number of operations. The solution of the decision problem is of fundamental importance for the theory of all areas whose sentences are capable of logical development from a finite number of axioms."*

> *"Das Entscheidungsproblem ist gelöst, wenn man ein Verfahren kennt, das bei einem vorgelegten logischen Ausdruck durch endlich viele Operationen die Entscheidung über die Allgemeingültigkeit bzw. Erfüllbarkeit erlaubt. [...] Das Entscheidungsproblem muss als das Hauptproblem der mathematischen Logik bezeichnet werden."*

David Hilbert, Wilhelm Ackermann (Figure 2.2) [96]

Section 2.1 elaborated on the elementary properties of formal systems through several examples. You might have noticed that we did not detail what a formal system's language, axioms, or inference rules must look like. The reason is we don't have to. For all of the subsequent considerations, it suffices that a formal system fulfills a few very elementary properties:

- The number of used alphabet symbols is finite.

- For each symbol string, it can be decided whether it is a formula.

- For each formula sequence, it can be decided whether it is a proof.

Wilhelm Friedrich Ackermann
(1896 – 1962)

Figure 2.2: The German mathematician Wilhelm Ackermann was one of David Hilbert's most renowned students. Today, his name is primarily associated with the *Ackermann function*, thoroughly discussed in Chapter 4.

It takes a good deal of destructive ingenuity to devise a formal system that does not fulfill these properties. Only obscure systems are excluded, such as those that permit infinitely long proof chains or infinitely many different symbols in the construction of formulas.

We are now well prepared to return to a central problem of Hilbert's program:

2.2 Decision Procedures

 Definition 2.3 (Syntactic Decision Problem)

The syntactic variant of the *decision problem* is defined as follows:

- Given: A formal system and a formula φ
- Asked: Does $\vdash \varphi$ hold?

To solve the syntactic variant of the decision problem, we must devise a systematic procedure that correctly determines for each formula φ whether φ is a theorem; that is, whether φ is derivable from the axioms by applying a finite number of inference steps. Such a procedure is called a *decision procedure*.

Hilbert considered the discovery of a decision procedure a central part of his program. However, Chapter 5 will show that Hilbert's vision must remain a dream for large parts of mathematics. But before discussing this negative result in detail, let us start with a positive one:

 Theorem 2.5

Every consistent, negation-complete formal system has a decision procedure.

Whether a presented formula φ is provable in a consistent and negation-complete formal system can be decided in different ways. Depending on its global strategy, we speak of a *bottom-up procedure* or a *top-down procedure*:

- **Bottom-up decision procedure (Figure 2.3)**

  To decide whether a given formula φ is provable, we first check whether φ or $\neg\varphi$ is among the axioms. If this is not the case, we generate new theorems by applying inference rules. Suppose the inference rules are always applied to all previously generated formulas. In that case, we obtain, after n inference steps, precisely those theorems deductible from the axioms in at most n steps. We continue the procedure until the formula φ or $\neg\varphi$ appears among the theorems. Since at least one of these formulas is derivable in a negation-complete calculus, the procedure will terminate after a finite number of steps.

  The bottom-up method appears to solve the decision problem straightforwardly, and for formal systems with a finite set of axioms

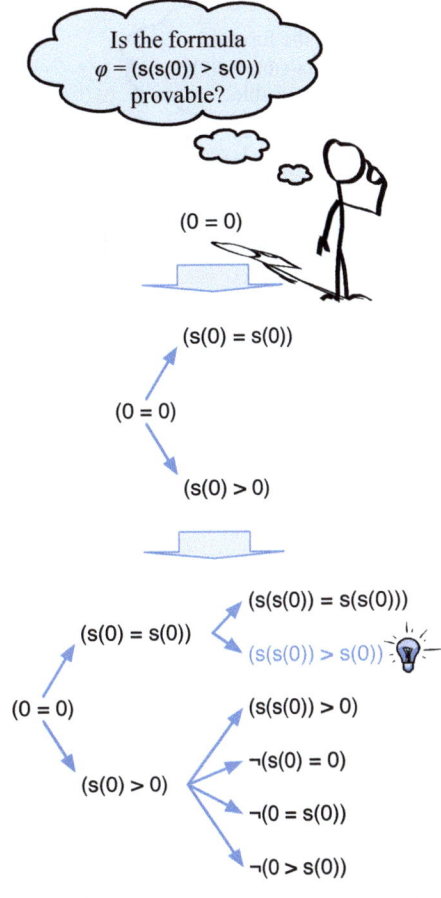

Figure 2.3: Bottom-up decision procedure for negation-complete and consistent formal systems. Starting from the axioms, new theorems are generated by successively applying inference rules until φ or $\neg\varphi$ is derived.

and a finite number of inference rules, this is indeed the case. For these systems, the mentioned procedure decides reliably whether a formula φ is provable or not. However, the situation becomes more intricate if the axioms come as schemas that can be instantiated with arbitrary subexpressions. In this case, we would have to start with infinitely many axioms. It is also conceivable that the inference rules were defined schematically. In this case, creating an infinite number of new theorems in a single step would be possible, and we could not apply the procedure described above without applying further modifications.

The top-down method provides a solution for such formal systems. This method is so general that it applies to any formal system that fulfills the abovementioned minimal properties.

■ **Top-down decision procedure (Figure 2.4)**

To decide whether a formula φ is provable, we proceed as follows:

- All symbol strings constructible with symbols of the system's formal language are systematically enumerated in sequence, e.g., by first enumerating the character strings of length 1, then the character strings of length 2, and so on. As we restrict ourselves to finite alphabets, each symbol string must appear at some point in the enumeration.

- All symbol strings that do not represent a formula are discarded, and the same applies to formula sequences that do not constitute a proof. Admittedly, the chance of a randomly formed symbol string to pass both checks is tiny. Nevertheless, the systematic enumeration ensures that every proof will eventually appear regardless of its complexity.

- For each string corresponding to a proof, it is checked whether the last formula equals φ or $\neg\varphi$. In the first case, we know that φ is a theorem; in the second case, that φ is not a theorem. Negation completeness ensures that a proof exists for at least one of the two formulas and that the algorithm terminates for each input.

Remember that consistency is essential for both procedures to function, as in a contradictory formal system, we cannot conclude the unprovability of φ from the provability of $\neg\varphi$. Hence, we can no longer abort the decision procedure when a proof for $\neg\varphi$ has come up, as a proof for φ can still appear later. In this case, φ would also be a theorem.

At this juncture, let's throw some sand in the gears and revisit the section's opening quote from Hilbert and Ackermann on page 76. Did you

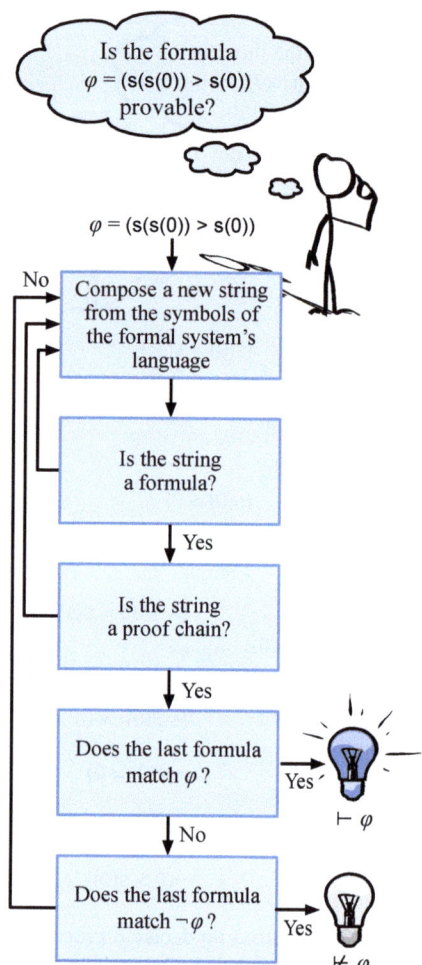

Figure 2.4: Top-down decision procedure for negation-complete and consistent formal systems.

2.2 Decision Procedures

notice that they do not mention the concept of provability at any point? In fact, Hilbert and Ackermann defined the decision problem on the semantic rather than the syntactic level.

Definition 2.4 (Semantic Decision Problem)

> The semantic variant of the *decision problem* is defined as follows:
> - Given: A formula φ
> - Asked: Does $\models \varphi$ hold?

This definition is in line with the historical formulation. Make sure to distinguish both variants, as they convey different meanings. Exceptions are formal systems being both consistent and complete. For those, the relation $\vdash \varphi \Leftrightarrow \models \varphi$ holds, making the syntactic decision problem and the semantic decision problem coincide.

Theorem 2.6

> For a correct and complete formal system, the syntactic decision problem and the semantic decision problem are equivalent.

We have seen that the differences between the syntactic and semantic levels disappear in correct and complete formal systems. A logic calculus with this property is first-order predicate logic, the formal system Hilbert and Ackermann's original wording refers to. In this system, thoroughly discussed in Section 2.4, whether we formulate the decision problem on the semantic or syntactic level is irrelevant. As a result, we will generally speak of (Hilbert's) decision problem from now on.

2.3 Propositional Logic

Propositional logic (PL0) deals with *atomic statements* that can be either true or false, and the relationships that exist between such statements (*"It is raining"*, *"The road is wet"*, *"If it is raining, then the road is wet"*). The significance of propositional logic is substantial. It is part of almost every logic calculus and can thus be considered the lowest common denominator among all calculi.

2.3.1 Syntax and Semantics

We repeat our methodology from Section 2.1 and approach propositional logic in two steps. We start by laying down the syntax and continue by defining the semantics.

Definition 2.5 (Syntax of Propositional Logic)

Let $V = \{A_1, A_3, \ldots\}$ be a set of variables. The set of *propositional formulas* over V is recursively defined:

- 0 and 1 are formulas.
- Each variable in V is a formula.
- If φ and ψ are formulas, so are
$$(\neg \varphi), (\varphi \wedge \psi), (\varphi \vee \psi), (\varphi \rightarrow \psi), (\varphi \leftrightarrow \psi), (\varphi \nleftrightarrow \psi).$$

The operator '$\neg$' stands for *negation*, '$\wedge$' for *conjunction* (AND operator), '$\vee$' for *disjunction* (OR operator), and '$\rightarrow$' for *implication*. Additionally, we refer to '$\leftrightarrow$' as the *equivalence operator* and to '$\nleftrightarrow$' as the *exclusive disjunction operator* (XOR operator).

A formula consisting solely of a truth value or a propositional variable is called *atomic*. It has the property of being indivisible. A formula φ that appears as part of another formula ψ is called a *subformula* of ψ, often informally denoted as $\varphi \in \psi$. Conversely, $\varphi \notin \psi$ indicates that φ is not a subformula of ψ. Variables will consistently be denoted by capital letters, though we will adjust the symbol set as needed, using A, B, C instead of A_1, A_2, A_3, for instance.

In Definition 2.5, we have introduced the logical operators in today's standard notation. Table 2.5 shows that the symbols trace back to the German mathematician Hans Hermes [90]. It comes as no surprise that the original notation has evolved, but the pace of change does. The reasons are manifold. On the one hand, the protagonists of the first hour were venturing into uncharted territory, and a suitable notation was not readily available. Secondly, many mathematicians believed only a specially created language would allow them to formulate their novel ideas. Unfortunately, the Babylonian confusion does not facilitate our lives today. For freshly trained mathematicians, historical publications on logic appear exotic and require a thorough familiarization with outdated terminology.

	Negation	Disjunction	Conjunction	Implication	Equivalence
Peano	$-\varphi$	$\cup$	$\cap$.	$\supset$	$=$
Russell	$\sim \varphi$	$\vee$	.	$\supset$	$\equiv$
Hilbert	$\overline{\varphi}$	$\vee$	&	$\rightarrow$	$\sim$
Hermes	$\neg \varphi$	$\vee$	$\wedge$	$\rightarrow$	$\leftrightarrow$

Table 2.5: Different notations of the propositional logic operators [136]

2.3 Propositional Logic

Furthermore, we will omit pairs of brackets whenever it improves the readability of a formula and does not cause ambiguity. As shown in Figure 2.5, we follow the standard convention that negation ('$\neg$') binds more strongly than conjunction ('$\wedge$'), which in turn binds more strongly than disjunction ('$\vee$'). The operators '$\rightarrow$', '$\leftrightarrow$', and '$\nleftrightarrow$' have the weakest binding strength. When these operators are mixed within an expression, the parentheses are placed left-associatively. The resulting chain is also considered left-associative if subformulas are linked with the same operator. The only exception is the negation operator, which is right-associative.

The binary operators '$\wedge$' and '$\vee$' can be generalized to n-ary operators. To this end, we adopt the following notation for a finite set of formulas $\varphi_1, \ldots, \varphi_n$:

$$\left(\bigwedge_{i=1}^{n} \varphi_i \right) := \varphi_1 \wedge \ldots \wedge \varphi_n$$

$$\left(\bigvee_{i=1}^{n} \varphi_i \right) := \varphi_1 \vee \ldots \vee \varphi_n$$

After establishing the syntactic structure of a formula, the next step is to define the semantics of propositional logic. We will begin by elucidating the term *interpretation*, which has been referenced multiple times, and continue by introducing the *model relation* '$\models$' afterward.

■ Binding rules (examples)

$$\neg A \wedge B \;\hat{=}\; ((\neg A) \wedge B)$$
$$A \vee B \wedge C \;\hat{=}\; (A \vee (B \wedge C))$$
$$A \rightarrow C \vee B \;\hat{=}\; (A \rightarrow (C \vee B))$$
$$A \rightarrow B \leftrightarrow C \;\hat{=}\; ((A \rightarrow B) \leftrightarrow C)$$
$$A \leftrightarrow B \rightarrow C \;\hat{=}\; ((A \leftrightarrow B) \rightarrow C)$$

■ Chaining rules (examples)

$$\neg \neg A \;\hat{=}\; (\neg(\neg A))$$
$$A \wedge B \wedge C \;\hat{=}\; ((A \wedge B) \wedge C)$$
$$A \vee B \vee C \;\hat{=}\; ((A \vee B) \vee C)$$
$$A \rightarrow B \rightarrow C \;\hat{=}\; ((A \rightarrow B) \rightarrow C)$$
$$A \leftrightarrow B \leftrightarrow C \;\hat{=}\; ((A \leftrightarrow B) \leftrightarrow C)$$
$$A \nleftrightarrow B \nleftrightarrow C \;\hat{=}\; ((A \nleftrightarrow B) \nleftrightarrow C)$$

Figure 2.5: Some parentheses may be omitted to simplify notation. Ambiguities are eliminated through binding and chaining rules. The former assign a binding priority to all operators, and the latter govern the handling of expressions containing the same operator several times in succession.

Definition 2.6 (Interpretation of Propositional Logic)

Let φ be a propositional logic formula. $A_1, \ldots, A_n$ denote the variables occurring in φ. Any mapping

$$I : \{A_1, \ldots, A_n\} \rightarrow \{0, 1\}$$

is called an *interpretation* of φ.

An interpretation, also known as an *assignment*, assigns one of the two truth values, 0 or 1, to each variable in a propositional formula.

The notion of interpretation allows us to define the semantics of propositional logic formally:

- Interpretation 1: $I(A) = 0, I(B) = 0$

  $(\underbrace{A \rightarrow B}_{I \not\models A \quad I \not\models B}) \rightarrow (\underbrace{B \rightarrow A}_{I \not\models B \quad I \not\models A})$
  $\underbrace{\qquad\qquad}_{I \models A \rightarrow B} \quad \underbrace{\qquad\qquad}_{I \models B \rightarrow A}$
  $\underbrace{\qquad\qquad\qquad\qquad\qquad}_{I \models (A \rightarrow B) \rightarrow (B \rightarrow A)}$

- Interpretation 2: $I(A) = 0, I(B) = 1$

  $(\underbrace{A \rightarrow B}_{I \not\models A \quad I \models B}) \rightarrow (\underbrace{B \rightarrow A}_{I \models B \quad I \not\models A})$
  $\underbrace{\qquad\qquad}_{I \models A \rightarrow B} \quad \underbrace{\qquad\qquad}_{I \not\models B \rightarrow A}$
  $\underbrace{\qquad\qquad\qquad\qquad\qquad}_{I \not\models (A \rightarrow B) \rightarrow (B \rightarrow A)}$

- Interpretation 3: $I(A) = 1, I(B) = 0$

  $(\underbrace{A \rightarrow B}_{I \models A \quad I \not\models B}) \rightarrow (\underbrace{B \rightarrow A}_{I \not\models B \quad I \models A})$
  $\underbrace{\qquad\qquad}_{I \not\models A \rightarrow B} \quad \underbrace{\qquad\qquad}_{I \models B \rightarrow A}$
  $\underbrace{\qquad\qquad\qquad\qquad\qquad}_{I \models (A \rightarrow B) \rightarrow (B \rightarrow A)}$

- Interpretation 4: $I(A) = 1, I(B) = 1$

  $(\underbrace{A \rightarrow B}_{I \models A \quad I \models B}) \rightarrow (\underbrace{B \rightarrow A}_{I \models B \quad I \models A})$
  $\underbrace{\qquad\qquad}_{I \models A \rightarrow B} \quad \underbrace{\qquad\qquad}_{I \models B \rightarrow A}$
  $\underbrace{\qquad\qquad\qquad\qquad\qquad}_{I \models (A \rightarrow B) \rightarrow (B \rightarrow A)}$

Figure 2.6: An interpretation I assigns one of the two truth values 0 (false) or 1 (true) to each propositional variable, demonstrated here using the example formula $\varphi = (A \rightarrow B) \rightarrow (B \rightarrow A)$.

 Definition 2.7 (Semantics of Propositional Logic)

Let φ and ψ be propositional logic formulas, and I an interpretation. The semantics of propositional logic is given by the *model relation* '$\models$', which is defined inductively over the structure of formulas:

$$I \models 1$$
$$I \not\models 0$$
$$I \models A_i :\Leftrightarrow I(A_i) = 1$$
$$I \models (\neg \varphi) :\Leftrightarrow I \not\models \varphi$$
$$I \models (\varphi \wedge \psi) :\Leftrightarrow I \models \varphi \text{ and } I \models \psi$$
$$I \models (\varphi \vee \psi) :\Leftrightarrow I \models \varphi \text{ or } I \models \psi$$
$$I \models (\varphi \rightarrow \psi) :\Leftrightarrow I \not\models \varphi \text{ or } I \models \psi$$
$$I \models (\varphi \leftrightarrow \psi) :\Leftrightarrow I \models \varphi \rightarrow \psi \text{ and } I \models \psi \rightarrow \varphi$$
$$I \models (\varphi \not\leftrightarrow \psi) :\Leftrightarrow I \not\models (\varphi \leftrightarrow \psi)$$

An interpretation I with $I \models \varphi$ is called a *model* for φ.

Figure 2.6 illustrates the concept through a specific example.

We can naturally view every propositional logic formula φ with n variables as a *Boolean function* $f_\varphi : \{0,1\}^n \rightarrow \{0,1\}$, mapping an assignment I to 1 precisely when I is a model for φ. More precisely, if I assigns the truth values $a_1, \ldots, a_n$ to the variables $A_1, \ldots, A_n$, respectively, then the value of $f_\varphi(a_1, \ldots, a_n)$ is given by:

$$f_\varphi(a_1, \ldots, a_n) := \begin{cases} 1 & \text{if } I \models \varphi \\ 0 & \text{if } I \not\models \varphi \end{cases}$$

Due to its finite domain, an n-ary Boolean function can be represented as a *truth table*, listing all possible combinations of the arguments along with the corresponding function value row by row (Figure 2.7).

Figure 2.8 demonstrates the generation of truth tables for compound expressions. Starting from the elementary statements, all subformulas are evaluated step by step, up to the entire expression. The three examples are chosen deliberately, as they exemplify three essential formula classes. φ_1 is structured to be true precisely if A is true or B is false. In the terminology of propositional logic, φ_1 is referred to as a *satisfiable* formula. φ_2 is also satisfiable, but in contrast to φ_1, it is always true regardless of the truth value of the elementary statements. Such

2.3 Propositional Logic

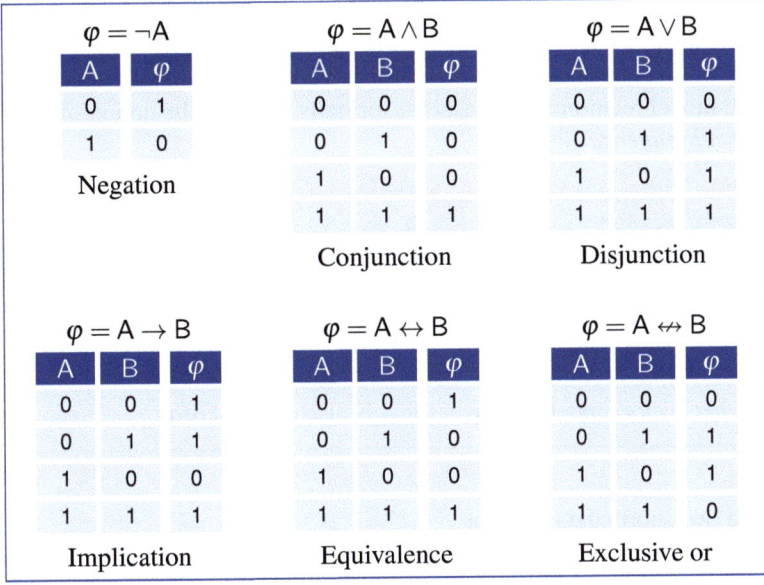

Figure 2.7: Truth tables of the propositional logic operators

formulas are called *universally valid*. Similarly, φ_3 is referred to as an *unsatisfiable formula*, as it never becomes true.

We summarize the above observations in the following definition:

 Definition 2.8 (Satisfiable, Unsatisfiable, Universally Valid)

A propositional logic formula φ is

- *satisfiable*, if φ has at least one model,
- *unsatisfiable*, if φ has no models,
- *universally valid*, if $\neg\varphi$ is unsatisfiable.

A universally valid formula is also called a *tautology*.

Figure 2.9 illustrates how the different formula classes relate to each other.

All notions introduced in Definition 2.8 can be extended to sets of formulas. A set $M = \{\varphi_1, \ldots, \varphi_n\}$ is satisfiable if some interpretation I is a model for all $\varphi_i \in M$. Unsatisfiable and universally valid sets of formulas are defined analogously. M is unsatisfiable if $\varphi_1, \ldots, \varphi_n$ have no

- $\varphi_1 := \underbrace{(A \rightarrow B)}_{\psi_1} \rightarrow \underbrace{(B \rightarrow A)}_{\psi_2}$

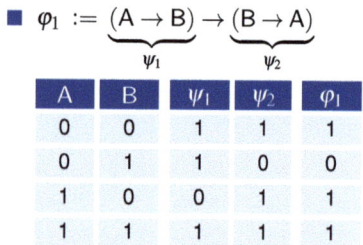

- $\varphi_2 := \underbrace{(A \vee B)}_{\psi_3} \rightarrow \underbrace{(B \vee A)}_{\psi_4}$

A	B	ψ_3	ψ_4	φ_2
0	0	0	0	1
0	1	1	1	1
1	0	1	1	1
1	1	1	1	1

- $\varphi_3 := \underbrace{(A \vee \neg A)}_{\psi_5} \rightarrow \underbrace{(B \wedge \neg B)}_{\psi_6}$

A	B	ψ_5	ψ_6	φ_3
0	0	1	0	0
0	1	1	0	0
1	0	1	0	0
1	1	1	0	0

Figure 2.8: Truth tables of several compound formulas

model in common. Conversely, if every interpretation is a model for all elements of M, we call M universally valid.

The model relation is the basis for formally defining the notion of *logical consequence*:

> **Definition 2.9** (Logical Consequence)
>
> Let $\varphi_1, \ldots, \varphi_n, \psi$ be propositional formulas. We write
>
> $$\varphi_1, \ldots, \varphi_n \models \psi,$$
>
> when every model of $\{\varphi_1, \ldots, \varphi_n\}$ is also a model of ψ.

The two shorthand notations

$$\models \psi \quad \text{for} \quad \emptyset \models \psi$$
$$\varphi \models \psi \quad \text{for} \quad \{\varphi\} \models \psi$$

allow us to establish the following relationships:

- $\models \psi$ holds if and only if ψ is universally valid.

- $\varphi \models \psi$ holds if and only if $\varphi \rightarrow \psi$ is universally valid.

- $\{\varphi_1, \varphi_2, \ldots, \varphi_n\} \models \psi$ holds if and only if $\{\varphi_2, \ldots, \varphi_n\} \models \varphi_1 \rightarrow \psi$.

In the considerations to come, the notion of *equivalence* will frequently crop up:

Figure 2.9: The principle of reflection. If φ is universally valid, then $\neg\varphi$ is unsatisfiable. If φ is not universally valid but satisfiable, the same applies to $\neg\varphi$. Thus, universal validity is an exclusive property only one of the two formulas, φ or $\neg\varphi$, can exhibit. In contrast, both φ and $\neg\varphi$ can be satisfiable.

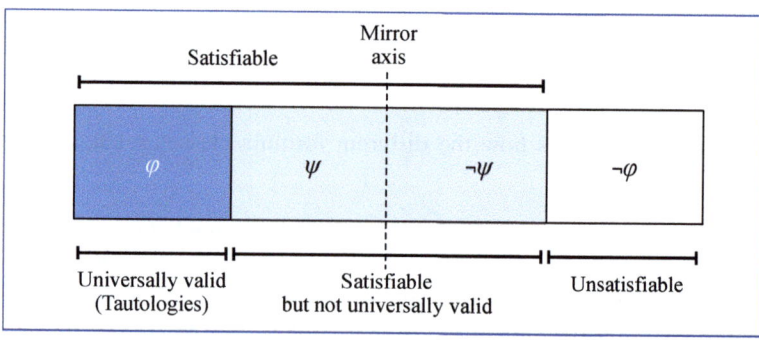

2.3 Propositional Logic

 Definition 2.10 (Equivalence)

Let φ and ψ be two propositional logic formulas. The relation '$\equiv$' is defined as follows:

$$\varphi \equiv \psi :\Leftrightarrow \varphi \models \psi \text{ and } \psi \models \varphi$$

Two formulas φ and ψ satisfying $\varphi \equiv \psi$ are called *equivalent*.

Put in words, two formulas φ and ψ are equivalent if they share the same models. Figure 2.10 summarizes important equivalences that are easy to verify via truth tables.

You might have noticed that Figure 2.10 solely features formulas containing the three propositional standard operators '$\neg$', '$\wedge$', and '$\vee$'. This choice is deliberate because these operators are sufficiently expressive to represent all others:

$$\varphi \rightarrow \psi \equiv \neg \varphi \vee \psi$$
$$\varphi \leftrightarrow \psi \equiv (\neg \varphi \vee \psi) \wedge (\varphi \vee \neg \psi)$$
$$\varphi \nleftrightarrow \psi \equiv (\neg \varphi \vee \neg \psi) \wedge (\varphi \vee \psi)$$

Alternatively, due to the equivalences

$$\varphi \wedge \psi \equiv \neg(\varphi \rightarrow \neg \psi)$$
$$\varphi \vee \psi \equiv \neg \varphi \rightarrow \psi$$
$$\varphi \wedge \psi \equiv \neg(\neg \varphi \vee \neg \psi)$$
$$\varphi \vee \psi \equiv \neg(\neg \varphi \wedge \neg \psi)$$

we could limit ourselves to one of the sets

$$\{\neg, \rightarrow\}, \{\neg, \vee\}, \text{ or } \{\neg, \wedge\},$$

and interpret the other operators as syntactic shortcuts for more complex formulas.

In the next section, we will exploit this fact and present a calculus in which only operators from the set $\{\neg, \rightarrow\}$ are utilized. The exclusion of other logical operators is no restriction in the strict sense since we have just shown that all propositional operators can be reduced to negation and implication.

- **Commutative laws**

$$\varphi \wedge \psi \equiv \psi \wedge \varphi$$
$$\varphi \vee \psi \equiv \psi \vee \varphi$$

- **Distributive laws**

$$\varphi \wedge (\psi \vee \chi) \equiv (\varphi \wedge \psi) \vee (\varphi \wedge \chi)$$
$$\varphi \vee (\psi \wedge \chi) \equiv (\varphi \vee \psi) \wedge (\varphi \vee \chi)$$

- **Identity laws**

$$\varphi \wedge 1 \equiv \varphi$$
$$\varphi \vee 0 \equiv \varphi$$

- **Complement laws**

$$\varphi \wedge \neg \varphi \equiv 0$$
$$\varphi \vee \neg \varphi \equiv 1$$

- **Associative laws**

$$(\varphi \wedge \psi) \wedge \chi \equiv \varphi \wedge (\psi \wedge \chi)$$
$$(\varphi \vee \psi) \vee \chi \equiv \varphi \vee (\psi \vee \chi)$$

- **Idempotence laws**

$$\varphi \wedge \varphi \equiv \varphi$$
$$\varphi \vee \varphi \equiv \varphi$$

- **Absorption laws**

$$(\varphi \wedge \psi) \vee \varphi \equiv \varphi$$
$$(\varphi \vee \psi) \wedge \varphi \equiv \varphi$$

- **De Morgan's laws**

$$\neg(\varphi \wedge \psi) \equiv \neg \varphi \vee \neg \psi$$
$$\neg(\varphi \vee \psi) \equiv \neg \varphi \wedge \neg \psi$$

- **Annulment laws**

$$\varphi \wedge 0 \equiv 0$$
$$\varphi \vee 1 \equiv 1$$

- **Double negation law**

$$\neg \neg \varphi \equiv \varphi$$

Figure 2.10: Fundamental laws of propositional logic

2.3.2 Propositional-Logic Calculus

In Section 2.3.1, we have defined the semantics of propositional logic via the model relation '$\models$'. At this juncture, we will introduce a formal system capable of deriving all universally valid formulas from the axioms. As customary in formal systems, the derivation happens exclusively on the syntactic level. Consequently, we can prove a proposition without knowledge about interpretations, models, or other semantic concepts.

Table 2.6 summarizes the axioms and inference rules of the propositional logic calculus. The first axiom is called the *weakening rule* as it lets us derive, for any statement φ, the weaker statement $\psi \to \varphi$. The second axiom expresses the *distributiveness* of the implication operator. The third and final axiom formalizes a well-known reasoning principle: the *logical contraposition*. It states that we can reverse the logical direction of inference when negating the arguments ("If it is raining, then the road is wet" is equivalent to "If the road is not wet, then it is not raining"). With the *modus ponens*, the logic calculus provides a single inference rule for deriving theorems. This rule is intuitively familiar to us. It guarantees that a proposition ψ must be true if we know that φ is true and φ implies ψ.

The following derivation shows how to derive the tautology $\varphi = A \to A$ from the axioms:

1. $\vdash (A \to ((A \to A) \to A)) \to ((A \to (A \to A)) \to (A \to A))$ (A2)

> The truth of the chosen example formula $A \to A$ follows immediately from the definition of the implication operator '$\to$'. Then why did we go through all the trouble of proving it? The reason is that provability and truth are two entirely different concepts. To show that the formula $A \to A$ is a theorem, we need to demonstrate its provability. In a formal sense, this does not mean that the formula is a true statement but merely that it is derivable from the axioms in a finite number of inference steps. In short, provability is a syntactic property, whereas truth is a semantic concept. It is the natural desire of mathematicians to align both concepts so that a formula's truth follows from its provability and vice versa. However, this is impossible for most parts of mathematics, as the following chapters will unmistakably demonstrate.

Axioms	
$\varphi \to (\psi \to \varphi)$	(A1)
$(\varphi \to (\psi \to \chi)) \to ((\varphi \to \psi) \to (\varphi \to \chi))$	(A2)
$(\neg\varphi \to \neg\psi) \to (\psi \to \varphi)$	(A3)
Inference rules	
$\dfrac{\varphi,\ \varphi \to \psi}{\psi}$	(MP)

Table 2.6: Axioms and inference rules of the propositional logic calculus

2.3 Propositional Logic

2. $\vdash A \to ((A \to A) \to A)$ (A1)
3. $\vdash (A \to (A \to A)) \to (A \to A)$ (MP, 1,2)
4. $\vdash A \to (A \to A)$ (A1)
5. $\vdash A \to A$ (MP, 3,4)

The first two members in the proof chain are instances of the axioms of distributivity and weakening, respectively. The third member is derived from applying the modus ponens to the previously generated formulas, and the fourth is again an instance of the weakening axiom. Now, φ can be derived from members 3 and 4 by applying the modus ponens a second time.

At the end of Section 2.1, we have introduced the notation $M \vdash \varphi$. It states that we can derive φ with a proof chain $\varphi_1, \ldots, \varphi_n$, which is structured according to the following scheme:

- φ_i is an axiom, or
- φ_i is a formula from M, or
- φ_i is obtained from the preceding members of the proof chain by applying an inference rule.

The formula sequence terminates with $\varphi_n = \varphi$. The set M adds several assumptions, allowing us to deal easily with statements of the form *"If M holds, it follows that …"*.

You may wonder if such an extension is truly necessary; after all, we can formulate any if-then relationship using the implication operator. However, the two constructs differ insofar as the operator '$\to$' exists *within* the formal system, while the inference relation $M \vdash \varphi$ makes a statement about the provability of φ. In other words, $M \vdash \varphi$ is a meta-statement *outside* the formal system.

Nevertheless, both constructs are closely related, as the following theorem underlines:

 Theorem 2.7 (Deduction Theorem of Propositional Logic)

> Let $\varphi, \varphi_1, \ldots, \varphi_n$, and ψ be propositional-logic formulas. Then, the following holds:
>
> $$\{\varphi_1, \ldots, \varphi_n\} \cup \{\varphi\} \vdash \psi \quad \Leftrightarrow \quad \{\varphi_1, \ldots, \varphi_n\} \vdash \varphi \to \psi$$

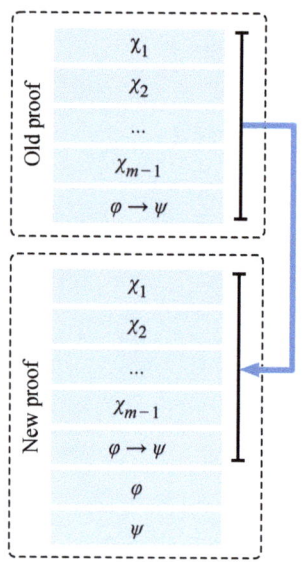

Figure 2.11: About the proof of the deduction theorem (direction from left to right)

Proof: The direction from right to left is almost trivial. If

$$\{\varphi_1, \ldots, \varphi_n\} \vdash \varphi \to \psi,$$

then there exists a formal proof for $\varphi \to \psi$ from $\{\varphi_1, \ldots, \varphi_n\}$. We can easily extend this proof to a proof for ψ from $\{\varphi_1, \ldots, \varphi_n, \varphi\}$ by first generating an instance of φ and applying the modus ponens to $\varphi \to \psi$ and φ afterward (Figure 2.11).

The direction from left to right is more elaborate, albeit it pursues a similar argument. Starting from a proof of ψ from $\{\varphi_1, \ldots, \varphi_n\} \cup \{\varphi\}$, we construct a proof of $\varphi \to \psi$ from $\{\varphi_1, \ldots, \varphi_n\}$.

Figure 2.12 sketches the general structure of the new proof. From the existing proof chain $\chi_1, \ldots, \chi_{m-1}, \psi$, we create a new one, which derives the formulas $\varphi \to \chi_i$ one after the other and ends with the assertion $\varphi \to \psi$.

After outlining the general structure of the proof chain, let us consider how to close the gaps. We distinguish three cases:

- χ_i is an axiom or an assumption

  $\vdash \chi_i$

  $\vdash \chi_i \to (\varphi \to \chi_i)$ (A1)

  $\vdash \varphi \to \chi_i$ (MP)

- χ_i matches the formula φ

  $\vdash (\varphi \to ((\varphi \to \varphi) \to \varphi)) \to ((\varphi \to (\varphi \to \varphi)) \to (\varphi \to \varphi))$ (A2)

  $\vdash \varphi \to ((\varphi \to \varphi) \to \varphi)$ (A1)

  $\vdash (\varphi \to (\varphi \to \varphi)) \to (\varphi \to \varphi)$ (MP)

  $\vdash \varphi \to (\varphi \to \varphi)$ (A1)

  $\vdash \varphi \to \varphi$ (MP)

- χ_i was generated via (MP) from χ_j and $\chi_j \to \chi_i$

  Then we know that the two lines

  $\vdash \varphi \to \chi_j$

  $\vdash \varphi \to (\chi_j \to \chi_i)$

  must appear somewhere in the proof chain, and we can proceed as follows:

  $\vdash (\varphi \to (\chi_j \to \chi_i)) \to ((\varphi \to \chi_j) \to (\varphi \to \chi_i))$ (A2)

  $\vdash (\varphi \to \chi_j) \to (\varphi \to \chi_i)$ (MP)

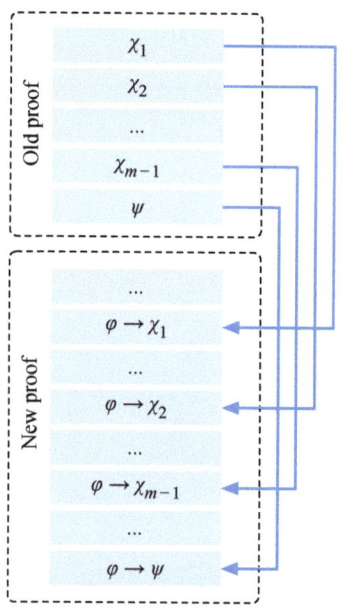

Figure 2.12: About the proof of the deduction theorem (direction from right to left)

2.3 Propositional Logic

Theorems		
■ Theorem T1 $\varphi \to \varphi$	■ Theorem T6 $(\varphi \to \psi) \to (\neg\psi \to \neg\varphi)$	■ Theorem T11 $\neg(\varphi \to \psi) \to \neg\psi$
■ Theorem T2 $(\varphi \to \psi) \to ((\psi \to \chi) \to (\varphi \to \chi))$	■ Theorem T7 $\varphi \to (\neg\psi \to \neg(\varphi \to \psi))$	■ Theorem T12 $(\varphi \to \neg\varphi) \to \neg\varphi$
■ Theorem T3 $\varphi \to ((\varphi \to \psi) \to \psi)$	■ Theorem T8 $\neg\varphi \to (\varphi \to \psi)$	■ Theorem T13 $(\neg\varphi \to \varphi) \to \varphi$
■ Theorem T4 $\neg\neg\varphi \to \varphi$	■ Theorem T9 $\varphi \to (\psi \to (\varphi \to \psi))$	■ Theorem T14 $(\varphi \to \psi) \to ((\neg\varphi \to \psi) \to \psi)$
■ Theorem T5 $\varphi \to \neg\neg\varphi$	■ Theorem T10 $\neg(\varphi \to \psi) \to \varphi$	■ Theorem T15 $\neg(\varphi \to \varphi) \to \psi$

Table 2.7: A small selection of formulas derivable in the propositional logic calculus

$$\vdash \varphi \to \chi_i \qquad (MP)$$

This concludes the proof. □

The deduction theorem is an invaluable tool for conducting formal proofs. On the one hand, it allows us to transition seamlessly between the object and the meta-level. Secondly, it offers the means to represent specific schemes of reasoning concisely, thus enabling us to shorten many proofs.

We will now bring the propositional logic calculus to life by proving the theorems in Table 2.7. They stem from [135] and [134] and vividly demonstrate how to carry out more complex proofs.

> Theorem T8 is of particular significance. Suppose a formal system has T8 among its theorems and the modus ponens among its inference rules. Then, the following applies: If a single contradiction is derivable, then every formula is a theorem. To see why, let us assume that the formulas φ and $\neg\varphi$ are both derivable from the axioms. From $\neg\varphi \to (\varphi \to \psi)$ we can deduce $\varphi \to \psi$ by modus ponens. According to our assumption, φ is also derivable, and another application of the modus ponens yields the theorem ψ. Since there is no restriction on the choice of ψ, we can substitute an arbitrary proposition. In short, in a contradictory formal system, all statements are provable without exception.
> We have thus unmasked consistency as an indispensable property of mathematical reasoning whose absence renders any formal system worthless.

Theorem T1

$\varphi \to \varphi$

1. $\vdash (\varphi \to ((\varphi \to \varphi) \to \varphi)) \to ((\varphi \to (\varphi \to \varphi)) \to (\varphi \to \varphi))$ (A2)
2. $\vdash \varphi \to ((\varphi \to \varphi) \to \varphi)$ (A1)
3. $\vdash (\varphi \to (\varphi \to \varphi)) \to (\varphi \to \varphi)$ (MP, 2,1)
4. $\vdash \varphi \to (\varphi \to \varphi)$ (A1)
5. $\vdash \varphi \to \varphi$ (MP, 3,4)

Theorem T2

$(\varphi \to \psi) \to ((\psi \to \chi) \to (\varphi \to \chi))$

1. $\{\varphi \to \psi, \psi \to \chi, \varphi\} \vdash \varphi$ (Theorem 2.4)
2. $\{\varphi \to \psi, \psi \to \chi, \varphi\} \vdash \varphi \to \psi$ (Theorem 2.4)
3. $\{\varphi \to \psi, \psi \to \chi, \varphi\} \vdash \psi$ (MP, 1,2)
4. $\{\varphi \to \psi, \psi \to \chi, \varphi\} \vdash \psi \to \chi$ (Theorem 2.4)
5. $\{\varphi \to \psi, \psi \to \chi, \varphi\} \vdash \chi$ (MP, 3,4)
6. $\{\varphi \to \psi, \psi \to \chi\} \vdash \varphi \to \chi$ (DT)
7. $\{\varphi \to \psi\} \vdash (\psi \to \chi) \to (\varphi \to \chi)$ (DT)
8. $\vdash (\varphi \to \psi) \to ((\psi \to \chi) \to (\varphi \to \chi))$ (DT)

Theorem T2 describes the *modus barbara*, which looks as follows in our symbolic notation:

$$\frac{\varphi \to \psi, \psi \to \chi}{\varphi \to \chi} \quad \text{(MB)}$$

To write down proof chains concisely, we will allow ourselves to utilize the modus barbara (MB) in addition to the already existing modus ponens (MP). The following derivation sequence demonstrates why it is safe to do so. It shows how the formula

$$\varphi \to \chi$$

can be derived from $\varphi \to \psi$ and $\psi \to \chi$ without applying the modus barbara at any point:

1. $\vdash \varphi \to \psi$
2. $\vdash \psi \to \chi$
3. $\vdash (\varphi \to \psi) \to ((\psi \to \chi) \to (\varphi \to \chi))$ (T2)
4. $\vdash (\psi \to \chi) \to (\varphi \to \chi)$ (MP, 1,3)
5. $\vdash \varphi \to \chi$ (MP, 2,4)

The derivation sequence demonstrates that theorem T2 is the key to eliminating the modus barbara from any proof chain.

2.3 Propositional Logic

			Theorem T3
1. $\{\varphi, \varphi \to \psi\} \vdash \varphi$	(Theorem 2.4)		$\varphi \to ((\varphi \to \psi) \to \psi)$
2. $\{\varphi, \varphi \to \psi\} \vdash \varphi \to \psi$	(Theorem 2.4)		
3. $\{\varphi, \varphi \to \psi\} \vdash \psi$	(MP, 1,2)		
4. $\{\varphi\} \vdash (\varphi \to \psi) \to \psi$	(DT)		
5. $\vdash \varphi \to ((\varphi \to \psi) \to \psi)$	(DT)		

			Theorem T4
1. $\vdash \neg\neg\varphi \to (\neg\neg\neg\neg\varphi \to \neg\neg\varphi)$	(A1)		$\neg\neg\varphi \to \varphi$
2. $\{\neg\neg\varphi\} \vdash \neg\neg\neg\neg\varphi \to \neg\neg\varphi$	(DT)		
3. $\vdash (\neg\neg\neg\neg\varphi \to \neg\neg\varphi) \to (\neg\varphi \to \neg\neg\neg\varphi)$	(A3)		
4. $\{\neg\neg\varphi\} \vdash \neg\varphi \to \neg\neg\neg\varphi$	(MP, 2,3)		
5. $\vdash (\neg\varphi \to \neg\neg\neg\varphi) \to (\neg\neg\varphi \to \varphi)$	(A3)		
6. $\{\neg\neg\varphi\} \vdash \neg\neg\varphi \to \varphi$	(MP, 4,5)		
7. $\{\neg\neg\varphi\} \vdash \varphi$	(DT)		
8. $\vdash \neg\neg\varphi \to \varphi$	(DT)		

			Theorem T5
1. $\vdash \neg\neg\neg\varphi \to \neg\varphi$	(T4)		$\varphi \to \neg\neg\varphi$
2. $\vdash (\neg\neg\neg\varphi \to \neg\varphi) \to (\varphi \to \neg\neg\varphi)$	(A3)		
3. $\vdash \varphi \to \neg\neg\varphi$	(MP, 1,2)		

			Theorem T6
1. $\vdash \neg\neg\varphi \to \varphi$	(T4)		$(\varphi \to \psi) \to (\neg\psi \to \neg\varphi)$
2. $\{\varphi \to \psi\} \vdash \varphi \to \psi$	(Theorem 2.4)		
3. $\{\varphi \to \psi\} \vdash \neg\neg\varphi \to \psi$	(MB, 1,2)		
4. $\vdash \psi \to \neg\neg\psi$	(T5)		
5. $\{\varphi \to \psi\} \vdash \neg\neg\varphi \to \neg\neg\psi$	(MB, 3,4)		
6. $\vdash (\neg\neg\varphi \to \neg\neg\psi) \to (\neg\psi \to \neg\varphi)$	(A3)		
7. $\{\varphi \to \psi\} \vdash \neg\psi \to \neg\varphi$	(MP, 5,6)		
8. $\vdash (\varphi \to \psi) \to (\neg\psi \to \neg\varphi)$	(DT)		

			Theorem T7
1. $\vdash \varphi \to ((\varphi \to \psi) \to \psi)$	(T3)		$\varphi \to (\neg\psi \to \neg(\varphi \to \psi))$
2. $\{\varphi\} \vdash (\varphi \to \psi) \to \psi$	(DT)		
3. $\vdash ((\varphi \to \psi) \to \psi) \to (\neg\psi \to \neg(\varphi \to \psi))$	(T6)		
4. $\{\varphi\} \vdash \neg\psi \to \neg(\varphi \to \psi)$	(MP, 2,3)		
5. $\vdash \varphi \to (\neg\psi \to \neg(\varphi \to \psi))$	(DT)		

Theorem T8
$\neg\varphi \to (\varphi \to \psi)$

1. $\vdash \neg\varphi \to (\neg\psi \to \neg\varphi)$ (A1)
2. $\{\neg\varphi\} \vdash \neg\psi \to \neg\varphi$ (DT)
3. $\vdash (\neg\psi \to \neg\varphi) \to (\varphi \to \psi)$ (A3)
4. $\{\neg\varphi\} \vdash (\varphi \to \psi)$ (MP, 2,3)
5. $\vdash \neg\varphi \to (\varphi \to \psi)$ (DT)

Theorem T9
$\varphi \to (\psi \to (\varphi \to \psi))$

1. $\{\varphi\} \vdash \psi \to (\varphi \to \psi)$ (A1)
2. $\vdash \varphi \to (\psi \to (\varphi \to \psi))$ (DT)

Theorem T10
$\neg(\varphi \to \psi) \to \varphi$

1. $\vdash \neg\varphi \to (\varphi \to \psi)$ (T8)
2. $\vdash (\neg\varphi \to (\varphi \to \psi)) \to (\neg(\varphi \to \psi) \to \neg\neg\varphi)$ (T6)
3. $\vdash \neg(\varphi \to \psi) \to \neg\neg\varphi$ (MP, 1,2)
4. $\vdash \neg\neg\varphi \to \varphi$ (T4)
5. $\vdash \neg(\varphi \to \psi) \to \varphi$ (MB, 3,4)

Theorem T11
$\neg(\varphi \to \psi) \to \neg\psi$

1. $\vdash \psi \to (\varphi \to \psi)$ (A1)
2. $\vdash (\psi \to (\varphi \to \psi)) \to (\neg(\varphi \to \psi) \to \neg\psi)$ (T6)
3. $\vdash \neg(\varphi \to \psi) \to \neg\psi$ (MP, 1,2)

Theorem T12
$(\varphi \to \neg\varphi) \to \neg\varphi$

1. $\vdash \varphi \to (\neg\neg\varphi \to \neg(\varphi \to \neg\varphi))$ (T7)
2. $\{\varphi\} \vdash \neg\neg\varphi \to \neg(\varphi \to \neg\varphi)$ (DT)
3. $\vdash \varphi \to \neg\neg\varphi$ (T5)
4. $\{\varphi\} \vdash \neg\neg\varphi$ (DT)
5. $\{\varphi\} \vdash \neg(\varphi \to \neg\varphi)$ (MP, 2,4)
6. $\vdash \varphi \to \neg(\varphi \to \neg\varphi)$ (DT)
7. $\vdash (\varphi \to \neg(\varphi \to \neg\varphi)) \to (\neg\neg(\varphi \to \neg\varphi) \to \neg\varphi)$ (T6)
8. $\vdash \neg\neg(\varphi \to \neg\varphi) \to \neg\varphi$ (MP, 6,7)
9. $\vdash (\varphi \to \neg\varphi) \to \neg\neg(\varphi \to \neg\varphi)$ (T5)
10. $\vdash (\varphi \to \neg\varphi) \to \neg\varphi$ (MB, 9,8)

Theorem T13
$(\neg\varphi \to \varphi) \to \varphi$

1. $\{\neg\varphi\} \vdash \varphi \to \neg\neg\varphi$ (T5)
2. $\vdash \neg\varphi \to (\varphi \to \neg\neg\varphi)$ (DT)
3. $\vdash (\neg\varphi \to (\varphi \to \neg\neg\varphi)) \to ((\neg\varphi \to \varphi) \to (\neg\varphi \to \neg\neg\varphi))$ (A2)

4. $\vdash (\neg\varphi \to \varphi) \to (\neg\varphi \to \neg\neg\varphi)$ (MP, 2,3)
5. $\{\neg\varphi \to \varphi\} \vdash (\neg\varphi \to \neg\neg\varphi)$ (DT)
6. $\vdash (\neg\varphi \to \neg\neg\varphi) \to \neg\neg\varphi$ (T12)
7. $\{\neg\varphi \to \varphi\} \vdash \neg\neg\varphi$ (MP, 5,6)
8. $\vdash \neg\neg\varphi \to \varphi$ (T4)
9. $\{\neg\varphi \to \varphi\} \vdash \varphi$ (MP, 7,8)
10. $\vdash (\neg\varphi \to \varphi) \to \varphi$ (DT)

Theorem T14
$(\varphi \to \psi) \to ((\neg\varphi \to \psi) \to \psi)$

1. $\{\varphi \to \psi\} \vdash \varphi \to \psi$ (Theorem 2.4)
2. $\vdash (\varphi \to \psi) \to (\neg\psi \to \neg\varphi)$ (T6)
3. $\{\varphi \to \psi\} \vdash (\neg\psi \to \neg\varphi)$ (MP, 1,2)
4. $\vdash (\neg\psi \to \neg\varphi) \to ((\neg\varphi \to \psi) \to (\neg\psi \to \psi))$ (T2)
5. $\{\varphi \to \psi\} \vdash (\neg\varphi \to \psi) \to (\neg\psi \to \psi)$ (MP, 3,4)
6. $\{\varphi \to \psi, \neg\varphi \to \psi\} \vdash \neg\psi \to \psi$ (DT)
7. $\vdash (\neg\psi \to \psi) \to \psi$ (T13)
8. $\{\varphi \to \psi, \neg\varphi \to \psi\} \vdash \psi$ (MP, 6,7)
9. $\{\varphi \to \psi\} \vdash (\neg\varphi \to \psi) \to \psi$ (DT)
10. $\vdash (\varphi \to \psi) \to ((\neg\varphi \to \psi) \to \psi)$ (DT)

Theorem T15
$\neg(\varphi \to \varphi) \to \psi$

1. $\{\neg\psi\} \vdash (\varphi \to \varphi)$ (T1)
2. $\vdash (\varphi \to \varphi) \to \neg\neg(\varphi \to \varphi)$ (T5)
3. $\{\neg\psi\} \vdash \neg\neg(\varphi \to \varphi)$ (MP, 1,2)
4. $\vdash \neg\psi \to \neg\neg(\varphi \to \varphi)$ (DT)
5. $\vdash (\neg\psi \to \neg\neg(\varphi \to \varphi)) \to (\neg(\varphi \to \varphi) \to \psi)$ (A3)
6. $\vdash \neg(\varphi \to \varphi) \to \psi$ (MP, 4,5)

Remember that the presented derivations are no genuine formal proofs due to the application of the modus barbara and the deduction theorem. Both rules operate on a meta-level and do not genuinely exist within the calculus. However, we have already shown how to simulate each step marked with (MB) or (DT) with a sequence that gets by with the native language elements. As a result, we can safely regard the derivation sequences as valid proofs. We may view the shown derivations as concise blueprints allowing us to generate native proofs.

Next, let us examine whether the propositional logic calculus fulfills all properties introduced in Definition 2.2. It is easy to see that the calculus

Frege (1879) [57]	
$\varphi \to (\psi \to \varphi)$	(F1)
$(\chi \to (\psi \to \varphi)) \to ((\chi \to \psi) \to (\chi \to \varphi))$	(F2)
$(\chi \to (\psi \to \varphi)) \to (\psi \to (\chi \to \varphi))$	(F3)
$(\psi \to \varphi) \to (\neg \varphi \to \neg \psi)$	(F4)
$\neg\neg\varphi \to \varphi$	(F5)
$\varphi \to \neg\neg\varphi$	(F6)

Russell and Whitehead (1910) [220]	
$\varphi \lor \varphi \to \varphi$	(P1)
$\psi \to \varphi \lor \psi$	(P2)
$\varphi \lor \psi \to \psi \lor \varphi$	(P3)
$\varphi \lor (\psi \lor \chi) \to \psi \lor (\varphi \lor \chi)$	(P4)
$(\psi \to \chi) \to (\varphi \lor \psi \to \varphi \lor \chi)$	(P5)

Kleene (1952) [118]	
$\varphi \to (\psi \to \varphi)$	(K1)
$(\varphi \to (\psi \to \chi)) \to ((\varphi \to \psi) \to (\varphi \to \chi))$	(K2)
$\varphi \land \psi \to \varphi$	(K3)
$\varphi \land \psi \to \psi$	(K4)
$\varphi \to (\psi \to \varphi \land \psi)$	(K5)
$\varphi \to \varphi \lor \psi$	(K6)
$\psi \to \varphi \lor \psi$	(K7)
$(\varphi \to \psi) \to ((\chi \to \psi) \to (\varphi \lor \psi \to \chi))$	(K8)
$(\varphi \to \psi) \to ((\varphi \to \neg\psi) \to \neg\varphi)$	(K9)
$\neg\neg\varphi \to \varphi$	(K10)

Rosser (1953) [175]	
$\varphi \to \varphi \land \varphi$	(R1)
$\varphi \land \psi \to \varphi$	(R2)
$(\varphi \to \psi) \to (\neg(\psi \land \chi) \to \neg(\chi \land \varphi))$	(R3)

Table 2.8: Alternative axiomatizations of propositional logic

is correct ($\vdash \varphi$ implies $\models \varphi$), as all axioms are universally valid, and the only inference rule, the modus ponens, preserves this property. A fortiori, the calculus is consistent according to Theorem 2.2.

The calculus is also complete, as all universally valid formulas are indeed derivable from the axioms ($\models \varphi$ implies $\vdash \varphi$). We do not want to delve deeper into the more involved details here and refer the interested reader to the elaborated proofs in [112] or [134] instead.

Note that the propositional logic calculus is not negation complete, as for some formulas φ, neither φ nor $\neg \varphi$ is derivable. The lack of negation completeness is not a flaw of the calculus; it is solely because the semantics of propositional logic is not tight to a single interpretation, as in the examples before. In propositional logic, we express $\models \varphi$ to mean that φ is universally valid; that is, φ is true under every possible interpretation. If a formula φ is true under some but not all interpretations, the same holds for $\neg \varphi$, which means that neither $\models \varphi$ nor $\models \neg \varphi$ holds. Hence, a correct calculus for propositional logic is never negation complete.

2.3 Propositional Logic

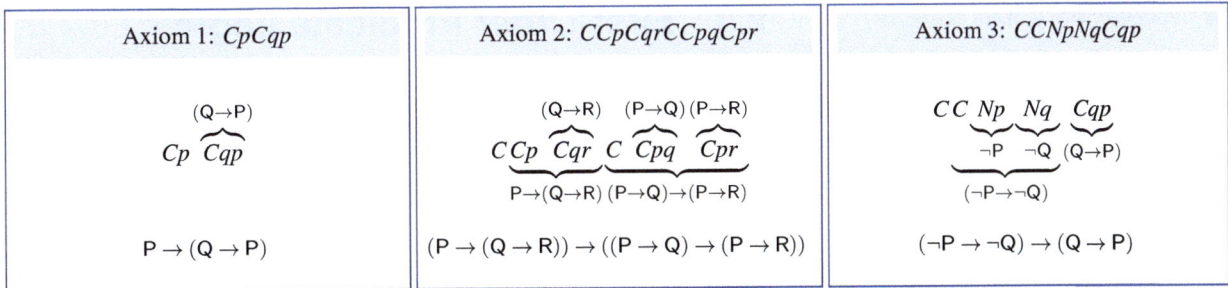

Figure 2.13: The axioms proposed by Łukasiewicz are the ones most commonly used in contemporary textbooks for the axiomatic foundation of propositional logic, with this book being no exception.

Further, note that the chosen axioms of our formal system are far from unambiguous. Table 2.8 summarizes several historically relevant axiomatizations, each constituting a correct and complete formal system for propositional logic.

Frege's axioms originate from his famous *Begriffsschrift* and, just like ours, utilize implication and negation as basic operators. Russell and Whitehead published their axioms in the first volume of *Principia Mathematica*. They based their logic on the logical primitives '¬' and '∨' and used the expression $\varphi \rightarrow \psi$ as an abbreviation for the formula $\neg \varphi \vee \psi$. Interestingly, Hilbert's student Paul Bernays proved in 1926 that the Principia's propositional axioms are not independent. The fourth axiom is redundant as it is deducable from the other axioms.

Frege's axioms are not minimal, either. The Polish mathematician Jan Łukasiewicz has shown that the number of axioms can be reduced without changing the derivable theorems. We read in a footnote from [128]:

> "*Frege's system is based upon the following six axioms: 'CpCqp', 'CCpCqrCCpqCpr', 'CCpCqrCqCpr', 'CCpqCNqNp', 'CNNpp', 'CpNNp'. Łukasiewicz has shown that in this system the third axiom is superfluous since it can be derived from the preceding two axioms, and that the last three axioms can be replaced by the single sentence 'CCNpNqCqp'.*"
>
> Jan Łukasiewicz, Alfred Tarski, 1930 [128]

Łukasiewicz's peculiar notation became known in the literature as *Polish notation*. Figure 2.13 demonstrates how the three axioms appear in contemporary notation. They match the axioms in Table 2.6, which reveals to whom we owe credit for the axioms used in this book.

2.4 First-Order Predicate Logic

Propositional logic allows us to capture logical relationships between elementary statements. Although many facts are describable in this manner, the expressive power of propositional logic is far from being strong enough to serve as a basis for the formalization of mathematics.

To formalize the terms and concepts of ordinary mathematics, we need to extend propositional logic by several building blocks. To identify the missing elements, let us examine a well-known example from ordinary analysis, the *continuity* of a real-valued function (cf. [224]).

Definition 2.11 (Continuity)

The function $f : D \to \mathbb{R}$ is *continuous* at $x_0 \in D$ if for every $\varepsilon > 0$, there exists a $\delta > 0$ with the following property:

$$x \in D \wedge |x - x_0| < \delta \Rightarrow |f(x) - f(x_0)| < \varepsilon$$

Using the universal quantifier '∀' and the existential quantifier '∃', we can write down the continuity condition as follows (Figure 2.14):

$$\forall (\varepsilon > 0) \, \exists (\delta > 0) \, \forall (x \in D) \, (|x - x_0| < \delta \to |f(x) - f(x_0)| < \varepsilon)$$

An in-depth analysis of the definition exposes the following entities besides the propositional logic connectives:

■ **Variables**

The formula contains the four variables x, x_0, ε, and δ, each representing an element of the *individual domain*, which in our example is the set of real numbers.

■ **Quantifiers**

Variables are bound by quantifiers to make quantitative statements about the elements of the individual domain. In our example, the variables ε and x are within the scope of a universal quantifier, while δ is *bound* by an existential quantifier. x_0 is not within the scope of any quantifier and is therefore referred to as *free* or *unbound*.

■ For every $\varepsilon > 0$, there exists …

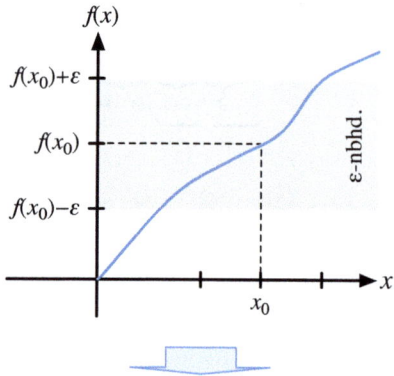

■ … a $\delta > 0$ such that …

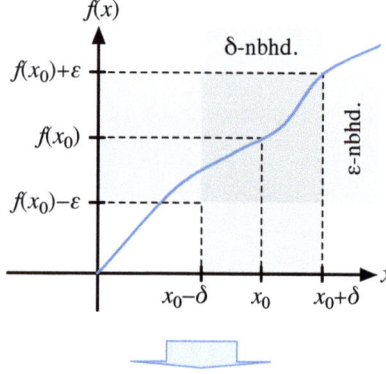

■ … $f(x)$ lies within the ε-neighborhood for all x from the δ-neighborhood.

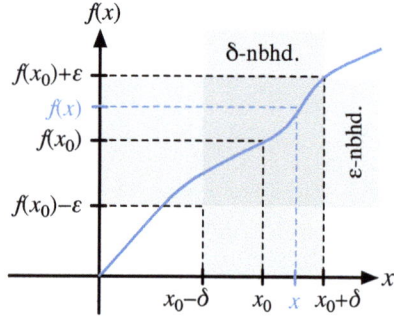

Figure 2.14: A real-valued function is continuous at the point x_0 if it satisfies the *epsilon-delta criterion*.

2.4 First-Order Predicate Logic

- **Functions**

  With f and $|\cdot|$ (absolute value function), the formula contains two function symbols of arity one. In general, an n-ary function symbol represents a mapping that assigns n elements of the individual domain to another element of the individual domain.

- **Predicates**

  With '$\in D$', the formula contains a predicate of arity one, and with '$<$', a predicate of arity two. In general, an n-ary predicate represents the existence or non-existence of a relationship between n elements of the individual domain.

The chosen example provides a first glimpse of predicate logic's nature and expressive power. In the next section, we will elaborate on the outlined ideas and work through the details with mathematical precision.

2.4.1 Syntax and Semantics

The syntax definition of predicate logic consists of three building blocks: *signatures*, *terms*, and *formulas*.

Definition 2.12 (Signature)

A signature Σ of predicate logic is a triple $(V_\Sigma, F_\Sigma, P_\Sigma)$, comprising

- a set V_Σ of *variables*, for example $\{x_1, x_2, \ldots\}$,
- a set F_Σ of *function symbols*, for example $\{f_1, f_2, \ldots\}$,
- a set P_Σ of *predicates*, for example $\{P_1, P_2, \ldots\}$.

Every function and every predicate has a fixed arity greater than or equal to 0.

Roughly speaking, a predicate logic signature defines the set of elementary symbols formulas are composed of. As in propositional logic, we allow us to customize the symbol pool from case to case and denote variables, e.g., by x, y, z, functions by f, g, h, and predicates by P, Q, R.

Predicate logic shares many similarities with propositional logic and even contains the latter as a subset. For this reason, we must keep certain terms apart. In particular, the term *variable* frequently presents a pitfall for many beginners, as it has different meanings in both logics.

In predicate logic, a variable is a placeholder for any element of a fixed set of individuals. A formula such as $P(x)$ only becomes a true or false statement through the specific choice of an individual element for the variable x.

In propositional logic, on the contrary, variables represent atomic statements that are either true or false. Thus, they are 0-ary predicates and only share the name with predicate-logic variables.

- Signature Σ

  $\Sigma = (V_\Sigma, F_\Sigma, P_\Sigma)$ with

  $V_\Sigma = \{x, y\}$
  $F_\Sigma = \{f \text{ (2-ary)}\}$
  $P_\Sigma = \{P \text{ (2-ary)}\}$

- Terms over Σ

  x
  y
  f(x,x)
  f(x,y)
  f(f(x,y),x)
  f(x,f(x,y))
  f(f(x,x),f(x,y))
  ...

- Atomic formulas over Σ

  P(x,x)
  P(x,y)
  P(f(x,y),x)
  P(x,f(x,y))
  P(x,f(f(x,y),x))
  P(f(f(x,y),x),y)
  ...

- Formulas over Σ

  $\forall x\, P(x,x)$
  $\exists x\, P(x,x)$
  $P(f(x,x),x) \leftrightarrow P(y,y)$
  $\forall y\, \exists x\, (P(f(x,x),x) \leftrightarrow P(y,y))$
  $\exists y\, \forall x\, (P(f(x,x),x) \leftrightarrow P(y,y))$
  ...

Figure 2.15: Step-by-step construction of predicate-logic expressions

 Definition 2.13 (Term)

> Let $\Sigma = (V_\Sigma, F_\Sigma, P_\Sigma)$ be a signature of predicate logic. The set of *predicate-logic terms* is defined inductively:
>
> - Every variable $\xi \in V_\Sigma$ is a term.
> - Every function symbol $f \in F_\Sigma$ of arity 0 is a term.
> - If $\sigma_1, \ldots, \sigma_n$ are terms and $f \in F_\Sigma$ is an n-ary function symbol, then $f(\sigma_1, \ldots, \sigma_n)$ is a term.

Figure 2.15 (above) demonstrates that the number of terms obtained from a signature's symbol set is usually infinite. Among the signature symbols, the function symbols of arity 0 play a distinct role. They act as *constants*, as they have no parameters.

The introduced terminology lets us precisely define the set of predicate-logic formulas:

Definition 2.14 (Syntax of Predicate Logic)

> Let Σ be a signature of predicate logic. The set of *atomic predicate-logic formulas* is defined as follows:
>
> - If $\sigma_1, \ldots, \sigma_n$ are terms and P is an n-ary predicate, then $P(\sigma_1, \ldots, \sigma_n)$ is an atomic formula.
>
> The set of *predicate-logic formulas* is defined inductively:
>
> - 0, 1, and every atomic formula are formulas.
> - Let $\xi \in V_\Sigma$. If φ and ψ are formulas, so are:
> $(\neg\varphi), (\varphi \wedge \psi), (\varphi \vee \psi), (\varphi \rightarrow \psi), (\varphi \leftrightarrow \psi), (\varphi \nleftrightarrow \psi)$
> $\forall \xi\, \varphi,\, \exists \xi\, \varphi$

Figure 2.15 (below) showcases a small selection of predicate-logic formulas. Note that not all variables necessarily have to be in the scope of a quantifier. For example, the variable x appears *free* or *unbound* in the formula P(x,x), whereas it is *bound* in the formula $\forall x\, P(x)$. The example in Figure 2.16 demonstrates that a variable may even occur freely and bound in the same formula. Formulas that do not contain free variables are called *closed*; all others are called *open*.

2.4 First-Order Predicate Logic

In the following, we write

$$\varphi(\xi_1, \ldots, \xi_n)$$

to indicate that the variables $\xi_1, \ldots, \xi_n$ may occur freely in φ. Accordingly, we write

$$\varphi(\cancel{\xi_1}, \ldots, \cancel{\xi_n})$$

to express that the mentioned variables do not occur freely in φ.

It is substantively easier to work with predicate-logic formulas if the variables in two independent sub-expressions are named differently. If the quantified variables of a closed formula φ are pairwise different, we speak of a *cleanst formula*. Figure 2.17 shows how to obtain such a formula by bounded renaming.

Similar to propositional logic, the semantics of predicate logic is given through a *model relation* '$\models$'. To provide a precise definition, we commence by extending the notion of interpretation to predicate-logic formulas:

Be cautious to correctly interpret the rules governing the syntactic structure of predicate-logic formulas. For instance, according to the stipulated formation rules, the formula ψ in

$$\forall x\, \varphi \to \psi$$

is unbound. To bring ψ into the scope of the quantifier, we need to parenthesize the subexpression $\varphi \to \psi$. Misinterpreting the formation rules is a common source of error. Thus, be advised to memorize the following carefully:

$$\forall x\, \varphi \to \psi \ne \forall x\, (\varphi \to \psi)$$

 Definition 2.15 (Interpretation of Predicate Logic)

Let $\Sigma = (V_\Sigma, F_\Sigma, P_\Sigma)$ be signature of predicate logic. An *interpretation* over Σ is a tuple (U, I) with the following properties:

- U is a non-empty set.
- I is a function mapping
  - every variable symbol $\xi \in V_\Sigma$ to an element $I(\xi) \in U$,
  - every n-ary function symbol $f \in F_\Sigma$ to a function
    $$I(f) : U^n \to U,$$
  - and every n-ary predicate symbol $P \in P_\Sigma$ to a relation
    $$I(P) \subseteq U^n.$$

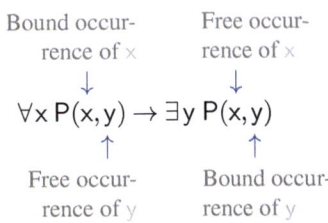

Figure 2.16: When a variable occurs within the range of a quantifier, we speak of a *bound* occurrence; otherwise, we speak of a *free* occurrence.

$$\forall x\, P(x,y) \to \exists x\, P(x,y)$$

Bounded renaming

$$\forall x\, P(x,y) \to \exists z\, P(z,y)$$

(Cleansed formula)

Figure 2.17: By renaming variables with multiple bindings, we obtain formulas in which all quantified variables are pairwise distinct.

The set U is referred to in the literature as the *set of individuals*, the *domain (of discourse)*, or the *universe*.

Note that the assignment of variables to the elements of U is only relevant for open formulas. In this case, it ensures that all free variables are assigned an individual element.

The notion of interpretation is not defined uniformly in the literature, as is the notion of model. In the definition used here, an interpretation is a tuple (U, I) with the property that the domain of the function I encompasses not only the function and predicate symbols but also the variable symbols of the underlying predicate-logic signature. This approach mimics the one in [183]. Alternative definitions employ a separate function b for assigning the free variables of a formula to elements of the individual domain and replace the expression $(U, I) \models \varphi$ by:

$$(U, I, b) \models \varphi \tag{2.5}$$

Taking this path opens up two possibilities for defining the concept of a model. One is to view a model as a triple (U, I, b) that satisfies the relation (2.5). This definition differs from ours primarily in the choice of notation and is adopted, for example, in [53] and [168]. Another is to define a model as a tuple (U, I) satisfying the relation (2.5) for all assignments b without exception [135]. Note that the two definitions only differ for formulas with free variables. All three definitions are equivalent if all variables are bound.

Also, note that the definition includes function and predicate symbols of arity 0. A 0-ary function symbol is formally assigned a function

$$U^0 \to U,$$

which conceals a single element from the individual range, hence, a constant. 0-ary predicate symbols represent relations over the set U^0. They are atomic statements that can be either true or false. They are thus nothing more than propositional variables, which are already well-known to us.

The mapping I, which assigns a function $I(f)$ to each function symbol f, can be expanded to complex terms in an obvious way. To do so, we apply I according to the following inductive scheme:

$$I(f(\sigma_1, \ldots, \sigma_n)) := I(f)(I(\sigma_1), \ldots, I(\sigma_n))$$

The expansion equips us with the necessary means to formally introduce the predicate-logic model relation '$\models$':

 Definition 2.16 (Semantics of Predicate Logic)

Let φ and ψ be predicate-logic formulas, and (U, I) an interpretation. The semantics of predicate logic is defined by the *model relation* '$\models$', which is defined inductively:

$$(U, I) \models 1$$
$$(U, I) \not\models 0$$
$$(U, I) \models P(\sigma_1, \ldots, \sigma_n) :\Leftrightarrow (I(\sigma_1), \ldots, I(\sigma_n)) \in I(P)$$
$$(U, I) \models (\neg \varphi) :\Leftrightarrow (U, I) \not\models \varphi$$
$$(U, I) \models (\varphi \wedge \psi) :\Leftrightarrow (U, I) \models \varphi \text{ and } (U, I) \models \psi$$
$$(U, I) \models (\varphi \vee \psi) :\Leftrightarrow (U, I) \models \varphi \text{ or } (U, I) \models \psi$$
$$(U, I) \models (\varphi \to \psi) :\Leftrightarrow (U, I) \not\models \varphi \text{ or } (U, I) \models \psi$$
$$(U, I) \models (\varphi \leftrightarrow \psi) :\Leftrightarrow (U, I) \models \varphi \to \psi \text{ and } (U, I) \models \psi \to \varphi$$
$$(U, I) \models (\varphi \leftrightarrow\!\!\!\!/\, \psi) :\Leftrightarrow (U, I) \not\models (\varphi \leftrightarrow \psi)$$
$$(U, I) \models \forall \xi\, \varphi :\Leftrightarrow \text{For all } u \in U \text{ it holds that } (U, I_{[\xi/u]}) \models \varphi$$
$$(U, I) \models \exists \xi\, \varphi :\Leftrightarrow \text{There is a } u \in U \text{ with } (U, I_{[\xi/u]}) \models \varphi$$

An interpretation (U, I) with $(U, I) \models \varphi$ is called a *model* of φ.

Definition 2.16 utilizes the term $I_{[\xi/u]}$ for the first time. If (U, I) is a predicate-logic interpretation, then $(U, I_{[\xi/u]})$ denotes the interpretation

2.4 First-Order Predicate Logic

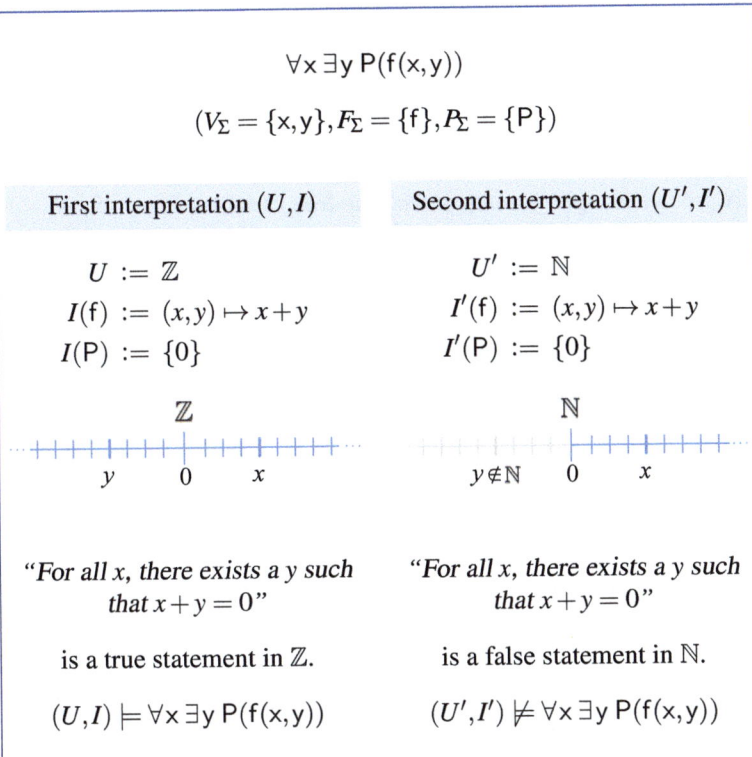

Figure 2.18: Two interpretations for the formula $\forall x\, \exists y\, P(f(x,y))$

that assigns the individual element u to ξ and is otherwise identical to (U,I).

Figure 2.18 presents two examples. Both interpretations associate the function symbol f with ordinary addition and the predicate symbol P with the set $\{0\}$, i.e., $P(x)$ is true precisely for the number 0. The underlying individual domains are different, though. The first interpretation draws from the range of integers, while the second considers the natural numbers. Under these conditions, the example formula

$$\varphi := \forall x\, \exists y\, P(f(x,y))$$

reads as follows:

"For all x, there exists a y such that $x+y=0$"

The statement is obviously true for the integers, but not for the natural numbers.

- Negation rules

$$\neg \exists \xi\, \varphi \equiv \forall \xi\, \neg\varphi$$
$$\exists \xi\, \neg\varphi \equiv \neg\forall \xi\, \varphi$$

- Scoping rules

$$\varphi(\bar{\xi}) \wedge (\exists \xi\, \psi) \equiv \exists \xi\, (\varphi \wedge \psi)$$
$$\varphi(\bar{\xi}) \wedge (\forall \xi\, \psi) \equiv \forall \xi\, (\varphi \wedge \psi)$$
$$\varphi(\bar{\xi}) \vee (\exists \xi\, \psi) \equiv \exists \xi\, (\varphi \vee \psi)$$
$$\varphi(\bar{\xi}) \vee (\forall \xi\, \psi) \equiv \forall \xi\, (\varphi \vee \psi)$$
$$\varphi(\bar{\xi}) \to (\exists \xi\, \psi) \equiv \exists \xi\, (\varphi \to \psi)$$
$$\varphi(\bar{\xi}) \to (\forall \xi\, \psi) \equiv \forall \xi\, (\varphi \to \psi)$$

Figure 2.19: Important equivalences. The expression $\varphi(\bar{\xi})$ states that the variable ξ does not occur freely in the formula φ.

The following notions, which we already learned about in the discussion of propositional logic, take this circumstance into account:

 Definition 2.17 (Satisfiable, Unsatisfiable, Universally Valid)

A predicate-logic formula φ with the free variables $\xi_1, \ldots, \xi_n$ is

- *satisfiable*, if φ has at least one model,
  ☞ there exists a (U,I) such that $(U,I) \models \varphi$,

- *unsatisfiable*, if φ has no models,
  ☞ there exists no (U,I) such that $(U,I) \models \varphi$,

- *universally valid*, if every interpretation is a model of φ,
  ☞ for all (U,I), $(U,I) \models \varphi$.

Just as in propositional logic, we can extend these notions to sets of formulas. A set

$$M = \{\varphi_1, \ldots, \varphi_n\}$$

is called satisfiable if some interpretation is a model for all $\varphi_i \in M$; it is called unsatisfiable if its elements have no model in common. If, on the other hand, each interpretation is a model for the formulas

$$\varphi_1, \ldots, \varphi_n,$$

we call the set M universally valid.

Now, it is almost self-evident how to define the concept of equivalence in predicate logic. We call two predicate-logic formulas φ and ψ equivalent, written as

$$\varphi \equiv \psi,$$

if they share the same models. Or, which is the same: if the formula

$$\varphi \leftrightarrow \psi$$

is universally valid.

Figure 2.19 summarizes relevant equivalences. The two negation laws are of particular significance, as they allow us to remove one of the two quantifiers from the symbol set without reducing the logic's expressive power. The following section will exploit this property in discussing a formal system that exclusively provides axioms and inference rules for the universal quantifier.

In propositional logic, we agreed on calling a formula a tautology if it is universally valid. We will use this exact definition in predicate logic, too, though we know that some books make a finer distinction. Some authors only label a predicate-logic formula a tautology if it is universally valid in the *propositional sense*.

We will illustrate the difference by considering the formula

$$(\forall x\, P(x)) \vee \neg(\forall x\, P(x)).$$

On the propositional level, this formula has the form $\varphi \vee \neg\varphi$, making it universally valid in the propositional sense. In contrast, the equivalent formula

$$\forall x\, P(x) \vee \exists x\, \neg P(x)$$

has the form $\varphi \vee \psi$ and is not universally valid in the propositional sense. In books following this terminology, not every universally valid predicate-logical formula is a tautology. However, the converse holds there as well.

2.4.2 Predicate-Logic Calculus

This section addresses the question of how to prove the universal validity of a predicate-logic formula. As we did for propositional logic, we will define a formal system capable of deriving universally valid formulas from a set of axioms by applying a finite number of inference steps. Table 2.9 summarizes the axioms and inference rules.

The axiom schemas (A1) to (A3) are familiar to us; they are identical to the axioms of propositional logic. The schema (A4) expresses that the formula $\forall \xi \, \varphi$ can be utilized to derive instances of the form $\varphi[\xi \leftarrow \sigma]$. The formula $\varphi[\xi \leftarrow \sigma]$ emerges from φ by replacing all free occurrences of ξ, and only those, with the predicate-logic term σ. We say that the formula $\varphi(\xi)$ has been *instantiated* with the term σ (Figure 2.20). The schema (A5) allows us to remove a subformula φ from the scope of a quantifier whenever the quantified variable does not occur in φ. Theorems can be deduced via the already-known modus ponens or the newly introduced generalization rule. The latter allows us to derive the generalization $\forall \xi \, \varphi$ from the formula φ.

As in the propositional case, the deduction theorem is an essential tool for conducting proofs. For closed formulas, we can adopt the formulation in Theorem 2.7 one-to-one. When dealing with free variables,

■ Example 1: $\mathfrak{S} := [x \leftarrow a]$

$(\forall x \, P(x,y))\mathfrak{S} = (\forall x \, P(x,y))$
$(\forall y \, P(x,y))\mathfrak{S} = (\forall y \, P(a,y))$
$(\forall x \forall y \, P(x,y))\mathfrak{S} = (\forall x \forall y \, P(x,y))$

💡 $\mathfrak{S}$ is a ground substitution

■ Example 2: $\mathfrak{S} := [x \leftarrow y, y \leftarrow f(y)]$

$(\forall x \, P(x,y))\mathfrak{S} = (\forall x \, P(x,f(y)))$
$(\forall y \, P(x,y))\mathfrak{S} = (\forall y \, P(y,y))$
$(\forall x \forall y \, P(x,y))\mathfrak{S} = (\forall x \forall y \, P(x,y))$

💡 $\mathfrak{S}$ is a substitution, but no ground substitution

Figure 2.20: A substitution of the form $[\xi \leftarrow \sigma]$ replaces all free occurrences of the variable ξ with the term σ. All bound occurrences remain unchanged. We speak of a *ground substitution* if the injected terms do not contain variables.

Axioms	
$\varphi \to (\psi \to \varphi)$	(A1)
$(\varphi \to (\psi \to \chi)) \to ((\varphi \to \psi) \to (\varphi \to \chi))$	(A2)
$(\neg \varphi \to \neg \psi) \to (\psi \to \varphi)$	(A3)
$\forall \xi \, \varphi \to \varphi[\xi \leftarrow \sigma]$ (for each collision-free substitution)	(A4)
$\forall \xi \, (\varphi(\xi) \to \psi) \to (\varphi \to \forall \xi \, \psi)$	(A5)

Inference Rules			
$\dfrac{\varphi, \varphi \to \psi}{\psi}$	(MP)	$\dfrac{\varphi}{\forall \xi \, \varphi}$	(G)

Table 2.9: Axioms and inference rules of the predicate-logic calculus.
Note that instances of (A4) can only be formed if the substitution of ξ by σ is carried out *collision-free*. A collision occurs when a variable from σ enters the scope of a quantifier due to substitution. This is the case, for example, with $\varphi = \exists y \, P(\xi, y)$ and $\sigma = y$. If we were to forego the requirement of collision-free substitution, the false statement $\forall x \exists y \, P(x,y) \to \exists y \, P(y,y)$ would be a theorem.
Caution is also required when instantiating (A5). Only formulas with no free occurrences of ξ may be substituted for φ.

Theorems
■ Theorem T16 $\forall \xi \forall \zeta\, \varphi(\xi,\zeta) \to \forall \zeta \forall \xi\, \varphi(\xi,\zeta)$
■ Theorem T17 $\forall \xi\, (\varphi(\xi) \to \psi(\xi)) \to (\forall \xi\, \varphi(\xi) \to \forall \xi\, \psi(\xi))$
■ Theorem T18 $\forall \xi \forall \zeta\, \varphi(\xi,\zeta) \to \forall \xi\, \varphi(\xi,\xi)$

Table 2.10: A small selection of theorems that can be proven in the predicate logic calculus

however, we need to pay particular attention, as the following derivation sequence demonstrates:

1. $\{P(x)\} \vdash P(x)$ (Theorem 2.4)
2. $\{P(x)\} \vdash \forall x\, P(x)$ (G. 1)

If the deduction theorem were applicable without restrictions, it would allow us to infer the formula

$$P(x) \to \forall x\, P(x)$$

which is not universally valid. To avoid deriving such formulas, we need to slightly alter the definition of the deduction theorem in the predicate-logic case.

Theorem 2.8 (Deduction Theorem of Predicate Logic)

Let $\varphi, \varphi_1, \ldots, \varphi_n$ and ψ be predicate-logic formulas.

■ If φ is closed, the following applies:

$$\{\varphi_1, \ldots, \varphi_n\} \cup \{\varphi\} \vdash \psi \Leftrightarrow \{\varphi_1, \ldots, \varphi_n\} \vdash \varphi \to \psi$$

■ If φ contains the free variables $\xi_1, \ldots, \xi_n$, the following applies:

$$\{\varphi_1, \ldots, \varphi_n\} \cup \{\varphi\} \vdash \psi \Leftrightarrow \{\varphi_1, \ldots, \varphi_n\} \vdash (\forall \xi_1 \ldots \forall \xi_n\, \varphi) \to \psi$$

Let us witness the calculus in action and prove the theorems in Table 2.10 (cf. [134]).

2.4 First-Order Predicate Logic

Theorem T16
$\forall\xi\,\forall\zeta\,\varphi(\xi,\zeta) \to \forall\zeta\,\forall\xi\,\varphi(\xi,\zeta)$

1. $\{\forall\xi\,\forall\zeta\,\varphi(\xi,\zeta)\} \vdash \forall\xi\,\forall\zeta\,\varphi(\xi,\zeta)$ (Theorem 2.4)
2. $\vdash \forall\xi\,\forall\zeta\,\varphi(\xi,\zeta) \to \forall\zeta\,\varphi(\xi,\zeta)$ (A4)
3. $\{\forall\xi\,\forall\zeta\,\varphi(\xi,\zeta)\} \vdash \forall\zeta\,\varphi(\xi,\zeta)$ (MP, 1,2)
4. $\vdash \forall\zeta\,\varphi(\xi,\zeta) \to \varphi(\xi,\zeta)$ (A4)
5. $\{\forall\xi\,\forall\zeta\,\varphi(\xi,\zeta)\} \vdash \varphi(\xi,\zeta)$ (MP, 3,4)
6. $\{\forall\xi\,\forall\zeta\,\varphi(\xi,\zeta)\} \vdash \forall\xi\,\varphi(\xi,\zeta)$ (G, 5)
7. $\{\forall\xi\,\forall\zeta\,\varphi(\xi,\zeta)\} \vdash \forall\zeta\,\forall\xi\,\varphi(\xi,\zeta)$ (G, 6)
8. $\vdash \forall\xi\,\forall\zeta\,\varphi(\xi,\zeta) \to \forall\zeta\,\forall\xi\,\varphi(\xi,\zeta)$ (DT)

Theorem T17
$\forall\xi\,(\varphi(\xi) \to \psi(\xi))$
$\to (\forall\xi\,\varphi(\xi) \to \forall\xi\,\psi(\xi))$

1. $\{\forall\xi\,(\varphi(\xi) \to \psi(\xi))\} \vdash \forall\xi\,(\varphi(\xi) \to \psi(\xi))$ (Theorem 2.4)
2. $\{\forall\xi\,\varphi(\xi)\} \vdash \forall\xi\,\varphi(\xi)$ (Theorem 2.4)
3. $\vdash \forall\xi\,(\varphi(\xi) \to \psi(\xi)) \to (\varphi(\xi) \to \psi(\xi))$ (A4)
4. $\{\forall\xi\,(\varphi(\xi) \to \psi(\xi))\} \vdash \varphi(\xi) \to \psi(\xi)$ (MP, 1,3)
5. $\vdash \forall\xi\,\varphi(\xi) \to \varphi(\xi)$ (A4)
6. $\{\forall\xi\,\varphi(\xi)\} \vdash \varphi(\xi)$ (MP, 2,5)
7. $\{\forall\xi\,(\varphi(\xi) \to \psi(\xi)), \forall\xi\,\varphi(\xi)\} \vdash \psi(\xi)$ (MP, 4,6)
8. $\{\forall\xi\,(\varphi(\xi) \to \psi(\xi)), \forall\xi\,\varphi(\xi)\} \vdash \forall\xi\,\psi(\xi)$ (G, 7)
9. $\{\forall\xi\,(\varphi(\xi) \to \psi(\xi))\} \vdash \forall\xi\,\varphi(\xi) \to \forall\xi\,\psi(\xi)$ (DT)
10. $\vdash \forall\xi\,(\varphi(\xi) \to \psi(\xi)) \to (\forall\xi\,\varphi(\xi) \to \forall\xi\,\psi(\xi))$ (DT)

Theorem T18
$\forall\xi\,\forall\zeta\,\varphi(\xi,\zeta) \to \forall\xi\,\varphi(\xi,\xi)$

for each collision-free substitution $\varphi[\zeta \leftarrow \xi]$

1. $\{\forall\xi\,\forall\zeta\,\varphi(\xi,\zeta)\} \vdash \forall\xi\,\forall\zeta\,\varphi(\xi,\zeta)$ (Theorem 2.4)
2. $\vdash \forall\xi\,\forall\zeta\,\varphi(\xi,\zeta) \to \forall\zeta\,\varphi(\xi,\zeta)$ (A4)
3. $\{\forall\xi\,\forall\zeta\,\varphi(\xi,\zeta)\} \vdash \forall\zeta\,\varphi(\xi,\zeta)$ (MP, 1,2)
4. $\vdash \forall\zeta\,\varphi(\xi,\zeta) \to \varphi(\xi,\xi)$ (A4)
5. $\{\forall\xi\,\forall\zeta\,\varphi(\xi,\zeta)\} \vdash \varphi(\xi,\xi)$ (MP, 3,4)
6. $\{\forall\xi\,\forall\zeta\,\varphi(\xi,\zeta)\} \vdash \forall\xi\,\varphi(\xi,\xi)$ (G, 5)
7. $\vdash \forall\xi\,\forall\zeta\,\varphi(\xi,\zeta) \to \forall\xi\,\varphi(\xi,\xi)$ (DT)

Finally, let us examine the predicate-logic calculus w.r.t. the properties outlined in Definition 2.2. The calculus' correctness is easy to verify. All instances of the axiom schemata (A1) to (A3) are universally valid, as seen in Section 2.3.2. Furthermore, the definition of the all-quantifier implies that all instances of the axiom schemata (A4) and (A5) are also universally valid, and both the modus ponens and the generalization rule preserve this property. For the letter, this follows directly from our definition of universally valid formulas with free variables (see Definition 2.17). Hence, the calculus is correct and a fortiori consistent according to Theorem 2.2.

As in propositional logic, the calculus of first-order predicate logic is complete; that is, every universally valid formula is derivable from the axioms by a finite number of inference steps. We know about the existence of a complete formal system for first-order predicate logic since 1929, the year Kurt Gödel proved his famous *completeness theorem* in his doctoral dissertation [71, 72]. The predicate-logic part of his formal system is identical to the one in Table 2.9. Different are the propositional axioms. Instead of the axiom schemas (A1) to (A3), Gödel used the four axioms schemata (P1), (P2), (P3), and (P5) in Table 2.8 on page 94, first introduced by Russell and Whitehead.

Be careful not to confuse Gödel's completeness theorem with the two incompleteness theorems you will become familiar with in Chapter 4. Even though their names sound confusingly similar, their substantive meanings are entirely different.

2.5 Predicate Logic with Equality

When attempting to formalize mathematical facts within predicate logic, you will quickly notice that in some cases, this can be done easily, while in others, it is cumbersome or even impossible. To demonstrate this phenomenon, let us construct a formula that is true if and only if the binary predicate symbol R is interpreted as a left-total, right-unique relation R. We can quickly write down the property of left-totality:

$$\forall x \, \exists y \, R(x,y)$$

However, the property of right-uniqueness poses difficulties for us. We need to express that for no x, two different elements, y and z, exist with $R(x,y)$ and $R(x,z)$. In other words, if both $R(x,y)$ and $R(x,z)$ hold, then y and z must be representatives of the same element. To formalize this relationship, we are missing a crucial concept: equality. One straightforward solution is to add the equality operator as a special predicate symbol with fixed semantics. Assuming access to such a predicate, we can quickly pen down the desired formula:

$$\forall x \, \exists y \, (R(x,y) \wedge \forall z \, (R(x,z) \rightarrow z \doteq y))$$

Only two definitions from Section 2.4 need to be adapted to obtain a formal definition of *predicate logic with equality*. The first one concerns the structure of atomic formulas:

2.5 Predicate Logic with Equality

Axioms
$\xi \doteq \xi$ (A6)
$\xi \doteq \zeta \rightarrow (\varphi[\zeta \leftarrow \xi] \rightarrow \varphi)$ (A7)
(for every collision-free substitution)

Table 2.11: Adding the axioms (A6) and (A7) to the formal system presented in Section 2.4.2 yields a complete formal system for first-order predicate logic with equality. The symbols ξ and ζ are placeholders for predicate-logic variables.

Theorems	
■ Theorem T19 $\sigma \doteq \sigma$	■ Theorem T20 $\sigma \doteq \tau \rightarrow \tau \doteq \sigma$
■ Theorem T21 $\sigma \doteq \tau \rightarrow (\tau \doteq \rho \rightarrow \sigma \doteq \rho)$	

Table 2.12: Selected theorems of first-order predicate logic with equality (cf. [134]). The placeholders σ, τ, and ρ represent arbitrary predicate-logic terms.

 Addendum to Definition 2.14

■ If σ and τ are terms, then $(\sigma \doteq \tau)$ is an atomic formula.

We also need to add a line to the definition of semantics, giving the symbol '$\doteq$' its intended meaning:

 Addendum to Definition 2.16

$(U, I) \models (\sigma \doteq \tau) :\Leftrightarrow I(\sigma) = I(\tau)$

Finally, adding further axioms will expand the formal system introduced in Sectio 2.4.2 into a calculus for predicate logic with equality. Table 2.11 summarizes the axioms involved. (A6) describes the reflexivity of the equality relation. It is an axiom schema where ξ is a placeholder for an arbitrary variable. (A7) is called the substitution axiom. It is also laid out as a schema and employs the symbols ξ and ζ as placeholders for arbitrary predicate-logic variables.

Informally speaking, the axiom of substitution states that no harm is made if we replace an element with another if both are identical. Make sure to apply the schema only if the substitution of ζ by ξ is collision-free, i.e., ξ does not fall within the scope of a quantifier at the inserted

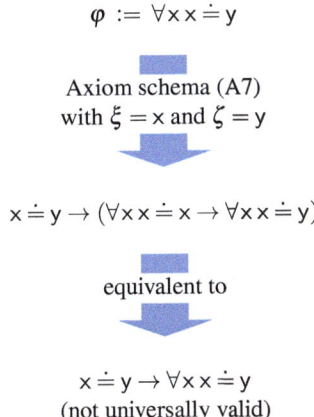

Figure 2.21: Not all instances were universally valid if (A7) allowed for arbitrary substitutions. In the example shown, variable x ends up in the scope of a quantifier and is thus no longer related to the variable x in the subexpression $x \doteq y$, appearing left to the implication operator.

position (Figure 2.21). This restriction ensures that the newly inserted variable ξ refers to the same individual element as the variable ξ that appears in (A7) on the left-hand side of the implication.

Theorem T19
$\sigma \doteq \sigma$

Let ξ be an arbitrary variable.

1. $\vdash \xi \doteq \xi$ (A6)
2. $\vdash \forall \xi\, \xi \doteq \xi$ (G, 1)
3. $\vdash \forall \xi\, \xi \doteq \xi \to \sigma \doteq \sigma$ (A4)
4. $\vdash \sigma \doteq \sigma$ (MP, 2,3)

Theorem T20
$\sigma \doteq \tau \to \tau \doteq \sigma$

Let ξ, ζ be arbitrary variables.

1. $\vdash \xi \doteq \zeta \to (\xi \doteq \xi \to \zeta \doteq \xi)$ (A7)
2. $\{\xi \doteq \zeta\} \vdash \xi \doteq \xi \to \zeta \doteq \xi$ (DT)
3. $\{\xi \doteq \zeta, \xi \doteq \xi\} \vdash \zeta \doteq \xi$ (DT)
4. $\{\xi \doteq \xi\} \vdash \xi \doteq \zeta \to \zeta \doteq \xi$ (DT)
5. $\vdash \xi \doteq \xi \to (\xi \doteq \zeta \to \zeta \doteq \xi)$ (DT)
6. $\vdash \xi \doteq \xi$ (A6)
7. $\vdash \xi \doteq \zeta \to \zeta \doteq \xi$ (MP, 5,6)
8. $\vdash \forall \xi\, (\xi \doteq \zeta \to \zeta \doteq \xi)$ (G, 7)
9. $\vdash \forall \xi\, (\xi \doteq \zeta \to \zeta \doteq \xi) \to (\sigma \doteq \zeta \to \zeta \doteq \sigma)$ (A4)
10. $\vdash \sigma \doteq \zeta \to \zeta \doteq \sigma$ (MP, 8,9)
11. $\vdash \forall \zeta\, (\sigma \doteq \zeta \to \zeta \doteq \sigma)$ (G, 10)
12. $\vdash \forall \zeta\, (\sigma \doteq \zeta \to \zeta \doteq \sigma) \to (\sigma \doteq \tau \to \tau \doteq \sigma)$ (A4)
13. $\vdash \sigma \doteq \tau \to \tau \doteq \sigma$ (MP, 11,12)

Theorem T21
$\sigma \doteq \tau \to (\tau \doteq \rho \to \sigma \doteq \rho)$

Let ξ, ζ, v be arbitrary variables.

1. $\vdash \xi \doteq \zeta \to \zeta \doteq \xi$ (T20)
2. $\vdash \zeta \doteq \xi \to (\zeta \doteq v \to \xi \doteq v)$ (A7)
3. $\vdash \xi \doteq \zeta \to (\zeta \doteq v \to \xi \doteq v)$ (MB, 1,2)
4. $\vdash \forall \xi\, (\xi \doteq \zeta \to (\zeta \doteq v \to \xi \doteq v))$ (G, 2)
5. $\vdash \forall \xi\, (\xi \doteq \zeta \to (\zeta \doteq v \to \xi \doteq v)) \to$
   $\quad (\sigma \doteq \zeta \to (\zeta \doteq v \to \sigma \doteq v))$ (A4)
6. $\vdash \sigma \doteq \zeta \to (\zeta \doteq v \to \sigma \doteq v)$ (MP, 4,5)
7. $\vdash \forall \zeta\, (\sigma \doteq \zeta \to (\zeta \doteq v \to \sigma \doteq v))$ (G, 6)

2.5 Predicate Logic with Equality

8. $\vdash \forall \zeta \, (\sigma \doteq \zeta \to (\zeta \doteq v \to \sigma \doteq v))$
   $\to (\sigma \doteq \tau \to (\tau \doteq v \to \sigma \doteq v))$ (A4)
9. $\vdash \sigma \doteq \tau \to (\tau \doteq v \to \sigma \doteq v)$ (MP, 7,8)
10. $\vdash \forall v \, (\sigma \doteq \tau \to (\tau \doteq v \to \sigma \doteq v))$ (G, 9)
11. $\vdash \forall v \, (\sigma \doteq \tau \to (\tau \doteq v \to \sigma \doteq v)) \to$
    $(\sigma \doteq \tau \to (\tau \doteq \rho \to \sigma \doteq \rho))$ (A4)
12. $\vdash \sigma \doteq \tau \to (\tau \doteq \rho \to \sigma \doteq \rho)$ (MP, 10,11)

You may have wondered why we have integrated equality as a dedicated symbol, which, in contrast to ordinary predicate symbols, has a fixed semantic. The answer is plain simple: In first-order predicate logic, there exists no formula $\varphi_=$ being true precisely when a particular predicate symbol, e.g., P, is interpreted as the equality relation:

$$(U,I) \models \varphi_= \Leftrightarrow I(\mathsf{P}) = \{(x,y) \in U^2 \mid x = y\} \quad (2.6)$$

We will outline why equality is not definable in first-order predicate logic. Specifically, we will show that if $\varphi_=$ has a model (U,I) that interprets P as equality, it has another model (U',I') that assigns P a different meaning, thus contradicting (2.6). If the formula $\varphi_=$ were to exist, all of its models had to interpret P as the equality relation without exception.

To create the new model, we take the set of individuals U and add a copy u' to each element $u \in U$. Furthermore, we extend the interpretation of the predicate and function symbols in a way that makes it irrelevant whether we deal with an original element from U or its copy.

Figure 2.22 demonstrates the construction with a binary predicate symbol P and an interpretation (U,I) with the domain $U = \{u_1, u_2\}$. (U,I) interprets P as the equality relation, as the relation $I(\mathsf{P})$ only comprises the two pairs (u_1,u_1) and (u_2,u_2). In the interpretation (U',I'), however, $I'(\mathsf{P})$ contains the combinations (u_1,u_1), (u'_1,u_1), (u_1,u'_1), (u'_1,u'_1) and (u_2,u_2), (u'_2,u_2), (u_2,u'_2), (u'_2,u'_2). The choice of $I'(\mathsf{P})$ makes it irrelevant whether we consider an original element $u \in U$ or its copy u'. As a result, we cannot distinguish between the original elements and their copies at the logic level. If (U,I) is a model for $\varphi_=$, so is (U',I'). In (U',I'), however, the predicate symbol P is no longer interpreted as the equality relation since each element is now also related to its copy. The example demonstrates that quantifying over the elements of the individual domain is not powerful enough to distinguish an element from its copy.

■ Transition from U to U'

$U = \{u_1, u_2\}$

$U' = \{u_1, u_2, u'_1, u'_2\}$

■ Transition from I to I'

$I(\mathsf{P}) = \{(u_1,u_1), (u_2,u_2)\}$

$I'(\mathsf{P}) = \{(u_1,u_1), (u_2,u_2),$
$(u'_1,u_1), (u'_2,u_2),$
$(u_1,u'_1), (u_2,u'_2),$
$(u'_1,u'_1), (u'_2,u'_2)\}$

Figure 2.22: First-order predicate logic lacks the expressive power to define equality. Every model (U,I) that interprets some predicate symbol as equality translates into a model (U',I') in which the symbol loses its intended meaning.

Revisiting the construction, you may notice that we have employed a simple mathematical trick here. The transition from U to U' relies on the construction of equivalence classes such that each element $x \in U$ and its copy x' constitute a distinct class $[x]_\sim$:

$$[x]_\sim = [x']_\sim = \{x, x'\}$$

This also reveals the meaning of the predicate symbol P under the interpretation (U', I'): It represents the equality relation between equivalence classes. In short, if an interpretation (U, I) fulfills the relationship

$$I(\mathsf{P}) = \{(x, y) \mid x = y\},$$

the following applies to the interpretation (U', I'):

$$I'(\mathsf{P}) = \{(x, y) \mid [x]_\sim = [y]_\sim\}$$

The discussion underlines that first-order predicate logic can capture the concept of equality at the level of equivalence classes. Yet, its expressive power does not suffice to define equality on the individual level.

2.6 Higher-Order Predicate Logic

In the previous sections, we agreed to apply the predicate-logic quantifiers '$\forall$' and '$\exists$' exclusively to variables. We can thus quantify over the elements of the individual domain without restriction, but not over functions and predicates. Such logics are called *first-order logics*. This section will free us from this shackle and pave the way to *higher-order logics*. The question concerning us the most is the following: Can we create a more expressive logic than PL1, first-order predicate logic, or will it turn out that quantification over predicates and functions is merely a matter of convenience? At the end of this section, we will realize that the expressive power does indeed increase. Nevertheless, there is no reason for euphoria, as there is a high price to pay.

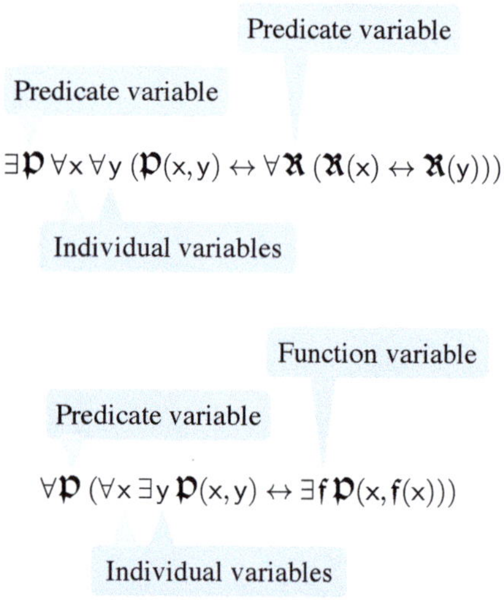

2.6.1 Syntax and Semantics

We start with the necessary syntactic modifications to extend first-order predicate logic to a higher-order logic. First, we add two new variable types to the set V_Σ of a predicate-logic signature. We refer to the variables of the first type as *predicate variables* and those of the second type as *function variables*. To distinguish the different variable types

Predicate variable

Predicate variable

$\exists \mathfrak{p} \, \forall x \, \forall y \, (\mathfrak{p}(x, y) \leftrightarrow \forall \mathfrak{A} \, (\mathfrak{A}(x) \leftrightarrow \mathfrak{A}(y)))$

Individual variables

Function variable

Predicate variable

$\forall \mathfrak{p} \, (\forall x \, \exists y \, \mathfrak{p}(x, y) \leftrightarrow \exists f \, \mathfrak{p}(x, f(x)))$

Individual variables

Figure 2.23: Second-order predicate-logic formulas

2.6 Higher-Order Predicate Logic

visually, we utilize the symbols $\mathfrak{P}, \mathfrak{Q}, \mathfrak{R}, \ldots$ for predicate variables and the symbols $\mathfrak{f}, \mathfrak{g}, \mathfrak{h}, \ldots$ for function variables. Each newly introduced variable has a fixed arity and may appear wherever a predicate sign or a function symbol with the same arity may appear in an expression of first-order predicate logic. Allowing the quantifiers '$\forall$' and '$\exists$' to bind variables of each type directly leads to second-order predicate logic, PL2 for short (Figure 2.23).

Next, we want to convey the quantifiers '$\forall$' and '$\exists$' an intuitive meaning, first in colloquial form:

$\forall \mathfrak{p} \ldots \;\widehat{=}\;$ "*For all predicates, it holds that* ..."
$\exists \mathfrak{p} \ldots \;\widehat{=}\;$ "*For some predicate, it holds that* ..."

$\forall \mathfrak{f} \ldots \;\widehat{=}\;$ "*For all functions, it holds that* ..."
$\exists \mathfrak{f} \ldots \;\widehat{=}\;$ "*For some function, it holds that* ..."

In the formal definition, we will utilize the expressions $I_{[\xi/P]}$ and $I_{[\xi/f]}$ in alignment with the expression $I_{[\xi/u]}$ from Definition 2.16. Suppose (U, I) is an interpretation and ξ is a predicate variable. Then, $(U, I_{[\xi/P]})$ refers to the interpretation that assigns the relation P to ξ and is otherwise identical to (U, I). The same applies to $I_{[\xi/f]}$ in the context of functions.

> Considering a one-character predicate sign P from a set-theoretic point of view, it is nothing more than a subset of the individual domain; P represents the subset comprising exactly those elements for which P is true. Doing so, a formula of the type $\forall \mathfrak{p}$ breaks down to a statement quantifying over subsets of the individual domain. Going one step further leads to the introduction of a new class of variables, allowing us to quantify over subsets of subsets. In this way, we arrive at *third-order predicate logic*. In this logic, we can quantify over individual elements, predicates, and functions, as well as over the properties of predicates and functions. Following this path, we arrive at ever-new logics, called fourth-order predicate logic, fifth-order predicate logic, and so on.

With the new terminology in hand, we can write down the semantics of second-order predicate logic in a few lines:

- ξ is a predicate variable of arity n

  $(U,I) \models \forall \xi\, \varphi \;:\Leftrightarrow\;$ For all $P \subseteq U^n$, it holds that $(U, I_{[\xi/P]}) \models \varphi$
  $(U,I) \models \exists \xi\, \varphi \;:\Leftrightarrow\;$ For some $P \subseteq U^n$, it holds that $(U, I_{[\xi/P]}) \models \varphi$

- ξ is a function variable of arity n

  $(U,I) \models \forall \xi\, \varphi \;:\Leftrightarrow\;$ For all $f: U^n \to U$, it holds that $(U, I_{[\xi/f]}) \models \varphi$
  $(U,I) \models \exists \xi\, \varphi \;:\Leftrightarrow\;$ For some $f: U^n \to U$, it holds that $(U, I_{[\xi/f]}) \models \varphi$

This is the *standard semantics* of second-order predicate logic.

Next, we want to explore the consequences of this logical extension. For this purpose, we first consider two formulas from first-order predicate logic. The first one reads like this:

$$\varphi_1 := \forall x\, \forall y\, (f(x) \doteq f(y) \to x \doteq y)$$

- Injective function

"*No element of the codomain has a preimage with more than one element.*"

- Surjective function

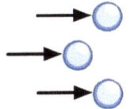

"*Every element of the codomain has a nonempty preimage.*"

Figure 2.24: Injective and surjective functions

The formula is true precisely when f is interpreted as an injective function (Figure 2.24 above). Surjectivity can be described similarly (Figure 2.24 below). Again, a formula from first-order predicate logic suffices:

$$\varphi_S := \forall y \, \exists x \, (y \doteq f(x))$$

Now, we employ the full expressive power of PL2 and combine φ_I and φ_S in the following way:

$$\varphi_{<\mathbb{N}} := \forall f \, (\varphi_I \to \varphi_S) \tag{2.7}$$
$$= \forall f \, (\forall x \, \forall y \, (f(x) \doteq f(y) \to x \doteq y) \to \forall y \, \exists x \, (y \doteq f(x)))$$

In a nutshell, $\varphi_{<\mathbb{N}}$ expresses that every injective self-mapping is surjective. That we can easily formulate this statement in second-order predicate logic illustrates the elegance all higher-order logics share. But what does this statement actually mean? A little knowledge of injective and surjective functions suffices to recognize that $\varphi_{<\mathbb{N}}$ is a true statement precisely when interpreted over a finite domain. In this case, and only in this, every injective function is necessarily surjective. Wrapping up:

$$(U,I) \models \varphi_{<\mathbb{N}} \iff U \text{ is finite} \tag{2.8}$$

Thus, we have successfully captured the concept of *finiteness* within second-order predicate logic. We say that $\varphi_{<\mathbb{N}}$ *defines* the notion of finiteness.

Of equal interest is this formula:

$$\varphi_{\leq \mathbb{N}} := \exists f \, \exists x \, \forall \mathfrak{p} \, (\mathfrak{p}(x) \wedge \forall y \, (\mathfrak{p}(y) \to \mathfrak{p}(f(y))) \to \forall x \, \mathfrak{p}(x))$$

The formula states that we can capture the elements of the domain in a sequence of the form:

$$x, f(x), f(f(x)), f(f(f(x))), f(f(f(f(x)))), \ldots \tag{2.9}$$

The variable x describes the first element, and the function f specifies how to proceed from one element to the next. To ensure that all elements of the individual range are covered, the formula states that all elements of the domain possess a property P if it is possessed by the elements enumerated in (2.9). If the domain of individuals were uncountable, this statement would not hold for all properties P, no matter how we choose x and f. On the contrary, if the domain is countable, an element x and a function f would exist such that all individual elements occur somewhere in the enumeration (2.9) (Figure 2.25). Thus, the following holds:

$$(U,I) \models \varphi_{\leq \mathbb{N}} \iff U \text{ is countable} \tag{2.10}$$

2.6 Higher-Order Predicate Logic

Combining $\varphi_{\leq \mathbb{N}}$ and $\neg \varphi_{< \mathbb{N}}$ conjunctively results in:

$$\varphi_\mathbb{N} := \exists f \, \exists x \, \forall \mathfrak{p} \, (\mathfrak{p}(x) \wedge \forall y \, (\mathfrak{p}(y) \to \mathfrak{p}(f(y)))) \to \forall x \, \mathfrak{p}(x)) \wedge$$
$$\neg \forall f \, (\forall x \, \forall y \, (f(x) \doteq f(y) \to x \doteq y) \to \forall x \, \exists y \, (x \doteq f(y)))$$

This formula is true precisely under those interpretations with a denumerable domain.

$$(U, I) \models \varphi_\mathbb{N} \; \Leftrightarrow \; U \text{ is denumerable} \qquad (2.11)$$

The formulas $\varphi_{<\mathbb{N}}$, $\varphi_{\leq\mathbb{N}}$, and $\varphi_\mathbb{N}$ share a noteworthy property; they describe notions impossible to capture within first-order predicate logic. Any attempt to come up with a PL1 formula satisfying (2.8), (2.10), or (2.11) is doomed to fail. The fact that only higher-order logic is expressive enough to formulate $\varphi_{<\mathbb{N}}$ follows from the *compactness theorem* of PL1, thoroughly discussed in Chapter 7. There, we will also present the famous *Löwenheim-Skolem-Tarski theorem*. It will explain why $\varphi_{\leq\mathbb{N}}$ and $\varphi_\mathbb{N}$ cannot be reduced to first-order formulas either.

Equality is another notion PL1 can't define. Section 2.5 has already outlined that no first-order formula is true precisely when a specific predicate symbol is interpreted as the equality relation. On the contrary, second-order predicate logic has little trouble defining this term. The key idea is that two distinct elements can always be separated from each other by some relation. In reverse, this means that two individual variables must conceal the same element if they cannot be distinguished by any relation:

$$\varphi_= := \forall x \, \forall y \, (P(x,y) \leftrightarrow \forall \mathfrak{A} \, (\mathfrak{A}(x) \leftrightarrow \mathfrak{A}(y)))$$

This formula is true precisely when the binary predicate P is interpreted as the equality relation '='.

The benefits of second-order predicate logic sound tempting, yet it plays a much smaller role than first-order predicate logic. Several reasons account for this:

- In comparison with first-order predicate logic, the semantic definition of PL2 relies heavily on the naive notion of a set. To establish the meaning of the constructs $\forall \mathfrak{p} \, \varphi$ and $\exists \mathfrak{p} \, \varphi$, we had to fall back on the notion of the power set, which is only possible if we assume this concept to be free of contradictions. In light of the set-theoretical paradoxes, some logicians reject this approach as illegitimate.

- The existence of concepts that are definable in PL2 but undefinable in PL1 does not render PL1 as weakly expressive. In Chapter 3, we

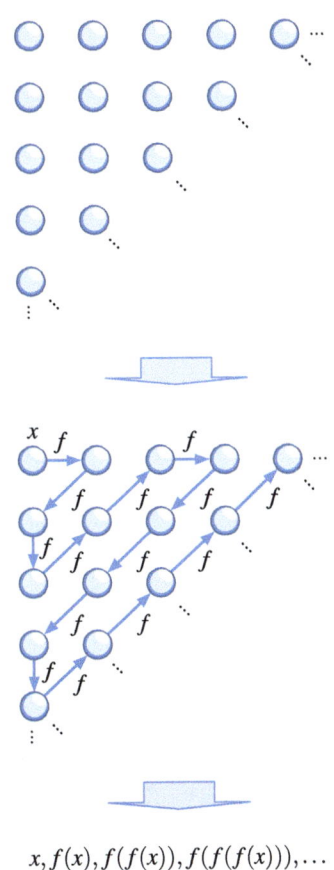

Figure 2.25: For each finite or countable set, there is an element x and a function f enumerating the elements in sequence by $x, f(x), f(f(x)), \ldots$.

will become acquainted with Zermelo-Fraenkel set theory, a *first-order theory* strong enough to formalize almost all branches of ordinary mathematics. Thus, there is usually no reason to leave the safe harbor of PL1.

- Given the standard semantics, PL2, unlike PL1, is no longer complete; that is, it is impossible to develop a correct formal system in which all universally valid PL2 formulas are derivable from the axioms. Chapter 4 will reveal why the incompleteness of PL2 is unavoidable. It is a direct consequence of *Gödel's first incompleteness theorem* and the expressive power of PL2, i.e., the ability of PL2 to axiomatize the natural numbers categorically.

2.6.2 Henkin Semantics

The definition of the quantifiers '∀' and '∃', as laid down in the standard semantics, is intuitively reasonable but not the only possible. In 1950, the US logician Leon Albert Henkin (Figure 2.26) proposed an alternative semantics, whose essentials we will now present [86].

Henkin semantics centers around the idea of letting the quantifiers no longer iterate over all possible relations or functions but only over a predefined selection. For that purpose, a Henkin interpretation (U, I) features two dedicated sets $\mathcal{R}(n)$ and $\mathcal{F}(n)$ for each positive natural number n. The set $\mathcal{R}(n)$ contains a selection of n-ary relations over the domain U, and the set $\mathcal{F}(n)$ contains a selection of n-ary functions. The semantics of the quantifiers '∀' and '∃' is defined as follows:

- ξ is a predicate variable of arity n

$$(U,I) \models \forall \xi\, \varphi \; :\Leftrightarrow \; (U, I_{[\xi/P]}) \models \varphi \text{ for all } P \in \mathcal{R}(n)$$
$$(U,I) \models \exists \xi\, \varphi \; :\Leftrightarrow \; (U, I_{[\xi/P]}) \models \varphi \text{ for some } P \in \mathcal{R}(n)$$

- ξ is a function variable of arity n

$$(U,I) \models \forall \xi\, \varphi \; :\Leftrightarrow \; (U, I_{[\xi/f]}) \models \varphi \text{ for all } f \in \mathcal{F}(n)$$
$$(U,I) \models \exists \xi\, \varphi \; :\Leftrightarrow \; (U, I_{[\xi/f]}) \models \varphi \text{ for some } f \in \mathcal{F}(n)$$

This is the *Henkin semantics* of second-order predicate logic.

There is a significant restriction, though: In a Henkin interpretation, the sets $\mathcal{R}(n)$ and $\mathcal{F}(n)$ may not be chosen freely. I.e., they must fulfill two conditions:

Leon Albert Henkin (1921 – 2006) [181]

Figure 2.26: The American logician Leon Albert Henkin has made significant discoveries in type theory. In 1950, he created Henkin semantics, an alternative to the standard semantics of higher-order predicate logic.

2.6 Higher-Order Predicate Logic

The American logician Leon Albert Henkin was born on April 19, 1921, in Brooklyn, New York. His academic career led him from Columbia College to the renowned University of Princeton, where he was awarded a doctorate in 1945 as a student of Alonzo Church. In 1953, Henkin accepted a position at the University of California in Berkeley, where he was appointed professor in 1958. He remained at his new academic home until his retirement in 1991.

Today, Henkin's name is primarily associated with a proof of the completeness of first-order predicate logic, published in 1949 [83, 85]. His result was not novel, as Kurt Gödel had already demonstrated the completeness of PL1 around 20 years earlier. Nevertheless, Henkin's proof is notable in two respects. Firstly, it is much simpler than Gödel's original proof, which is why many contemporary textbooks follow Henkin's approach today. Secondly, his line of reasoning also applies to higher-order logic. Henkin had already recognized this fact during his doctoral thesis and explicitly pointed it out in [85]: *"In the second place the proof suggests a new approach to the problem of completeness for functional calculi of higher order. Both of these matters will be taken up in future papers."*

The promised paper appeared just one year later. In [86], Henkin introduced what we now refer to as Henkin interpretations, quickly becoming an essential alternative to the standard semantics of higher-order logic.

Leon Albert Henkin died on November 1, 2006, aged 85.

- All instances of the *comprehension schema*

$$\exists \xi \, (\forall x_1 \ldots \forall x_n \, (\xi(x_1,\ldots,x_n) \leftrightarrow \varphi(x_1,\ldots,x_n)))$$

are true. In this schema, the placeholder ξ stands for any predicate variable of arity n, and φ for a formula with n free variables and no occurrence of ξ. The comprehension schema ensures that all relations in $\mathcal{R}(n)$ can be described by a formula φ with n free variables.

- All instances of the *function definition schema*

$$\forall \xi \, (\forall x_1 \ldots \forall x_n \, \exists_1 y \, \xi(x_1,\ldots,x_n,y) \to \\ \exists v \, (\forall x_1 \ldots \forall x_n \, \xi(x_1,\ldots,x_n,v(x_1,\ldots,x_n))))$$

are true. Herein, ξ denotes a predicate variable of arity $n+1$ and v a function variable of arity n. We use the expression $\exists_1 y$ as shorthand to express that the subsequent formula is true for *exactly one* assignment of y. In simple terms, the function definition schema states that the set $\mathcal{F}(n)$ contains every n-ary function that is describable with a left-total, right-unique relation from $\mathcal{R}(n+1)$. Consequently, the sets $\mathcal{R}(n+1)$ and $\mathcal{F}(n)$ of a Henkin interpretation are not independent of each other.

Standard semantics and Henkin semantics are closely related. Since we can choose $\mathcal{R}(n)$ and $\mathcal{F}(n)$ in such a way that these sets contain all n-ary relations or functions, every interpretation of the standard semantics is also a Henkin interpretation, but not vice versa. As a result, every formula being true under all Henkin interpretations is also true under

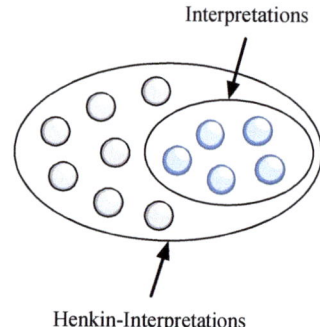

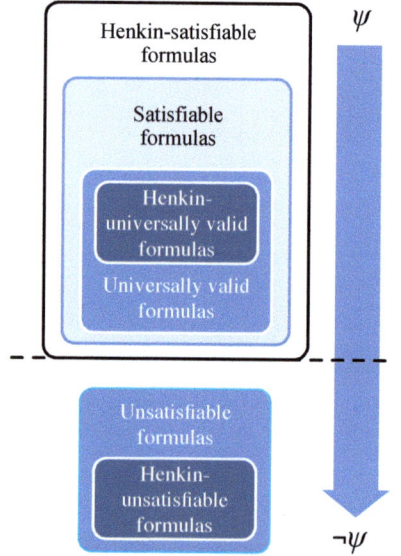

Figure 2.27: Standard semantics and Henkin semantics are closely interconnected. Every Henkin-universally valid formula is universally valid, and every satisfiable formula is Henkin-satisfiable.

all standard-semantic interpretations. Likewise, any formula being true under at least one standard-semantic interpretation is also true under at least one Henkin interpretation. Overall, the following relationships hold (Figure 2.27):

$$\varphi \text{ is Henkin-universally valid} \Rightarrow \varphi \text{ is universally valid}$$
$$\varphi \text{ is satisfiable} \Rightarrow \varphi \text{ is Henkin-satisfiable}$$

The converse of these statements does not apply.

Henkin semantics lets us regain a property already thought to be lost. In contrast to standard semantics, it allows for the definition of complete formal systems, that is, formal systems in which all Henkin-universally valid formulas, and only those, are derivable from the axioms. Yet despite all the euphoria, keeping sight of two aspects is crucial. For one thing, the quantifiers in Henkin semantics no longer carry the intuitive meaning we are inclined to give them. For another, second-order predicate logic loses its expressive power if we replace standard semantics with Henkin semantics. For example, it is no longer possible to uniquely characterize simple structures such as natural numbers.

At this point, we start to sense the pull of Gödel's incompleteness theorems, which will continuously intensify from chapter to chapter. But before we investigate the details behind these baffling theorems in Chapter 4, let us first draw our attention to the formal foundation of modern mathematics in the next chapter.

2.7 Exercises

Through the formal systems $E, E_2, \ldots, E_6$, we have explored the basic properties of formal systems.

Exercise 2.1

a) Recap the characteristics of the different formal systems by completing the table below:

	E	E_2	E_3	E_4	E_5	E_6
Consistent				✔		
Negation complete				✔		
Correct			✘	✔		
Complete				✔		

b) Suppose the symbol '>' is no longer interpreted as "greater" but as "greater or equal". Does the system E_4 remain consistent, negation complete, correct, and complete?

c) Is it possible to change the interpretations of the symbols '>' and '=' to turn E_3 into a correct formal system?

Exercise 2.2

Consider the following five axioms defining a set of properties of two classes, K and L. They are taken from [142] and retained in their colloquial form:

1. Any two members of K are contained in just one member of L.

2. No member of K is contained in more than two members of L.

3. The members of K are not all contained in a single member of L.

4. Any two members of L contain just one member of K.

5. No member of L contains more than two members of K.

Various consequences can be drawn from the axioms using the standard mathematical rules of inference. Is it possible to derive a contradiction? How could a formal proof of consistency be conducted?

Exercise 2.3

In this exercise, we consider four formal systems over a rudimentary language. In total, 42 formulas can be formed by instantiating the formulas $\varphi_1(\xi)$, $\varphi_2(\xi)$, $\varphi_3(\xi)$, $\neg\varphi_1(\xi)$, $\neg\varphi_2(\xi)$, and $\neg\varphi_3(\xi)$ with the terms $\bar{0},\ldots,\bar{6}$. The following matrix indicates which of the 42 statements are true and which are false:

	$\bar{0}$	$\bar{1}$	$\bar{2}$	$\bar{3}$	$\bar{4}$	$\bar{5}$	$\bar{6}$
$\varphi_1(\xi)$	$\models$	$\not\models$	$\models$	$\not\models$	$\models$	$\not\models$	$\not\models$
$\varphi_2(\xi)$	$\models$	$\models$	$\models$	$\models$	$\models$	$\models$	$\models$
$\varphi_3(\xi)$	$\models$	$\models$	$\not\models$	$\models$	$\models$	$\not\models$	$\models$
$\neg\varphi_1(\xi)$	$\not\models$	$\models$	$\not\models$	$\models$	$\not\models$	$\models$	$\models$
$\neg\varphi_2(\xi)$	$\not\models$	$\not\models$	$\not\models$	$\not\models$	$\not\models$	$\not\models$	$\not\models$
$\neg\varphi_3(\xi)$	$\not\models$	$\not\models$	$\models$	$\not\models$	$\not\models$	$\models$	$\not\models$

The axioms and inference rules of the four calculi are unknown. However, we possess a matrix for each formal system, indicating which formulas can be derived from the axioms and which cannot. State, for each calculus, whether it is complete, correct, consistent, and negation complete.

K_1	$\bar{0}$	$\bar{1}$	$\bar{2}$	$\bar{3}$	$\bar{4}$	$\bar{5}$	$\bar{6}$
$\varphi_1(\xi)$	$\vdash$	$\not\vdash$	$\vdash$	$\not\vdash$	$\vdash$	$\not\vdash$	$\not\vdash$
$\varphi_2(\xi)$	$\vdash$	$\vdash$	$\vdash$	$\vdash$	$\vdash$	$\vdash$	$\vdash$
$\varphi_3(\xi)$	$\vdash$	$\vdash$	$\not\vdash$	$\vdash$	$\vdash$	$\not\vdash$	$\vdash$
$\neg\varphi_1(\xi)$	$\not\vdash$	$\vdash$	$\not\vdash$	$\vdash$	$\not\vdash$	$\vdash$	$\vdash$
$\neg\varphi_2(\xi)$	$\not\vdash$	$\not\vdash$	$\not\vdash$	$\not\vdash$	$\not\vdash$	$\not\vdash$	$\not\vdash$
$\neg\varphi_3(\xi)$	$\not\vdash$	$\not\vdash$	$\vdash$	$\not\vdash$	$\not\vdash$	$\vdash$	$\not\vdash$

K_2	$\bar{0}$	$\bar{1}$	$\bar{2}$	$\bar{3}$	$\bar{4}$	$\bar{5}$	$\bar{6}$
$\varphi_1(\xi)$	$\vdash$	$\not\vdash$	$\vdash$	$\not\vdash$	$\vdash$	$\not\vdash$	$\not\vdash$
$\varphi_2(\xi)$	$\vdash$	$\vdash$	$\vdash$	$\vdash$	$\not\vdash$	$\vdash$	$\vdash$
$\varphi_3(\xi)$	$\vdash$	$\vdash$	$\not\vdash$	$\vdash$	$\vdash$	$\not\vdash$	$\vdash$
$\neg\varphi_1(\xi)$	$\not\vdash$	$\vdash$	$\not\vdash$	$\vdash$	$\not\vdash$	$\vdash$	$\vdash$
$\neg\varphi_2(\xi)$	$\not\vdash$	$\not\vdash$	$\not\vdash$	$\not\vdash$	$\not\vdash$	$\not\vdash$	$\not\vdash$
$\neg\varphi_3(\xi)$	$\not\vdash$	$\not\vdash$	$\vdash$	$\not\vdash$	$\not\vdash$	$\vdash$	$\not\vdash$

K_3	$\bar{0}$	$\bar{1}$	$\bar{2}$	$\bar{3}$	$\bar{4}$	$\bar{5}$	$\bar{6}$
$\varphi_1(\xi)$	$\vdash$	$\not\vdash$	$\vdash$	$\not\vdash$	$\vdash$	$\not\vdash$	$\vdash$
$\varphi_2(\xi)$	$\vdash$	$\vdash$	$\vdash$	$\vdash$	$\vdash$	$\vdash$	$\vdash$
$\varphi_3(\xi)$	$\vdash$	$\vdash$	$\not\vdash$	$\vdash$	$\vdash$	$\not\vdash$	$\vdash$
$\neg\varphi_1(\xi)$	$\not\vdash$	$\vdash$	$\not\vdash$	$\vdash$	$\not\vdash$	$\vdash$	$\vdash$
$\neg\varphi_2(\xi)$	$\not\vdash$	$\not\vdash$	$\not\vdash$	$\not\vdash$	$\not\vdash$	$\not\vdash$	$\not\vdash$
$\neg\varphi_3(\xi)$	$\not\vdash$	$\not\vdash$	$\vdash$	$\not\vdash$	$\not\vdash$	$\vdash$	$\not\vdash$

K_4	$\bar{0}$	$\bar{1}$	$\bar{2}$	$\bar{3}$	$\bar{4}$	$\bar{5}$	$\bar{6}$
$\varphi_1(\xi)$	$\vdash$	$\not\vdash$	$\vdash$	$\not\vdash$	$\vdash$	$\not\vdash$	$\not\vdash$
$\varphi_2(\xi)$	$\vdash$	$\vdash$	$\vdash$	$\vdash$	$\vdash$	$\vdash$	$\vdash$
$\varphi_3(\xi)$	$\vdash$	$\vdash$	$\vdash$	$\vdash$	$\vdash$	$\not\vdash$	$\vdash$
$\neg\varphi_1(\xi)$	$\not\vdash$	$\vdash$	$\not\vdash$	$\vdash$	$\not\vdash$	$\vdash$	$\vdash$
$\neg\varphi_2(\xi)$	$\not\vdash$	$\not\vdash$	$\not\vdash$	$\not\vdash$	$\not\vdash$	$\not\vdash$	$\not\vdash$
$\neg\varphi_3(\xi)$	$\not\vdash$	$\not\vdash$	$\not\vdash$	$\not\vdash$	$\not\vdash$	$\vdash$	$\not\vdash$

2.7 Exercises

Exercise 2.4

Let the logic calculi P_1 and P_2 be defined by the following axioms and inference rules:

Axioms (calculus P_1)		Axioms (calculus P_2)	
(11,110)	(A1)	(01,011)	(A1')

Inference rules (calculus P_1)		Inference rules (calculus P_2)	
$\dfrac{(\varphi,\psi)}{(\varphi 011, \psi 100)}$	(S1)	$\dfrac{(\varphi,\psi)}{(\varphi 001, \psi 0)}$	(S1')
$\dfrac{(\varphi,\psi)}{(\varphi 11, \psi 110)}$	(S2)	$\dfrac{(\varphi,\psi)}{(\varphi 01, \psi 011)}$	(S2')
$\dfrac{(\varphi,\psi)}{(\varphi 010, \psi 011)}$	(S3)	$\dfrac{(\varphi,\psi)}{(\varphi 01, \psi 101)}$	(S3')
		$\dfrac{(\varphi,\psi)}{(\varphi 10, \psi 001)}$	(S4')

Both calculi operate on the same basic principle. The axiom specifies a pair of binary strings, and the inference rules govern how to extend those strings successively. Both calculi differ only in the binary substrings hard-coded into the axiom and the inference rules.

a) To which calculus belongs the following proof? On the right-hand side, indicate the applied inference rule for each derivation step.

1. ⊢ (01,011)   (   )
2. ⊢ (0110,011001)   (   )
3. ⊢ (011001,011001101)   (   )
4. ⊢ (01100110,011001101001)   (   )
5. ⊢ (0110011010,011001101001001)   (   )

b) Can a theorem of the form (φ, φ) be derived in P_1 and P_2?

c) Is there a decision procedure for P_1 and P_2?

d) Does a procedure exist that decides for all calculi of the above type whether a theorem of the form (φ, φ) can be derived?

Exercise 2.5 Complete the truth tables of the following propositional logic formulas. Are the formulas satisfiable, universally valid or unsatisfiable?

- $\varphi_1 = (\neg A \vee B) \wedge (\neg B \vee C) \wedge (\neg C \vee A)$

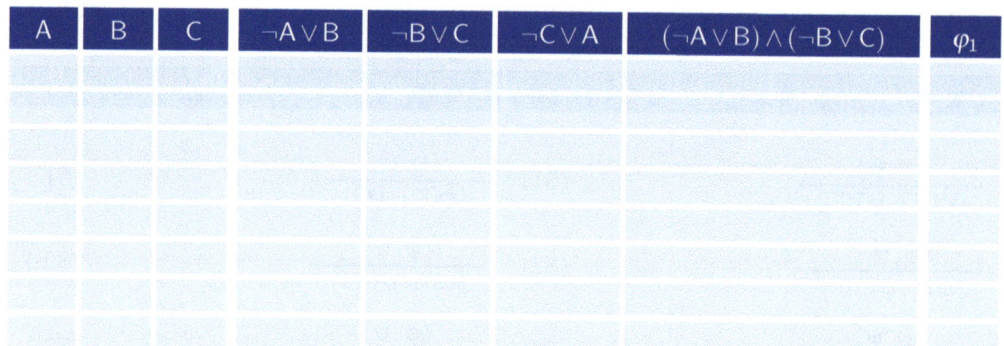

- $\varphi_2 = (A \rightarrow B) \wedge (B \rightarrow C) \rightarrow (A \rightarrow C)$

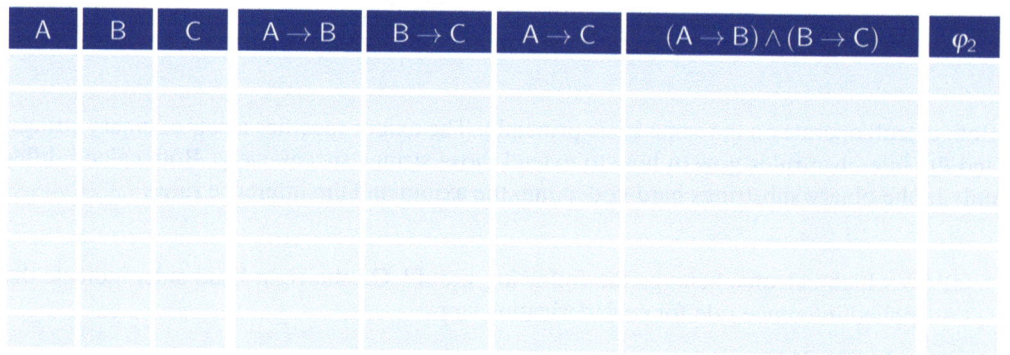

- $\varphi_3 = (A \leftrightarrow B) \wedge (B \leftrightarrow C) \wedge (A \leftrightarrow C)$

A	B	C	$A \leftrightarrow B$	$B \leftrightarrow C$	$A \leftrightarrow C$	$(A \leftrightarrow B) \wedge (B \leftrightarrow C)$	φ_3

2.7 Exercises

Exercise 2.6

Dirichlet's drawer principle is named after the German mathematician Peter Gustav Lejeune Dirichlet. It states that a finite set M cannot be mapped injectively to a set N if N contains fewer elements than M. We are all familiar with the drawer principle from everyday life. If $m > n$ and we distribute m objects to n drawers, then at least one drawer must contain more than one object. Dirichlet's drawer principle is usually called the *pigeonhole principle* in the English-speaking world. The reasoning stays the same: If m pigeons are distributed over n pigeonholes and $m > n$, at least one pigeonhole must be occupied more than once.

Peter Gustav Lejeune Dirichlet
(1805 – 1859)

Formalize Dirichlet's drawer principle for n objects and $n-1$ drawers. For this purpose, introduce a propositional variable A_{ij} for each possible combination of objects and drawers, which equals 1 if and only if the i-th object is in the j-th drawer.

Exercise 2.7

In Section 2.3.1, we have shown that every propositional logic formula can be rewritten to contain no other operators than '$\neg$' or '$\rightarrow$'. Because of this property, we refer to the set $\{\neg, \rightarrow\}$ as a *complete operator set*.

In this exercise, we consider the binary operators '$\overline{\wedge}$' (nand) and '$\overline{\vee}$' (nor) with:

$$I \models (\varphi \overline{\wedge} \psi) :\Leftrightarrow I \not\models \varphi \text{ or } I \not\models \psi$$
$$I \models (\varphi \overline{\vee} \psi) :\Leftrightarrow I \not\models \varphi \text{ and } I \not\models \psi$$

a) Prove that $\{\overline{\wedge}\}$ and $\{\overline{\vee}\}$ are complete operator sets.

b) Prove that $\{\overline{\wedge}\}$ and $\{\overline{\vee}\}$ are the only complete operator sets with a single element.

Exercise 2.8

The propositional quantifiers '$\forall$' and '$\exists$' are defined as follows:

$$\forall x\, \varphi := \varphi[x \leftarrow 0] \wedge \varphi[x \leftarrow 1]$$
$$\exists x\, \varphi := \varphi[x \leftarrow 0] \vee \varphi[x \leftarrow 1]$$

a) How do these quantifiers differ from those used in predicate logic?

b) Let φ be a PL0 formula with x being the only variable. Check the veracity of the following statements:

$\exists x\, \varphi \equiv 1 \Leftrightarrow \varphi$ is satisfiable
$\neg \exists x\, \varphi \equiv 1 \Leftrightarrow \varphi$ is unsatisfiable

$\forall x\, \varphi \equiv 1 \Leftrightarrow \varphi$ is universally true
$\neg \forall x\, \varphi \equiv 1 \Leftrightarrow \varphi$ ist unsatisfiable

c) Which of the following equivalences are valid?

$\forall x\, (\varphi \wedge \psi) \equiv \forall x\, \varphi \wedge \forall x\, \psi$
$\forall x\, (\varphi \vee \psi) \equiv \forall x\, \varphi \vee \forall x\, \psi$

$\exists x\, (\varphi \wedge \psi) \equiv \exists x\, \varphi \wedge \exists x\, \psi$
$\exists x\, (\varphi \vee \psi) \equiv \exists x\, \varphi \vee \exists x\, \psi$

Exercise 2.9

Some time ago, I asked the following question in an exam:

Consider φ as a propositional logic formula containing the variable x, with no other variables present. Verify whether the following assertion holds true: "φ is either equivalent to the formula x or to the formula $\neg x$."

Several students have solved the task roughly like this:

> The statement is correct. Proof: Symbolically, the statement "φ is either equivalent to the formula x or to the formula $\neg x$" is equivalent to:
>
> $$(\varphi \leftrightarrow x) \vee (\varphi \leftrightarrow \neg x) \equiv 1$$
>
> The statement can be proven with a few elementary transformations:
>
> $$(\varphi \leftrightarrow x) \vee (\varphi \leftrightarrow \neg x) \equiv \neg \varphi \neg x \vee \varphi x \vee \neg \varphi x \vee \varphi \neg x$$
> $$\equiv \varphi(x \vee \neg x) \vee \neg \varphi(x \vee \neg x)$$
> $$\equiv \varphi \vee \neg \varphi$$
> $$\equiv 1$$

The students have missed the existence of formulas such as $\varphi = x \vee \neg x$ or $\varphi = x \wedge \neg x$, which are neither equivalent to x nor to $\neg x$. Can you point out the error in their proof?

2.7 Exercises

Exercise 2.10

Russell and Whitehead built the propositional calculus of the *Principia Mathematica* upon the following five axioms:

1. $\varphi \vee \varphi \rightarrow \varphi$ (Taut)
2. $\psi \rightarrow \varphi \vee \psi$ (Add)
3. $\varphi \vee \psi \rightarrow \psi \vee \varphi$ (Perm)
4. $\varphi \vee (\psi \vee \chi) \rightarrow \psi \vee (\varphi \vee \chi)$ (Assoc)
5. $(\psi \rightarrow \chi) \rightarrow (\varphi \vee \psi \rightarrow \varphi \vee \chi)$ (Sum)

Hilbert's student Paul Bernays has shown that the propositional axioms of the *Principia* are not independent, as the fourth axiom (Assoc) is derivable from the others.

The following derivation sequence is an adaptation of the original proof from the 1926 paper *Axiomatische Untersuchung des Aussagen-Kalküls der Principia Mathematica* [9], in which Paul Bernays summarized the results of his habilitation thesis from 1918.

1. $\vdash \chi \rightarrow \varphi \vee \chi$ (▓)
2. $\vdash$ ▓ (Sum)
3. $\vdash \psi \vee \chi \rightarrow \psi \vee (\varphi \vee \chi)$ (MP, 1,2)
4. $\vdash$ ▓ (▓)
5. $\vdash \varphi \vee (\psi \vee \chi) \rightarrow \varphi \vee (\psi \vee (\varphi \vee \chi))$ (MP, 3,4)
6. $\vdash \varphi \vee (\psi \vee (\varphi \vee \chi)) \rightarrow (\psi \vee (\varphi \vee \chi)) \vee \varphi$ (▓)
7. $\vdash$ ▓ (MB, 5,6)
8. $\vdash \varphi \rightarrow \chi \vee \varphi$ (▓)
9. $\vdash$ ▓ (▓)
10. $\vdash \varphi \rightarrow \varphi \vee \chi$ (MB, 8,9)
11. $\vdash \varphi \vee \chi \rightarrow \psi \vee (\varphi \vee \chi)$ (▓)
12. $\vdash$ ▓ (MB, 10,11)
13. $\vdash$ ▓ (Sum)
14. $\vdash (\psi \vee (\varphi \vee \chi)) \vee \varphi \rightarrow (\psi \vee (\varphi \vee \chi)) \vee (\psi \vee (\varphi \vee \chi))$ (MP, 12,13)
15. $\vdash$ ▓ (▓)
16. $\vdash (\psi \vee (\varphi \vee \chi)) \vee \varphi \rightarrow \psi \vee (\varphi \vee \chi)$ (MB, 14,15)
17. $\vdash \varphi \vee (\psi \vee \chi) \rightarrow \psi \vee (\varphi \vee \chi)$ (▓)

Try to reconstruct the partially printed proof.

Exercise 2.11

Prove the following formulas in the propositional logic calculus:

a) $(\varphi \to (\psi \to \chi)) \to (\psi \to (\varphi \to \chi))$

b) $\varphi \to (\psi \to \neg(\varphi \to \neg\psi))$

Exercise 2.12

Section 2.5 has introduced first-order predicate logic with equality. Does this logic permit the construction of formulas with the following meanings?

$$(U,I) \models \varphi_{\geq n} \Leftrightarrow U \text{ contains at least } n \text{ elements}$$
$$(U,I) \models \varphi_{\leq n} \Leftrightarrow U \text{ contains at most } n \text{ elements}$$
$$(U,I) \models \varphi_{=n} \Leftrightarrow U \text{ contains precisely } n \text{ elements}$$

Exercise 2.13

We call a relation R

- *reflexive*, if $R(x,x)$ holds for all x,
- *left-comparative*, if $R(x,y) \wedge R(x,z)$ implies $R(y,z)$,
- *symmetric*, if $R(x,y)$ implies $R(y,x)$.

a) Formalize the statement *"Every reflexive, left-comparative relation is symmetric"* in first-order predicate logic.

b) Try to derive a formal proof using the predicate-logic calculus. For the sake of simplicity, consider all propositional tautologies as already proven.

Exercise 2.14

Assume that the inference rules of a formal system ensure that the generated theorems become longer in each derivation step; that is, the conclusion always consists of more symbols than the premises.

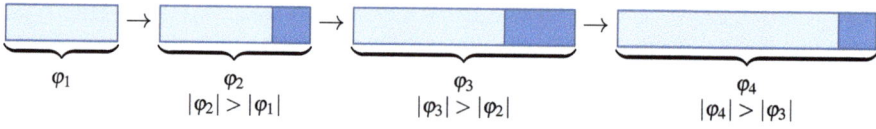

$\varphi_1 \quad\quad \varphi_2 \quad\quad \varphi_3 \quad\quad \varphi_4$
$\quad\quad |\varphi_2| > |\varphi_1| \quad |\varphi_3| > |\varphi_2| \quad |\varphi_4| > |\varphi_3|$

Does such a system always have a decision procedure?

Exercise 2.15

Section 2.6 has demonstrated that second-order predicate logic is sufficiently expressive to define the concept of finiteness. The definition exploited the idea of claiming that every injective function is surjective. Is it also possible to define the concept by claiming that every surjective function is injective?

Exercise 2.16

Name the mathematical concepts defined by the following formulas:

a) $\forall f \, (\forall x \, \exists y \, (x \doteq f(y)) \rightarrow \forall x \, \forall y \, (f(x) \doteq f(y) \rightarrow x \doteq y))$

b) $\exists \mathfrak{A} \, (\forall x \, \forall y \, \forall z \, (\mathfrak{A}(x,y) \wedge \mathfrak{A}(y,z) \rightarrow \mathfrak{A}(x,z)) \wedge \forall x \, (\neg \mathfrak{A}(x,x) \wedge \exists y \, \mathfrak{A}(x,y)))$

3 Foundations of Mathematics

> *"No one shall drive us out of the paradise which Cantor has created for us."*
>
> David Hilbert [94]

Chapter 2 introduced propositional and predicate logic and showed how to model the reasoning principles of ordinary mathematics within a formal system. This chapter will extend predicate logic to so-called *theories* by adding new axioms. In a nutshell, a mathematical theory is a formal system whose axioms are divided into two groups: *non-logical* and *logical axioms*.

- The *non-logical axioms* define the relationships between the modeled objects and give a theory its face. For example, the non-logical axioms of Peano arithmetic describe the characteristic properties of natural numbers and ensure that the symbols '+' and '×' adopt their familiar arithmetic meaning. Analogously, the non-logical axioms of Zermelo-Fraenkel set theory determine the relationships between the objects we commonly call sets. The non-logical axioms are also called the *proper axioms* of a theory, its *theory axioms*, or simply its *axioms*.

- The logical axioms define the mathematical reasoning apparatus of the formal system. Though frequently ignored in the discussion of a theory, they are no less important. Only these axioms enable us to derive logical conclusions from the theory axioms. If the logical axioms are the axioms of predicate logic introduced in Section 2.4, we speak of a first-order theory. If, on the other hand, the logical axioms are those from second-order predicate logic, we speak of a second-order theory and so on.

Let's suit the action to the word and examine two essential first-order theories in detail: Peano arithmetic and Zermelo-Fraenkel set theory.

> A glimpse at other books reveals that some authors distinguish between the concept of a *theory* and the concept of an *axiomatized theory*. At first glance, this approach is puzzling since we need axioms to formulate a theory. Doesn't this trivially make every theory an axiomatized theory? And if so, why does such a distinction exist at all? The apparent contradiction is due to different definitions of the term theory. Some authors use it as a synonym for a set of formulas M closed under the logical inference relation ($M \models \varphi$ implies $\varphi \in M$) rather than a synonym for a formal system. In books adopting this approach, the notion of an axiomatizable theory takes on a very natural meaning. A theory M is axiomatizable if a formal system exists in which the formulas from M, and only these, are derivable from the axioms in a finite number of steps.

3.1 Peano Arithmetic

Peano arithmetic, or PA for short, is the theory of the natural numbers, including addition and multiplication. As in the previous chapter, we approach PA in multiple steps. In Section 3.1.1, we introduce the syntax of Peano arithmetic; that is, we agree on the rules governing the construction of *arithmetic terms* and *arithmetic formulas*. Section 3.1.2 defines the semantics by giving the formula components a substantive meaning. After that, in Section 3.1.3, we present the axioms and the logical reasoning apparatus of Peano arithmetic and provide several examples to demonstrate how arithmetic statements can be formally proven.

Arithmetic term	Shorthand form
0	
x_1	
$(x_1 \times x_2)$	
$s(0)$	$\overline{1}$
$s(s(s(s(0))))$	$\overline{5}$
$(x_1 \times s(s(0)))$	$x_1 \times \overline{2}$
$(s(0)+s(s(0)))$	$\overline{1}+\overline{2}$
$(s(0) \times (x_1+x_2))$	$\overline{1} \times (x_1+x_2)$

Table 3.1: Examples of arithmetic terms. The shorthand notation

$$\overline{n} := \underbrace{s(s(\ldots s(0)\ldots))}_{n \text{ times}}$$

is familiar to us from Section 2.1. Once again, it will serve as a convenient writing aid.

Arithmetic formulas
$\exists x_1\, \overline{2} \times x_1 = \overline{6}$
$\exists x_1\, x_1 \times x_1 = \overline{9}$
$\exists x_1\, \overline{3} \times x_1 = \overline{9}$
$\exists x_1\, \overline{7} = \overline{6} + x_1$
$\forall x_1\, \exists x_2\, ($
$\quad \exists x_5\, (x_2 = x_1 + x_5 + s(0)) \wedge$
$\quad \neg(x_2 = s(0)) \wedge$
$\quad \forall x_3\, (\exists x_4\, (x_3 \times x_4 = x_2)$
$\quad\quad \rightarrow (x_3 = s(0) \vee x_3 = x_2)))$

Table 3.2: Examples of arithmetic formulas. To improve readability, we will frequently refrain from writing down pairs of parentheses as long as ambiguities are avoided.

3.1.1 Syntax

Definition 3.1 (Syntax of Peano Arithmetic)

The set of *arithmetic terms* is inductively defined:

- $0, x_1, x_2, x_3, \ldots$ are arithmetic terms.

- If σ and τ are arithmetic terms, so are

$$s(\sigma), (\sigma + \tau), (\sigma \times \tau).$$

The set of *arithmetic formulas* is inductively defined:

- If σ and τ are arithmetic terms, then

$$(\sigma = \tau) \text{ is an arithmetic formula.}$$

- If φ and ψ are arithmetic formulas, so are

$$(\neg\varphi), (\varphi \wedge \psi), (\varphi \vee \psi), (\varphi \rightarrow \psi), (\varphi \leftrightarrow \psi), (\varphi \nleftrightarrow \psi).$$

- If φ is an arithmetic formula, so are

$$\forall \xi\, \varphi, \exists \xi\, \varphi \quad \text{where} \quad \xi \in \{x_1, x_2, x_3, \ldots\}$$

Tables 3.1 and 3.2 list several terms and formulas that can be formed according to the stipulated rules.

On the surface, the language of Peano arithmetic differs little from the

3.1 Peano Arithmetic

language of predicate logic, and a closer look at Definition 2.14 confirms this assumption. The two languages differ only in their stock of constant and function symbols, which Peano arithmetic limits to

'0', 's', '+', and '×',

and the stock of predicate symbols, which Peano arithmetic limits to the symbol '='. Therefore, we can safely apply all syntactical simplifications from predicate logic to arithmetic formulas. In particular, we will allow ourselves to alter the notation of variables from case to case and dispense with one or another pair of brackets as long as the penned-down expression remains unambiguous.

3.1.2 Semantics

Does the appearance of the arithmetic terms remind you of the calculi from Section 2.1, which we employed as examples to work out the basic properties of formal systems? The similarities are not accidental, and we can, in fact, regard Peano arithmetic as a generalization of these systems. Hence, the definition of the model relation '$\models$' almost writes itself:

Definition 3.2 (Semantics of Peano Arithmetic)

Let φ and ψ be closed arithmetic formulas. The semantics of Peano arithmetic is given by the *model relation* '$\models$', which is inductively defined:

$$\models (\sigma_1 = \sigma_2) :\Leftrightarrow I(\sigma_1) = I(\sigma_2)$$
$$\models (\neg \varphi) :\Leftrightarrow \not\models \varphi$$
$$\models (\varphi \land \psi) :\Leftrightarrow \models \varphi \text{ and } \models \psi$$
$$\models (\varphi \lor \psi) :\Leftrightarrow \models \varphi \text{ or } \models \psi$$
$$\models (\varphi \to \psi) :\Leftrightarrow \not\models \varphi \text{ or } \models \psi$$
$$\models (\varphi \leftrightarrow \psi) :\Leftrightarrow \models \varphi \to \psi \text{ and } \models \psi \to \varphi$$
$$\models (\varphi \nleftrightarrow \psi) :\Leftrightarrow \not\models (\varphi \leftrightarrow \psi)$$
$$\models \forall \xi\, \varphi :\Leftrightarrow\, \models \varphi[\xi \leftarrow \overline{n}] \text{ for all } n \in \mathbb{N}$$
$$\models \exists \xi\, \varphi :\Leftrightarrow\, \models \varphi[\xi \leftarrow \overline{n}] \text{ for some } n \in \mathbb{N}$$

In Peano arithmetic, each term σ represents a natural number, denoted by $I(\sigma)$ in the definition above. If σ contains no variables, $I(\sigma)$ can be

$$I(0) := 0$$
$$I(\mathsf{s}(\sigma)) := I(\sigma) + 1$$
$$I(\sigma_1 + \sigma_2) := I(\sigma_1) + I(\sigma_2)$$
$$I(\sigma_1 \times \sigma_2) := I(\sigma_1) \times I(\sigma_2)$$

■ Example: $\sigma := \mathsf{s}(\mathsf{s}(0)) + \mathsf{s}(0)$

$$\begin{aligned}
I(\sigma) &= I(\mathsf{s}(\mathsf{s}(0)) + \mathsf{s}(0)) \\
&= I(\mathsf{s}(\mathsf{s}(0))) + I(\mathsf{s}(0)) \\
&= I(\mathsf{s}(0)) + 1 + I(0) + 1 \\
&= I(0) + 1 + 1 + 1 \\
&= 1 + 1 + 1 \\
&= 3
\end{aligned}$$

Figure 3.1: If an arithmetic term σ contains no variables, its assigned numerical value $I(\sigma)$ can be calculated recursively along the term structure.

Various nomenclatures exist in the literature to name the successor of a natural number. This book describes the natural numbers $0, 1, 2, 3, \ldots$ using the arithmetic terms

$$0, \mathsf{s}(0), \mathsf{s}(\mathsf{s}(0)), \mathsf{s}(\mathsf{s}(\mathsf{s}(0))), \ldots$$

Other books diverge from this path and adopt the following notation:

$$0, S0, SS0, SSS0, \ldots$$

The latter is more compact but deviates from the prevailing convention of writing function symbols in lowercase. Yet other books use an apostrophe instead of 'S', thus listing the above number series as

$$0, 0', 0'', 0''', \ldots$$

calculated recursively along the term structure, as shown in Figure 3.1. As expected, the symbols 's', '+', and '×' refer to the successor function, addition, and multiplication, respectively.

Always be aware that the model relation has a different meaning in Peano arithmetic than in propositional or predicate logic. In the latter two, $\models \varphi$ expresses that the formula φ is universally valid, i.e., it becomes a true statement under all possible interpretations. In Peano arithmetic, however, we have a specific interpretation in mind: the natural numbers. We call it the *standard interpretation* and agree on the shorthand notation

$$(\mathbb{N}, \{s, +, \times\}).$$

In Peano arithmetic, $\models \varphi$ thus expresses that φ is a true statement if we interpret the symbols according to the standard interpretation. More concisely, we can express this relationship as such:

$$\models \varphi \ :\Leftrightarrow \ (\mathbb{N}, \{s, +, \times\}) \text{ is a model of } \varphi$$

By linking the model relation in Peano arithmetic to a single interpretation, either $\models \varphi$ or $\models \neg \varphi$ applies to every formula φ. In propositional and predicate logic, this is not the case.

Next, we will employ several examples to illustrate that Peano arithmetic is sufficiently expressive to formalize typical number-theoretical statements. Figure 3.2 lists the chosen statements in plain language.

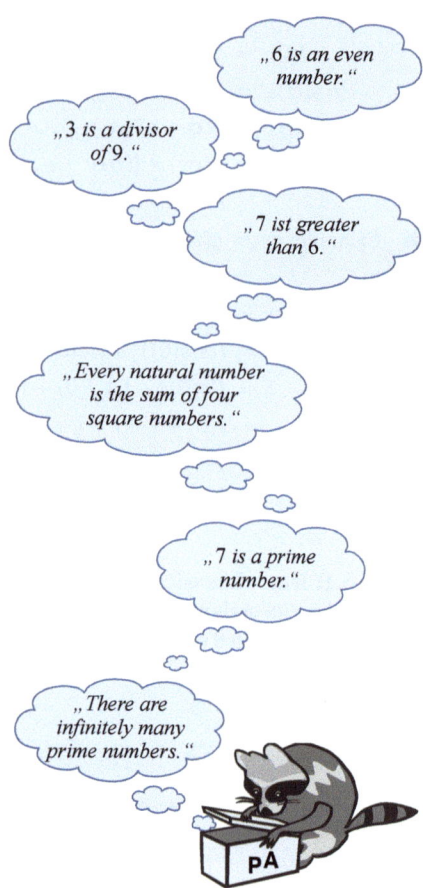

Figure 3.2: A selection of number theoretical statements

- $\exists x_1 \ \overline{2} \times x_1 = \overline{6}$

  The formula states the existence of a natural number x_1, which, when multiplied by 2, yields 6. Hence, it represents the equivalent statement

  "6 is an even number."

- $\exists x_1 \ \overline{3} \times x_1 = \overline{9}$

  The formula postulates the existence of a natural number x_1, which, multiplied by 3, yields 9. Therefore, it is equivalent to the statement

  "3 is a divisor of 9."

To write down formulas of this kind concisely, we agree on the following abbreviation:

$$\sigma \mid \tau := \exists \xi \ \sigma \times \xi = \tau$$

In this formula and all following formulas introducing an abbreviation, ξ is a placeholder for any variable that does not occur freely in σ and τ.

3.1 Peano Arithmetic

- $\exists x_1 \, \overline{7} = \overline{6} + x_1$

  The formula stipulates the relationship $7 \geq 6$. It demonstrates that the comparison relation '$\geq$' can be easily reduced to addition and thus be described within Peano arithmetic. As in the case of the divisibility relation, we permit the symbols '$\geq$', '$\leq$', '$>$', and '$<$' to be used as abbreviations within arithmetic formulas. The following definitions formally justify these language extensions:

$$(\sigma \geq \tau) := \exists \xi \, \sigma = \tau + \xi$$
$$(\sigma \leq \tau) := \exists \xi \, \sigma + \xi = \tau$$
$$(\sigma > \tau) := \exists \xi \, \sigma = \tau + \xi + \overline{1}$$
$$(\sigma < \tau) := \exists \xi \, \sigma + \xi + \overline{1} = \tau$$

  Furthermore, we introduce *conditional quantifiers* by agreeing on the following:

$$\exists (\xi > \sigma) \, \varphi := \exists \xi \, (\xi > \sigma \wedge \varphi)$$
$$\forall (\xi > \sigma) \, \varphi := \forall \xi \, (\xi > \sigma \to \varphi)$$

  An analogous definition applies to the operators '$<$', '$\geq$', and '$\leq$'.

- $\forall z \, \exists x_1 \, \exists x_2 \, \exists x_3 \, \exists x_4 \, z = x_1 \times x_1 + x_2 \times x_2 + x_3 \times x_3 + x_4 \times x_4$

  The formula formally describes *Lagrange's four-square theorem*, which states that every natural number is the sum of four square numbers. In Section 5.4.3, we will reencounter this theorem and clarify its role in Hilbert's tenth problem.

- $\forall z \, (z \mid \overline{7}) \to (z = \overline{1} \vee z = \overline{7})$

  This formula states that all natural numbers z dividing 7 must satisfy the relationship $z = 1$ or $z = 7$. The formula is thus equivalent to the statement

  "7 is a prime number."

  Again, we want to facilitate writing by using a corresponding definition:

$$\text{prime}(\sigma) := \neg(\sigma = \overline{1}) \wedge \forall \xi \, (\xi \mid \sigma \to (\xi = \overline{1} \vee \xi = \sigma))$$

- $\forall x_1 \, \exists (x_2 > x_1) \, \text{prime}(x_2)$

  In plain language, the formula reads: For every natural number x_1, a larger number x_2 exists, with x_2 being a prime number. The formula is a formalization of Euclid's famous theorem about the existence of infinitely many prime numbers.

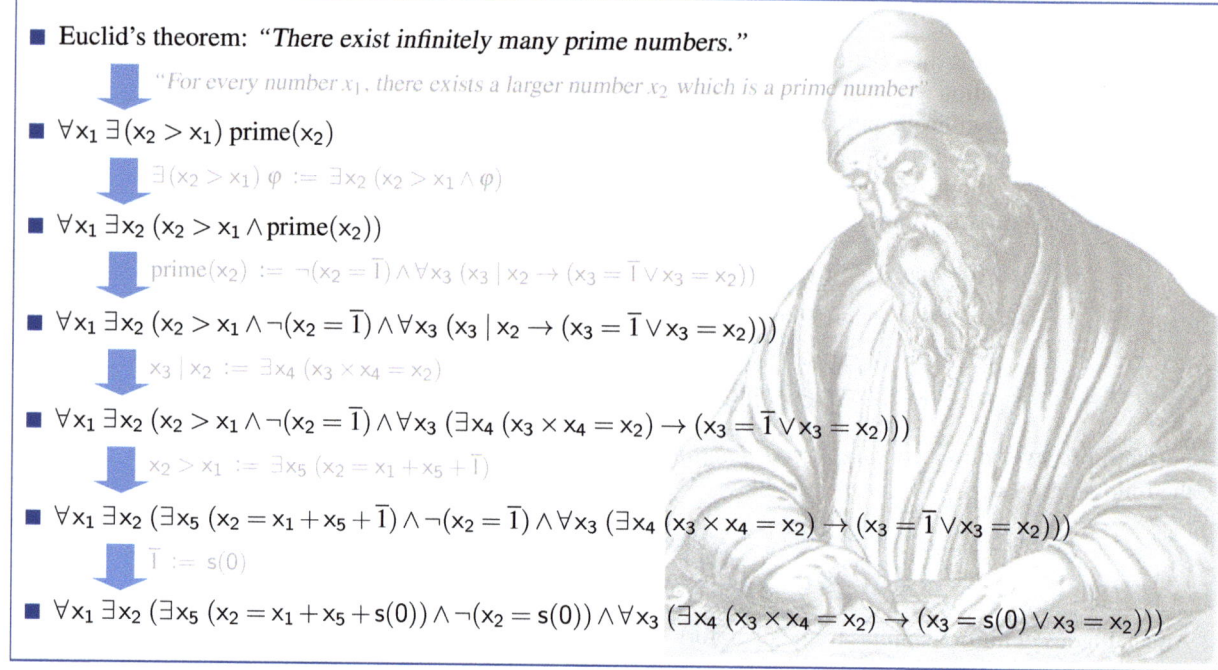

Figure 3.3: Formalization of Euclid's theorem in Peano arithmetic

Remember that the newly introduced operators are merely syntactic abbreviations and do not increase the expressive power of Peano arithmetic. Though they are dispensable in principle, they serve us well. Besides helping us write down formulas more concisely, the abbreviations vastly contribute to a formula's comprehensibility. For example, Figure 3.3 demonstrates what Euclid's theorem would look like if we only used the basic syntactic building blocks from Definition 3.1. After the transformation, recognizing the formula's true meaning becomes almost impossible.

Finally, we introduce one last writing aid: the expanded existential quantifier $\exists_1$:

$$\exists_1 \xi\, \varphi(\xi) := \exists \xi\, (\varphi(\xi) \wedge \forall \zeta\, (\varphi(\zeta) \to \zeta = \xi)).$$

The expanded existential quantifier concisely expresses that $\varphi(x)$ becomes true for *precisely one* assignment of x. In contrast,

$$\exists x\, \varphi(x)$$

merely states that *at least one* such assignment exists.

3.1.3 Axioms and Inference Rules

Having established the syntax and semantics of Peano arithmetic in the previous section, we want to move our attention to the axioms and inference rules. To understand the axioms' deeper meaning, let us first try to characterize the natural numbers by their properties.

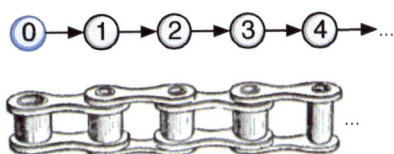

Figure 3.4: Chain-like structure of the natural numbers

The first two axioms almost write themselves:

- "0 is a natural number." (P1)
- "Each number x has a unique successor $s(x)$." (P2)

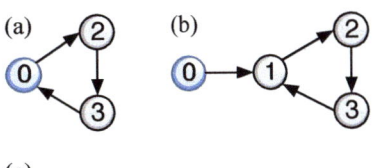
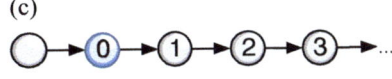
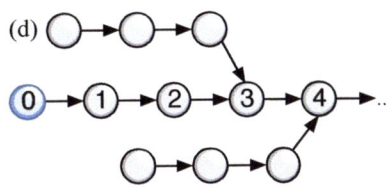

At first glance, the two properties seem to accurately describe the structure of the natural numbers, as shown in Figure 3.4. However, the examples in Figure 3.5 reveal that other structures also fulfill these properties, structures that do not align with our idea of the natural numbers at all.

The uninvited guests disappear by demanding two additional properties:

- "0 is not the successor of any number." (P3)
- "Different numbers have different successors." (P4)

The first property eliminates structures (a) and (c), and the second eliminates structures (b) and (d). Although we are already close to unambiguously characterizing the natural numbers, a fundamental problem has yet to be solved. The formulated properties do not rule out that the strain of natural numbers is accompanied by another strain of numbers with the same chain-line structure (Figure 3.6). To rule out the existence of such shadow numbers, we must resort to a more profound property than those mentioned so far. The solution comes in the form of the *induction axiom*:

Figure 3.5: Axioms (P1) and (P2) are not sufficient to exclude the structures (a) to (d).

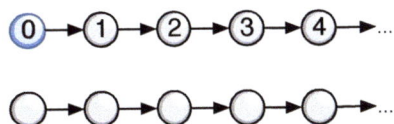

Figure 3.6: The depicted structure satisfies axioms (P1) to (P4) but contradicts the induction axiom.

- "If a set M contains the number 0 and $x \in M$ implies $s(x) \in M$, then M contains all natural numbers."

Since we can interpret the set membership of a number as a property and, at the same time, express every property in the form of a set membership, we can also formulate the induction axiom as such:

- "If the number 0 has the property φ and $\varphi(x)$ implies $\varphi(s(x))$, then all natural numbers have the property φ." (P5)

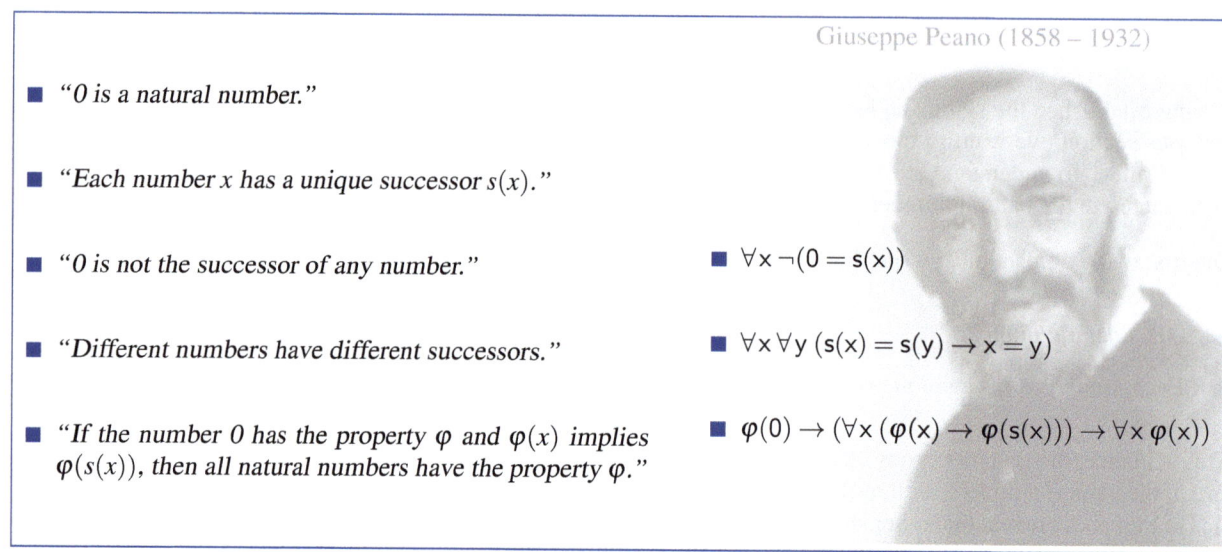

Figure 3.7: The Peano axioms, informally noted on the left and formally expressed on the right

(P1) to (P5) are the famous *Peano axioms*, published by Giuseppe Peano in 1889 in his *Arithmetices principia*. In Figure 1.30 on page 32, we have already seen an excerpt from the English translation.

The example structures depicted in Figures 3.4 to 3.6 showed that no Peano axiom is dispensable. If we remove just one, other structures apart from the natural numbers fulfill the remaining axioms. The opposite consideration, however, is equally important: Do the Peano axioms fully characterize the natural numbers, or do we need to add further axioms to attain an unambiguous description? Richard Dedekind's famous isomorphism theorem gives a reassuring answer. It states that the axioms (P1) to (P5) uniquely characterize the natural numbers up to isomorphism, which is, in essence, how we name or write down the numbers.

At this point, we have gathered the necessary knowledge to understand the axioms and inference rules of Peano arithmetic in all their richness. Figure 3.7 shows how to write down the colloquially formulated axioms as arithmetic formulas; all three of them become theory axioms of Peano arithmetic. Note that we do not need to translate the first two axioms, as they are already formalized by the fact that 0 and s are integral parts of the language of Peano arithmetic in the form of a constant symbol and a function symbol, respectively.

3.1 Peano Arithmetic

Non-logical axioms		Logical axioms	
$\sigma = \tau \to (\sigma = \rho \to \tau = \rho)$	(S1)	$\varphi \to (\psi \to \varphi)$	(A1)
$\sigma = \tau \to s(\sigma) = s(\tau)$	(S2)	$(\varphi \to (\psi \to \chi)) \to ((\varphi \to \psi) \to (\varphi \to \chi))$	(A2)
$\neg(0 = s(\sigma))$	(S3)	$(\neg \varphi \to \neg \psi) \to (\psi \to \varphi)$	(A3)
$s(\sigma) = s(\tau) \to \sigma = \tau$	(S4)	$\forall \xi\, \varphi \to \varphi[\xi \leftarrow \sigma]$ ($[\xi \leftarrow \sigma]$ collision-free)	(A4)
$\sigma + 0 = \sigma$	(S5)	$\forall \xi\, (\varphi \to \psi) \to (\varphi \to \forall \xi\, \psi)$ ($\xi \notin \varphi$)	(A5)
$\sigma + s(\tau) = s(\sigma + \tau)$	(S6)	Inference rules	
$\sigma \times 0 = 0$	(S7)	$\dfrac{\varphi,\, \varphi \to \psi}{\psi}$	(MP)
$\sigma \times s(\tau) = (\sigma \times \tau) + \sigma$	(S8)	$\dfrac{\varphi}{\forall \xi\, \varphi}$	(G)
$\varphi(0) \to (\forall x\, (\varphi(x) \to \varphi(s(x))) \to \forall x\, \varphi(x))$	(S9)		

Table 3.3: The axioms and inference rules of Peano arithmetic

The Peano axioms alone do not constitute a formal system. To obtain such a system, we must supplement them with further axioms. The result is a longer list of axioms and inference rules, summarized in Table 3.3. Axioms (S5) to (S8) describe the elementary properties of addition and multiplication and give the operators '+' and '×' their intended meaning. (S9) is the axiom of induction, which we have shown to be necessary above. The right half of the table lists the logic axioms and the inference rules. They are taken one-to-one from predicate logic, turning Peano arithmetic into a first-order theory.

All axioms are laid out as schemata, with the placeholders σ and τ representing terms, φ, ψ, χ representing formulas, and ξ representing a variable.

Let's pause shortly and turn our attention to the induction axiom again. A closer look at the different formulations shows that the formalized variant (S9) is a first-order formula. However, the colloquial variant (P5) is of second order, as it makes a statement about any property of the natural numbers. In other words, the axiom of induction quantifies over

Theorems
■ Theorem PA1 $\sigma = \sigma$
■ Theorem PA2 $\sigma = \tau \rightarrow \tau = \sigma$
■ Theorem PA3 $\sigma = \tau \rightarrow (\tau = \rho \rightarrow \sigma = \rho)$
■ Theorem PA4 $\sigma = \tau \rightarrow (\rho = \tau \rightarrow \sigma = \rho)$
■ Theorem PA5 $\forall x \, (\sigma = \tau \rightarrow \sigma + x = \tau + x)$
■ Theorem PA6 $\forall x \, (x = 0 + x)$
■ Theorem PA7 $\sigma + \bar{1} = s(\sigma)$
■ Theorem PA8 $\sigma \times \bar{1} = \sigma$

Table 3.4: A small selection of arithmetic formulas provable in the formal system of Peano arithmetic.

predicates. Peano arithmetic works around the situation by laying out (S9) as an axiom schema. Since we can replace the placeholder φ with any formula, there is not a single induction axiom in Peano arithmetic but infinitely many.

At first glance, this does the trick. Laying out the induction axiom as a schema appears to formally capture the colloquial formulation of the induction axiom. At second glance, however, it becomes clear that we only came close to an exact formalization. Specifically, we are facing the problem of an axiom schema not being able to guarantee the induction principle for all properties. The reason is simple: There are uncountably many properties but only a countable number of formulas. As a result, an axiom schema can only cover a fraction of all existing properties. For the moment, we will ignore this subtle flaw. In due course, in Section 7.2, we will revisit the topic and show that this phenomenon does indeed have far-reaching consequences.

The time has come to bring the formal system to life by proving the theorems listed in Table 3.4. When revisiting the proofs, you will notice that they hardly differ from those in Section 2.4; after all, Peano arithmetic is a first-order theory and utilizes the same logical reasoning apparatus as the first-order predicate-logic calculus. As a result, we can employ not only the pool of previously proven tautologies but also all the tools developed before. Above all, the deduction theorem will be faithfully at our side again.

Calculemus – let us calculate!

Theorem PA1
$\sigma = \sigma$

1. $\vdash \sigma + 0 = \sigma$ (S5)
2. $\vdash \sigma + 0 = \sigma \rightarrow (\sigma + 0 = \sigma \rightarrow \sigma = \sigma)$ (S1)
3. $\vdash \sigma + 0 = \sigma \rightarrow \sigma = \sigma$ (MP, 1,2)
4. $\vdash \sigma = \sigma$ (MP, 1,3)

Theorem PA2
$\sigma = \tau \rightarrow \tau = \sigma$

1. $\vdash \sigma = \tau \rightarrow (\sigma = \sigma \rightarrow \tau = \sigma)$ (S1)
2. $\{\sigma = \tau\} \vdash \sigma = \sigma \rightarrow \tau = \sigma$ (DT)
3. $\{\sigma = \tau, \sigma = \sigma\} \vdash \tau = \sigma$ (DT)
4. $\{\sigma = \sigma\} \vdash \sigma = \tau \rightarrow \tau = \sigma$ (DT)
5. $\vdash \sigma = \sigma \rightarrow (\sigma = \tau \rightarrow \tau = \sigma)$ (DT)
6. $\vdash \sigma = \sigma$ (PA1)
7. $\vdash \sigma = \tau \rightarrow \tau = \sigma$ (MP, 5,6)

3.1 Peano Arithmetic

Theorem PA3
$\sigma = \tau \to (\tau = \rho \to \sigma = \rho)$

1. $\vdash \tau = \sigma \to (\tau = \rho \to \sigma = \rho)$ (S1)
2. $\vdash \sigma = \tau \to \tau = \sigma$ (PA2)
3. $\{\sigma = \tau\} \vdash \tau = \sigma$ (DT)
4. $\{\sigma = \tau\} \vdash \tau = \rho \to \sigma = \rho$ (MP, 1,3)
5. $\vdash \sigma = \tau \to (\tau = \rho \to \sigma = \rho)$ (DT)

Theorem PA4
$\sigma = \tau \to (\rho = \tau \to \sigma = \rho)$

1. $\vdash \sigma = \tau \to (\tau = \rho \to \sigma = \rho)$ (PA3)
2. $\{\sigma = \tau\} \vdash \tau = \rho \to \sigma = \rho$ (DT)
3. $\vdash \rho = \tau \to \tau = \rho$ (PA2)
4. $\{\rho = \tau\} \vdash \tau = \rho$ (DT)
5. $\{\sigma = \tau, \rho = \tau\} \vdash \sigma = \rho$ (MP, 2,4)
6. $\{\sigma = \tau\} \vdash \rho = \tau \to \sigma = \rho$ (DT)
7. $\vdash \sigma = \tau \to (\rho = \tau \to \sigma = \rho)$ (DT)

Theorem PA5
$\forall x\, (\sigma = \tau \to \sigma + x = \tau + x)$
$x \notin \sigma, x \notin \tau$

Proof by induction. Let $\psi(x) := (\sigma = \tau \to (\sigma + x = \tau + x))$

1. $\vdash \sigma + 0 = \sigma$ (S5)
2. $\vdash \tau + 0 = \tau$ (S5)
3. $\vdash \sigma + 0 = \sigma \to (\sigma = \tau \to \sigma + 0 = \tau)$ (PA3)
4. $\vdash \sigma = \tau \to \sigma + 0 = \tau$ (MP, 1,3)
5. $\{\sigma = \tau\} \vdash \sigma + 0 = \tau$ (DT)
6. $\vdash \sigma + 0 = \tau \to (\tau + 0 = \tau \to \sigma + 0 = \tau + 0)$ (PA4)
7. $\{\sigma = \tau\} \vdash \tau + 0 = \tau \to \sigma + 0 = \tau + 0$ (MP, 5,6)
8. $\{\sigma = \tau\} \vdash \sigma + 0 = \tau + 0$ (MP, 2,7)
9. $\vdash \sigma = \tau \to \sigma + 0 = \tau + 0$ (DT)
10. $\vdash \psi(0)$ (Definition)
11. $\{\psi(x)\} \vdash \sigma = \tau \to (\sigma + x = \tau + x)$ (Theorem 2.4)
12. $\{\psi(x), \sigma = \tau\} \vdash \sigma + x = \tau + x$ (DT)
13. $\vdash \sigma + s(x) = s(\sigma + x)$ (S6)
14. $\vdash \tau + s(x) = s(\tau + x)$ (S6)
15. $\vdash \sigma + x = \tau + x \to s(\sigma + x) = s(\tau + x)$ (S2)
16. $\{\psi(x), \sigma = \tau\} \vdash s(\sigma + x) = s(\tau + x)$ (MP, 12,15)
17. $\vdash \sigma + s(x) = s(\sigma + x) \to$
 $(s(\sigma + x) = s(\tau + x) \to \sigma + s(x) = s(\tau + x))$ (PA3)

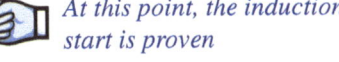
At this point, the induction start is proven

18. ⊢ s(σ+x) = s(τ+x) → σ+s(x) = s(τ+x)		(MP, 13,17)
19. {ψ(x), σ = τ} ⊢ σ+s(x) = s(τ+x)		(MP, 16,18)
20. ⊢ σ+s(x) = s(τ+x) → (τ+s(x) = s(τ+x) → σ+s(x) = τ+s(x))		(PA4)
21. {ψ(x), σ = τ} ⊢ τ+s(x) = s(τ+x) → σ+s(x) = τ+s(x)		(MP, 19,20)
22. {ψ(x), σ = τ} ⊢ σ+s(x) = τ+s(x)		(MP, 14,21)
23. {ψ(x)} ⊢ σ = τ → σ+s(x) = τ+s(x)		(DT)
24. {ψ(x)} ⊢ ψ(s(x))		(Definition)
25. ⊢ ψ(x) → ψ(s(x))		(DT)
26. ⊢ ∀x (ψ(x) → ψ(s(x)))		(G, 25)
27. ⊢ ψ(0) → (∀x (ψ(x) → ψ(s(x))) → ∀x ψ(x))		(S9)
28. ⊢ ∀x (ψ(x) → ψ(s(x))) → ∀x ψ(x)		(MP, 10,27)
29. ⊢ ∀x ψ(x)		(MP, 26,28)
30. ⊢ ∀x (σ = τ → σ+x = τ+x)		(Definition)

Theorem PA6
∀x (x = 0+x)

At this point, the induction start is proven

At this point, the induction step is proven

Proof by induction. Let ψ(x) := (x = 0+x)

1. ⊢ 0+0 = 0		(S5)
2. ⊢ 0+0 = 0 → 0 = 0+0		(PA2)
3. ⊢ 0 = 0+0		(MP, 1,2)
4. ⊢ ψ(0)		(Definition)
5. {ψ(x)} ⊢ x = 0+x		(Theorem 2.4)
6. ⊢ 0+s(x) = s(0+x)		(S6)
7. ⊢ x = 0+x → s(x) = s(0+x)		(S2)
8. {ψ(x)} ⊢ s(x) = s(0+x)		(MP, 5,7)
9. ⊢ s(x) = s(0+x) → (0+s(x) = s(0+x) → s(x) = 0+s(x))		(PA4)
10. {ψ(x)} ⊢ 0+s(x) = s(0+x) → s(x) = 0+s(x)		(MP, 8,9)
11. {ψ(x)} ⊢ s(x) = 0+s(x)		(MP, 6,10)
12. {ψ(x)} ⊢ ψ(s(x))		(Definition)
13. ⊢ ψ(x) → ψ(s(x))		(DT)
14. ⊢ ∀x (ψ(x) → ψ(s(x)))		(G, 13)
15. ⊢ ψ(0) → (∀x (ψ(x) → ψ(s(x))) → ∀x ψ(x))		(S9)
16. ⊢ ∀x (ψ(x) → ψ(s(x))) → ∀x ψ(x)		(MP, 4,15)
17. ⊢ ∀x ψ(x)		(MP, 14,16)

3.1 Peano Arithmetic

18. ⊢ ∀x (x = 0+x)   (Definition)

		Theorem PA7
		$\sigma + \bar{1} = s(\sigma)$

1. ⊢ $\sigma + s(0) = s(\sigma + 0)$ (S6)
2. ⊢ $\sigma + 0 = \sigma$ (S5)
3. ⊢ $\sigma + 0 = \sigma \to s(\sigma + 0) = s(\sigma)$ (S2)
4. ⊢ $s(\sigma + 0) = s(\sigma)$ (MP, 2,3)
5. ⊢ $\sigma + s(0) = s(\sigma + 0) \to$
   $(s(\sigma + 0) = s(\sigma) \to \sigma + s(0) = s(\sigma))$ (PA3)
6. ⊢ $s(\sigma + 0) = s(\sigma) \to \sigma + s(0) = s(\sigma)$ (MP, 1,5)
7. ⊢ $\sigma + s(0) = s(\sigma)$ (MP, 4,6)
8. ⊢ $\sigma + \bar{1} = s(\sigma)$ (Definition)

		Theorem PA8
		$\sigma \times \bar{1} = \sigma$

1. ⊢ $\sigma \times s(0) = (\sigma \times 0) + \sigma$ (S8)
2. ⊢ $\sigma \times 0 = 0$ (S7)
3. ⊢ $\forall x\, (\sigma \times 0 = 0 \to (\sigma \times 0) + x = 0 + x)$ (PA5)
4. ⊢ $\forall x\, (\sigma \times 0 = 0 \to (\sigma \times 0) + x = 0 + x) \to$
   $(\sigma \times 0 = 0 \to (\sigma \times 0) + \sigma = 0 + \sigma)$ (A4)
5. ⊢ $\sigma \times 0 = 0 \to (\sigma \times 0) + \sigma = 0 + \sigma$ (MP, 3,4)
6. ⊢ $(\sigma \times 0) + \sigma = 0 + \sigma$ (MP, 2,5)
7. ⊢ $\sigma \times s(0) = (\sigma \times 0) + \sigma \to$
   $((\sigma \times 0) + \sigma = 0 + \sigma \to \sigma \times s(0) = 0 + \sigma)$ (PA3)
8. ⊢ $(\sigma \times 0) + \sigma = 0 + \sigma \to \sigma \times s(0) = 0 + \sigma$ (MP, 1,7)
9. ⊢ $\sigma \times s(0) = 0 + \sigma$ (MP, 6,8)
10. ⊢ $\forall x\, x = 0 + x$ (PA6)
11. ⊢ $\forall x\, x = 0 + x \to \sigma = 0 + \sigma$ (A4)
12. ⊢ $\sigma = 0 + \sigma$ (MP, 10,11)
13. ⊢ $\sigma = 0 + \sigma \to 0 + \sigma = \sigma$ (PA2)
14. ⊢ $0 + \sigma = \sigma$ (MP, 12,13)
15. ⊢ $\sigma \times s(0) = 0 + \sigma \to (0 + \sigma = \sigma \to \sigma \times s(0) = \sigma)$ (PA3)
16. ⊢ $0 + \sigma = \sigma \to \sigma \times s(0) = \sigma$ (MP, 9,15)
17. ⊢ $\sigma \times s(0) = \sigma$ (MP, 14,16)
18. ⊢ $\sigma \times \bar{1} = \sigma$ (Definition)

After a long journey, we have finally reached our goal and formally derived all theorems from Table 3.4 in the formal system of Peano arithmetic. You might have wondered why we went through the examples in such detail. The reason is simple: The mere listing of axioms and inference rules does not convey any sense or feeling of how easy or difficult it is to derive a theorem. We can't gain such a sense without bringing the axioms and inference rules to life, which we have been doing by proving the example theorems.

Let us consider a particular case before turning away from the proven theorems. By substituting σ with $\bar{1}$ in (PA7) and (PA8), we obtain two of the most frequently quoted theorems about the natural numbers with a single stroke.

 Corallary 3.1

> The following formulas are theorems of Peano arithmetic:
> $$\bar{1} + \bar{1} = \bar{2} \qquad \bar{1} \times \bar{1} = \bar{1}$$

3.2 Axiomatic Set Theory

This section centers on set theory, one of the most significant areas of modern mathematics. The concept of a set gains value mainly due to its expressive power. Set theory not only contains Peano arithmetic as a sub-theory; it is expressive enough to formalize all the concepts of ordinary mathematics. Moreover, the antinomies that began to surface at the beginning of the twentieth century drew the attention of many scholars to this branch of mathematics. They not only helped set theory to gain unintentional popularity but also revealed that the foundation of mathematics is a fragile one.

Today, many theories gather under the umbrella of axiomatic set theory, all sharing the common goal of closing the cracks in the foundation of mathematics. One of the oldest is *ramified type theory*. It is a centerpiece of the *Principia Mathematica*, the monumental work put forward by Russell and Whitehead as a supposed panacea against the antinomies of set theory. Type theory centers around the idea of imposing a hierarchical level to each set, a so-called *type*. Considering only those sets with a higher type than their elements as legitimate eliminates self-referential constructs such as the set of all sets from the outset. The American logician Willard Van Orman Quine simplified type theory in 1937 [163, 164]. That year, he published an axiomatized variant named *New Foundations* that eliminated many of its predecessor's shortcomings. Since then, type theory in its original form has been widely considered obsolete.

Today, the concept of set is usually formalized with theories that fall into one of the following two categories:

- **Theories based on sets**

    Theories belonging to this category exclusively deal with *sets*. Its best-known representatives are *Zermelo-Fraenkel set theory*, ZF for short, and ZFC (*Zermelo-Fraenkel with Choice*), which also contains the axiom of choice. Both are first-order theories formed by nine respective ten axioms, which Ernst Zermelo and Abraham Fraenkel formulated between 1908 and 1921. Another theory in this category is the lesser-known *Kripke-Platek set theory* (KP) [81].

- **Theories based on sets and classes**

    Theories in this category distinguish between *sets* and *classes*. For example, while Russell's set of all sets does not exist in ZF and ZFC, it is contained in such theories as a class. We can think of a class as

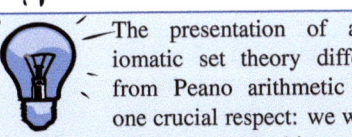
The presentation of axiomatic set theory differs from Peano arithmetic in one crucial respect: we will refrain from defining a standard interpretation for set theory. There is a good reason for deviating from our usual path. While in Peano arithmetic, it was straightforward to translate our intuitive conception of the natural numbers into the model relation '$\models$', doing the same in set theory is far more challenging. First and foremost, to define a standard interpretation, we would have to commit ourselves to a domain of individuals. In Peano arithmetic, this was easy – the range of individuals corresponded to the set of natural numbers, posing no issue. But in set theory? The domain of individuals had to be, dare we say it, the set of all sets. Forming a standard interpretation in such a naive way would reopen the door to Russell's antinomy, highlighting the need for extreme caution in set theory to avoid paradoxes.

a collection of objects too large to exist as a self-contained whole. Thus, these theories do not overcome antinomies by excluding the disputed objects; instead, they move them from the realm of sets to the realm of classes. Classes are subject to several restrictions, though. For instance, they must never be an element of another set or class.

Among these theories are the set theory of Wilhelm Ackermann [3] and the lesser-known *Morse-Kelley set theory* [115, 140]. The best-known representative of this category is the *Neumann-Bernays-Gödel set theory*, or NBG for short, developed around 1940. In contrast to ZF, it is *finitely axiomatizable* as it does not depend on axiom schemata. Despite their different focus, ZF and NBG are closely intertwined. On the one hand, all theorems of ZF are also provable in NBG. On the other hand, the reverse holds, too, if we restrict ourselves to theorems that solely deal with sets. All of these NBG theorems are also provable in ZF.

Section 3.2.1 will focus on Zermelo-Fraenkel set theory, which is dominant in axiomatic set theory today. Among others, we will demonstrate how to interpret the natural numbers as sets and utilize ZF or ZFC to formalize number-theoretic statements. In Section 3.2.2, the picked route will guide us directly into the realm of ordinal numbers. In this astounding world of thought, we will overcome alleged limits and become acquainted with giant numbers far beyond the imaginable. Armed with the theory of ordinal numbers, we will finally explain the concept of cardinality with mathematical precision in Section 3.2.3.

The time has come to look behind the scenes at one of the most exciting areas of mathematics. The path we are about to take is undoubtedly challenging. But rest assured: The insights gained at the end of the chapter will amply reward us for our efforts.

> A glimpse at Zermelo's 1908 paper reveals that his original set theory contains not only sets but also other objects. These objects are so-called *urelements*, which, unlike proper sets, never contain elements. In a sense, urelements and classes are two complementary concepts. While classes are too large to exist within a set or another class, urelements are too small to contain elements of their own. Based on their nature, they are also called individuals or atoms.
> Urelements fit our intuitive idea of a set's elements. However, we can easily represent them by other sets, which explains why Zermelo's set theory is frequently counted among the theories that only recognize sets as objects. Accordingly, it came as no surprise that the later-developed Zermelo-Fraenkel set theory dropped the distinction. ZF no longer mentions the concept of urelements and recognizes no entities other than sets.

3.2.1 Zermelo-Fraenkel Set Theory

We introduce the building blocks of Zermelo-Fraenkel set theory in two steps. First, we define the rules governing the construction of set-theoretic formulas. Afterward, we examine the axioms of this theory in depth.

3.2 Axiomatic Set Theory

 Definition 3.3 (Syntax of Zermelo-Fraenkel Set Theory)

> The set of *ZF-formulas* over the stock of variables $\{x_1, x_2, x_3, \ldots\}$ is inductively defined:
>
> - If ξ and ν are variables, then
> $$(\xi = \nu) \text{ and } (\xi \in \nu)$$
> are formulas.
>
> - If φ and ψ are formulas, so are
> $$(\neg \varphi), (\varphi \wedge \psi), (\varphi \vee \psi), (\varphi \to \psi), (\varphi \leftrightarrow \psi), (\varphi \nleftrightarrow \psi).$$
>
> - If φ is a formula, so are
> $$\forall \xi\, \varphi, \exists \xi\, \varphi \quad \text{with} \quad \xi \in \{x_1, x_2, x_3, \ldots\}.$$

In addition to the propositional operators and the predicate-logic quantifiers, ZF set theory is limited to two predicate symbols: '$=$' and '$\in$'. The intuitive meaning of these symbols is straightforward: $x_1 = x_2$ expresses the equality between the sets x_1 and x_2, while $x_1 \in x_2$ states that x_1 is an element of x_2. As before, we will adjust the pool of variables from time to time and write, for instance, x, y, z instead of x_1, x_2, x_3. We will also refrain from writing down one or the other pair of parentheses as long as no ambiguities arise.

To further simplify matters, we allow the following language constructs:

$$
\begin{aligned}
x \neq y &:= \neg(x = y) \\
x \notin y &:= \neg(x \in y) \\
x \subseteq y &:= \forall z\, (z \in x \to z \in y) \\
x \subset y &:= x \subseteq y \wedge x \neq y \\
\forall (x \in y)\, \varphi &:= \forall x\, (x \in y \to \varphi) \\
\exists (x \in y)\, \varphi &:= \exists x\, (x \in y \wedge \varphi)
\end{aligned}
$$

3.2.1.1 ZF Axioms

This section details the substantive meaning of the various non-logical axioms, thus shedding a brighter light on Zermelo-Fraenkel set theory.

> Different authors note set-theoretic formulas differently, especially the variables. The dilemma: In mathematics, we are used to denoting sets with capital letters, just as we are used to denoting the variables of a logic formula with lowercase letters. In axiomatic set theory, variables represent sets, so there are good reasons to both capitalize and lowercase them. This book follows the convention of predicate logic and denotes variables in lowercase.
>
> Some theories, including the NBG set theory, enforce a mixed notation to distinguish sets and classes already at the syntactic level. For example, in the NBG formula
>
> $$\exists X\, \forall y\, y \in X,$$
>
> X denotes a class and y a set. Although each notation fulfills its purpose, it is advisable always to check the notation when reviewing other literature to avoid misunderstandings in advance.

Axiom of extensionality
(Axiom of definitness)

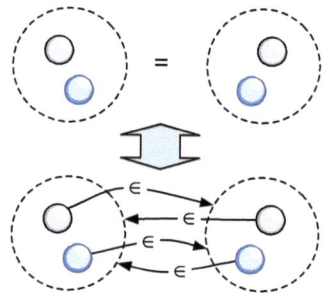

- $\forall x \forall y \, (x = y \leftrightarrow \forall z \, (z \in x \leftrightarrow z \in y))$

> "If every element of a set M is also an element of N and vice versa, if, therefore, both $M \subset N$ and $N \subset M$, then always $M = N$; or, more briefly: Every set is determined by its elements."
>
> Ernst Zermelo, 1908

Axiomatic set theory applies the principle of *extensionality* (*extensional equality*). It states that the meaning of an expression is determined solely by its scope, that is, by the objects it names or describes. In terms of sets, it means that two sets, x and y, are equal if and only if they contain the same elements ($z \in x$ implies $z \in y$ and vice versa).

Using the abbreviations introduced above, we can write down the axiom in the following form:

$$\forall x \forall y \, (x = y \leftrightarrow x \subseteq y \land y \subseteq x)$$

Axiom of the empty set

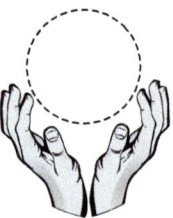

- $\exists x \forall y \, y \notin x$

> "There exists a (fictitious) set, the *null set*, $\emptyset$, that contains no element at all."
>
> Ernst Zermelo, 1908

This axiom postulates the existence of the empty set. It is the only constructive axiom that permits us to create a set out of nothing. All others will only allow us to create a new set from existing ones.

To facilitate the construction of formulas, we will use ordinary mathematical notation and denote the empty set with $\emptyset$. Remember that we do not integrate a new symbol into the formal language but merely agree on a syntactic abbreviation. Specifically, every formula φ in which the symbol $\emptyset$ occurs represents the expression

$$\exists x \, (\forall y \, y \notin x \land \varphi[\emptyset \leftarrow x]).$$

Herein, x and y are two variables, of which x must not occur in φ. Furthermore, the expression

$$\varphi[\emptyset \leftarrow x]$$

3.2 Axiomatic Set Theory

denotes the formula φ with each occurrence of the symbol $\emptyset$ substituted with the newly introduced variable x. For instance, applying the substitution to the expression

$$\emptyset \in z$$

yields the formula

$$\exists x \, (\forall y \, y \notin x \land x \in z). \tag{3.1}$$

The resulting expression emphasizes that set-theoretical formulas quickly become unmanageable if only the native language elements are permitted.

There is a simple reason why replacing $\emptyset$ is rather cumbersome: While all abbreviations introduced so far resembled predicates, $\emptyset$ is an artificial constant symbol. In the cleansed formula (3.1), the newly introduced variable x describes this constant, and the variable y is needed to ensure that x corresponds to the proper set, in this case, the empty set.

■ $\forall x \, \forall y \, \exists z \, \forall u \, (u \in z \leftrightarrow u = x \lor u = y)$ **Axiom of pairing**

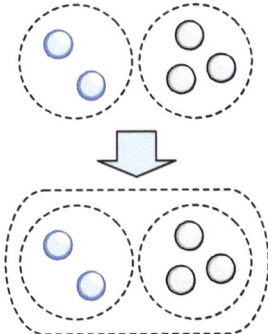

> "[...]; if a and b are any two objects of the domain, there always exists a set $\{a,b\}$ containing as elements a and b but no object x distinct from both."
>
> Ernst Zermelo, 1908

For any two sets, x and y, the pairing axiom guarantees the existence of a set z that contains x and y as sole elements. In ordinary mathematical notation, it states that we can always construct the set $\{x,y\}$ from any two sets x and y. If $x = y$, the pairing set is $\{x\}$.

Together with the axiom of the empty set, the pairing axiom allows the construction of a large number of sets, namely:

$$\emptyset, \{\emptyset\}, \{\emptyset, \{\emptyset\}\}, \{\emptyset, \{\emptyset, \{\emptyset\}\}\}, \{\{\emptyset\}, \{\emptyset, \{\emptyset\}\}\}, \{\{\emptyset\}\}, \ldots$$

Axiom of the union

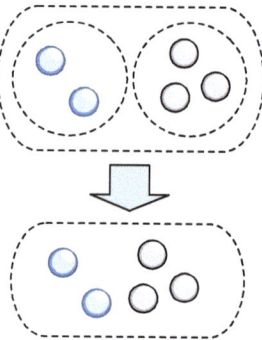

- $\forall x \, \exists y \, \forall z \, (z \in y \leftrightarrow \exists (w \in x) \, z \in w)$

> "To every set T there corresponds a set $\mathfrak{S}T$, the union of T, that contains as elements precisely all elements of the elements of T."
>
> Ernst Zermelo, 1908

In ordinary mathematical notation, the axiom states that for every set x, the set

$$y = \bigcup x := \bigcup_{w \in x} w$$

also exists. For example, for

$$x = \{\, \emptyset, \{\emptyset, \{\emptyset\}\}, \{\{\emptyset\}\} \,\},$$

the axiom guarantees the existence of the set

$$y = \emptyset \cup \{\emptyset, \{\emptyset\}\} \cup \{\{\emptyset\}\} = \{\emptyset, \{\emptyset\}\}.$$

The union set is figuratively formed from x by flattening the set hierarchy for one level and removing duplicates.

As usual, we will denote the union of two sets with the symbol '$\cup$'. If a formula φ contains the expression $\xi \cup \nu$, we shall treat it as an abbreviation for the following formula:

$$\exists x \, (\forall y \, (y \in x \leftrightarrow y \in \xi \vee y \in \nu) \wedge \varphi[\xi \cup \nu \leftarrow x]) \qquad (3.2)$$

The replacement schema is very similar to that of the empty set. x and y are two new variables, where x must not occur in φ, and the expression $\varphi[\xi \cup \nu \leftarrow x]$ denotes the formula φ, in which each occurrence of the string $\xi \cup \nu$ was substituted by the newly introduced variable x. The reason why we need an additional variable is similar to the case of the empty set. Having created a new function symbol with '$\cup$', we need the variable x to reference the function value.

In addition to the symbol '$\cup$', we will also use the symbol '$\cap$' to describe the intersection of two sets. Analogous to (3.2), we can rewrite any formula containing the expression $\xi \cap \nu$ as follows:

$$\exists x \, (\forall y \, (y \in x \leftrightarrow y \in \xi \wedge y \in \nu) \wedge \varphi[\xi \cap \nu \leftarrow x])$$

3.2 Axiomatic Set Theory

- $\forall x \, \exists y \, \forall z \, (z \in y \leftrightarrow z \in x \land \varphi(z))$

Axiom of separation

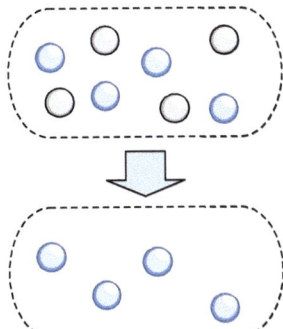

"Whenever the propositional function $\mathfrak{E}(x)$ is definite for all elements of a set M, M possesses a subset $M_\mathfrak{E}$ containing as elements precisely those elements x of M for which $\mathfrak{E}(x)$ is true."

Ernst Zermelo, 1908

This axiom stipulates the existence of the set

$$y = \{z \in x \mid P(z)\}$$

for every set x and for every property P that can be described by a formula φ. The set y contains exactly those elements of x that satisfy P. The axiom is laid out as an axiom schema, with the placeholder φ representing the property P. φ is an arbitrary formula with z being its sole free variable.

- $\exists x \, (\emptyset \in x \land \forall (y \in x) \, \{y\} \in x)$

Axiom of infinity

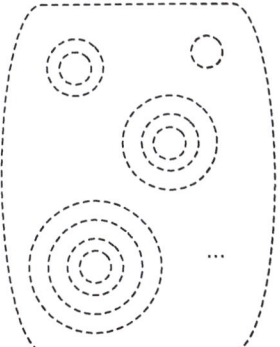

"There exists in the domain at least one set Z that contains the null set as an element and is so constituted that to each of its elements a there corresponds a further element of the form $\{a\}$, in other words, that with each of its elements a it also contains the corresponding set $\{a\}$ as an element."

Ernst Zermelo, 1908

The axiom of infinity guarantees the existence of a set x,

- such that the empty set is an element of x,
  ☞ $\emptyset \in x$

- and for every $y \in x$, the set $\{y\}$ is also an element of x.
  ☞ $\forall (y \in x) \, \{y\} \in x$

Note that the expression $\{y\}$ is not a native language element. We use it in anticipation of Section 3.2.1.3, where we will show how to rebuild the expression with native language elements.

Substantively, the axiom of infinity states that x contains the elements

$$\emptyset, \{\emptyset\}, \{\{\emptyset\}\}, \{\{\{\emptyset\}\}\}, \{\{\{\{\emptyset\}\}\}\}, \{\{\{\{\{\emptyset\}\}\}\}\}, \ldots$$

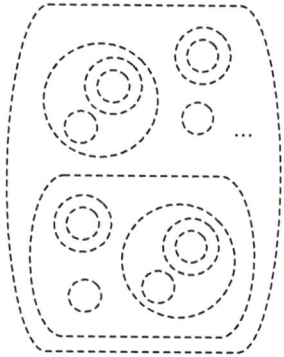

The axiom's name indicates that no finite set can fulfill this property. Consequently, in Zermelo-Fraenkel set theory, we can always assume the existence of a set with infinitely many elements.

Zermelo's formulation of the axiom of infinity is one of many possible. Modern treatises on set theory prefer to formulate the axiom as such:

- $\exists x \, (\emptyset \in x \land \forall (y \in x) \, y \cup \{y\} \in x)$

This variant requires

- the empty set to be a member of x and
  ☞ $\emptyset \in x$

- the set $y \cup \{y\}$ to be a member of x for every $y \in x$.
  ☞ $\forall (y \in x) \, y \cup \{y\} \in x$

A set with this property is, for example:

$$x = \{\emptyset, \{\emptyset\}, \{\emptyset, \{\emptyset\}\}, \{\emptyset, \{\emptyset\}, \{\emptyset, \{\emptyset\}\}\},$$
$$\{\emptyset, \{\emptyset\}, \{\emptyset, \{\emptyset\}\}, \{\emptyset, \{\emptyset\}, \{\emptyset, \{\emptyset\}\}\}\}, \ldots$$

Both variants of the infinity axiom are equivalent, as the other ZF axioms let us derive the first variant from the second and vice versa.

The discussion of Zermelo's and von Neumann's number series in Section 3.2.1.4 will uncover why the formulation of the infinity axiom has changed over time: Both series are alternative set representations of the natural numbers. It will become evident that the axiom has an intuitive substantive meaning, as it simply claims that the natural numbers form a set. The first axiom carries this substantive meaning if we base the natural numbers on Zermelo's number representation, and the second if we follow von Neumann's approach.

Axiom of the power set

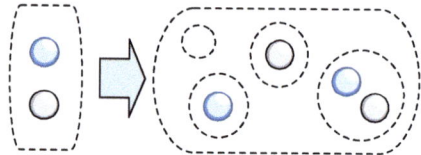

- $\forall x \, \exists y \, \forall z \, (z \in y \leftrightarrow z \subseteq x)$

> "To every set T there corresponds a set $\mathfrak{U}T$, the power set of T, that contains as elements precisely all subsets of T."
>
> Ernst Zermelo, 1908

For every set x, the axiom guarantees the existence of the the power set $\mathcal{P}(x)$, the set of all subsets. For example, the power set of

$$x = \{x_1, x_2, x_3\}$$

is the following set:

$$y = \{\emptyset, \{x_1\}, \{x_2\}, \{x_3\}, \{x_1, x_2\}, \{x_1, x_3\}, \{x_2, x_3\}, \{x_1, x_2, x_3\}\}$$

■ $(\forall (a \in x)\ \exists_1 b\ \varphi(a,b)) \rightarrow (\exists y\ \forall b\ (b \in y \leftrightarrow \exists (a \in x)\ \varphi(a,b)))$

Axiom of replacement

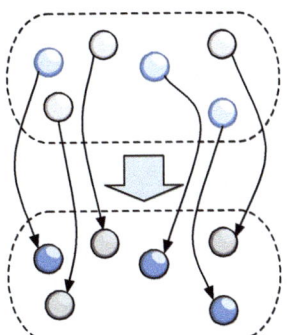

> "If M is a set and each element of M is replaced by a 'thing of the domain $\mathfrak{B}$', then M again becomes a set."
>
> Abraham Fraenkel, 1922

For every function f describable by a formula φ and every set

$$x = \{x_1, x_2, x_3, \ldots\},$$

the axiom postulates the existence of

$$y = \{f(x_1), f(x_2), f(x_3), \ldots\}.$$

The axiom is a schema in which the placeholder φ may be substituted with any formula containing a and b as free variables. The partial expression $\forall (a \in x)\ \exists_1 b\ \varphi(a,b)$ ensures that φ models a function that assigns each element of the domain precisely one image.

Abraham Fraenkel introduced the axiom of replacement as a substitute for Ernst Zermelo's axiom of separation. Interestingly, we can derive the separation axiom from the replacement axiom in ZF, but not vice versa, making Zermelo's set theory a proper sub-theory of Zermelo-Fraenkel set theory. As the axiom of separation is, in principle, dispensable, some books such as [224] consistently introduce ZF set theory with only eight axioms. In most modern expositions, however, the axiom of separation is still listed as an axiom, and we do not want to oppose this practice in this book.

■ $\forall x\ (x \neq \emptyset \rightarrow \exists (y \in x)\ x \cap y = \emptyset)$

Axiom of foundation
(Axiom of regularity)

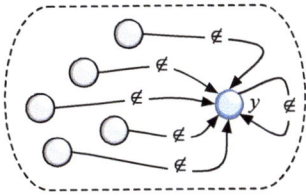

> "Every (decreasing) chain of elements, in which each term is an element of the preceding term, breaks off with finite index at an urelement. Or equally: Every partial domain T [$\neq \emptyset$] contains at least one element t_0 none of whose elements are in T."
>
> Ernst Zermelo, 1930

- Self-inclusion

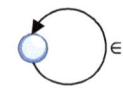

- Ring inclusion

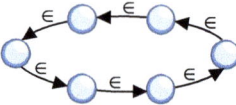

- Infinite descent

Figure 3.8: The axiom of foundation prevents sets from directly or indirectly containing themselves as elements or forming an infinitely descending chain of inclusion.

The axioms formulated so far allow us to construct sets explicitly; however, they do not exclude the possibility of other sets existing side by side. To keep such uninvited guests at bay, Fraenkel formulated the so-called *axiom of restriction*, "that imposes on the concept of set, or more appropriately the domain [of sets], the smallest extension compatible with the remaining axioms." [182].

John von Neumann was dissatisfied with the formulation, as it relates substantively to the other axioms. In his set theory from 1925, he replaced it with the *axiom of foundation*, which prohibits infinitely descending inclusion chains. The axiom was adopted by Zermelo in 1930 and put into the form quoted above. It states that every non-empty set x contains an element y whose elements are all different from the elements of x (☞ $x \cap y = \emptyset$).

The consequences of the foundation axiom are far-reaching. First, we note that no set that contains itself as an element can ever exist (Figure 3.8). To see why, let us assume the existence of a set x_1 with $x_1 \in x_1$. In this case, the set $x = \{x_1\}$ immediately violated the foundation axiom, since x and x_1 share a common element (because of $x_1 \in x_1$, $x \cap x_1 = \{x_1\} \neq \emptyset$). The foundation axiom even prevents self-inclusion if it occurs as a ring inclusion extending over several hierarchy levels. If there were sets $x_1, x_2, x_3, \ldots x_n$ satisfying

$$x_1 \in x_2 \in x_3 \in \ldots \in x_n \in x_1,$$

the set $x = \{x_1, x_2, x_3, \ldots, x_n\}$ would violate the foundation axiom, as each element of this set would contain an element also contained in x.

Likewise, infinitely descending inclusion chains of the form

$$x_1 \ni x_2 \ni x_3 \ldots$$

are eliminated. In this case, the set $\{x_1, x_2, x_3, \ldots\}$ contradicts the statement of the foundation axiom. Note that the converse does not apply. Infinitely increasing sequences

$$x_1 \in x_2 \in x_3 \ldots$$

are perfectly in line with the axiom.

As an aside, the foundation axiom demonstrates the importance of the notational simplifications introduced so far. Figure 3.9 depicts how the foundation axiom would look like if we solely used native language constructs. The axiom's true meaning would barely be recognizable and disappear almost entirely amid formal nomenclature.

3.2 Axiomatic Set Theory

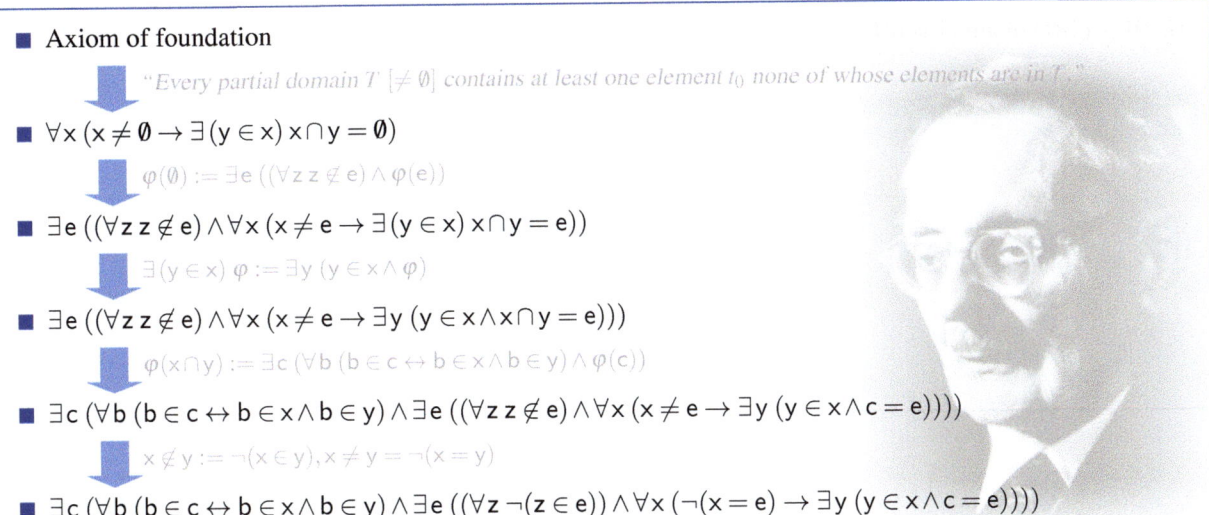

- **Axiom of foundation**

  "Every partial domain T [$\neq \emptyset$] contains at least one element t_0 none of whose elements are in T."

- $\forall x \, (x \neq \emptyset \rightarrow \exists (y \in x) \, x \cap y = \emptyset)$

  $\varphi(\emptyset) := \exists e \, ((\forall z \, z \notin e) \wedge \varphi(e))$

- $\exists e \, ((\forall z \, z \notin e) \wedge \forall x \, (x \neq e \rightarrow \exists (y \in x) \, x \cap y = e))$

  $\exists (y \in x) \, \varphi := \exists y \, (y \in x \wedge \varphi)$

- $\exists e \, ((\forall z \, z \notin e) \wedge \forall x \, (x \neq e \rightarrow \exists y \, (y \in x \wedge x \cap y = e)))$

  $\varphi(x \cap y) := \exists c \, (\forall b \, (b \in c \leftrightarrow b \in x \wedge b \in y) \wedge \varphi(c))$

- $\exists c \, (\forall b \, (b \in c \leftrightarrow b \in x \wedge b \in y) \wedge \exists e \, ((\forall z \, z \notin e) \wedge \forall x \, (x \neq e \rightarrow \exists y \, (y \in x \wedge c = e))))$

  $x \notin y := \neg (x \in y), x \neq y := \neg (x = y)$

- $\exists c \, (\forall b \, (b \in c \leftrightarrow b \in x \wedge b \in y) \wedge \exists e \, ((\forall z \, \neg (z \in e)) \wedge \forall x \, (\neg (x = e) \rightarrow \exists y \, (y \in x \wedge c = e))))$

Figure 3.9: Without notational simplifications, the foundation axiom is hardly recognizable.

3.2.1.2 Axiom of Choice

The time has come to turn our attention to the tenth and final axiom of Zermelo-Fraenkel set theory: The *axiom of choice* (AC). In the past, there was a controversial debate about whether AC should be included among the other axioms of set theory. Although most mathematicians accept AC today, we are, strictly speaking, dealing with two different theories: the ZF set theory, consisting of the nine axioms from Section 3.2.1.1, and the ZFC set theory (*Zermelo-Fraenkel with Choice*), which also includes the axiom of choice. Before we look closely at the consequences of adding this in some respects mysterious axiom, let us first reveal its substantive meaning:

Axiom of choice

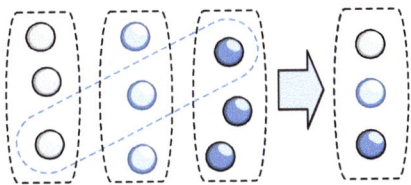

- $(\forall (u,v \in x) \, (u \neq v \rightarrow u \cap v = \emptyset) \wedge \forall (u \in x) \, u \neq \emptyset) \rightarrow$
  $\exists y \, \forall (z \in x) \, \exists_1 (w \in z) \, w \in y$

"If T is a set whose elements all are sets that are different from $\emptyset$ and mutually disjoint, its union $\cup T$ includes at least one subset S_1 having one and only one element in common with each element of T."

Ernst Zermelo, 1908

The Axiom of choice makes a statement about all sets x,

- whose elements are pairwise disjoint sets,
  ☞ $\forall (u, v \in x)\, (u \neq v \rightarrow u \cap v = \emptyset)$

- and none of these elements is the empty set.
  ☞ $\forall (u \in x)\, u \neq \emptyset$

For such sets, the axiom of choice guarantees the existence of a set y that contains a single element from each set $z \in x$.
☞ $\exists y\, \forall (z \in x)\, \exists_1 (w \in z)\, w \in y$

Note that the axiom of choice only guarantees the existence of a choice set but does not tell us anything about its construction. In other words, the axiom is non-constructive. Besides, it appears more like a theorem; its particular claim makes it look like a consequence of the other axioms. Yet, you have already learned about the opposite in Chapter 1 on Page 57. The axiom of choice is independent of the other axioms and is thus neither provable nor disprovable in ZF [37, 75]. Just as in the case of the continuum hypothesis, it is safe to add the axiom of choice or its negation as another axiom without the risk of contradiction.

Despite its unprovability, the axiom of choice appears to express an intuitive truth. Why should taking an element out of a non-empty set not be possible? After all, the selection itself is not subject to any restriction, which means that an arbitrary element fulfills our purpose. And what if a set contains an infinite number of elements? All the better! There are then more than enough elements to choose from. Well, sometimes, it is advisable to distrust our intuition, especially when we are getting perilously close to the concept of infinity.

Let's set aside the axiom of choice for a moment and recapitulate a familiar concept from ordinary mathematics: the *ordering relation* (Figure 3.10 and 3.11).

3.2 Axiomatic Set Theory

Example 1	Example 2	Example 3
$(\mathbb{Z} \setminus \{0\}, \prec)$ with $x \prec y :\Leftrightarrow x < y \wedge \|x\| = \|y\|$	$(\mathbb{Z}, <)$	$(\mathbb{N}, <)$
-1 ≺ 1 -2 ≺ 2 -3 ≺ 3	-2 < -1 < 0 < 1 < 2 < 3	0 < 1 < 2 < 3 < 4 < 5 < 6
Order: ✔ Yes Total order: ✘ No Well-order: ✘ No	Order: ✔ Yes Total order: ✔ Yes Well-order: ✘ No	Order: ✔ Yes Total order: ✔ Yes Well-order: ✔ Yes

Figure 3.10: Illustration of the different ordering types. The set $\mathbb{Z} \setminus \{0\}$ is ordered by '$\prec$' but not totally ordered as for all elements x and y with $|x| \neq |y|$, neither $x \prec y$ nor $y \prec x$ holds. The set $\mathbb{Z}$ is totally ordered by '$<$' but not well-ordered. For example, the set $\mathbb{Z}$ itself has no minimal element. In contrast, the set $\mathbb{N}$ is well-ordered by '$<$' as every non-empty subset always contains a minimal element.

Definition 3.4 (Order)

Let M be a set and '$<$' a binary relation on M.

- '$<$' is a *partial order* or *order* on M, if '$<$'
  1. is *irreflexive*, i.e., $x < x$ never holds,
  2. is *asymmetric*, i.e., $x < y$ implies $y \not< x$,
  3. is *transitive*, i.e., $x < y$ and $y < z$ imply $x < z$.

- '$<$' is a *linear order* or *total order* on M, if
  1. '$<$' is an order on M, and
  2. for all $x, y \in M$ with $x \neq y$, either $x < y$ or $y < x$ holds.

- '$<$' is a *well-ordering* on M, if
  1. '$<$' is a total order on M, and
  2. every subset $N \subseteq M$ with $N \neq \emptyset$ has a least element; that is, there exists an $x \in N$ such that $x < y$ for all $y \in N \setminus \{x\}$.

The term *partial order* has different definitions in the literature. Some books, like this one, require the property of *irreflexivity*. Others assume *reflexivity* and demand *antisymmetry* instead of asymmetry. In mathematics, a relation is antisymmetric if the relations $x < y$ and $y < x$ only hold simultaneously if $x = y$. Some authors emphasize the difference by speaking of irreflexive or reflexive orders explicitly.

Both variants have their justification. For example, the subset relation $\subseteq$ is a reflexive order ($x \subseteq x$ always holds), whereas the element relation $\in$ is an irreflexive order ($x \in x$ does not hold for any set x). These examples highlight that the decisive property of an order relation is not whether it is reflexive or irreflexive but the asymmetric or antisymmetric arrangement of the elements in combination with transitivity. Whether the examined order relations are irreflexive or reflexive is irrelevant at most places in this book. Hence, we will largely ignore the distinction and simply refer to orderings in the subsequent chapters.

If an order relation is total, two elements x and y always relate to each other, that is, either $x < y$ or $y < x$. Total orders are also called *linear orders*, as we can imagine the elements arranged in a sequence. Always

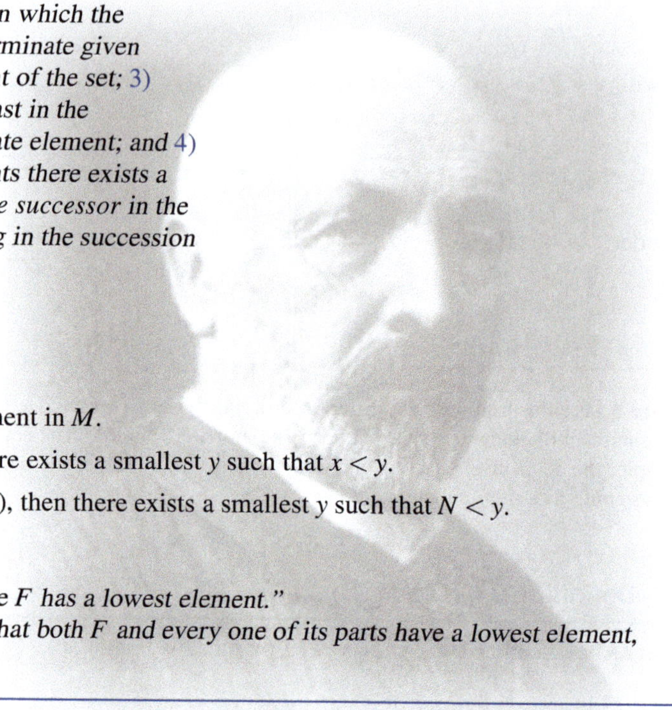

- Georg Cantor, 1883 [21]

    "A well-ordered set is a well-defined set 1) in which the elements are bound to one another by a determinate given succession such that 2) there is a first element of the set; 3) every single element (provided it is not the last in the succession) is followed by another determinate element; and 4) for any desired finite or infinite set of elements there exists a determinate element which is their immediate successor in the succession (unless there is absolutely nothing in the succession following all of them)."

- Contemporary formulation

    1) M is totally ordered with respect to '<'.
    2) If $M \neq \emptyset$, then there exists a smallest element in M.
    3) If there exists a y such that $x < y$, then there exists a smallest y such that $x < y$.
    4) If there exists a y such that $N < y$ ($N \subset M$), then there exists a smallest y such that $N < y$.

- Georg Cantor, 1897 [23]

    "A. Every part F_1 of a well-ordered aggregate F has a lowest element."
    "B. If a simply ordered aggregate F is such that both F and every one of its parts have a lowest element, then F is a well-ordered aggregate."

Figure 3.11: A piece of history. In 1883, Georg Cantor characterized the notion of a well-order with four properties, which, at first glance, clearly differ from those in Definition 3.4. In 1897, however, Cantor proved that both definitions are equivalent.

be aware that such an arrangement does not necessarily imply that every element x has an immediate successor. For example, '<' totally orders the set $\mathbb{Q}$, but between two elements x and y, there is always another element z with $x < z$ and $z < y$.

From the concept of a total order, the idea of a well-order is just a few steps away. A totally ordered set M becomes well-ordered if every non-empty subset $N \subseteq M$ has a minimal element. It is easy to identify the natural numbers as well-ordered with respect to '<' as every subset of $\mathbb{N}$ contains an element smaller than all others. The situation differs for the integers where the set $\mathbb{Z}$ already violates the well-ordering principle. The set of integers has no lower bound and, thus, no minimum element.

Similarly, the set of positive rational numbers $\mathbb{Q}^+$ is not well-ordered with respect to '<', as every left-open interval provides a counterexample. Nevertheless, there is a simple trick to well-order the elements

3.2 Axiomatic Set Theory

from $\mathbb{Q}^+$ by introducing an order relation '$\prec$' that is defined as follows for reduced fractions (Figure 3.12):

$$\frac{p_1}{q_1} \prec \frac{p_2}{q_2} :\Leftrightarrow \begin{cases} p_1 < p_2 & \text{if } p_1 \neq p_2 \\ q_1 < q_2 & \text{if } p_1 = p_2 \end{cases}$$

To determine the smaller of two rational numbers, $\frac{p_1}{q_1}$ and $\frac{p_2}{q_2}$, the two numerators p_1 and p_2 are compared first. A smaller numerator is synonymous with a smaller number in the new comparison relation. Only if the numerators are the same do the denominators q_1 and q_2 decide which number is smaller. Thanks to the new arrangement, we have indeed managed to define a well-ordering of the rational numbers. We can pick out any non-empty subset and always identify a unique minimal element with respect to '$\prec$'.

The considerations above might make us believe every set can be ordered without exception. Expressed as a theorem, our assumption reads as follows:

Theorem 3.1 (Well-Ordering Theorem)

> Every set can be well-ordered.

As an example, let us consider the real numbers. The ordinary comparison operation '<' does not well-order $\mathbb{R}$ since, e.g., every left-open interval violates the well-ordering property. However, assuming the well-ordering theorem holds, we should also be able to well-order the real numbers. At this point, please try to find a suitable arrangement of the real numbers yourself.

You have probably noticed that the same trick that allowed us to order rational numbers fails in the case of real numbers. However, it is no disgrace that you have failed in your attempts. To this day, nobody has succeeded in constructively defining an ordering relation '$\prec$' that leads to a well-ordering of the real numbers.

The vast number of failed attempts feeds the suspicion that the well-ordering of the real numbers cannot succeed. But can we really trust our intuition this time, trying to convince us of the falsity of the well-ordering theorem with increasing insistence? Is the axiom of choice true and the well-ordering theorem false?

Now is the time to lift the veil and let the facts speak out. A fundamental result of modern set theory states that the well-ordering theorem and the

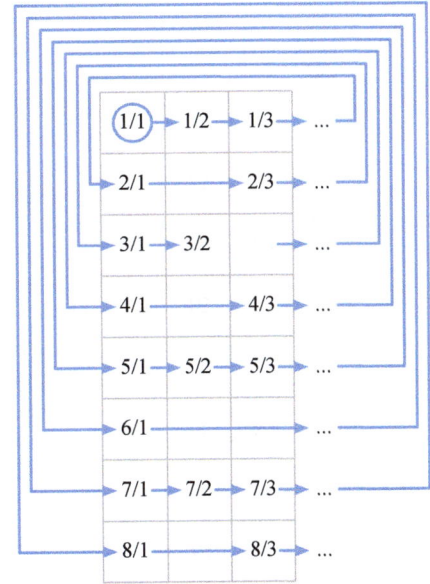

Figure 3.12: Rearranging the rational numbers lets us order the set $\mathbb{Q}^+$. In the new enumeration, the fractions with the numerator 1 appear first, followed by the fractions with the numerator 2, and so on. Note that every rational number has infinitely many representations besides its reduced standard representation. The latter are grayed out and play no role in the order definition.

> **Proof that every set can be well-ordered**
> E. ZERMELO (1904)
>
> 1) Let M be an arbitrary set [...]
>
> 2) *Imagine that with every subset M' [of M] there is associated an arbitrary elememt m'_1 that occurs in M' itself; let m'_1 be called the 'distinguished' element of M'. This yields a 'covering' γ of the set* $\mathbf{M}$ $[\mathcal{P}(M) - \{0\}]$ *by certain elements of the set M.* [...] In what follows we take an arbitrary covering γ and derive from it a definite well-ordering of the elements of M.
>
> 3) *Definition.* Let us apply the term 'γ-set' to any well-ordered set M_γ that consists entirely of elements of M and has the following property: whenever a is an arbitrary element of M_γ and A is the 'associated' segment, which consists of the elements x of M such that $x \prec a$, a is the distinguished element of $M - A$.
>
> 4) *There are γ-sets included in M.* Thus, for example, m_1, the distinguished element of M' when $M' = M$, is itself a γ-set. [...]
>
> 5) Whenever M'_γ and M''_γ are any too distinct γ-sets (associated, however, with the same covering γ chosen once and for all!), *one of the two is identical with a segment of the other*.
>
> [...]
>
> 6) *Consequences.* If two γ-sets have an element a in common, they also have the segment A of the preceding elements in common. If they have *two* elements a and b in common, then either in *both* sets $a \prec b$ or in *both* sets $b \prec a$.
>
> 7) If we call any element of M that occurs in some γ-set a 'γ-element', the following theorem holds: *The totality L_γ of all γ-elements can be so ordered that it will itself be a γ-set, and it contains all elements of the original set M. M itself is thereby well-ordered.* [...] Accordingly, to every covering γ there corresponds a definite well-ordering of the set M. [...]
>
> The present proof rests upon the assumption that coverings γ actually do exist, hence upon the principle that even for an infinite totality of sets there are always mappings that associate with every set one of its elements, [...]
>
> I owe to Mr. Erhard Schmidt the idea that, by invoking this principle, we can take an *arbitrary* covering γ as a basis for the well-ordering; the proof as I carried it through, then rests upon the fusion of various possible 'γ-sets', that is, of the well-ordered segments resulting from the ordering principle.

Figure 3.13: Proof of the well-ordering theorem by Ernst Zermelo [51, 228]. He first formulated his line of argument in a 1904 letter to David Hilbert, which is quoted here. The passages highlighted in color show that Zermelo used the axiom of choice in a slightly different form than is common today. For one thing, he did not postulate the existence of a choice set but the existence of a selection function. For another, he dispensed with today's convention that the sets from which a selection is made must be pairwise disjoint.

axiom of choice are equivalent. We can prove the well-ordering theorem within ZFC and derive the axiom of choice in ZF if we accept the well-ordering theorem as an axiom (Figure 3.13). However, the proof does not provide a specific ordering. In the same way the axiom of choice merely postulates the existence of a choice set, the proof of the well-ordering theorem establishes the existence of a well-order. Still, it does not tell us how to arrange the elements in concrete terms.

The independence of the axiom of choice has far-reaching consequences. It follows that the ZF axioms are compatible with both the well-ordering theorem and its negation. A fortiori, the ZF axioms are compatible with the more specific assumption that the real numbers can

3.2 Axiomatic Set Theory

be well-ordered, as well as with the assumption that this is impossible. This is why all your attempts to constructively define a well-ordering of the real numbers were doomed to fail. If you had succeeded, you would have shown that the ZF axioms are incompatible with the assumption that the real numbers cannot be well-ordered, which is not the case. A truly astonishing result of modern set theory!

A third characterization of the axiom of choice is worth mentioning. It comes in the form of *Zorn's lemma*:

Theorem 3.2 (Zorn's lemma)

> Every non-empty partially ordered set, in which every totally ordered subset has an upper bound, contains at least one maximal element.

Zorn's lemma, the axiom of choice, and the well-ordering principle are equivalent. If we add one of them to the ZF axioms, we can derive the other two as theorems. Even though Zorn's lemma makes the least intuitive statement, it has several applications in ordinary mathematics. It allows for concise proofs of essential theorems in linear algebra and functional analysis.

3.2.1.3 Set Theory as the Foundation of Mathematics

This section will further develop the language of Zermelo-Fraenkel set theory and demonstrate how concepts of ordinary mathematics can be formalized. To this end, we will introduce several new language elements that act as syntactic abbreviations in the usual way. They will contribute to writing down the formulas of ZF set theory even more concisely and comprehensibly.

Above all, we want to allow ordinary set notation inside ZF formulas. This is easily doable for finite sets since for any formula φ containing an expression of the form $\{\xi_1, \ldots, \xi_n\}$, there is an equivalent formula of the following type:

$$\exists x \, (\psi(x) \wedge \varphi[\{\xi_1, \ldots, \xi_n\} \leftarrow x])$$

In this formula, $\psi(x)$ is chosen such that the newly introduced variable x uniquely describes the set $\{\xi_1, \ldots, \xi_n\}$. Naturally, any other variable not occurring in φ may be used instead of x.

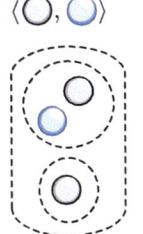

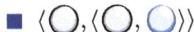

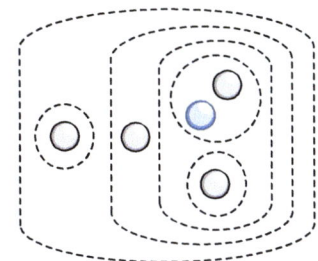

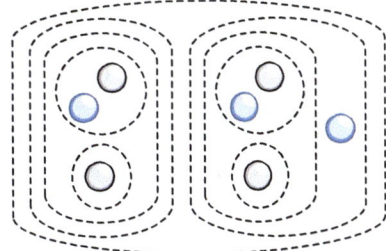

Figure 3.14: The construction $\langle \xi, \nu \rangle := \{\{\xi\}, \{\xi, \nu\}\}$ imposes an order on the formerly unordered elements of a set. Thus, the concept of an ordered pair can be reduced to sets, as can the more general concept of an ordered n-tuple.

- *"x is a binary relation"*

$$\forall (y \in x)\, \exists u\, \exists v\; y = \langle u, v \rangle$$

- Definition of *ordered pair*

$$\forall (y \in x)\, \exists u\, \exists v\; y = \{\{u\}, \{u,v\}\}$$

- Resolving $\{\{u\}, \{u,v\}\}$

$$\forall (y \in x)\, \exists u\, \exists v$$
$$\exists z\, (\{u\} \in z \wedge \{u,v\} \in z\, \wedge$$
$$\forall (z' \in z)\, (z' = \{u\} \vee z' = \{u,v\})\, \wedge$$
$$y = z)$$

- Resolving $\{u\}$

$$\forall (y \in x)\, \exists u\, \exists v$$
$$\exists z_1\, ((u \in z_1\, \wedge$$
$$\forall (z' \in z_1)\, z' = u)\, \wedge$$
$$\exists z\, (z_1 \in z \wedge \{u,v\} \in z\, \wedge$$
$$\forall (z' \in z)\, (z' = z_1 \vee z' = \{u,v\})\, \wedge$$
$$y = z))$$

- Resolving $\{u,v\}$

$$\forall (y \in x)\, \exists u\, \exists v$$
$$\exists z_1\, ((u \in z_1\, \wedge$$
$$\forall (z' \in z_1)\, z' = u)\, \wedge$$
$$\exists z_2\, ((u \in z_2 \wedge v \in z_2\, \wedge$$
$$\forall (z' \in z_2)\, (z' = u \vee z' = v))\, \wedge$$
$$\exists z\, (z_1 \in z \wedge z_2 \in z\, \wedge$$
$$\forall (z' \in z)\, (z' = z_1 \vee z' = z_2)\, \wedge$$
$$y = z)))$$

- Eliminating the conditional quantifiers

$$\forall y\, (y \in x \to \exists u\, \exists v$$
$$\exists z_1\, ((u \in z_1\, \wedge$$
$$\forall z'\, (z' \in z_1 \to z' = u))\, \wedge$$
$$\exists z_2\, ((u \in z_2 \wedge v \in z_2\, \wedge$$
$$\forall z'\, (z' \in z_2 \to (z' = u \vee z' = v)))\, \wedge$$
$$\exists z\, (z_1 \in z \wedge z_2 \in z\, \wedge$$
$$\forall z'\, (z' \in z \to (z' = z_1 \vee z' = z_2))\, \wedge$$
$$y = z))))$$

Figure 3.15: Formalizing relations in the language of Zermelo-Fraenkel set theory

For example, if $y = \{y_1\}$ and $x = \{x_1, x_2\}$, we can describe the set y by the formula

$$\psi_1(y) := y_1 \in y \wedge \forall (z \in y)\, z = y_1,$$

and the set x by

$$\psi_2(x) := x_1 \in x \wedge x_2 \in x \wedge \forall (z \in x)\, (z = x_1 \vee z = x_2).$$

This allows us to express the formula

$$\{y_1\} = \{x_1, x_2\} \to (x_1 = y_1 \wedge x_2 = y_1) \tag{3.3}$$

in the original language of ZF set theory, too. In pure form, it reads as follows:

$$\exists y\, (y_1 \in y \wedge \forall (z \in y)\, z = y_1\, \wedge$$
$$\exists x\, (x_1 \in x \wedge x_2 \in x \wedge \forall (z \in x)\, (z = x_1 \vee z = x_2)\, \wedge$$
$$(y = x \to (x_1 = y_1 \wedge x_2 = y_1))))$$

3.2 Axiomatic Set Theory

Building on the established set notation, we can formalize the notion of an *ordered pair*. Unlike in an ordinary set $\{\xi, v\}$, the elements ξ and v in an ordered pair $\langle \xi, v \rangle$ have a fixed position, urging us to encode the position somehow using standard set notation. For this purpose, we employ a simple but elegant trick, depicted in Figure 3.14. We set:

$$\langle \xi, v \rangle := \{\, \{\xi\}, \{\xi, v\} \,\}$$

The definition fulfills its purpose: Although the set $\{\, \{\xi\}, \{\xi, v\} \,\}$ itself is unordered, we can clearly extract the position of ξ and v from the structure of its elements $\{\xi\}$ and $\{\xi, v\}$.

The proposed set representation for ordered pairs was introduced in 1921 by Kazimierz Kuratowski [122] and can be extended straightforwardly to ordered tuples of any size:

$$\langle \xi_1 \rangle := \xi_1$$
$$\langle \xi_1, \ldots, \xi_{n+1} \rangle := \langle \langle \xi_1, \ldots, \xi_n \rangle, \xi_{n+1} \rangle$$

The chosen path leads us directly to the concept of *relations*. Formally, a relation is a set ξ of ordered pairs, allowing us to describe the statement "ξ is a relation" as such:

$$\mathfrak{R}(\xi) := \forall (y \in \xi) \, \exists u \, \exists v \, (y = \langle u, v \rangle)$$

In this notation, the formula looks compact and elegant. What it looks like in native terms becomes apparent by gradually expanding all abbreviations. Figure 3.15 reveals the result; the formula is a true monster and demonstrates again the importance of syntactic simplifications. Without them, set-theoretical formulas degenerate into massive symbol arrangements whose content is no longer recognizable to us.

Finally, we aim to formally capture the concept of a *partial function* (Figure 3.16). We understand such a function as a binary relation ξ of the following form:

$$\xi = \{\langle u, v \rangle \mid v = f(u)\}$$

More precisely, a binary relation ξ is a partial function if it is *right-unique*; that is, if, for every u, there is at most one v with $(u, v) \in \xi$. Thus, we can formalize the statement "ξ is a partial function" as such:

$$\mathfrak{F}(\xi) := \mathfrak{R}(\xi) \wedge \forall u \, \forall v \, \forall w \, (\langle u, v \rangle \in \xi \wedge \langle u, w \rangle \in \xi \to v = w)$$

■ Partial function

"No element of the domain is mapped to more than one element of the codomain."

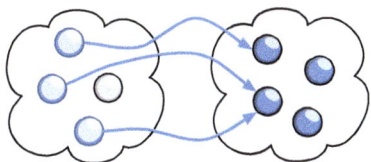

■ Total function

"Every element of the domain is mapped to one and only one element of the codomain."

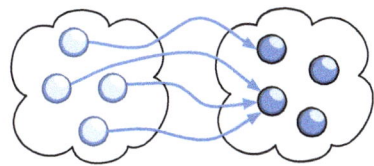

Figure 3.16: The notion of a *partial function* is a generalization of the notion of a function. We already speak of a partial function if each domain element has at most one image element, thus leveraging the requirement that there must be exactly one. The latter are sometimes referred to as *total functions* to explicitly distinguish partial functions from ordinary functions.

Formal Proofs

In several places in this book, we emphasized that Zermelo-Fraenkel set theory can serve as a formal foundation of mathematics, and we already have an early justification in our hands. The examples of ordered pairs, relations, and functions vividly show that set theory is capable of describing ordinary concepts of mathematics. Furthermore, Section 3.2.2 will elaborate on how to represent numbers as sets. After formalizing the natural numbers, we can expand on them and define the integers. Once the integers have been defined, we can erect the theory of rational numbers, followed by the theory of real numbers. If we continue this process of abstraction, we reach all areas of ordinary mathematics; they unfold in front of us like the branches of a tree whose root has found a firm hold in the foundation of set theory.

However, our approach is still lacking an essential element of justification. We must ensure that the theory of sets is expressive enough to recreate not only the concepts but also the proofs of ordinary mathematics. Using the following theorem as an example, we want to demonstrate how this is feasible:

 Theorem 3.3 (Equality of Ordered Pairs)

> For any ordered pairs $\langle x, y \rangle$ and $\langle u, v \rangle$, the following holds:
> $$\langle x, y \rangle = \langle u, v \rangle \text{ implies } x = u \text{ and } y = v$$

To see how to prove this theorem in ordinary mathematics, let us refer to the standard literature. In [134], for instance, the original wording of the proof reads as follows:

> "Assume $\langle x, y \rangle = \langle u, v \rangle$. Then $\{\{x\}, \{x, y\}\} = \{\{u\}, \{u, v\}\}$. Since $\{x\} \in \{\{x\}, \{x, y\}\}$, $\{x\} \in \{\{u\}, \{u, v\}\}$. Hence, $\{x\} = \{u\}$ or $\{x\} = \{u, v\}$. In either case, $x = u$. Now, $\{u, v\} \in \{\{u\}, \{u, v\}\}$; so, $\{u, v\} \in \{\{x\}, \{x, y\}\}$. Then, $\{u, v\} = \{x\}$ or $\{u, v\} = \{x, y\}$. Similarly, $\{x, y\} = \{u\}$ or $\{x, y\} = \{u, v\}$. If $\{u, v\} = \{x\}$ and $\{x, y\} = \{u\}$, then $x = y = u = v$; if not, $\{u, v\} = \{x, y\}$. Hence, $\{u, v\} = \{u, y\}$. So, if $v \neq u$, then $y = v$; if $v = u$, then $y = v$. Thus, in all cases, $y = v$."

To prove the theorem within ZF, we start translating it into formal notation:
$$\forall x\, \forall y\, \forall u\, \forall v\, (\langle x, y \rangle = \langle u, v \rangle \rightarrow x = u \wedge y = v)$$

Table 3.5: Auxiliary Theorems for the proof of Theorem 3.3

Set-theoretical theorems	
$\xi \in \{\xi, \nu\}$	(H1)
$\xi \in \{\nu, \xi\}$	(H2)
$\xi \in \{\nu, \mu\} \rightarrow (\xi = \nu \vee \xi = \mu)$	(H3)
$\xi \in \nu \rightarrow (\nu = \mu \rightarrow \xi \in \mu)$	(H4)
$\xi = \nu \rightarrow \nu = \xi$	(H5)
$\xi = \nu \rightarrow (\nu = \mu \rightarrow \xi = \mu)$	(H6)
$\nu = \xi \rightarrow (\{\xi, \mu\} = \{\nu, \nu\} \rightarrow \{\xi, \mu\} = \{\xi, \nu\})$	(H7)
$\{\xi\} = \{\nu\} \rightarrow \xi = \nu$	(H8)
$\{\xi, \nu\} = \{\mu\} \rightarrow \mu = \xi$	(H9)
$\{\xi, \nu\} = \{\mu\} \rightarrow \nu = \mu$	(H10)
$\{\xi, \nu\} = \{\xi, \mu\} \rightarrow \mu = \nu$	(H11)
$\xi = \nu \rightarrow \{\mu, \nu\} = \{\mu, \xi\}$	(H12)
Propositional theorems	
$\varphi \rightarrow (\psi \rightarrow (\varphi \wedge \psi))$	(TA1)
$(\varphi \vee \psi) \rightarrow ((\chi \vee \psi) \rightarrow ((\varphi \wedge \chi) \vee \psi))$	(TA2)
$(\varphi \rightarrow \psi) \rightarrow ((\neg \varphi \rightarrow \psi) \rightarrow \psi)$	(TA3)
$(\varphi \vee \psi) \rightarrow ((\varphi \rightarrow \chi) \rightarrow ((\psi \rightarrow \chi) \rightarrow \chi))$	(TA4)
$(\varphi \rightarrow (\psi \rightarrow \chi)) \rightarrow (\varphi \wedge \psi \rightarrow \chi)$	(TA5)

3.2 Axiomatic Set Theory

To keep the proof as straightforward as possible, we will fall back on several auxiliary theorems that are either propositional tautologies or make elementary statements about the element relation '$\in$' and the equality relation '='. Table 3.5 summarizes the theorems. Further, we assume that the variables ξ, ν, and μ may be substituted with both variables and sets such as $\{x,y\}$.

A word of warning: Resist the temptation to underestimate the role of the auxiliary theorems! Even if they resemble a collection of trivialities in substantive terms, their proofs are not at all trivial in every case. If we were also to list the proofs, the following derivation would lengthen by several pages.

---

1. $\langle x,y \rangle = \langle u,v \rangle \vdash$
   $\{\{x\},\{x,y\}\} = \{\{u\},\{u,v\}\}$ (Def)   *Assume $\langle x,y \rangle = \langle u,v \rangle$.*
       Then $\{\{x\},\{x,y\}\} = \{\{u\},\{u,v\}\}$.
2. $\vdash \{x\} \in \{\{x\},\{x,y\}\}$ (H1)   *Since $\{x\} \in \{\{x\},\{x,y\}\}$,*
3. $\vdash \{x\} \in \{\{x\},\{x,y\}\} \to$
   $(\{\{x\},\{x,y\}\} = \{\{u\},\{u,v\}\} \to \{x\} \in \{\{u\},\{u,v\}\})$ (H4)
4. $\vdash \{\{x\},\{x,y\}\} = \{\{u\},\{u,v\}\} \to \{x\} \in \{\{u\},\{u,v\}\}$ (MP, 2,3)
5. $\langle x,y \rangle = \langle u,v \rangle \vdash \{x\} \in \{\{u\},\{u,v\}\}$ (MP, 1,4) $\{x\} \in \{\{u\},\{u,v\}\}$.
6. $\vdash \{x\} \in \{\{u\},\{u,v\}\} \to (\{x\} = \{u\} \lor \{x\} = \{u,v\})$ (H3)
7. $\langle x,y \rangle = \langle u,v \rangle \vdash \{x\} = \{u\} \lor \{x\} = \{u,v\}$ (MP, 5,6)   *Hence, $\{x\} = \{u\}$ or $\{x\} = \{u,v\}$.*
8. $\vdash \{x\} = \{u\} \to x = u$ (H8)
9. $\vdash \{u,v\} = \{x\} \to x = u$ (H9)
10. $\vdash \{x\} = \{u,v\} \to \{u,v\} = \{x\}$ (H5)
11. $\{x\} = \{u,v\} \vdash \{u,v\} = \{x\}$ (DT)
12. $\{x\} = \{u,v\} \vdash x = u$ (MP, 9,11)
13. $\vdash \{x\} = \{u,v\} \to x = u$ (DT)
14. $\vdash (\{x\} = \{u\} \lor \{x\} = \{u,v\}) \to ((\{x\} = \{u\} \to x = u) \to$
    $((\{x\} = \{u,v\} \to x = u) \to x = u))$ (TA4)
15. $\langle x,y \rangle = \langle u,v \rangle \vdash (\{x\} = \{u\} \to x = u) \to$
    $((\{x\} = \{u,v\} \to x = u) \to x = u)$ (MP, 7,14)
16. $\langle x,y \rangle = \langle u,v \rangle \vdash (\{x\} = \{u,v\} \to x = u) \to x = u$ (MP, 8,15)
17. $\langle x,y \rangle = \langle u,v \rangle \vdash x = u$ (MP, 13,16)   *In either case, $x = u$.*
18. $\vdash \{u,v\} \in \{\{u\},\{u,v\}\}$ (H2)   *Now, $\{u,v\} \in \{\{u\},\{u,v\}\}$;*
19. $\vdash \{u,v\} \in \{\{u\},\{u,v\}\} \to$
    $(\{\{u\},\{u,v\}\} = \{\{x\},\{x,y\}\} \to \{u,v\} \in \{\{x\},\{x,y\}\})$ (H4)

20. $\vdash \{\{u\},\{u,v\}\} = \{\{x\},\{x,y\}\} \to$
    $\{u,v\} \in \{\{x\},\{x,y\}\}$ (MP, 18,19)
21. $\vdash \{\{x\},\{x,y\}\} = \{\{u\},\{u,v\}\} \to$
    $\{\{u\},\{u,v\}\} = \{\{x\},\{x,y\}\}$ (H5)
22. $\langle x,y\rangle = \langle u,v\rangle \vdash \{\{u\},\{u,v\}\} = \{\{x\},\{x,y\}\}$ (MP, 1,21)
23. $\langle x,y\rangle = \langle u,v\rangle \vdash \{u,v\} \in \{\{x\},\{x,y\}\}$ (MP, 20,22)
24. $\vdash \{u,v\} \in \{\{x\},\{x,y\}\} \to (\{u,v\} = \{x\} \lor \{u,v\} = \{x,y\})$ (H3)

so, $\{u,v\} \in \{\{x\},\{x,y\}\}$.

25. $\langle x,y\rangle = \langle u,v\rangle \vdash \{u,v\} = \{x\} \lor \{u,v\} = \{x,y\}$ (MP, 23,24)

Then, $\{u,v\} = \{x\}$ *or* $\{u,v\} = \{x,y\}$.

26. $\vdash \{x,y\} \in \{\{x\},\{x,y\}\}$ (H2)

Similarly,

27. $\vdash \{x,y\} \in \{\{x\},\{x,y\}\} \to$
    $(\{\{x\},\{x,y\}\} = \{\{u\},\{u,v\}\} \to \{x,y\} \in \{\{u\},\{u,v\}\})$ (H4)
28. $\vdash \{\{x\},\{x,y\}\} = \{\{u\},\{u,v\}\} \to$
    $\{x,y\} \in \{\{u\},\{u,v\}\}$ (MP, 26,27)
29. $\langle x,y\rangle = \langle u,v\rangle \vdash \{x,y\} \in \{\{u\},\{u,v\}\}$ (MP, 1,28)
30. $\vdash \{x,y\} \in \{\{u\},\{u,v\}\} \to (\{x,y\} = \{u\} \lor \{x,y\} = \{u,v\})$ (H3)

$\{x,y\} = \{u\}$ *or* $\{x,y\} = \{u,v\}$.

31. $\langle x,y\rangle = \langle u,v\rangle \vdash \{x,y\} = \{u\} \lor \{x,y\} = \{u,v\}$ (MP, 29,30)
32. $\langle x,y\rangle = \langle u,v\rangle \vdash \{x,y\} \neq \{u\} \to \{x,y\} = \{u,v\}$ (31, Def. $\lor$)
33. $\langle x,y\rangle = \langle u,v\rangle, \{x,y\} \neq \{u\} \vdash \{x,y\} = \{u,v\}$ (DT)
34. $\vdash \{x,y\} = \{u,v\} \to \{u,v\} = \{x,y\}$ (H5)
35. $\langle x,y\rangle = \langle u,v\rangle, \{x,y\} \neq \{u\} \vdash \{u,v\} = \{x,y\}$ (MP, 33,34)
36. $\langle x,y\rangle = \langle u,v\rangle \vdash \{x,y\} \neq \{u\} \to \{u,v\} = \{x,y\}$ (DT)
37. $\langle x,y\rangle = \langle u,v\rangle \vdash \{x,y\} = \{u\} \lor \{u,v\} = \{x,y\}$ (36, Def. $\lor$)
38. $\vdash \{x,y\} = \{u\} \to u = x$ (H9)
39. $\{x,y\} = \{u\} \vdash u = x$ (DT)
40. $\vdash \{x,y\} = \{u\} \to y = u$ (H10)
41. $\{x,y\} = \{u\} \vdash y = u$ (DT)
42. $\vdash y = u \to (u = x \to y = x)$ (H6)
43. $\{x,y\} = \{u\} \vdash u = x \to y = x$ (MP, 41,42)
44. $\{x,y\} = \{u\} \vdash y = x$ (MP, 38,43)
45. $\vdash y = x \to x = y$ (H5)
46. $\{x,y\} = \{u\} \vdash x = y$ (MP, 44,45)
47. $\vdash \{u,v\} = \{x\} \to x = u$ (H9)
48. $\{u,v\} = \{x\} \vdash x = u$ (DT)
49. $\vdash \{u,v\} = \{x\} \to v = x$ (H10)
50. $\{u,v\} = \{x\} \vdash v = x$ (DT)

3.2 Axiomatic Set Theory

51. $\vdash v = x \to (x = u \to v = u)$ (H6)
52. $\{u,v\} = \{x\} \vdash x = u \to v = u$ (MP, 50,51)
53. $\{u,v\} = \{x\} \vdash v = u$ (MP, 48,52)
54. $\vdash v = u \to u = v$ (H5)
55. $\{u,v\} = \{x\} \vdash u = v$ (MP, 53,54)
56. $\vdash y = u \to (u = v \to y = v)$ (H6)
57. $\{x,y\} = \{u\} \vdash u = v \to y = v$ (MP, 41,56)
58. $\{x,y\} = \{u\}, \{u,v\} = \{x\} \vdash y = v$ (MP, 55,57)
59. $\vdash x = y \to (y = u \to (x = y \land y = u))$ (TA1)
60. $\{x,y\} = \{u\} \vdash y = u \to (x = y \land y = u)$ (MP, 46,59)
61. $\{x,y\} = \{u\} \vdash x = y \land y = u$ (MP, 41,60)
62. $\vdash (x = y \land y = u) \to (u = v \to (x = y \land y = u \land u = v))$ (TA1)
63. $\{x,y\} = \{u\} \vdash u = v \to (x = y \land y = u \land u = v)$ (MP, 61,62)
64. $\{x,y\} = \{u\}, \{u,v\} = \{x\} \vdash$
    $x = y \land y = u \land u = v$ (MP, 55,63)   *If $\{u,v\} = \{x\}$ and $\{x,y\} = \{u\}$, then $x = y = u = v$;*
65. $\vdash (\{x,y\} = \{u\} \lor \{u,v\} = \{x,y\}) \to$
    $((\{u,v\} = \{x\} \lor \{u,v\} = \{x,y\}) \to$
    $((\{x,y\} = \{u\} \land \{u,v\} = \{x\}) \lor \{u,v\} = \{x,y\}))$ (TA2)
66. $\langle x,y \rangle = \langle u,v \rangle \vdash (\{u,v\} = \{x\} \lor \{u,v\} = \{x,y\}) \to$
    $((\{x,y\} = \{u\} \land \{u,v\} = \{x\}) \lor \{u,v\} = \{x,y\})$ (MP, 37,65)
67. $\langle x,y \rangle = \langle u,v \rangle \vdash$
    $(\{x,y\} = \{u\} \land \{u,v\} = \{x\}) \lor \{u,v\} = \{x,y\}$ (MP, 25,66)
68. $\langle x,y \rangle = \langle u,v \rangle \vdash$
    $\neg(\{x,y\} = \{u\} \land \{u,v\} = \{x\}) \to \{u,v\} = \{x,y\}$ (67, Def. $\lor$)
69. $\langle x,y \rangle = \langle u,v \rangle, \neg(\{x,y\} = \{u\} \land \{u,v\} = \{x\}) \vdash$ *if not, $\{u,v\} = \{x,y\}$.*
    $\{u,v\} = \{x,y\}$ (DT)
70. $\vdash x = u \to (\{u,v\} = \{x,y\} \to \{u,v\} = \{u,y\})$ (H7)
71. $\langle x,y \rangle = \langle u,v \rangle \vdash \{u,v\} = \{x,y\} \to \{u,v\} = \{u,y\}$ (MP, 17,70)
72. $\langle x,y \rangle = \langle u,v \rangle, \neg(\{x,y\} = \{u\} \land \{u,v\} = \{x\}) \vdash$
    $\{u,v\} = \{u,y\}$ (MP, 69,71)
73. $\langle x,y \rangle = \langle u,v \rangle \vdash$
    $\neg(\{x,y\} = \{u\} \land \{u,v\} = \{x\}) \to \{u,v\} = \{u,y\}$ (DT)
74. $\vdash y = v \to \{u,v\} = \{u,y\}$ (H12)
75. $\{x,y\} = \{u\}, \{u,v\} = \{x\} \vdash \{u,v\} = \{u,y\}$ (MP, 58,74)
76. $\{x,y\} = \{u\} \vdash \{u,v\} = \{x\} \to \{u,v\} = \{u,y\}$ (DT)

	77. ⊢ $\{x,y\} = \{u\} \to (\{u,v\} = \{x\} \to \{u,v\} = \{u,y\})$		(DT)
	78. ⊢ $(\{x,y\} = \{u\} \to (\{u,v\} = \{x\} \to \{u,v\} = \{u,y\})) \to$		
	$\quad (\{x,y\} = \{u\} \wedge \{u,v\} = \{x\} \to \{u,v\} = \{u,y\})$		(TA5)
	79. ⊢ $\{x,y\} = \{u\} \wedge \{u,v\} = \{x\} \to \{u,v\} = \{u,y\}$		(MP, 77,78)
	80. ⊢ $((\{x,y\} = \{u\} \wedge \{u,v\} = \{x\}) \to \{u,v\} = \{u,y\}) \to$		
	$\quad ((\neg(\{x,y\} = \{u\} \wedge \{u,v\} = \{x\}) \to \{u,v\} = \{u,y\}) \to$		
	$\quad \{u,v\} = \{u,y\})$		(TA3)
	81. ⊢ $(\neg(\{x,y\} = \{u\} \wedge \{u,v\} = \{x\}) \to \{u,v\} = \{u,y\})$		
	$\quad \to \{u,v\} = \{u,y\}$		(MP, 79,80)
Hence, $\{u,v\} = \{u,y\}$.	82. $\langle x,y \rangle = \langle u,v \rangle$ ⊢ $\{u,v\} = \{u,y\}$		(MP, 73,81)
	83. $v \neq u$ ⊢ $\{u,v\} = \{u,y\} \to y = v$		(H11)
So, if $v \neq u$, then $y = v$;	84. $\langle x,y \rangle = \langle u,v \rangle, v \neq u$ ⊢ $y = v$		(MP, 82,83)
	85. $\langle x,y \rangle = \langle u,v \rangle$ ⊢ $v \neq u \to y = v$		(DT)
	86. $v = u$ ⊢ $\{u,v\} = \{u,y\} \to y = v$		(H11)
if $v = u$, then $y = v$.	87. $\langle x,y \rangle = \langle u,v \rangle, v = u$ ⊢ $y = v$		(MP, 82,86)
	88. $\langle x,y \rangle = \langle u,v \rangle$ ⊢ $v = u \to y = v$		(DT)
	89. ⊢ $(v = u \to y = v) \to ((v \neq u \to y = v) \to y = v)$		(TA3)
	90. $\langle x,y \rangle = \langle u,v \rangle$ ⊢ $(v \neq u \to y = v) \to y = v$		(MP, 88,89)
Thus, in all cases, $y = v$.	91. $\langle x,y \rangle = \langle u,v \rangle$ ⊢ $y = v$		(MP, 85,90)
	92. ⊢ $x = u \to (y = v \to (x = u \wedge y = v))$		(TA1)
	93. $\langle x,y \rangle = \langle u,v \rangle$ ⊢ $y = v \to (x = u \wedge y = v)$		(MP, 17,92)
	94. $\langle x,y \rangle = \langle u,v \rangle$ ⊢ $x = u \wedge y = v$		(MP, 91,93)
	95. ⊢ $\langle x,y \rangle = \langle u,v \rangle \to x = u \wedge y = v$		(DT)
	96. ⊢ $\forall x \forall y \forall u \forall v (\langle x,y \rangle = \langle u,v \rangle \to x = u \wedge y = v)$		(G, 4 mal)

A great deal of work is behind us! Even if the content of the proven statement is unspectacular, the complexity of the proof is awe-inspiring. Without any doubt, the example highlights the difficulty of proving even seemingly apparent propositions. But rest assured. Nobody will ask you to derive complex theorems on such a deep level of abstraction, let alone come up with the idea of proclaiming the symbolic level of Zermelo-Fraenkel set theory as the tool of a new kind of mathematics. In their daily work, mathematicians from all disciplines will continue to carry out proofs on the level of abstraction we are familiar with, where formulas and colloquial formulations exist side by side. As humans, this is our only way to master the complexity of sophisticated theorems. The critical point here is the assurance that all conclusions of ordinary mathematics can ultimately be broken down to the symbolic manipulation of strings at the lowest level.

3.2.1.4 Embedding of the Natural Numbers

The previous section demonstrated the formalization of mathematical concepts within Zermelo-Fraenkel set theory through the example of relations and functions. This time, we aim to illustrate the expressive power of set theory by showing how to represent natural numbers. Fortunately, such a construction is relatively easy, as the following encoding already serves our needs (Figure 3.17):

 Definition 3.5 (Zermelo, 1908)

> The natural numbers are defined as the following sets:
> $$0 := \emptyset \qquad n+1 := \{n\}$$

Ernst Zermelo first investigated this natural number representation in 1908. It gave rise to Zermelo's famous number series Z_0:

> "The set Z_0 contains the elements 0, $\{0\}$, $\{\{0\}\}$, and so forth, and it may be called the *number sequence*, because its elements can take the place of the numerals. It is the simplest example of a denumerably infinite set."
>
> Ernst Zermelo, 1908 [232]

A more modern embedding of the natural numbers traces back to John von Neumann (Figure 3.18). Because Ernst Zermelo came up with this number series almost simultaneously, some authors speak of the Neumann-Zermelo number series.

 Definition 3.6 (von Neumann, 1923)

> The natural numbers are defined as the following sets:
> $$0 := \emptyset \qquad n+1 := n \cup \{n\}$$

Figure 3.19 compares the initial parts of both number series. At first glance, Zermelo's representation appears superior, as it embeds the natural numbers more concisely. A second glance, however, reveals the

■ $n = 0$

■ From n to $n+1$

Figure 3.17: Construction of Zermelo's number sequence

■ $n = 0$

■ From n to $n+1$

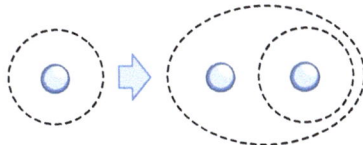

Figure 3.18: Construction of von Neumann's number sequence

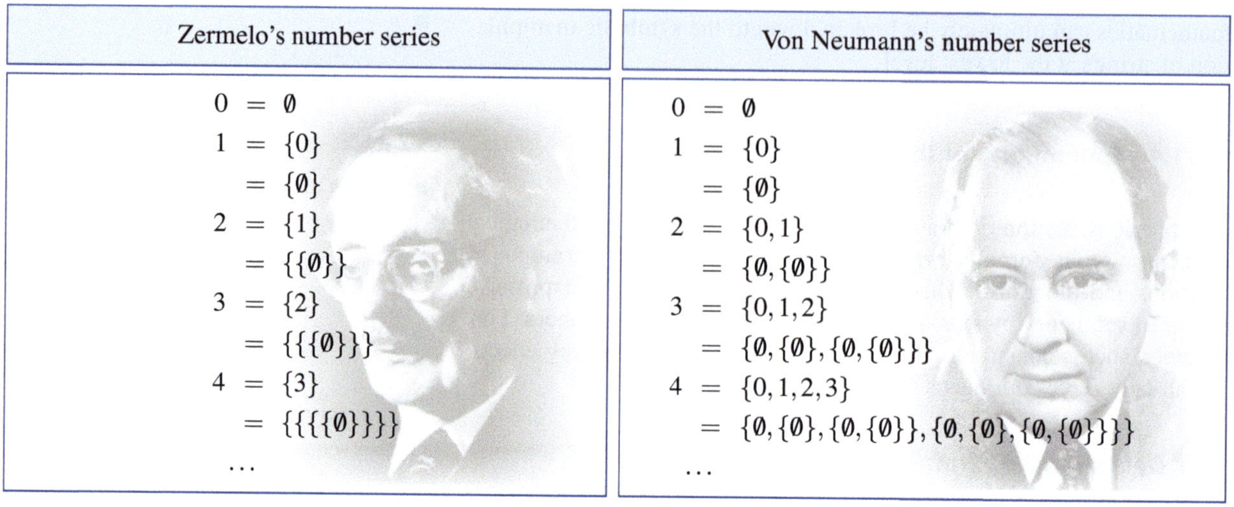

Zermelo's number series	Von Neumann's number series
$0 = \emptyset$ $1 = \{0\}$ $ = \{\emptyset\}$ $2 = \{1\}$ $ = \{\{\emptyset\}\}$ $3 = \{2\}$ $ = \{\{\{\emptyset\}\}\}$ $4 = \{3\}$ $ = \{\{\{\{\emptyset\}\}\}\}$ $\ldots$	$0 = \emptyset$ $1 = \{0\}$ $ = \{\emptyset\}$ $2 = \{0, 1\}$ $ = \{\emptyset, \{\emptyset\}\}$ $3 = \{0, 1, 2\}$ $ = \{\emptyset, \{\emptyset\}, \{\emptyset, \{\emptyset\}\}\}$ $4 = \{0, 1, 2, 3\}$ $ = \{\emptyset, \{\emptyset\}, \{\emptyset, \{\emptyset\}\}, \{\emptyset, \{\emptyset\}, \{\emptyset, \{\emptyset\}\}\}\}$ $\ldots$

Figure 3.19: Representing natural numbers as sets according to Ernst Zermelo and John von Neumann

elegance of von Neumann's number series. Two properties are particularly noteworthy:

- The set representation of a number n is contained in the set representation of a number m if and only if $n < m$. As a result, von Neumann's number series is totally ordered with respect to the element relation '$\in$'. In the same way, it is totally ordered with respect to '$\subseteq$'. The subset property is equivalent to $n \leq m$.

- The set representation of a number n comprises precisely n elements, ensuring that its cardinality corresponds to the number it represents. Consequently, von Neumann's number sequence elucidates both the ordinal and the cardinal aspects of natural numbers.

Von Neumann's construction of natural numbers is a particular case of a more general construction scheme examined thoroughly in the next section. It opens the door to the realm of *ordinal numbers*, a fascinating world of numbers so immense that we can conceive only the tiniest of them with our limited human imagination.

3.2.2 Ordinal Numbers

> *"Every simply ordered aggregate M has a definite ordinal type $\overline{M}$; this type is the general concept which results from M if we abstract from the nature of its elements while retaining their order of precedence, [...]. We call the ordinal type of a well-ordered aggregate F its 'ordinal number'."*
>
> Georg Cantor, 1897 [23]

The previous section has paved the way for representing natural numbers as sets, and, as we have seen, von Neumann's number series has proven particularly fruitful for this purpose. Here, the empty set represents the number 0, and $n \cup \{n\}$ denotes the successor of n. This section will generalize the idea of von Neumann's number series and lead us straight into the world of ordinal numbers. We will become familiar with a new type of number that allows us to count far beyond the limits of natural numbers.

The journey into the realms of ordinal numbers is a journey towards infinity, and we do not dare to enter without a warning. The path we are about to take leads right into darkness. Although our imagination stands faithfully by our side at the beginning, we will soon experience it slowly fading and finally failing. The numbers awaiting us are so complex that our limited human intuition does not give us the slightest chance of grasping them, even rudimentarily.

3.2.2.1 Definition and Properties

Let our journey begin. We gently step forward and start with the definition of *transitive sets*:

 Definition 3.7 (Transitive Set)

A set z is called *transitive* if the following holds:

- From $x \in y$ and $y \in z$, it follows that $x \in z$.

Or, equivalently:

- From $y \in z$, it follows that $y \subseteq z$.

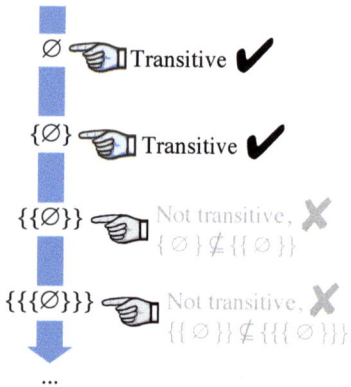

Figure 3.20: Only the first two elements of Zermelo's number series are transitive sets.

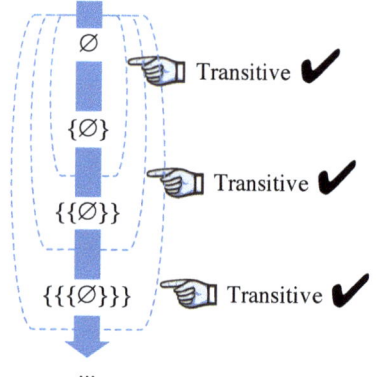

Figure 3.21: Combining the elements of any initial piece of Zermelo's number series yields a transitive set.

The first defining property has a familiar face. If we replaced the element operator '$\in$', for example, with the symbol '$\subseteq$', the wording of this definition would reduce to the well-known transitivity of the subset relation. However, the second and equivalent definition quickly clarifies that transitive sets are truly peculiar entities; they have the strange property that each element is simultaneously a subset.

To get a glimpse of what such sets look like, let us revisit Zermelo's number series Z_0. It will help us recognize how transitive sets differ from their non-transitive counterparts. The series starts with the empty set $\emptyset$. The empty set trivially fulfills the relationship required in Definition 3.7 and is thus transitive. The set $\{\emptyset\}$ is at the second place. It is also transitive, as its only element $\emptyset$ is a subset of every other set and thus a subset of $\{\emptyset\}$. The set $\{\{\emptyset\}\}$, the Zermelo representation of the number 2, on the other hand, is not transitive since the element $\{\emptyset\}$ is not a subset of $\{\{\emptyset\}\}$. The same applies to Zermelo's set representations of other natural numbers (Figure 3.20).

On the other hand, we once again arrive at a transitive set by considering any initial piece of Z_0, that is, a set of the following form:

$$\{\emptyset, \{\emptyset\}, \{\{\emptyset\}\}, \{\{\{\emptyset\}\}\}, \ldots, \{\ldots\{\{\{\emptyset\}\}\}\ldots\}\}$$

Each of its elements is a subset of itself.

This first impression shall suffice for the moment. We will now work out some elementary relationships, starting with the proof of several closure properties of transitive sets:

Theorem 3.4

1. If x is a transitive set, so is $x \cup \{x\}$.

2. If x is a set of transitive sets, then $\bigcup x$ is also a transitive set.

3. If x is a set of transitive sets, then $\bigcap x$ is also a transitive set.

4. If x is a transitive set, so is $\mathcal{P}(x)$.

Proof: 1. Let $y \in x \cup \{x\}$. Then either $y = x$ or $y \in x$. In the first case, $y \subseteq x$ trivially holds. In the second case, $y \subseteq x$ also holds due to the transitivity of x. In both cases, it follows that $y \subseteq x$, and thus $y \subseteq x \cup \{x\}$.
2. Let $y \in \bigcup x$. Then y is contained in at least one set x' with $x' \in x$. Since x' is transitive, $y \subseteq x'$ holds. Consequently, $y \subseteq \bigcup x$. 3. Let $y \in \bigcap x$. Then y is contained in all sets x' with $x' \in x$. Since x' is transitive, $y \subseteq x'$ holds.

Consequently, $y \subseteq \bigcap x$. 4. Every element $y \in \mathcal{P}(x)$ is, by definition, a subset of x. Hence, every element $y' \in y$ is also an element of x. Since x is transitive, $y' \subseteq x$ follows, and therefore $y' \in \mathcal{P}(x)$. Consequently, $y \subseteq \mathcal{P}(x)$, which identifies the power set as transitive. □

From this theorem, we can conclude that Zermelo's number series Z_0 is a transitive set.

We are now well-prepared to formally define the concept of an ordinal number:

 Definition 3.8 (Ordinal Number)

A transitive set x, whose elements are also transitive, is called an *ordinal number*, or *ordinal* for short.

In the upcoming considerations, we will follow the convention of denoting ordinal numbers with the lowercase Greek letters α, β, etc. Remember that ordinal numbers are not numbers in the traditional sense, even if their name provokes this misunderstanding. Definition 3.8 undoubtedly makes it clear: Ordinal numbers are sets!

Figures 3.22 and 3.23 emphasize that only the first two elements of Zermelo's number series are ordinal numbers. Before we turn to larger numbers, we want to elaborate on some consequences of Definition 3.8. First of all, we note that most assertions of Theorem 3.4 transfer one-to-one to ordinal numbers:

 Theorem 3.5

1. If α is an ordinal number, so is $\alpha \cup \{\alpha\}$.
2. If x is a set of ordinal numbers, then $\bigcup x$ is an ordinal number.
3. If x is a set of ordinal numbers, then $\bigcap x$ is an ordinal number.

Proof: The transitivity of $\alpha \cup \{\alpha\}$, $\bigcup x$, and $\bigcap x$ follows directly from Theorem 3.4. It remains to show that these sets contain only elements that are themselves transitive, which follows directly from the definition of an ordinal number.

The subsequent theorems highlight several properties that give us an intuitive understanding of ordinal numbers. Their proofs are technical and

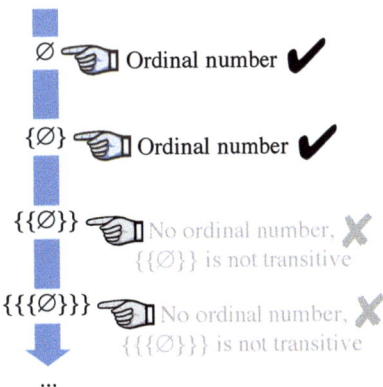

Figure 3.22: Only the first two elements of Zermelo's number series are ordinal numbers.

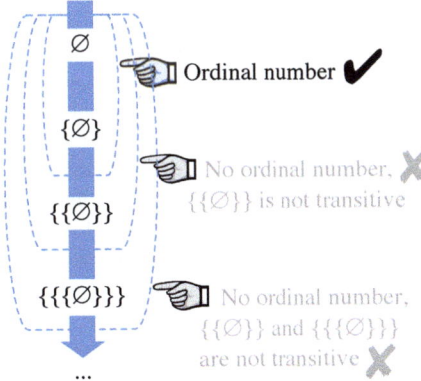

Figure 3.23: Although all initial segments of Zermelo's number series are transitive, they violate the ordinal number property if they contain more than two elements.

> All roads lead to Rome. This idiom applies to ordinal numbers probably more than to any other mathematical concept. The literature contains a multitude of definitions, all of which appear very different from the outset but ultimately turn out to be equivalent.
>
> The first to define an ordinal number as a transitive set of transitive sets was Kurt Gödel in 1937. In 1941, Paul Bernays presented an alternative characterization. Following his definition, an ordinal number α is a transitive set with the property that every transitive proper subset of α is itself an element of α [10].
>
> Other authors define ordinal numbers as transitive sets well-ordered under set inclusion. This characterization comes closer to Georg Cantor's original definition, which saw ordinal numbers as canonical representatives of well-ordered sets [22]. In Cantor's sense, ordinal numbers emerge from a process that reduces a set to its ordering structure by abstracting away the concrete nature of the elements. The historical definition of the ordinal number is significant, and we will return to it later in this chapter.

undoubtedly require a certain amount of hard work. However, there is a simple reason why we are proceeding so formally here. The world of ordinal numbers extends beyond our intuition, and we would soon find ourselves on unstable ground if we dispense with a mathematically precise proof. Nevertheless, you can safely skip the details at first reading and return to them later.

 Theorem 3.6 (Trichotomy Theorem)

> Any two ordinal numbers α and β satisfy:
> $$\alpha \in \beta \quad \text{or} \quad \alpha = \beta \quad \text{or} \quad \beta \in \alpha$$

Proof: We write $\mathfrak{T}(\alpha,\beta)$ for $(\alpha \in \beta \vee \alpha = \beta \vee \beta \in \alpha)$ and prove the assumption by contradiction. So let us assume that $\mathfrak{T}(\alpha,\beta)$ does not hold for all ordinal numbers α and β. In this case, the axiom of foundation guarantees the existence of a smallest α with respect to '$\in$' (we call it α_0), for which $\mathfrak{T}(\alpha,\beta)$ is false for at least one β. Now, analogously, we can conclude that there must be a smallest β (we call it β_0), for which $\mathfrak{T}(\alpha_0,\beta)$ becomes false. Then, the following holds:

a) $\neg\mathfrak{T}(\alpha_0,\beta_0)$ (due to the specific choice of α_0 and β_0).

b) $\alpha \in \alpha_0$ implies $\mathfrak{T}(\alpha,\beta_0)$ (due to the minimality of α_0).

c) $\beta \in \beta_0$ implies $\mathfrak{T}(\alpha_0,\beta)$ (due to the minimality of β_0).

We will now show that $\mathfrak{T}(\alpha_0,\beta_0)$ must nevertheless be true, contrary to a). We distinguish three cases:

1. $\alpha_0 = \beta_0$. Then $\mathfrak{T}(\alpha_0,\beta_0)$ is trivially true.

2. There exists an $\alpha \in \alpha_0$ with $\alpha \notin \beta_0$. Due to b), $\mathfrak{T}(\alpha,\beta_0)$ holds, thus $\beta_0 = \alpha$ or $\beta_0 \in \alpha$. In both cases, $\beta_0 \in \alpha_0$ (in the latter case due to the transitivity of α_0) and thus $\mathfrak{T}(\alpha_0,\beta_0)$.

3. There exists a $\beta \in \beta_0$ with $\beta \notin \alpha_0$. Due to c), $\mathfrak{T}(\alpha_0,\beta)$ holds, thus $\alpha_0 = \beta$ or $\alpha_0 \in \beta$. In both cases, $\alpha_0 \in \beta_0$ (in the latter case due to the transitivity of β_0) and thus $\mathfrak{T}(\alpha_0,\beta_0)$. □

The trichotomy of ordinal numbers is essential. It guarantees that any two ordinal numbers can be compared, thus ensuring that ordinals are totally ordered with respect to '$\in$'. From now on, we also express this property symbolically:

3.2 Axiomatic Set Theory

 Definition 3.9 (Ordinal Ordering)

On the ordinal numbers, we establish the following order:

$$\alpha < \beta :\Leftrightarrow \alpha \in \beta$$

The subsequent theorem reveals another unique property of ordinal numbers. It states that the subset relation and the element relation coincide:

 Theorem 3.7 (Equivalence of '$\in$' and '$\subset$')

For any two ordinal numbers α and β, it holds that:

$$\alpha \in \beta \text{ if and only if } \alpha \subset \beta$$

Proof: If $\alpha \in \beta$, then the transitivity of β implies $\alpha \subseteq \beta$. Since a set cannot contain itself, $\alpha \neq \beta$ follows, thus $\alpha \subset \beta$. Now, suppose $\alpha \subset \beta$. This implies that neither $\alpha = \beta$ nor $\beta \in \alpha$ (in the latter case, we would have $\beta \in \beta$). According to Theorem 3.6, it follows that $\alpha \in \beta$, which was to be proven. □

The next theorem uncovers an essential closure property. Ordinal numbers remain among themselves!

 Theorem 3.8 (Closure under '$\in$')

For any ordinal number β, it holds that:

Every element $\alpha \in \beta$ is also an ordinal number.

Proof: Let $\alpha \in \beta$. All elements of an ordinal number are transitive, so α is transitive, too. Due to the transitivity of β, every element $x \in \alpha$ is also contained in β. Therefore, every element of α is also transitive, identifying α as an ordinal number. □

 Be careful never to speak of the set of all ordinal numbers! The Italian mathematician Cesare Burali-Forti discovered as early as 1897 that combining all ordinal numbers into a single set opens the door to paradoxes.
To evoke the *Burali-Forti paradox*, let us assume that the set of all ordinal numbers exists. Since this set, we call it O, contains only ordinal numbers, it would be a set of transitive sets. According to Theorem 3.8, all elements of an ordinal number are ordinal numbers, too, so every element of O would be a subset of O. In short, the set of all ordinal numbers would be transitive and thus an ordinal number itself. It would be a number satisfying $O \in O$, which is impossible according to the regularity axiom. Overall, the union of all ordinal numbers proves too large to exist as a totality.
Though the ordinal numbers do not constitute a set, they form a proper class, referred to in the literature with the capital Greek letter omega:

$$\Omega = \text{Class of all ordinal numbers}$$

Be careful not to confuse the symbol Ω with *Chaitin's constant* Ω, which you will become familiar with in Chapter 6. In both cases, the same letter denotes something completely different.

3.2.2.2 Towards Infinity

> "Moria... You fear to go into those mines. The dwarves delved too greedily and too deep. You know what they awoke in the darkness of Khazad-dûm... shadow and flame."
>
> From Lord of the Rings

We begin this section by constructing several ordinal numbers in real terms. The simplest ordinal is the one with no elements, the empty set ∅. The appearance of ordinal numbers with one element is also evident. According to the trichotomy theorem, ∅ must be an element of these numbers, so only {∅} comes into question. We have already identified this set as an ordinal number above. A similar argument exposes the ordinal numbers with more elements. Due to trichotomy, all ordinals constructed so far must be elements of the new set. Hence, for each $n \in \mathbb{N}$, there is a uniquely determined ordinal number with n elements. Figure 3.24 reveals that the constructed sets are old acquaintances: The finite ordinal numbers are precisely the elements of von Neumann's number series presented in Section 3.2.1.4.

Likewise, the construction highlights that counting within ordinal numbers is possible. Starting from an ordinal number x, we get to the next by augmenting x with all previous ordinals using the construction rule $x \cup \{x\}$. Counting within the ordinals is so central that we provide a separate definition for this term:

Definition 3.10 (Successor of an Ordinal Number)

Let α be an ordinal number. The ordinal number

$$s(\alpha) := \alpha \cup \{\alpha\}$$

is called the *successor* of α.

Theorem 3.5 guarantees that for each ordinal α, the set $\alpha \cup \{\alpha\}$ is again an ordinal number. In other words, we can continue counting ad infinitum. The transition from α to $s(\alpha)$ is a specific case of *ordinal addition*, which we will introduce below. In anticipation of the formal definiton, we already write $\alpha + 1$ for $s(\alpha)$ and $\alpha + 2$ for $s(s(\alpha))$, etc.

At this point, we want to leave the finite ordinals behind and construct the first infinite set satisfying all requirements for being an ordinal num-

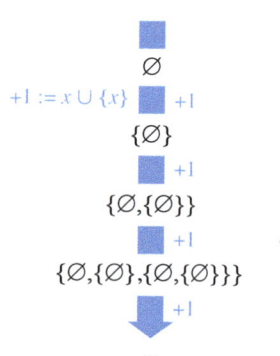

Figure 3.24: The systematic construction of all finite ordinal numbers yields the von Neumann number series.

3.2 Axiomatic Set Theory

ber. Theorem 3.5 guides the way. Forming the union of all finite ordinals, as shown in Figure 3.25, lets us arrive at another ordinal, the famous number:

$$\omega := \{\, \emptyset, \{\emptyset\}, \{\emptyset, \{\emptyset\}\}, \{\emptyset, \{\emptyset\}, \{\emptyset, \{\emptyset\}\}\}, \ldots \,\}$$

ω is relevant in several respects. For one thing, it is the smallest *transfinite ordinal*, that is, the smallest number comprising an infinite number of elements. For another, it has the unique property of not being the successor of any other ordinal; no α satisfies $s(\alpha) = \omega$. Ordinal numbers bearing this property are so significant that they carry their own name:

Definition 3.11 (Limit Ordinal)

An ordinal number is a *limit ordinal*, if it is not the successor of any other ordinal number.

Another aspect is just as remarkable. ω is identical to von Neumann's number series, which, according to what we detailed in Section 3.2.1.4, corresponds to the set of natural numbers. Accordingly, we can write down ω in the following form (Figure 3.26):

$$\omega = \{0, 1, 2, 3, \ldots\}$$

Consequently, we can treat not only every natural number as an ordinal number but also the set of natural numbers itself.

Now comes the decisive step: By applying the successor operation, we can form the new ordinals

$$\omega + 1, \omega + 2, \ldots,$$

which means nothing less than having counted beyond the natural numbers. This is, in fact, a remarkable achievement: the theory of ordinal numbers enables us to transcend the limits of natural numbers. The chosen path leads us straight into a universe that extends far beyond the boundaries of the natural numbers. Let's take a moment to glimpse the immense scale of this newly discovered world.

Starting from ω, we can create the following numbers by repeated application of the successor operation:

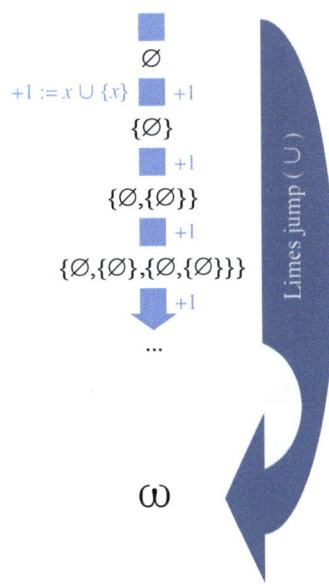

Figure 3.25: We reach the first infinite ordinal number ω by taking the union of all finite ordinals. It is identical to the von Neumann ordinal sequence and corresponds to the set $\mathbb{N}$ of natural numbers.

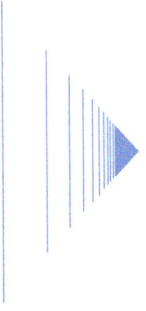

Figure 3.26: Visualization of the first transfinite ordinal ω. Each bar represents a natural number. The sequence of bars extends to infinity.

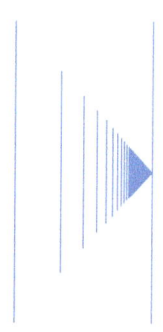

Figure 3.27: Visualization of $\omega+1$

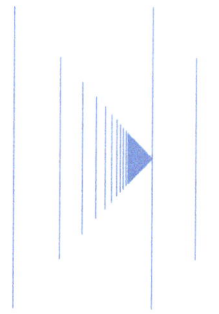

Figure 3.28: Visualization of $\omega+2$

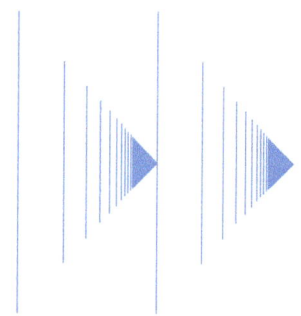

Figure 3.29: Visualization of $\omega+\omega$

$$\omega = \{0,1,2,\ldots\}$$
$$\omega+1 = \{0,1,2,\ldots,\omega\}$$
$$\omega+2 = \{0,1,2,\ldots,\omega,\omega+1\}$$
$$\ldots$$
$$\omega+(n+1) = \{0,1,2,\ldots,\omega,\omega+1,\omega+2,\ldots,\omega+n\}$$
$$\ldots$$

Figures 3.27 and 3.28 visualize the result.

Another limit jump leads us to the number $\omega+\omega$, depicted in Figure 3.29:

$$\omega+\omega = \bigcup_{n\in\mathbb{N}} (\omega+n)$$

Denoting the new number as $\omega+\omega$ seems a natural choice but still lacks a formal foundation. The following definitions will provide such a basis by extending the familiar construction rule to a general scheme for *ordinal addition*:

$$\alpha+0 := \alpha$$
$$\alpha+s(\beta) := s(\alpha+\beta)$$
$$\alpha+\gamma := \bigcup_{\beta<\gamma} \alpha+\beta \quad \text{(for all limit ordinals } \gamma\text{)}$$

This scheme allows us to continue our journey and aim for even larger numbers. These are the numbers $\omega+\omega+\ldots+\omega$, abbreviated by $\omega\cdot n$. In particular, we obtain the following ordinals:

$$\omega\cdot 2 = \{0,1,\ldots,\omega,\omega+1,\ldots\}$$
$$\omega\cdot 3 = \{0,1,\ldots,\omega,\omega+1,\ldots,\omega\cdot 2,\omega\cdot 2+1,\ldots\}$$
$$\ldots$$
$$\omega\cdot(n+1) = \omega\cdot n \cup \{\omega\cdot n,\omega\cdot n+1,\omega\cdot n+2,\ldots\}$$
$$\ldots$$

Another limit jump takes us to

$$\omega\cdot\omega = \bigcup_{n\in\mathbb{N}} (\omega\cdot n).$$

Again, we lack the formal justification to refer to the obtained number as $\omega\cdot\omega$. The following definition of *ordinal multiplication* closes the gap:

$$\alpha \cdot 0 := 0$$
$$\alpha \cdot s(\beta) := \alpha \cdot \beta + \alpha$$
$$\alpha \cdot \gamma := \bigcup_{\beta < \gamma} \alpha \cdot \beta \quad \text{(for all limit ordinals } \gamma\text{)}$$

Figure 3.30 visualizes the set $\omega \cdot \omega$. Inside the resulting structure, the natural numbers repeat infinitely often.

Let's keep going and aim for even larger numbers. Next, we come across the numbers $\omega \cdot \omega \cdot \ldots \cdot \omega$, which we call ω^n:

$$\omega^2 = \omega \cdot \omega$$
$$\omega^3 = \omega^2 \cdot \omega$$
$$\omega^4 = \omega^3 \cdot \omega$$
$$\ldots$$
$$\omega^{n+1} = \omega^n \cdot \omega$$
$$\ldots$$

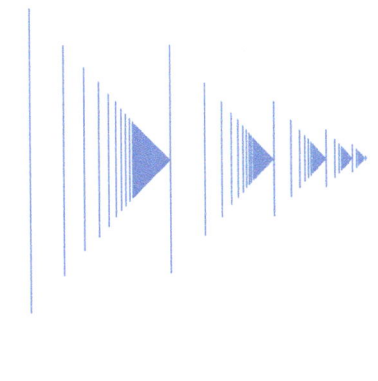

Figure 3.30: Visualization of $\omega \cdot \omega$

In almost familiar fashion, let's perform the next limit jump. Doing so yields the incredibly huge number

$$\omega^\omega = \bigcup_{n \in \mathbb{N}} \omega^n.$$

Figure 3.31 suggests a way to visualize this ordinal.

After establishing addition and multiplication, the time has come to introduce ordinal exponentiation. Based on what we have worked out above, the definition almost writes itself:

$$\alpha^0 := 1$$
$$\alpha^{s(\beta)} := \alpha^\beta \cdot \alpha$$
$$\alpha^\gamma := \bigcup_{\beta < \gamma} \alpha^\beta \quad \text{(for all limit ordinals } \gamma\text{)}$$

Even if the ordinal numbers constructed so far have already strained our intuition, our journey is not over. We can readily form further ordinal numbers that leave ω^ω far behind. These are:

$$\omega^{\omega^\omega}, \omega^{\omega^{\omega^\omega}}, \omega^{\omega^{\omega^{\omega^\omega}}}, \omega^{\omega^{\omega^{\omega^{\omega^\omega}}}}, \omega^{\omega^{\omega^{\omega^{\omega^{\omega^\omega}}}}}, \omega^{\omega^{\omega^{\omega^{\omega^{\omega^{\omega^\omega}}}}}}, \ldots,$$

Let's make the next limit jump by joining all ω-towers:

$$\varepsilon_0 := \bigcup_{n \in \mathbb{N}} \left. \omega^{\omega^{\cdot^{\cdot^{\cdot^\omega}}}} \right\} n \text{ times}$$

Resist the temptation to apply familiar arithmetic rules to the addition and multiplication of ordinal numbers! Contrary to what we are used to when dealing with natural, rational, or real numbers, neither the addition nor the multiplication of ordinal numbers is commutative. In general, the following holds:

$$\alpha + \beta \neq \beta + \alpha, \quad \alpha \cdot \beta \neq \beta \cdot \alpha$$

A simple calculation shows why. For instance, $\omega + 1 \neq \omega$, but at the same time,

$$1 + \omega = \bigcup_{\beta < \omega} 1 + \beta = \omega.$$

The situation is similar for ordinal multiplication. On the one hand, $\omega \cdot 2 \neq \omega$, on the other hand, we have

$$2 \cdot \omega = \bigcup_{\beta < \omega} 2 \cdot \beta = \omega.$$

Figure 3.31: Visualization of the ordinal number ω^ω with an infinitely rotating spiral [223]. To understand its structure, let's first look at the first two revolutions:

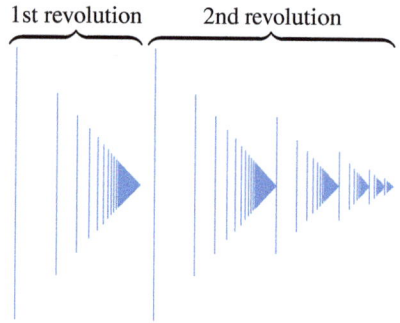

We are already familiar with both structures; they are identical to those from Figures 3.26 and 3.30, corresponding to the ordinal numbers ω and ω^2, respectively. The rest of the structure is now obvious: Each revolution represents an ordinal number of the form ω^n. The third revolution corresponds to ω^3, the fourth to ω^4, and so on. As the spiral turns infinitely often, it corresponds to the union of all ordinal numbers ω^n and therefore to the number ω^ω.

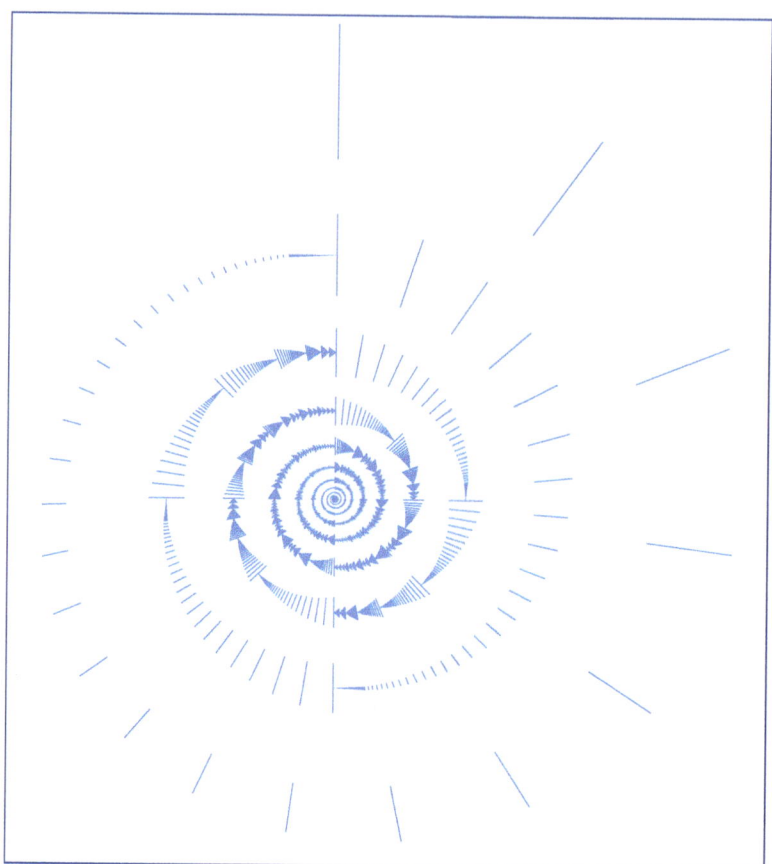

By doing so, we arrive at a truly colossal number that is fascinating in many respects. For example, it is the first ordinal satisfying the *fixed point property*

$$\omega^{\varepsilon_0} = \varepsilon_0.$$

The property is almost impossible to grasp with human intuition. Calculating with natural numbers has taught us that exponentiation with large numbers leads to ever larger numbers, and the size increase is dramatic if we raise the exponent even slightly. The number ε_0 breaks this rule. In a sense, it is so incomprehensibly large that it also includes its powers. Note that the inclusion does not conceal an antinomy of the Russell type. ε_0 is a well-defined set of ordinal numbers and does not contain itself. It is even possible to consecutively number the elements of ε_0, which means that the set ε_0, unimaginably complex as it is, is still countable.

ε_0 has another striking property. All numbers between ω and ε_0 can be reached through a finite number of additions and exponentiations of ω. More specifically, we can express any ordinal number α with $\omega < \alpha < \varepsilon_0$ in the form

$$\alpha = \omega^{\beta_0} + \omega^{\beta_1} + \ldots + \omega^{\beta_n} \tag{3.4}$$

where $\beta_1, \ldots, \beta_n$ are ordinal numbers smaller than α. By additionally permitting constant factors, (3.4) translates into the so-called *Cantor normal form*:

Theorem 3.9 (Cantor Normal Form)

> Every ordinal number α with $0 < \alpha < \varepsilon_0$ can be written in the form
>
> $$\alpha = c_0 \cdot \omega^{\beta_0} + c_1 \cdot \omega^{\beta_1} + \ldots + c_n \cdot \omega^{\beta_n}$$
>
> with $\alpha > \beta_0 > \beta_1 > \ldots > \beta_n$ and $c_0, \ldots, c_n \in \mathbb{N}$.

Waiving the restriction $\alpha > \beta_0 > \beta_1 > \ldots > \beta_n$ lets us write ε_0 in Cantor normal form, too. The result is the fixed point representation

$$\varepsilon_0 = \omega^{\varepsilon_0}$$

encountered above. In this case, however, the Cantor normal form does not yield any simplification. In fact, it is impossible to represent ε_0 in the form (3.4). ε_0 is the smallest number losing this property.

Cantor had grouped ordinal numbers into classes according to their cardinality, upon which he later built cardinal number theory. For instance, all finite ordinal numbers belong to the first number class, and all countably infinite numbers to the second. ω is the smallest ordinal number in the second class. Cantor observed that addition, multiplication, and exponentiation do not break out of the respective class. These operations map finite sets to finite sets, countable sets to countable sets, and so on. Surprisingly, though, is the fact that we cannot reach every element of the respective class with these operations. Starting from ω, we always remain within ε_0 by applying a finite number of additions, multiplications, and exponentiations. Nevertheless, it is easy to see that there must be other countable ordinal numbers beyond ε_0. We get such numbers with $\varepsilon_0 + 1, \varepsilon_0 + 2$, etc. by simply counting on.

Resist the temptation to visualize these numbers! The set ε_0 is already beyond our imagination, so we will only sketch how the rest of our journey unfolds. As Figure 3.32 shows, numerous other ε-numbers exist

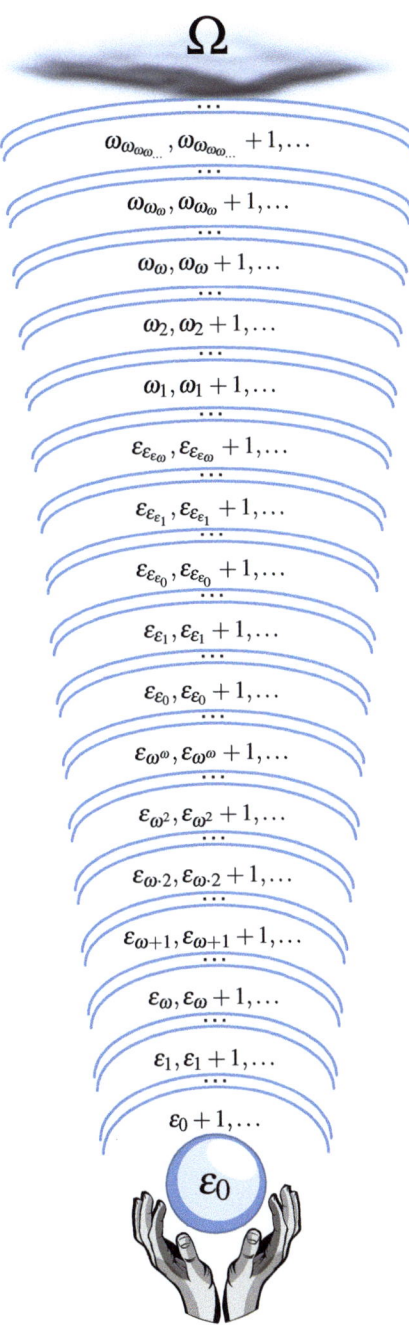

Figure 3.32: Some ordinals beyond ε_0

beyond ε_0, which are also a solution of the fixed point equation $\omega^x = x$. These numbers belong to the second number class, which means they are still countable sets. In general, every number ε_α whose ordinal index is itself countable is indeed a countable set.

Moving on to ever larger numbers, we eventually reach ω_1, the smallest representative of the third number class. It is the first uncountable ordinal number and arises from the union of all countable ordinals. It is followed by the numbers $\omega_2, \omega_3, \ldots$, etc. As unimaginably large as the number ε_0 is, it appears like a single drop in the ocean of infinity against the numbers waiting at the horizon. We don't want to stretch our imaginations any further. The ordinal numbers have already lured us so far into the realm of the infinite that we can hardly grasp even a tiny bit of what we encounter.

Our strength is also exhausted in other respects. Between ε_0 and ω_1 there can be numbers whose existence can neither be proven nor disproven in Zermelo-Fraenkel set theory. It seems as if we lose our footing beyond ε_0; without sight, we wander in a world in which we can not even be sure of the existence of the objects we touch.

3.2.2.3 Order Types and Well-Orders

In Section 3.2.2.1, we outlined that the element relation '$\in$' imposes an ordering on the ordinal numbers. In particular, we agreed on the following definition:

$$\alpha < \beta \ :\Leftrightarrow\ \alpha \in \beta$$

It is a fundamental principle of ordinal number theory that this definition leads to a well-ordering. Let us briefly explain why. First, the ordering is total, as we can always compare two ordinal numbers, α and β, due to trichotomy. Furthermore, every non-empty set of ordinals contains a minimal element. This property follows immediately from the fact that we can rewrite every infinitely descending chain of the form

$$\alpha_1 > \alpha_2 > \alpha_3 > \alpha_4 \ldots$$

into an equivalent chain of the form

$$\alpha_1 \ni \alpha_2 \ni \alpha_3 \ni \alpha_4 \ldots$$

The well-ordering property is thus a direct consequence of the axiom of foundation, which prevents the construction of infinitely descending $\in$-chains.

3.2 Axiomatic Set Theory

The following theorem, which we will accept without formal proof, emphasizes the strong interrelation between well-orderings and ordinal numbers even further:

Theorem 3.10 (Isomorphism Theorem for Ordinal Numbers)

> Every well-ordered set is order-isomorphic to exactly one ordinal.

First, we need to elucidate the term *order isomorphism*. As outlined in Figure 3.33, two orderings, $(M_1, <_1)$ and $(M_2, <_2)$, are called *order-isomorphic* if there is an order-preserving bijective function

$$f : M_1 \to M_2.$$

We speak of an *order-preserving function* if it satisfies the following:

$$x <_1 y \Leftrightarrow f(x) <_2 f(y) \quad \text{for all } x, y \in M_1$$

Accordingly, two sets are order isomorphic if they have the same cardinality and their elements have the same order structure. The specific nature of the elements in M_1 and M_2 is irrelevant, as the isomorphism function f abstracts from them.

Now, we are in a position to understand the isomorphism theorem in its full glory. First, it states that there is an ordinal number with the same order structure for every well-ordered set. Second, it states that this number is unique. Consequently, we may regard every ordinal number as the representative of a specific well-ordering. Conversely, we can assign an order type to each set M in the form of a unique ordinal. This order type is "the general concept which results from M if we abstract from the nature of its elements while retaining their order of precedence." I did not end the last sentence with my own words. They are the words of Georg Cantor, taken from the opening quotation of Section 3.2.2. The original idea Cantor had associated with the ordinal numbers is now evident. He approached them not through the concept of transitive sets, as we did. It was his meticulous studies on the nature of orderings and well-orderings that led him to discover this truly fascinating world of giant numbers.

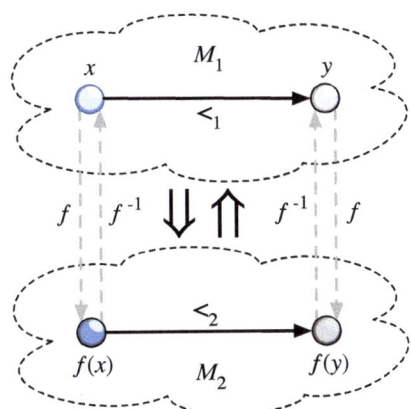

Figure 3.33: Two sets, M_1 and M_2, are order-isomorphic if there exists a bijective function that assigns all elements of M_1 to M_2 in an order-preserving manner: $x <_1 y$ implies $f(x) <_2 f(y)$ and vice versa.

Examples

Let us examine some specific arrangements of natural numbers and determine which ordinal numbers describe their order types.

0, 1, 2, 3, 4, 5, 6, 7, ...

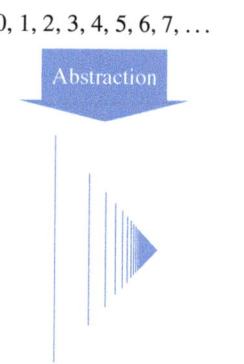

Figure 3.34: Well-ordering of the natural numbers with order type ω

1, 2, 3, 4, 5, 6, 7, ..., 0

Figure 3.35: Well-ordering of the natural numbers with order type $\omega + 1$

0, 2, 4, 6, ..., 1, 3, 5, 7, ...

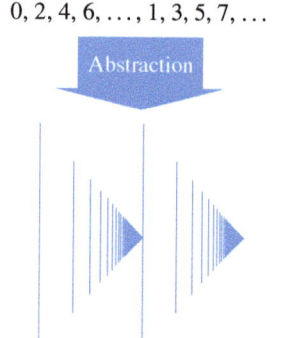

Figure 3.36: Well-ordering of the natural numbers with order type $\omega \cdot 2$

■ **Example 1** (Figure 3.34)

This example leaves the elements of $\mathbb{N}$ in their natural order. We already know the order type of this number series; it is the smallest transfinite ordinal ω.

We can rearrange the natural numbers in various ways without losing the well-ordering property. The following examples show how.

■ **Example 2** (Figure 3.35)

By definition, the number 0 is larger than all other natural numbers in this arrangement and, therefore, does not appear at the beginning but at the end of the number sequence. All other natural numbers retain their normal position. It takes little time to find the order type of this arrangement. It is described by $\omega + 1$, the direct successor of ω.

■ **Example 3** (Figure 3.36)

This example divides the natural numbers into two sequences. The first includes all even numbers, and the second all odd numbers. Each subset has a minimal element, so we still have a well-ordered sequence. Once again, we find a unique representative in the series of ordinal numbers matching this arrangement: The ordinal number $\omega + \omega$ or, in another notation, the ordinal number $\omega \cdot 2$.

■ **Example 4** (Figure 3.37)

The arrangement builds upon the idea of enumerating the natural numbers as multiples of prime numbers. The numbers 0 and 1 come first, followed by all multiples of the prime number 2, then all multiples of the prime number 3, and so on. Each number is only listed once to avoid duplicates. This procedure divides the natural numbers into an infinite number of sequences, each with an infinite number of elements. We already know the associated order type: It is described by the ordinal number $\omega \cdot \omega = \omega^2$.

■ **Example 5** (Figure 3.38)

As in the previous example, we start with the number 0. The rest of the arrangement pursues the idea of deriving the position of a natural number from its prime factorization. The first element in the enumeration is 0, followed by all powers of two of the form $2^n (n \geq 0)$:

$$1, 2, 2^2, 2^3, 2^4, 2^5, \ldots \tag{3.5}$$

We continue the enumeration by multiplying all elements generated so far by all powers of the form $3^n (n \geq 1)$. In this way, we obtain the following numbers:

$$3, 3^2, 3^3, \ldots, 2 \cdot 3, 2 \cdot 3^2, 2 \cdot 3^3, \ldots, 2^2 \cdot 3, 2^2 \cdot 3^2, 2^2 \cdot 3^3, \ldots$$

Note that each element of the initial segment (3.5) has become a separate sequence with an infinite number of elements.

In the next step, we continue the number series by multiplying the previously generated elements by all powers of the form 5^n ($n \geq 1$):

$$5, 5^2, \ldots, 2 \cdot 5, 2 \cdot 5^2, \ldots, 2^2 \cdot 5, 2^2 \cdot 5^2, \ldots,$$
$$3 \cdot 5, 3 \cdot 5^2, \ldots, 3^2 \cdot 5, 3^2 \cdot 5^2, \ldots, 3^3 \cdot 5, 3^3 \cdot 5^2, \ldots,$$
$$2 \cdot 3 \cdot 5, 2 \cdot 3 \cdot 5^2, \ldots, 2 \cdot 3^2 \cdot 5, 2 \cdot 3^2 \cdot 5^2, \ldots, 2 \cdot 3^3 \cdot 5, 2 \cdot 3^3 \cdot 5^2, \ldots,$$
$$2^2 \cdot 3 \cdot 5, 2^2 \cdot 3 \cdot 5^2, \ldots, 2^2 \cdot 3^2 \cdot 5, 2^2 \cdot 3^2 \cdot 5^2, \ldots, 2^2 \cdot 3^3 \cdot 5, 2^2 \cdot 3^3 \cdot 5^2, \ldots,$$

So far, we have only recorded those numbers with no prime factors other than 2, 3, and 5. We will eventually reach every natural number if we continue along these lines and multiply the generated initial piece with more prime powers. The resulting arrangement is an adventurous construction of rows and sub-rows that branch out further and further. Nevertheless, we are already familiar with the resulting order type. It has the structure from Figure 3.31, represented by the ordinal number ω^ω.

3.2.2.4 Transfinite Induction

This section will generalize the well-known proof principle of mathematical induction to arbitrary well-ordered sets. The considerations will lead us directly to the principle of *transfinite induction*, which allows us to prove many statements about well-ordered sets with elegance.

Let us first recapitulate what we mean by the principle of mathematical induction. Section 3.1 introduced it in the form of the induction axiom:

$$\varphi(0) \to (\forall x \, (\varphi(x) \to \varphi(s(x)))) \to \forall x \, \varphi(x) \qquad (3.6)$$

The principle of mathematical induction is applicable whenever a parameterized statement $\varphi(x)$ is to be proven for all natural numbers $x \in \mathbb{N}$. In this case, it is sufficient to prove that the statement is true for $x = 0$ and that its validity inherits from any $x \in \mathbb{N}$ to its successor.

Sometimes, the principle is utilized in an alternative formulation, which we will now derive. It follows directly from the *principle of minimality*, which in its colloquial formulation reads as follows:

$$0, 1, 2, 4, \ldots, 3, 9, 15, \ldots, 5, 25, 35, \ldots$$

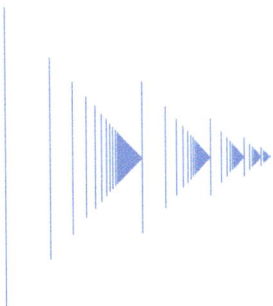

Figure 3.37: Well-ordering of the natural numbers with order type $\omega \cdot \omega$

$$0, 1, 2, 4, 8, 16, 32, \ldots,$$
$$3, 9, 27, \ldots, 6, 18, 54, \ldots, 12, 36, 108, \ldots,$$
$$5, 25, \ldots, 10, 50, \ldots, 20, 100, \ldots,$$
$$15, 75, \ldots, 45, 225, \ldots, 135, 675, \ldots,$$
$$30, 150, \ldots, 90, 450, \ldots, 270, 1350, \ldots,$$
$$60, 300, \ldots, 180, 900, \ldots, 540, 2700, \ldots$$

Figure 3.38: Well-ordering of the natural numbers with order type ω^ω

"*If there exists an x with* $\neg\varphi(x)$,
then there exists a smallest x with $\neg\varphi(x)$."

In formal notation, the minimality principle reads like this:

$$\exists x \, \neg\varphi(x) \to \exists x \, (\neg\varphi(x) \land \forall (y < x) \, \varphi(y))$$

Reversing the direction of the conclusion by negating the left and right sides of the implication yields an equivalent statement of the following form:

$$\forall x \, (\varphi(x) \lor \neg\forall (y < x) \, \varphi(y)) \to \forall x \, \varphi(x)$$

Slightly reshaping the expression lets us reach our goal:

$$\forall x \, (\forall (y < x) \, \varphi(y) \to \varphi(x)) \to \forall x \, \varphi(x) \tag{3.7}$$

This is the second frequently used variant of mathematical induction. To prove a statement $\varphi(x)$ for all $x \in \mathbb{N}$, we assume for an arbitrary x that $\varphi(y)$ is true for all $y < x$. If, under this assumption, $\varphi(x)$ is also a true statement, then $\varphi(x)$ is proven for all $x \in \mathbb{N}$.

There is no doubt that mathematical induction is a powerful proof method. It is all the more astonishing that we could legitimize it by the mere fact that the natural numbers always contain a smallest counterexample, if there is one at all. However, this property applies not only to the natural numbers but to any well-ordered set (Figure 3.39).

At this point, we are ready to state the principle of transfinite induction for well-ordered sets:

Theorem 3.11 (Transfinite Induction for Well-Ordered Sets)

For every well-ordered set $(\xi, <)$ and every formula $\varphi(v)$, the following holds:

$$\forall (x \in \xi) \, (\forall (y \in \xi \land y < x) \, \varphi(y) \to \varphi(x)) \to \forall (x \in \xi) \, \varphi(x)$$

The principle of transfinite induction is frequently utilized for proving properties about the ordinal numbers. The formula in Theorem 3.11 then becomes:

$$\forall \alpha \, (\forall (\beta < \alpha) \, \varphi(\beta) \to \varphi(\alpha)) \to \forall \alpha \, \varphi(\alpha) \tag{3.8}$$

We can even go one step further and distinguish three cases for every ordinal number α:

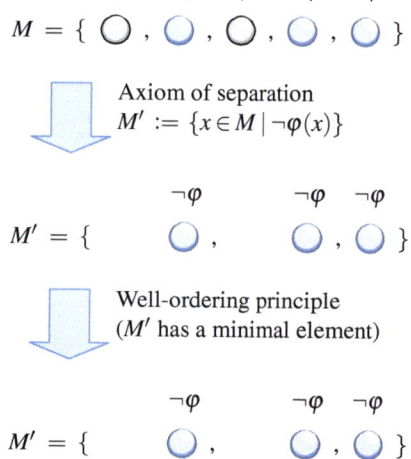

"*If there is an x with* $\neg\varphi(x)$, *then there is a smallest x with* $\neg\varphi(x)$."

Figure 3.39: Every well-ordered set M is subject to the *principle of minimality*. It states that M contains a *smallest* counterexample in M for every property φ, provided one exists at all.

3.2 Axiomatic Set Theory

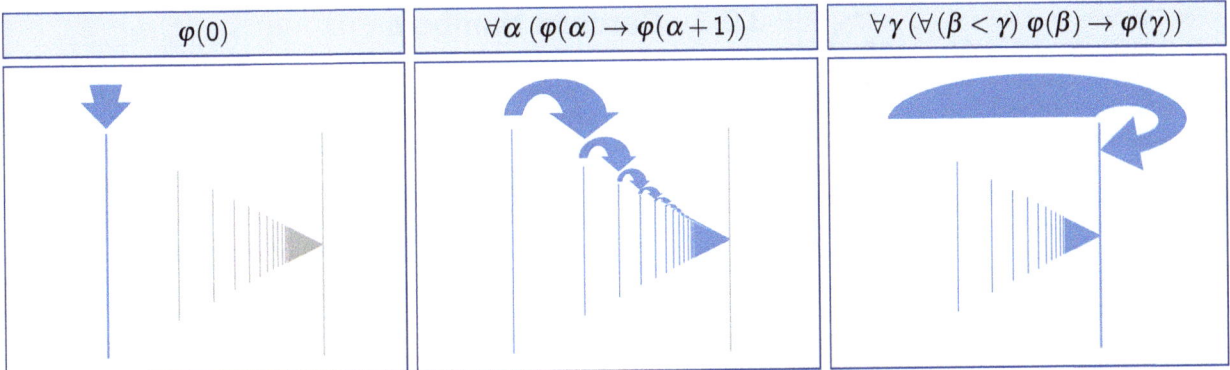

Figure 3.40: A transfinite induction proof consists of three parts. The proof commences with the induction start. After that, it is shown that the validity of φ passes to all direct successors and all limit ordinals.

- $\alpha = 0$.
- α is the successor of another ordinal.
- α is a limit ordinal.

This allows us to translate (3.8) into a form we can look up in many text books under the keyword transfinite induction:

$$\begin{pmatrix} \varphi(0) \wedge \\ \forall \alpha \, (\varphi(\alpha) \to \varphi(\alpha+1)) \wedge \\ \forall \gamma \, (\forall (\beta < \gamma) \, \varphi(\beta) \to \varphi(\gamma)) \end{pmatrix} \to \forall \alpha \, \varphi(\alpha)$$

In this formula, the expression $\forall \gamma$ is to be understood as a conditional quantifier that quantifies exclusively over limit ordinals.

For proving a property of ordinal numbers, it is thus sufficient to verify the individual statements outlined in Figure 3.40. Specifically, this involves the following three proof steps:

- Induction start
  - ☞ $\varphi(0)$

- First induction step: Inheritance to the successor ordinal
  - ☞ $\forall \alpha \, (\varphi(\alpha) \to \varphi(\alpha+1))$

- Second induction step: Inheritance to the next limit ordinal
  - ☞ $\forall \gamma \, (\forall (\beta < \gamma) \, \varphi(\beta) \to \varphi(\gamma))$

If the proof succeeds for all three cases, the principle of transfinite induction ensures that φ holds for all ordinals.

Let's take a moment to revisit formula (3.7) for a closer inspection. At first glance, the base case seems to have vanished in the alternative formulation of mathematical induction. To expose this impression as deceptive, let's resolve the universal quantifier and take a look at the left-hand side of the formula (3.7) for $x = 0$:

$$\forall (y < 0) \, \varphi(y) \to \varphi(0)$$

Since 0 is the smallest natural number, the subformula

$$\forall (y < 0) \, \varphi(y)$$

is always true, and we can simplify the left side as such:

$$\varphi(0)$$

Voilà. The induction start has not disappeared; it hides as a particular case in the induction formula.

> For Cantor, ordinal and cardinal numbers were a different kettle of fish. The opening quotations of Sections 3.2.2 and 3.2.3 expose his conceptual distinction between the two. For Cantor, ordinal numbers arose from an abstraction that focused solely on the order of the elements rather than their concrete nature. With the introduction of cardinal numbers, he carried out a second abstraction that neither distinguished sets by the nature of their elements nor their order (Figure 3.41).
>
> Even though it suits human intuition to view ordinal numbers and cardinal numbers as descriptions of sets at different levels of abstraction, this separation is unnecessary. In fact, modern set theory dispenses with this distinction and traces the concept of cardinal numbers directly back to the concept of ordinal numbers. Cantor's idea is nevertheless valuable as a mental aid. Compared with the formal formulation in Definition 3.12, it sheds bright light on the rationale behind ordinal and cardinal numbers.

3.2.3 Cardinal Numbers

> *"We will call by the name 'power' or 'cardinal number' of M the general concept which, by means of our active faculty of thought, arises from the aggregate M when we make abstraction of the nature of its various elements m and of the order in which they are given. We denote the result of this double act of abstraction, the cardinal number or power of M, by $\overline{\overline{M}}$."*
>
> Georg Cantor, 1895 [24]

Now that we have extensively dealt with ordinal numbers, let's take a closer look at *cardinal numbers*, which are closely related. These numbers are not entirely alien to us; we have already used them to describe the cardinality of sets in Chapter 1. For finite sets, this was straightforward, as the cardinality of a finite set is simply the number of its elements. For infinite sets, on the other hand, we had resorted to Cantor's conceptual framework, which denoted the smallest infinity by the *cardinal number* $\aleph_0$, the next largest by $\aleph_1$, and so on.

So far, we have only used the term cardinal number informally. Now, the time has come to put the concept on a solid basis. As the following definition shows, we can trace the concept of cardinal numbers back to the concept of ordinal numbers, which is already familiar to us.

Definition 3.12 (Cardinal Number)

> An ordinal number α is called a *cardinal number* if
> $$\beta < \alpha \text{ implies } |\beta| < |\alpha|$$
> for all ordinal numbers β.

The definition clarifies that cardinal numbers are nothing but special ordinal numbers. Specifically, they are those ordinal numbers that are minimal in cardinality. In other words, a cardinal number has the property that there is no smaller ordinal number with the same cardinality. In the rest of this book, we follow the convention of denoting cardinal numbers with the lowercase Greek letters κ, λ, or μ.

Note that for two finite ordinal numbers, α and β, the relations $\beta < \alpha$ and $|\beta| < |\alpha|$ are equivalent. Therefore, as long as we do not leave the trusted ground of the finite, every ordinal is also a cardinal. However,

3.2 Axiomatic Set Theory

once we enter the realm of the infinite, we must be cautious. First of all, the smallest infinite cardinal number $\aleph_0$ is equivalent to the smallest infinite ordinal number; that is, the following equation applies:

$$\aleph_0 = \omega$$

Though their paths diverge from here, structural similarities between the two numerical worlds remain. The Aleph series also extends to infinity and exhibits a structure similar to the ordinal numbers. In particular, for every ordinal number α, a cardinal number $\aleph_\alpha$ exists. Consequently, the cardinal numbers also form a proper class.

Just as in the case of ordinal numbers, we can work our way up to ever larger cardinal numbers, but now we are pushing forward at an even more dizzying speed. Remember the astronomically large number ω_1, the smallest uncountable ordinal number? On the infinite Aleph scale, ω_1 is not far away; it already appears in second place! It is immediately followed by the ordinal number ω_2 and so on.

The formal definition of cardinal numbers enables us to precisely capture the symbolism $|M|$ even for infinite sets. So far, we have mainly used the notation in comparisons of the form $|M| = |N|$, expressing that a bijective mapping exists between the sets M and N. However, $|M|$ had no precise meaning without the equals sign. With the concept of cardinal numbers in hand, we can become more accurate and assign the symbol a specific meaning:

 Definition 3.13 (Cardinality)

> Let M be any set. The *cardinality* of M, denoted as $|M|$, is the cardinal number κ which can be mapped bijectively to M.

This settles the debt we have burdened ourselves with the informal use of $|M|$ and the Aleph numbers in Chapter 1. With Definition 3.13 in mind, the previously symbolic notation $|M| = \aleph_n$ has become a proper equation with precisely defined sets on both sides.

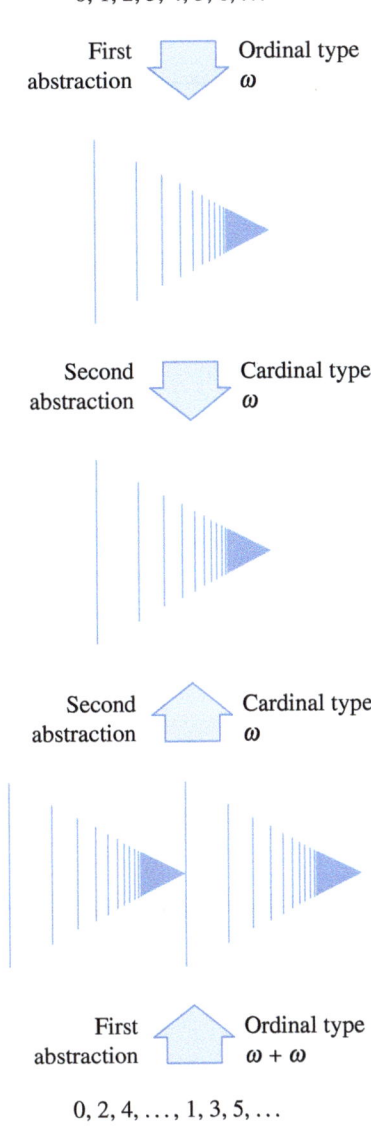

Figure 3.41: Ordinal numbers abstract from the nature of the elements of a set. Cardinal numbers perform a second abstraction in which the order of elements plays no role either. This example shows two orders of natural numbers of different ordinal types that share the same cardinal type.

3.3 Exercises

Exercise 3.1 Formalize the following propositions within Peano arithmetic:

a) "There exist natural numbers x and y satisfying $x^2 + y^2 = 9$."

b) "$x^3 + y^3 = z^3$ has no solution in the positive natural numbers."

c) "x is a power of two."

Exercise 3.2 Which of the following formulas corresponds to the statement "7 is a prime number"?

a) $\forall z \, (z \mid \overline{7} \rightarrow (z = \overline{1} \vee z = \overline{7}))$

b) $\neg \exists (y > 1) \, \exists (z > 1) \, \overline{7} = y \times z$

Which of the following formulas corresponds to the statement "x is a prime number"?

c) $\forall z \, ((z \mid x) \rightarrow (z = \overline{1} \vee z = x))$

d) $\neg \exists (y > 1) \, \exists (z > 1) \, x = y \times z$

Exercise 3.3 Formalize the following statements using Peano arithmetic:

a) "Every even natural number $n > 2$ can be expressed as the sum of two prime numbers."

b) "For infinitely many numbers n, both n and $n+2$ are prime numbers."

Both statements are familiar to us from Chapter 1. The first is Goldbach's famous conjecture; the second is the conjecture about the existence of infinitely many prime twins.

Exercise 3.4 Some textbooks introduce the axiom of induction

$$\varphi(0) \rightarrow (\forall x \, (\varphi(x) \rightarrow \varphi(s(x)))) \rightarrow \forall x \, \varphi(x)$$

in a slightly different form:

$$(\varphi(0) \wedge \forall x \, (\varphi(x) \rightarrow \varphi(s(x)))) \rightarrow \forall x \, \varphi(x)$$

Demonstrate the equivalence of both definitions.

3.3 Exercises

Exercise 3.5

At the top of the ZF axiom list, we introduced the axiom of extensionality. In formal notation, it reads as follows:

$$\forall x \forall y \, (x = y \leftrightarrow \forall z \, (z \in x \leftrightarrow z \in y))$$

Suppose we remove the equals sign from the language and understand the axiom of extensionality not as a proper axiom but as the equal sign's definition. In doing so, we can only treat the equals sign as a syntactic abbreviation rather than a native language element. Can we still derive the same theorems as before?

Exercise 3.6

At the fourth position of the ZF axiom list, we have introduced the axiom of the union:

$$\forall x \, \exists y \, \forall z \, (z \in y \leftrightarrow \exists (w \in x) \, z \in w)$$

For two sets, x and y, it guarantees the existence of the union set $x \cup y$. In connection with this axiom, we have also introduced the notation $x \cap y$ as a syntactic abbreviation for the intersection. Is the existence of the intersection also guaranteed by the axiom of the union? If not, which axioms are additionally required?

Exercise 3.7

The axiom of foundation is an indispensable part of ZF set theory. It states that every non-empty set x contains an element y that has no elements in common with x.

a) What is the significance of the axiom for ZF set theory?

b) Is the axiom compatible with Russell's ramified type theory?

Exercise 3.8

Section 3.2.1.3 has shown that the concept of an ordered pair can be captured with the following formula:

$$\langle \xi, \nu \rangle := \{\, \{\xi\}, \{\xi, \nu\} \,\}$$

However, this representation is only one of many possibilities. For example, Norbert Wiener suggested in 1914 to represent ordered pairs as follows [150, 221, 222]:

$$\langle \xi, \nu \rangle := \{\, \{\emptyset, \{\xi\}\}, \{\{\nu\}\} \,\}$$

a) Which of the following set diagrams visualizes the given definitions of the ordered pair?

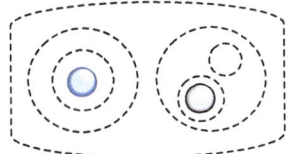 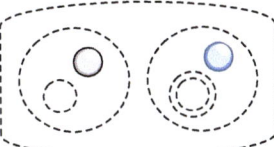 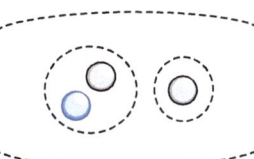

b) What is the construction pattern of the remaining diagram? Is it suitable for representing ordered pairs as well?

Exercise 3.9

We know from Section 3.2.1.2 that the relation '<' does not well-order $\mathbb{Z}$, the set of integers.

Redefine the ordering relation '<' such that $\mathbb{Z}$ becomes a well-ordered set.

Exercise 3.10

The *Cartesian product* of two sets is defined as follows:

$$v \times \mu := \{\langle x,y \rangle \mid x \in v \land y \in \mu\}$$

a) Formalize the Cartesian product in the system of ZF set theory by specifying a formula $\mathfrak{K}(\xi, v, \mu)$ that is true if and only if ξ, v, μ satisfy $\xi = v \times \mu$.

b) Provide a formula $\mathfrak{R}(\xi, v)$ that is true precisely if ξ is a relation over the set v. Base your definition on the formula $\mathfrak{K}(\xi, v, \mu)$ derived in Part a).

c) On page 159, you learned about the formula $\mathfrak{R}(\xi)$, which is true precisely if ξ is a relation over some set. Try to find an alternative definition for $\mathfrak{R}(\xi)$ that utilizes the notion of the Cartesian product.

d) Building on the concept of relations, we have shown how to describe partial functions within Zermelo-Fraenkel set theory. Can total, injective, and surjective functions be formalized accordingly?

Exercise 3.11

In this exercise, we take on the formal proof of Theorem 3.3, establishing the component equality of ordered pairs.

a) Consider the proof steps in lines 59 - 64. Was it essential to prove the entire statement $x = y = u = v$?

b) The colloquially formulated original proof ends by distinguishing two cases. First, it considered the case $v \neq u$, then the case $v = u$. Could we have dispensed with this distinction in the formal proof?

3.3 Exercises

Exercise 3.12 Which of the following sets are transitive? Which are ordinal numbers? Which are cardinal numbers?

a) $\emptyset$

b) $\{\emptyset\}$

c) $\{\{\emptyset\},\{\{\emptyset\}\}\}$

d) $\{\emptyset,\{\{\emptyset\}\}\}$

e) $\{\emptyset,\{\emptyset\},\{\{\emptyset\}\}\}$

f) $\{\emptyset,\{\emptyset\},\{\emptyset,\{\emptyset\}\}\}$

Exercise 3.13 Prove or refute the following assertions:

a) If x is an ordinal number, so is $\mathcal{P}(x)$.

b) The addition of ordinal numbers is commutative, that is, $\alpha + \beta = \beta + \alpha$.

c) Ordinal multiplication is commutative, that is, $\alpha \cdot \beta = \beta \cdot \alpha$.

Exercise 3.14 Which of the following statements are equivalent?

a) X is empty

b) X is finite

c) X is countable

d) X is denumerable

e) X is infinite

f) X is uncountable

g) $|X| = \emptyset$

h) $|X| = \aleph_0$

i) $|X| < \aleph_0$

j) $|X| > \aleph_0$

k) $|X| \leq \aleph_0$

l) $|X| \geq \aleph_0$

m) $|X| \in \aleph_0$

n) $|X| \subseteq \aleph_0$

o) $|X| \subset \aleph_0$

p) $\aleph_0 \in |X|$

q) $\aleph_0 \subseteq |X|$

r) $\aleph_0 \subset |X|$

4 Proof Theory

"Man kann – unter Voraussetzung der Widerspruchsfreiheit der klassischen Mathematik – sogar Beispiele für Sätze (und zwar solche von der Art des Goldbach'schen oder Fermat'schen) angeben, die zwar inhaltlich richtig, aber im formalen System der klassischen Mathematik unbeweisbar sind."

"One can – assuming the consistency of classical mathematics – even give examples of sentences (namely those of Goldbach's or Fermat's type) that are indeed correct in content, but unprovable in the formal system of classical mathematics."

Kurt Gödel [187]

This chapter introduces proof theory, which most logicians consider the primary building block of mathematical logic. It pursues the idea of interpreting proofs as mathematical objects, thereby making them accessible for rigorous analysis. The development of proof theory was in full swing in the first half of the twentieth century, yielding profound insights into the nature of mathematical reasoning. However, it also revealed fundamental limitations inherent in mathematics, which are the subject of this chapter.

Sections 4.1 to 4.4 will derive Gödel's incompleteness theorems and discuss their far-reaching implications. Afterward, we will show that the phenomenon of incompleteness is omnipresent in mathematics. For instance, Section 4.5 will present the Goodstein sequence, demonstrating that even seemingly harmless statements of ordinary mathematics are affected.

4.1 Gödel's Incompleteness Theorems

Gödel's incompleteness theorems are at the heart of modern proof theory. Their substantive content is dark, yet they shed such a bright

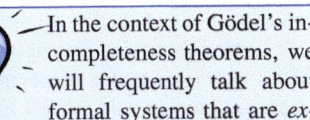 In the context of Gödel's incompleteness theorems, we will frequently talk about formal systems that are *expressive enough to formalize Peano arithmetic*. What exactly does this mean? In his original paper, Gödel proved the incompleteness theorem for a specific formal system, which he called P. In his own words, *"P is essentially the system which one obtains by building the logic of PM [Principia Mathematica] around Peano's axioms."* [77] He also states in his work that his result is not limited to P but covers all formal systems that are expressive enough to talk about the additive and multiplicative properties of the natural numbers. In addition to Peano arithmetic, this includes all theories representing the natural numbers in the form of other objects. With Zermelo-Fraenkel set theory, we have already become acquainted with such a formal system. Although the natural numbers are not independent objects in ZF and ZFC, they are representable as specific sets, with addition and multiplication translating into suitable set operations. This is meant when we say a formal system is expressive enough to formalize Peano arithmetic.

light on the nature of the mathematical method that they have captivated countless mathematicians and scientists since their discovery in 1931. Speaking for myself, I first read about them in Douglas Hofstadter's masterpiece *Gödel, Escher, Bach* [105] shortly before entering university. Even though many years have passed since then, my fascination with Gödel's work remains unbroken. It is undoubtedly the incompleteness theorems that motivated me to write this book.

Two leitmotifs shape the following pages. First, I aim to derive the incompleteness theorems along Gödel's original line of reasoning from 1931. That way, I will try not only to prove the substantive content of the incompleteness theorems but also, as far as possible, to provide an insight into Gödel's thinking. However, there is no room for too much euphoria because his work will remain difficult to grasp even after finishing this chapter. Gödel's original work contains countless formulas and definitions that initially obscure the essentials. Nevertheless, his meticulous precision in proving his results is anything but a flaw; without it, the theorems would never have found the necessary acceptance among his critics. For almost all of his contemporaries, Gödel's incompleteness theorems were a heavy blow, and many were highly skeptical for this reason alone.

This book does not aim to defend Gödel's results against criticism; instead, it attempts to expose the core ideas of his compelling proofs. Therefore, I have deliberately tried not to overburden the following sections with technical details. This applies particularly to several auxiliary theorems, which are not spectacular in content but sometimes require a lengthy technical justification. Instead of presenting the proofs, the relevant passages indicate where to look them up.

4.2 The First Incompleteness Theorem

Gödel's first incompleteness theorem is the most frequently quoted in mathematical logic. Roughly speaking, it states that the concepts of truth and provability cannot coincide in sufficiently expressive formal systems. These systems must inevitably be incomplete; there must be substantively true statements that are unprovable within the system.

Let's start by ruling out a common misconception: not every formal system is incomplete. Gödel's first incompleteness theorem only affects formal systems with a certain expressiveness. In particular, it affects all systems expressive enough to formalize Peano arithmetic, that is, the natural numbers together with addition and multiplication. There

---

Among Gödel's harshest opponents was the famous set theorist Ernst Zermelo, whose name appears numerous times in this book. In September 1931, the two came face to face at the meeting of the German Mathematical Society in Bad Elster. In appearance, the reserved Gödel had little to oppose his 60-year-old antagonist. Zermelo was known for his eloquence and quick-tempered, sometimes irascible manner [50]. In Bad Elster, he left no doubt about his attitude towards the young mathematician and initially refused to engage in any discussion. Nevertheless, a personal conversation did take place, which ended unexpectedly peacefully. Just six days later, however, Zermelo informed Gödel in writing that he had found an error in his proof. Gödel tried in several letters to clear up the misunderstanding, but the arguments did not sway Zermelo. Finally, Zermelo made his criticism public in 1932 [231]. Gödel was not a man of confrontation and gave up any further attempt to explain his incompleteness theorems to the aging rival. Rudolf Carnap later commented on the correspondence that Zermelo had completely misunderstood Gödel's explanations [50].

4.2 The First Incompleteness Theorem

> **Satz V:** Zu jeder rekursiven Relation $R(x_1 \ldots x_n)$ gibt es ein n-stelliges *Relationszeichen* r (mit den *freien Variablen*[38] $u_1, u_2 \ldots u_n$), so daß für alle Zahlen-n-tupel $(x_1 \ldots x_n)$ gilt:
>
> $$R(x_1 \ldots x_n) \rightarrow \text{Bew}\left[Sb\left(r \begin{smallmatrix} u_1 & \ldots & u_n \\ Z(x_1) & \ldots & Z(x_n) \end{smallmatrix}\right)\right] \quad (3)$$
>
> $$\overline{R}(x_1 \ldots x_n) \rightarrow \text{Bew}\left[\text{Neg } Sb\left(r \begin{smallmatrix} u_1 & \ldots & u_n \\ Z(x_1) & \ldots & Z(x_n) \end{smallmatrix}\right)\right] \quad (4)$$
>
> Wir begnügen uns hier damit, den Beweis dieses Satzes, da er keine prinzipiellen Schwierigkeiten bietet und ziemlich umständlich ist, in Umrissen anzudeuten[39]. Wir beweisen den Satz für alle
>
> ...
>
> Das allgemeine Resultat über die Existenz unentscheidbarer Sätze lautet:
>
> **Satz VI:** Zu jeder ω-widerspruchsfreien rekursiven Klasse $\varkappa$ von *Formeln* gibt es rekursive *Klassenzeichen* r, so daß weder v Gen r noch Neg (v Gen r) zu Flg ($\varkappa$) gehört (wobei v die *freie Variable* aus r ist).
>
> Beweis: Sei $\varkappa$ eine beliebige rekursive ω-widerspruchsfreie Klasse von *Formeln*. Wir definieren:

Figure 4.1: Two theorems from Gödel's original work [73]. Theorem V is a technical milestone in the proof of the first incompleteness theorem. It corresponds to our Theorem 4.5, discussed in Section 4.2.3. Instead of providing a detailed proof of this theorem, Gödel limited himself to delivering a brief outline. For a fully worked out proof, see, e.g., [195].

Theorem VI is the main incompleteness result. Later in his work, Gödel utilizes it to derive a corollary substantively equivalent to our Theorem 4.2. Gödel's theorems are deliberately shown here in their original form. They highlight how the terminology used at the time differs from what we use today. Their substantive meaning is not easy to recognize, even at a second glance.

is no doubt that the natural numbers belong to the vital core of mathematics; without them, this science would pretty much vaporize. Thus, the incompleteness theorem attests nothing less than the impossibility of constructing a formal system capable of proving precisely the true statements of ordinary mathematics.

Much has been published about the first incompleteness theorem, and comparing the various presentations highlights two peculiarities. Firstly, different notations and terminologies often obscure the fact that we are dealing with the same sentence in substantive terms (cf. Figure 4.1). Secondly, the lengths of the presented proofs differ drastically. For example, the author in [184] arrives at the desired result after just a few paragraphs, while Gödel's original proof from 1931 spans many pages. How can that be?

Two main reasons account for this. First of all, many more recent proofs utilize the notion of *computability*. By formalizing this concept in 1936, Alan Turing paved the way to reach Gödel's result comparatively quickly. However, the main reason is another one. Several variants of the first incompleteness theorem exist that differ not only in terminology but also in their exact substantive meaning. The literature rarely points this out, yet it's recognition is crucial to avoid misunderstandings. One frequently used variant is this:

- Semantic variant

    "Any correct formal system expressive enough to formalize Peano arithmetic is incomplete."

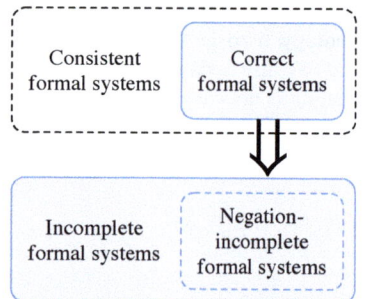

- Syntactic variant

    "Any consistent formal system expressive enough to formalize Peano arithmetic is negation incomplete."

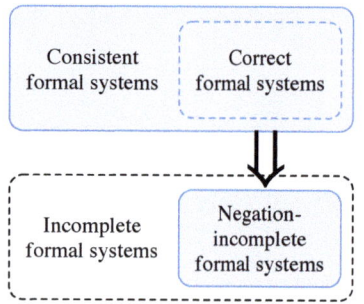

Figure 4.2: The semantic variant of the first incompleteness theorem is substantively weaker than the syntactic variant.

Theorem 4.1 (First Incompleteness Theorem, Semantical)

> Any correct formal system expressive enough to formalize Peano arithmetic is incomplete.

This is the *semantic variant* of Gödel's first incompleteness theorem. It makes a statement about correct formal systems, systems in which all derivable statements are true ($\vdash \varphi$ implies $\models \varphi$). If such a system comprises Peano arithmetic, i.e., if it is expressive enough to talk about the additive and multiplicative properties of the natural numbers, the first incompleteness theorem asserts its incompleteness. In such a system, at least one formula represents a true statement but is unprovable within the system. Comparatively short proofs exist for the semantic variant of Gödel's first incompleteness theorem, which we will encounter in Chapter 5.

In addition to the semantic variant, another formulation circumvents the concept of correctness. It reads as follows:

Theorem 4.2 (First Incompleteness Theorem, Syntactical)

> Any consistent formal system expressive enough to formalize Peano arithmetic is negation incomplete.

This is the *syntactic variant* of Gödel's first incompleteness theorem. It makes a statement about a broader class of formal systems since it only assumes consistency rather than correctness. Because every negation-incomplete formal system that formalizes Peano arithmetic is also incomplete, and the correctness of a formal system always implies its consistency, the semantic version is a direct consequence of the syntactic variant (Figure 4.2).

The substantive meaning of Gödel's first incompleteness theorem is undoubtedly impressive; even more stunning, however, is how Gödel proved the theorem. In simple terms, he managed to construct a formula with the following meaning:

$$\text{"I am unprovable within the formal system."} \quad (4.1)$$

The self-referential nature of this proposition is reminiscent of the *Barber paradox* from Section 1.2.5 and the central element in Gödel's proof. For this proposition, we will show that in a formal system satisfying the conditions of the first incompleteness theorem, neither the

4.2 The First Incompleteness Theorem

proposition itself nor its negation is derivable from the axioms. In other words, Gödel's statement is undecidable within the system. Section 4.2.4 will demonstrate how to prove the undecidability of this theorem in a few lines. The actual difficulty lies elsewhere, namely in constructing the theorem itself.

How did Gödel manage to construct a theorem that postulates its own unprovability? This theorem is unlike any theorem we are familiar with in analysis, algebra, or any other branch of ordinary mathematics. It is a meta-theorem, as it makes a statement about the formal system it is formulated in. Talking about itself requires the theorem to somehow step out of the formal system, but how can something like this possibly happen?

In fact, Gödel discovered a back door in all formal systems with sufficient expressiveness. The core idea of his approach is the construction of arithmetic statements that carry two substantive meanings simultaneously (Figure 4.3).

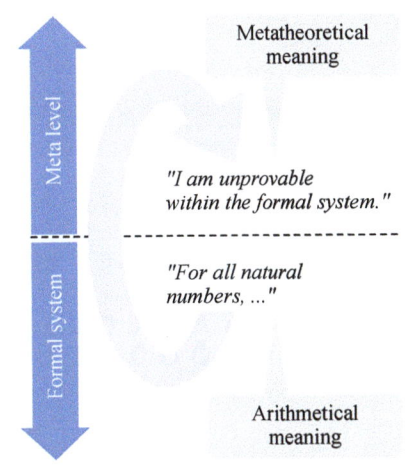

Figure 4.3: Gödel constructed an arithmetic statement with a second metatheoretical meaning in addition to its number-theoretical one. It is the result of an invisible isomorphism that establishes a connection between the symbolic manipulation of character strings and the arithmetic properties of natural numbers. With a clever trick, Gödel devised an arithmetic formula that postulated its own unprovability.

- First, these theorems have an arithmetical meaning. Viewed within the formal system, they appear as ordinary theorems of Peano arithmetic and, as such, make statements about natural numbers.

- Viewed from outside the formal system, the propositions bear a second, metatheoretical meaning. It arises through a hidden isomorphism whose unveiling is a paramount moment of mathematical logic. Gödel discovered how to represent the rules and axioms of a formal system arithmetically and replicate the symbolic manipulation of strings on the arithmetic level. In this way, he managed to encode metatheoretical statements, such as the question about the existence of a proof, into arithmetic formulas.

All Gödel needed was part of Peano arithmetic, namely the natural numbers in combination with addition and multiplication. He thus unveiled a surprising phenomenon: Every formal system that includes Peano arithmetic is expressive enough to formulate metatheoretical statements and, therefore, can speak about itself.

4.2.1 Arithmetization of Syntax

It is time to examine more closely how arithmetic formulas can talk about the properties of formal systems. To achieve this goal, we will relate formulas and proofs with natural numbers by defining a mapping

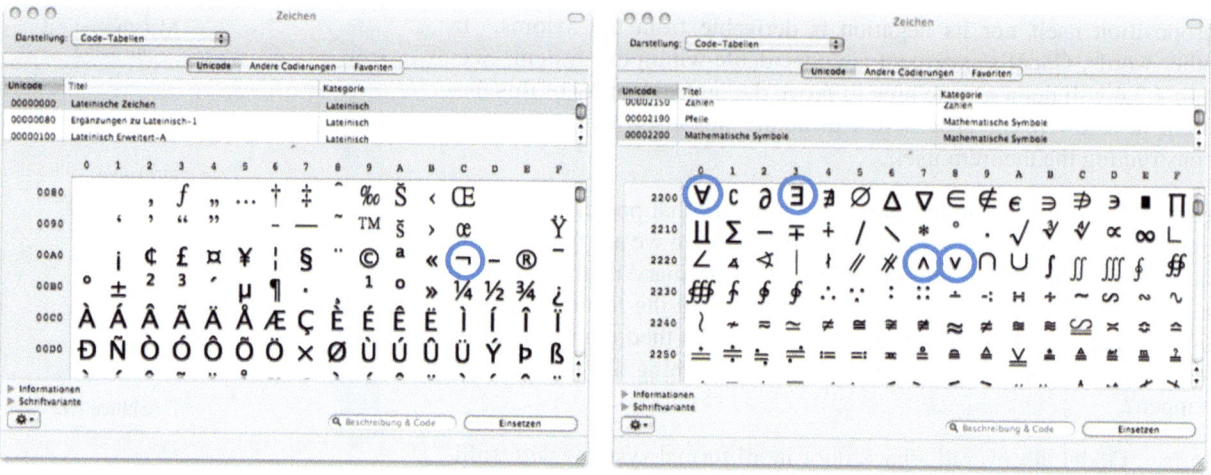

Figure 4.4: Unicode comprises a total of 16 planes, each capable of holding 65536 different characters. The result is a universal symbol table assigning a unique binary code to each known character [4]. This code is always the same, on any hardware, under any operating system, and in any programming language.

> Which variant of the first incompleteness theorem did Gödel prove in 1931? In fact, his incompleteness result is a weakened variant of Theorem 4.2. At the time of publication, Gödel did not manage to prove his result under the assumption of consistency alone; instead, he had to assume an ω-consistent formal system. It was not until 1936 that Barkley Rosser proved that Gödel's assumption of ω-consistent is replaceable by ordinary consistency [46, 173]. This much in advance: Every ω-consistent formal system is consistent, but not vice versa.
>
> Gödel's decision to not prove the semantic but the more difficult syntactic variant must be viewed in its historical context. He considered it crucial not to base his proof on the semantic concept of truth, as he wrote his work when the aftershocks of set-theoretic paradoxes were still present, and many of his contemporaries were skeptical or even hostile towards the concept of truth.

that translates every formula and every proof of a formal system into a unique natural number. The calculated number will then serve as a substitute for the original formula or the original proof. We will carry out our plan inside Peano arithmetic as introduced in Section 3.1. However, the developed methodology is so general that it applies to any other sufficiently expressive formal system.

The syntax of Peano arithmetic can be arithmetized in many different ways. A straightforward approach is to type the formulas on a PC and interpret the internally stored bit sequence as a natural number. The conversion becomes particularly easy using Unicode (Figure 4.4). This standardized character table contains all common logic symbols, so we do not even have to change the formula's syntax.

For example, to translate the formula

$$\forall x \, x = x$$

in macOS into the Unicode-based UTF-16 format, it is sufficient to type the following command sequence into the command shell:

```
echo -n "∀xx=x" | iconv -t UTF-16 | hexdump
```

The result is a 12-element byte sequence or, equivalently, a 24-digit

4.2 The First Incompleteness Theorem

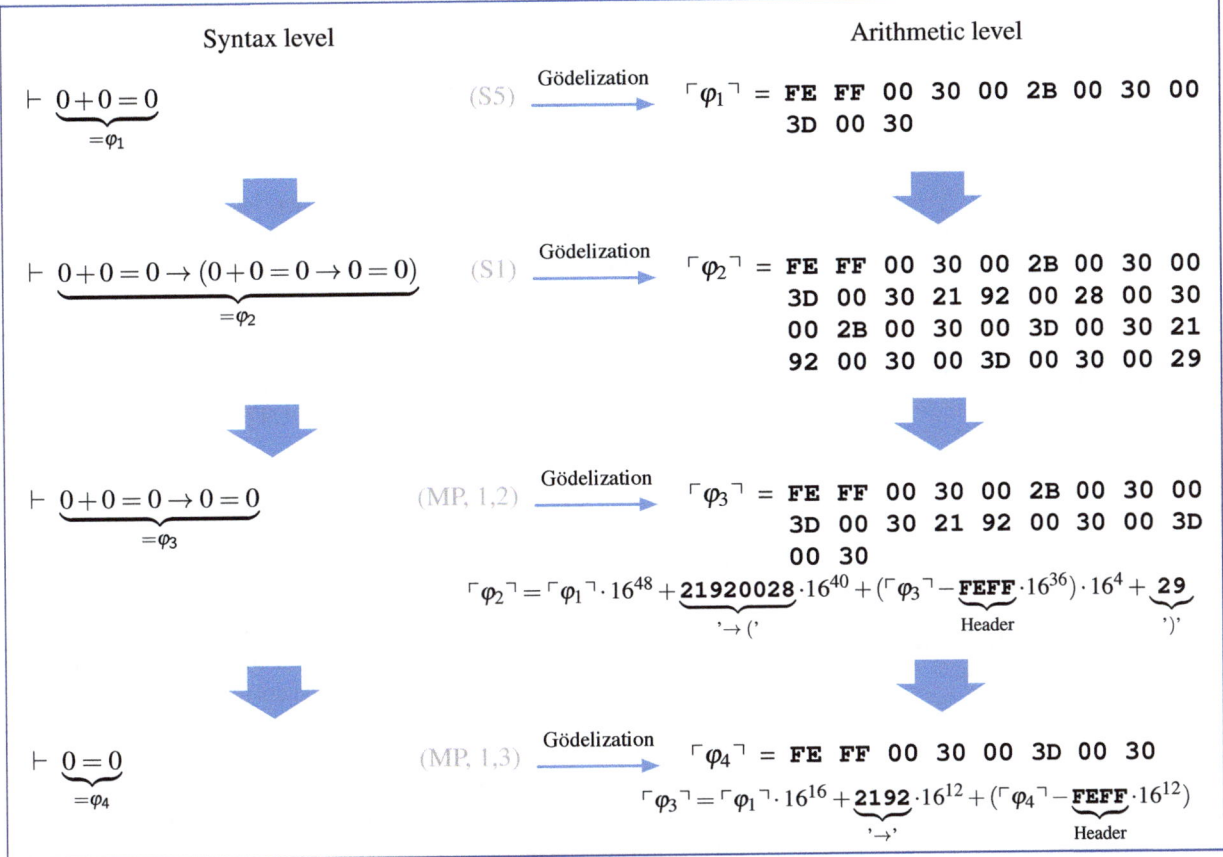

Figure 4.5: Any syntactic manipulation that extends a formal proof chain $\varphi_0, \ldots, \varphi_i$ by applying an inference rule, can be interpreted as an arithmetic relationship between the Gödel numbers $\ulcorner \varphi_0 \urcorner, \ldots, \ulcorner \varphi_i \urcorner, \ulcorner \varphi_{i+1} \urcorner$.

hexadecimal number, which we denote with $\ulcorner \varphi \urcorner$:

$$\ulcorner \varphi \urcorner = \underbrace{\mathbf{FE\ FF}}_{\text{Header}}\ \underbrace{22\ 00}_{\text{'}\forall\text{'}}\ \underbrace{00\ 78}_{\text{'}x\text{'}}\ \underbrace{00\ 78}_{\text{'}x\text{'}}\ \underbrace{00\ 3D}_{\text{'='}}\ \underbrace{00\ 78}_{\text{'}x\text{'}}$$

The sequence starts with two bytes forming the UTF-16 header, followed by two-byte packets containing the Unicode of the respective formula character. We refer to the number $\ulcorner \varphi \urcorner$ as the *Gödel number* of the formula φ and the coding process as *Gödelization*.

The coding presented is just one of many, and the assigned values play a subordinate role. Every encoding that fulfills three minimum requirements is suitable for our purposes:

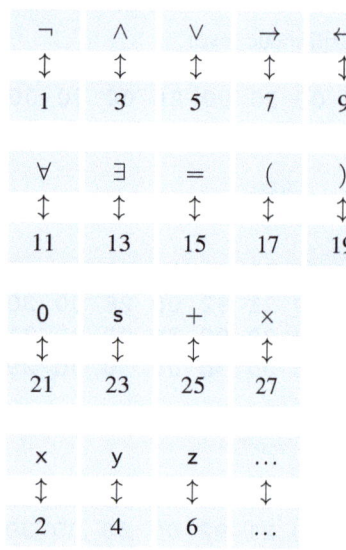

Table 4.1: The arithmetization of syntax starts by translating every elementary symbol of the formal system's artificial language into a natural number.

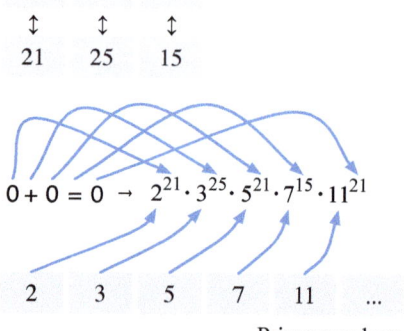

Figure 4.6: A formula is Gödelized by using the numerical value of the i-th formula character as the exponent of the i-th prime number and joining all expressions into a single product.

- The encoding must injectively embed the formulas in the set of natural numbers, i.e., assigning different Gödel numbers to different formulas. The UTF-16 encoding fulfills this requirement, as different text fragments always have different UTF-16 representations.

- The Gödel numbers must be calculable, i.e., a procedure must exist that systematically determines the number $\ulcorner\varphi\urcorner$ for each formula φ. The UTF-16 encoding fulfills this requirement, as we can generate it programmatically on any PC.

- For each natural number, we must be able to determine whether it encodes a symbol string that complies with the syntax rules of our formal system's artificial language. In short, we must be able to decide whether it represents a formula. UTF-16 encoded numbers satisfy this requirement with ease.

Although the UTF-16 encoding meets all listed requirements, it is of limited value for our purposes. To understand why, consider the following scenario: Let $\varphi_0, \ldots, \varphi_{i+1}$ be formulas of Peano arithmetic, where φ_{i+1} arose from the previous formulas by applying an inference rule. On the one hand, there is a syntactic relationship between the formulas $\varphi_0, \ldots, \varphi_{i+1}$ as applying an inference rule corresponds to a symbolic manipulation of strings. On the other hand, there is an arithmetic relationship between the Gödel numbers $\ulcorner\varphi_0\urcorner, \ldots, \ulcorner\varphi_{i+1}\urcorner$. Figure 4.5 demonstrates the relationships using a proof from Section 3.1.3, the proof of Theorem PA1 with the instantiation $\sigma = 0$. Encoded in UTF-16, the arithmetic relationships between the Gödel numbers become cumbersome to describe. Yet, this comes as no surprise, as the genuine purpose of Unicode is entirely different.

For this reason, we will adopt the approach presented in [195], which is more suitable for our purposes due to its mathematical nature. The coding is closely related to that in Gödel's original work and performs the translation into natural numbers step by step:

- Like the UTF-16 encoding, we assign a natural number to each elementary symbol of the formal system's language but use the numerical values from Table 4.1 instead of the Unicode table. The values are chosen such that all logic symbols are assigned odd numbers. The even numbers are reserved for coding variables.

- To encode a single formula, we do no longer write down the numerical values of the formula characters one after the other. Instead, we use the numerical value of the i-th formula character as the exponent

4.2 The First Incompleteness Theorem

of the i-th prime number and combine all expressions into a joint product, as shown in Figure 4.6. If we denote the numerical value of the i-th formula character by c_i and the i-th prime number by π_i, we can write down the Gödel number $\ulcorner \varphi \urcorner$ as follows:

$$\ulcorner \varphi \urcorner := \pi_1^{c_1} \cdot \pi_2^{c_2} \cdot \pi_3^{c_3} \cdots$$

Using prime numbers is essential. Since every natural number is uniquely described by its prime factors, two distinct formulas are guaranteed to map to different Gödel numbers. Figure 4.7 summarizes how the four formulas from the proof of Theorem PA1 are Gödelized this way.

■ Since, according to Definition 2.1, a formal proof is a sequence of formulas, we can translate it into a natural number according to the same scheme. To encode a sequence of the form $\varphi_1, \varphi_2, \varphi_3, \ldots$, we use the Gödel number of the i-th formula as the exponent of the i-th prime number and again combine all expressions into a joint product:

$$\ulcorner \varphi_1, \varphi_2, \varphi_3, \ldots \urcorner := \pi_1^{\ulcorner \varphi_1 \urcorner} \cdot \pi_2^{\ulcorner \varphi_2 \urcorner} \cdot \pi_3^{\ulcorner \varphi_3 \urcorner} \cdots$$

Our example proof translates into

$$2^{2^{21} 3^{25} 5^{21} 7^{15} 11^{21}} \cdot$$
$$3^{2^{21} 3^{25} 5^{21} 7^{15} 11^{21} 13^{7} 17^{17} 19^{21} 23^{25} 29^{21} 31^{15} 37^{21} 41^{7} 43^{21} 47^{15} 53^{21} 59^{19}} \cdot$$
$$5^{2^{21} 3^{25} 5^{21} 7^{15} 11^{21} 13^{7} 17^{21} 19^{15} 23^{21}} \cdot$$
$$7^{2^{21} 3^{15} 5^{21}},$$

which is a number with around 2^{10383} decimal digits. No book in the world comprises enough pages even to come close to accommodating its decimal notation. We are thus well advised to leave the number in its factorized representation.

Even though the two presented encodings differ significantly, they share a common disadvantage: both map formulas injectively into the natural numbers but not surjectively. Consequently, not all natural numbers are Gödel numbers. Sometimes, however, it is convenient to assume a one-to-one relationship between the set of formulas and the set of natural numbers, which raises the question of whether bijective Gödelizations exist. The answer is a resounding yes since we can enumerate all syntactically correct symbol sequences one after the other and thus associate the Gödel number i with the i-th formula in the enumeration. However, this type of Gödelization is far from being practical. Just like in the case of UTF-16 encoding, the syntactical relationships of formulas are cumbersome to describe at the arithmetic level.

■ Gödelization of φ_1

$\ulcorner 0 + 0 = 0 \urcorner$
$= 2^{21} \cdot 3^{25} \cdot 5^{21} \cdot 7^{15} \cdot 11^{21}$
$= 2976791086050777886254142258705\ldots$
$4735259615108039000000000000000\ldots$
000000
$\approx 3 \cdot 10^{67}$

■ Gödelization of φ_2

$\ulcorner 0 + 0 = 0 \rightarrow (0 + 0 = 0 \rightarrow 0 = 0) \urcorner$
$= 2^{21} \cdot 3^{25} \cdot 5^{21} \cdot 7^{15} \cdot 11^{21} \cdot 13^{7} \cdot 17^{17} \cdot$
$19^{21} \cdot 23^{25} \cdot 29^{21} \cdot 31^{15} \cdot 37^{21} \cdot 41^{7} \cdot$
$43^{21} \cdot 47^{15} \cdot 53^{21} \cdot 59^{19}$
$= 4254009852517873300162885099095\ldots$
$2062912177152225723412983561076\ldots$
$4204241788115952166723818682709\ldots$
$1838340314531482866349985859639\ldots$
$6267146087126501265378899938492\ldots$
$1198219578838439107499451558520\ldots$
$6839301168107657439662602002788\ldots$
$1200381075268878821015628074667\ldots$
$9122187572659828211350474489248\ldots$
$5934282167896560823266182229402\ldots$
$7587626403589167148247045777416\ldots$
$0442969912665803065843000000000\ldots$
000000000000
$\approx 4{,}3 \cdot 10^{383}$

■ Gödelization of φ_3

$\ulcorner 0 + 0 = 0 \rightarrow 0 = 0 \urcorner$
$= 2^{21} \cdot 3^{25} \cdot 5^{21} \cdot 7^{15} \cdot 11^{21} \cdot 13^{7} \cdot 17^{21} \cdot$
$19^{15} \cdot 23^{21}$
$= 7733351355658080332438994260291\ldots$
$6040167200248925434737188323592\ldots$
$2061839580272843306988370847036\ldots$
$4426273272154163855898916739004\ldots$
$77670000000000000000000000$
$= \approx 7{,}7 \cdot 10^{148}.$

■ Gödelization of φ_4

$\ulcorner 0 = 0 \urcorner$
$= 2^{21} \cdot 3^{15} \cdot 5^{21}$
$= 14348907000000000000000000000$
$\approx 1{,}4 \cdot 10^{28}$

Figure 4.7: Gödelization of the proof steps of Theorem PA1

It's important to note that the terminology we use in the field of recursive functions today emerged only after 1931. Hence, the term *primitive-recursive function* does not appear in Gödel's original work. What we now refer to as *primitive-recursive*, Gödel simply called *recursive*.

The oldest known paper mentioning the term *primitive recursion* was published by the Hungarian mathematician Rózsa Péter (Figure 4.8) in 1934 [154], and the term *primitive-recursive function* first appeared in a 1936 paper by Stephen Cole Kleene [117].

Nevertheless, the term *recursive function* is still in use today as an abbreviation for the larger class of so-called μ-*recursive functions*, which comprises all computable functions.

4.2.2 Primitiv-Recursive Functions

This section sheds light on specific arithmetic functions central to Gödel's proof. In the literature, they are aptly referred to as *primitive-recursive functions*, as we can obtain them recursively from a small number of primitive functions. The following definition clarifies the details:

 Definition 4.1 (Primitive-Recursive Function)

- The following functions are primitive recursive:
  - The zero function $z(n) := 0$
  - The successor function $s(n) := n+1$
  - The projection $p_i^n(x_1, \ldots, x_n) := x_i$

- If $g : \mathbb{N}^k \to \mathbb{N}$ and $h_1, \ldots, h_k : \mathbb{N}^n \to \mathbb{N}$ are primitive recursive, so is $f(x_1, \ldots, x_n)$ defined by:

$$f(x_1, \ldots, x_n) = g(h_1(x_1, \ldots, x_n), \ldots, h_k(x_1, \ldots, x_n))$$

- If $g : \mathbb{N}^n \to \mathbb{N}$ and $h : \mathbb{N}^{n+2} \to \mathbb{N}$ are primitive recursive, so is $f(m, x_1, \ldots, x_n)$ defined by:

$$f(0, x_1, \ldots, x_n) = g(x_1, \ldots, x_n),$$
$$f(m+1, x_1, \ldots, x_n) = h(f(m, x_1, \ldots, x_n), m, x_1, \ldots, x_n).$$

The first rule defines the elementary primitive-recursive functions, namely the zero function, the successor function, and the projections. The other rules specify how to create new primitive-recursive functions from existing ones. According to the definition above, we are dealing with two different construction schemes:

- **Composition**

  The composition rule permits the usage of primitive-recursive functions as parameters in other primitive-recursive functions. For example, if $g(x_1, x_2, x_3)$ is primitive-recursive, so is the following function:

  $$f(x_1, x_2) := g(x_2, x_1, x_1) = g(p_2^2(x_1, x_2), p_1^2(x_1, x_2), p_1^2(x_1, x_2))$$

  In passing, the example demonstrates a valuable property of the projection functions. They allow us to select or swap certain variables.

Rózsa Péter
(1905 – 1977)

Figure 4.8: The Hungarian mathematician Rózsa Péter was one of the leading figures in the field of recursion theory. As the author of several popular science books, she also succeeded in inspiring an audience far beyond the scientific community [156,157].

4.2 The First Incompleteness Theorem

■ **Primitive recursion**

This construction scheme embodies the very essence of primitive-recursive functions. A closer look reveals that the function value f is calculated in a loop with m acting as the iteration variable. If $m = 0$, the function value is given by the function g. If $m > 0$, the function value is determined by applying the function h to the function value $f(m-1, x_1, \ldots, x_n)$ and the parameters $m-1$ and $x_1, \ldots, x_n$.

The primitive-recursion scheme is capable of expressing all standard arithmetic operations. For example, to come up with a primitive-recursive definition of the addition, multiplication, and exponentiation of natural numbers, we start with the following representation:

$$\text{add}(m,n) = \begin{cases} n & \text{if } m = 0 \\ s(\text{add}(m-1,n)) & \text{if } m > 0 \end{cases} \quad (4.2)$$

$$\text{mult}(m,n) = \begin{cases} 0 & \text{if } m = 0 \\ \text{add}(\text{mult}(m-1,n),n) & \text{if } m > 0 \end{cases} \quad (4.3)$$

$$\text{pow}(m,n) = \begin{cases} 1 & \text{if } m = 0 \\ \text{mult}(\text{pow}(m-1,n),n) & \text{if } m > 0 \end{cases} \quad (4.4)$$

Through the clever application of the projection function, we can bring (4.2) to (4.4) into the desired form:

$$\text{add}(0,n) = p_1^1(n),$$
$$\text{add}(m+1,n) = s(p_1^3(\text{add}(m,n),m,n))$$

$$\text{mult}(0,n) = 0,$$
$$\text{mult}(m+1,n) = \text{add}(p_1^3(\text{mult}(m,n),m,n), p_3^3(\text{mult}(m,n),m,n))$$

$$\text{pow}(0,n) = s(0),$$
$$\text{pow}(m+1,n) = \text{mult}(p_1^3(\text{pow}(m,n),m,n), p_3^3(\text{pow}(m,n),m,n))$$

It is straightforward to transfer the concept of primitive-recursive functions to relations. For this purpose, we associate the existence or non-existence of a relation to the existence of a corresponding *characteristic function*:

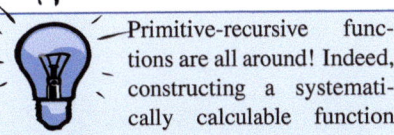

Primitive-recursive functions are all around! Indeed, constructing a systematically calculable function that cannot be defined using primitive recursion alone is quite challenging. In 1926, David Hilbert even conjectured that all computable functions might be primitive recursive [92]. He had been disproved the same year by Wilhelm Ackermann, who successfully constructed a function that was not primitive-recursive but still calculable using nested recursion. Ackermann published his function in 1928 [2]. In 1935, Rózsa Péter simplified the construction and came up with the following well-known form [155]:

$$\mathfrak{A}(0,n) := 2 \cdot n + 1$$
$$\mathfrak{A}(m+1,0) := \mathfrak{A}(m,1)$$
$$\mathfrak{A}(m+1,n+1) := \mathfrak{A}(m, \mathfrak{A}(m+1,n))$$

At first glance, the function appears innocuous. However, due to the clever choice of the recursion scheme, it grows faster than any primitive-recursive function. The function value $\mathfrak{A}(4,2)$ already corresponds to a number with approximately 20,000 decimal places, and for larger values of m and n, we can hardly calculate $\mathfrak{A}(m,n)$.

In honor of Wilhelm Ackermann, this function is called the *Ackermann function*. Some authors are more precise and refer to it as the *Ackermann-Péter function* in recognition of its origin.

> **Definition 4.2** (Primitive-Recursive Relation)
>
> A relation R between the natural numbers $x_1, \ldots, x_n$ is called *primitive recursive*, if there exists a primitive-recursive function f with the following property:
>
> $$R(x_1, \ldots, x_n) \Leftrightarrow f(x_1, \ldots, x_n) = 0$$
>
> We call f the *characteristic function* of R.

4.2.3 Arithmetic Representability

In this section, we will utilize Peano arithmetic (PA) to talk about primitive-recursive functions. In the preceding sections, we have repeatedly stressed that PA can formalize manifold properties of numbers and functions. But how exactly was this meant? For instance, how can we formally express that a natural number is even? Peano arithmetic knows no other operations apart from the successor function, addition, and multiplication, so how can it talk about anything that is not even present in its language reservoir?

Once again, the solution comes in the form of the *extensionality principle*, a concept we have already encountered in the context of set theory. According to this principle, the meaning of an expression is given by its scope, its extension, that is, the objects it names or describes. In our case, these objects are sets of natural numbers. The principle points the way to capture the following statement extensionally:

> "x is an even natural number"

It is described by the set of even numbers.

The expressive power of Peano arithmetic makes it easy to characterize the set of even numbers by a formula $\varphi(\xi)$ with a free variable ξ. In particular, we need to choose $\varphi(\xi)$ such that the formulas

$$\varphi(\overline{0}), \varphi(\overline{2}), \varphi(\overline{4}), \varphi(\overline{6}), \varphi(\overline{8}), \varphi(\overline{10}), \ldots$$

are true, and the formulas

$$\varphi(\overline{1}), \varphi(\overline{3}), \varphi(\overline{5}), \varphi(\overline{7}), \varphi(\overline{9}), \varphi(\overline{11}), \ldots$$

are false. The following formula suits our needs:

$$\varphi(\mathsf{x}) = (\exists \mathsf{z}\, \mathsf{x} = \mathsf{z} \times \overline{2})$$

4.2 The First Incompleteness Theorem

Upon closer inspection, it becomes apparent that we can not only represent properties of natural numbers, i.e., single-digit relations. It is also possible to arithmetically represent relations between two or more natural numbers (Figure 4.9). All we need to do is generalize the outline methodology with the following definition:

 Definition 4.3 (Semantically Representable Relation)

Let $R \subseteq \mathbb{N}^n$ be a relation and φ be a formula with n free variables. R is *semantically represented* by φ, if the following holds:

$$(x_1,\ldots,x_n) \in R \;\Rightarrow\; \models \varphi(\overline{x_1},\ldots,\overline{x_n})$$
$$(x_1,\ldots,x_n) \notin R \;\Rightarrow\; \models \neg\varphi(\overline{x_1},\ldots,\overline{x_n})$$

Functions are also arithmetically representable by treating a function of arity n as a relation of arity $n+1$:

 Definition 4.4 (Semantically Representable Function)

Let $f : \mathbb{N}^n \to \mathbb{N}$ be a function and φ be a formula with $n+1$ free variables. f is *semantically represented* by φ, if the following holds:

$$f(x_1,\ldots,x_n) = y \;\Rightarrow\; \models \varphi(\overline{x_1},\ldots,\overline{x_n},\overline{y})$$
$$f(x_1,\ldots,x_n) \neq y \;\Rightarrow\; \models \neg\varphi(\overline{x_1},\ldots,\overline{x_n},\overline{y})$$

Let's breathe life into the definition and explore how to arithmetically represent the function $\mathrm{pow}(x,y)$ introduced in Section 4.2.2. We commence by constructing a formula with a single free variable z, structured according to the following scheme:

$$\exists u_0 \ldots \exists u_y \, (\psi_0(x,u_0) \wedge \ldots \wedge \psi_y(x,u_y) \wedge z = u_y) \qquad (4.5)$$

For each natural number i with $0 \leq i \leq y$, this formula contains a bound variable u_i and a subformula ψ_i. If we choose ψ_i such that $\psi_i(x,u_i)$ is true precisely for $u_i = x^i$, variable u_y represents the function value z. The construction of the subformulas ψ_i presents no difficulty for us, as we can extract their wording directly from the primitive-recursion scheme of exponentiation:

$$\psi_0(x,u_0) := (u_0 = \overline{1})$$
$$\psi_{i+1}(x,u_{i+1}) := (\forall w \, (\psi_i(x,w) \to u_{i+1} = w \times x))$$

Arithmetically representable relations

- "x is an even natural number."
$$\varphi(x) := (\exists z \, x = z \times \overline{2})$$

- "x is a square number."
$$\varphi(x) := (\exists z \, x = z \times z)$$

- "x divides y."
$$\varphi(x,y) := (\exists z \, x \times z = y)$$

- "x is greater than or equal to y."
$$\varphi(x,y) := (\exists z \, x = y + z)$$

- "x is greater than y."
$$\varphi(x,y) := (\exists z \, x = y + z + \overline{1})$$

- "x is a prime number."
$$\varphi(x) :=$$
$$(\neg(x = \overline{1}) \wedge$$
$$\forall z \, (z \mid x \to (z = \overline{1} \vee z = x)))$$

- "x and y are twin primes."
$$\varphi(x,y) :=$$
$$(\neg(x = \overline{1}) \wedge \neg(y = \overline{1}) \wedge$$
$$\forall z \, (z \mid x \to (z = \overline{1} \vee z = x)) \wedge$$
$$\forall z \, (z \mid y \to (z = \overline{1} \vee z = y)) \wedge$$
$$y = x + \overline{2})$$

Figure 4.9: A small selection of arithmetically representable relations

- $\varphi_\alpha(u,0,\overline{1})$

    "Value 1 is at position 0."

    u = | 1 | | | ... | |

- $\forall v \, \forall w \, (v < y \wedge \varphi_\alpha(u,v,w) \to$
  $\varphi_\alpha(u,v+\overline{1}, w \times x))$

    "If value w is at position v, then value $w \cdot x$ is at position $v+1$."

    u = | 1 | x | x^2 | ... | x^y |

- $\varphi_\alpha(u,y,z)$

    "Value z is at position y."

    u = | 1 | x | x^2 | ... | z |
     y

Figure 4.10: If there were a function with the properties of φ_α, we would be able to represent the exponential function $z = x^y$ arithmetically.

However, a serious problem remains: Because we have employed an indeterminate number of quantifiers and the free variable y also appears as the index of the variable u_y, (4.5) is not a genuine formula of Peano arithmetic. Hence, we must find a way to translate it into a proper arithmetic formula.

To do so, we will pick up the idea from Chapter 1 that enabled us to identify the sets $\mathbb{N}$ and $\mathbb{N}^n$ as equipotent. We showed how to uniquely encode finite sequences of natural numbers into a single natural number, and we can apply a similar trick to our number sequence $x^0, \ldots, x^y$.

For this purpose, we assume the existence of a function $\alpha : \mathbb{N}^2 \to \mathbb{N}$ such that for every finite sequence $a_0, \ldots, a_y$, there is a number b satisfying

$$\alpha(b,0) = a_0, \alpha(b,1) = a_1, \ldots, \alpha(b,y) = a_y.$$

If we find a way to represent α arithmetically with a formula φ_α, formula (4.5) could be rewritten as follows:

$$\exists u \, (\varphi_\alpha(u,0,\overline{1}) \wedge$$
$$\forall v \, \forall w \, (v < y \wedge \varphi_\alpha(u,v,w) \to \varphi_\alpha(u,v+\overline{1}, w \times x)) \wedge \quad (4.6)$$
$$\varphi_\alpha(u,y,z))$$

Figure 4.10 illustrates the meaning of the individual formula components.

We are already close to solving our problem. The number of quantifiers in (4.6) is now constant, and the variable y no longer occurs as an index of another variable. Yet, we must still develop a function with the property of α.

Thanks to Gödel, we can now call such a function our own. In contrast to our fictitious function α with two variables, he introduced a function β with three variables:

$$\beta(x,y,z) := x \bmod (1 + y \cdot (z+1))$$

The following theorem reveals that this function does the trick:

 Theorem 4.3

> For every finite sequence of natural numbers $a_0, \ldots, a_{k-1}$, there exist two natural numbers, b and c, such that
>
> $$a_i = \beta(b,c,i) = b \bmod (1 + c \cdot (i+1))$$

4.2 The First Incompleteness Theorem

Proof: We prove the theorem in two steps:

■ **Step 1:** We show that the numbers

$$1+l! \cdot 1, \ 1+l! \cdot 2, \ 1+l! \cdot 3, \ldots, \ 1+l! \cdot l$$

are pairwise coprime for every $l \in \mathbb{N}$. We can conduct the proof with elementary number-theoretical arguments. If there were a prime number p dividing both

$$(1+l! \cdot i) \text{ and } (1+l! \cdot j), \qquad (1 \leq i < j \leq l)$$

then p would also divide the difference

$$(1+l! \cdot j) - (1+l! \cdot i) \ = \ l! \cdot (j-i).$$

Consequently, at least one of the numbers $l!$ or $(j-i)$ must be divisible by p. We now show that both assumptions lead to a contradiction:

- Suppose $p \mid l!$. Then p is also a divisor of $l! \cdot i$, which contradicts the assumption that p divides $1 + l! \cdot i$.

- Suppose $p \mid (j-i)$. Since $(j-i) < l$, $(j-i)$ is a divisor of $l!$. However, this implies that p also divides $l!$, which we have just disproven.

■ **Step 2:** We define the number l as

$$l := \max\{k, a_0, a_1, \ldots, a_{k-1}\}$$

and consider the congruence

$$\begin{aligned} x &\equiv a_0 & \mod \ (1+l! \cdot 1) \\ x &\equiv a_1 & \mod \ (1+l! \cdot 2) \\ & \cdots \\ x &\equiv a_{k-1} & \mod \ (1+l! \cdot k) \end{aligned}$$

Since the modules $(1 + l! \cdot i)$ are pairwise coprime, we can apply the *Chinese remainder theorem*, which guarantees the simultaneous congruence to have a solution. Let's denote this solution as b and set $c := l!$. Then, for all a_i, it holds that

$$a_i = b \bmod (1 + c \cdot (i+1)),$$

which was to be proven. □

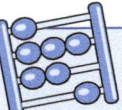

The *Sunzi Suanjing* is one of the most influential works of early Chinese mathematics. It was composed by the arithmetic master Sunzi in the first half of the first century, probably between 280 and 473 AD [217]. We find the most famous passage of the Sunzi Suanjing in the third and last chapter, where the master asks in Task 26 for the solution to the following riddle [113]:

"There are certain things whose number is unknown. If we count them by threes, we have two left over; by fives, we have three left over; and by sevens, two are left over. How many things are there?"

In modern parlance, solving the riddle means to solve the following system of *linear congruences*:

$$\begin{aligned} x &\equiv 2 & \mod 3 \\ x &\equiv 3 & \mod 5 \\ x &\equiv 2 & \mod 7 \end{aligned}$$

We also call such a system a *simultaneous congruence*. Today, we can fall back on several theorems that make statements about their solvability. Due to their origin, these theorems are called *Chinese remainder theorems*. The variant used in the proof of Theorem 4.3 states the following: If $m_0, \ldots, m_n$ are natural numbers that are pairwise coprime, and $a_0, \ldots, a_n$ are arbitrary integers, then the simultaneous congruence

$$\begin{aligned} x &\equiv a_0 & \mod m_0 \\ x &\equiv a_1 & \mod m_1 \\ & \cdots \\ x &\equiv a_n & \mod m_n \end{aligned}$$

has a unique solution modulo $m_0 \cdot \ldots \cdot m_n$.

- $a_0 = 1, a_1 = 2$

  For $b = 5$ and $c = 1$, we have:

  5 mod $(1 + 1 \cdot 1) = 1$
  5 mod $(1 + 1 \cdot 2) = 2$
  5 mod $(1 + 1 \cdot 3) = 1$
  5 mod $(1 + 1 \cdot 4) = 0$
  5 mod $(1 + 1 \cdot 5) = 5$

$\beta(b,c,0)$	$\beta(b,c,1)$	$\beta(b,c,2)$	$\beta(b,c,3)$	$\beta(b,c,4)$	
1	2	1	0	5	...

- $a_0 = 1, a_1 = 2, a_2 = 4$

  For $b = 67$ and $c = 2$, we have:

  67 mod $(1 + 2 \cdot 1) = 1$
  67 mod $(1 + 2 \cdot 2) = 2$
  67 mod $(1 + 2 \cdot 3) = 4$
  67 mod $(1 + 2 \cdot 4) = 4$
  67 mod $(1 + 2 \cdot 5) = 1$

$\beta(b,c,0)$	$\beta(b,c,1)$	$\beta(b,c,2)$	$\beta(b,c,3)$	$\beta(b,c,4)$	
1	2	4	4	1	...

- $a_0 = 1, a_1 = 2, a_2 = 4, a_3 = 8$

  For $b = 43058$ and $c = 6$, we have:

  43058 mod $(1 + 6 \cdot 1) = 1$
  43058 mod $(1 + 6 \cdot 2) = 2$
  43058 mod $(1 + 6 \cdot 3) = 4$
  43058 mod $(1 + 6 \cdot 4) = 8$
  43058 mod $(1 + 6 \cdot 5) = 30$

$\beta(b,c,0)$	$\beta(b,c,1)$	$\beta(b,c,2)$	$\beta(b,c,3)$	$\beta(b,c,4)$	
1	2	4	8	30	...

Figure 4.11: Coding of number sequences with Gödel's β function

Figure 4.11 demonstrates how Gödel's β function can represent the initial segments of the sequence of all powers of two.

Numerous functions exist to encode natural number sequences into a single number. Nevertheless, it is essential to represent the function arithmetically, which is readily feasible for Gödel's β function:

$$\varphi_\beta(b,c,i,a) := \exists d\, b = s(c \times s(i)) \times d + a \wedge a < s(c \times s(i)) \quad (4.7)$$

At this point, we have crossed the finish line, as we can arithmetically represent the exponential function as follows:

$$\exists b\, \exists c\, (\varphi_\beta(b,c,0,\overline{1}) \wedge$$
$$\forall v\, \forall w\, (v < y \wedge \varphi_\beta(b,c,v,w) \rightarrow \varphi_\beta(b,c,v+\overline{1},w \times x)) \wedge$$
$$\varphi_\beta(b,c,y,z))$$

Replacing φ_β by its definition, we obtain:

$$\exists b\, \exists c\, (\exists d\, b = s(c \times s(0)) \times d + \overline{1} \wedge \overline{1} < s(c \times s(0)) \wedge$$
$$\forall v\, \forall w\, (v < y \wedge \exists d\, b = s(c \times s(v)) \times d + w \wedge w < s(c \times s(v)) \rightarrow$$
$$\exists d\, b = s(c \times s(v+\overline{1})) \times d + (w \times x) \wedge (w \times x) < s(c \times s(v+\overline{1}))) \wedge$$
$$\exists d\, b = s(c \times s(y)) \times d + z \wedge z < s(c \times s(y)))$$

At first glance, the formula may appear to be a random assortment of arithmetic expressions. However, a closer inspection makes it shine brightly. On the inside, it conceals a mathematical jewel in the form of Gödel's β function.

Can we represent other functions arithmetically in the same way? The answer is yes! Gödel's β function is of such a general nature that we can represent any primitive-recursive function or relation arithmetically according to the same scheme. Generalizing the formula construction shown above yields the following theorem:

Theorem 4.4

Every primitive-recursive relation $R(x_1, \ldots, x_n)$ is semantically representable within Peano Arithmetic.

So far, we have been using a semantic type of representation since we have distinguished between true and false formulas. We now add a second type of arithmetic representability by replacing the concept of truth with provability. Since proofs in formal systems are carried out entirely on the syntactic level, we speak of a *syntactic representation*.

4.2 The First Incompleteness Theorem

Definition 4.5 (Syntactically Representable Relation)

Let $R \subseteq \mathbb{N}^n$ be a relation and φ be a formula with n free variables. R is *syntactically represented* by φ, if the following holds:

$$(x_1, \ldots, x_n) \in R \;\Rightarrow\; \vdash \varphi(\overline{x_1}, \ldots, \overline{x_n})$$
$$(x_1, \ldots, x_n) \notin R \;\Rightarrow\; \vdash \neg\varphi(\overline{x_1}, \ldots, \overline{x_n})$$

The formulas in Figure 4.9 represent the corresponding relations and functions not only semantically but also syntactically. We refrain from proving this result, as it would take substantial effort. In particular, for all natural numbers, we had to show that the corresponding formula instances of φ are provable or unprovable in the formal system of Peano arithmetic.

Gödel demonstrated that the statement of Theorem 4.4 remains valid on the syntactic level. In his original work, this is the statement of his famous Theorem V, which we quoted in its original wording in Figure 4.1. In a modern formulation, it reads as follows:

Theorem 4.5 (Gödel, 1931)

Every primitive-recursive relation $R(x_1, \ldots, x_n)$ is syntactically representable within Peano Arithmetic.

This theorem is a milestone on the way to the incompleteness results. Its proof is so technical that even Gödel confined himself to a rough sketch. A detailed elaboration can be found in [195].

Next, we will extend the concept of syntactic representability to functions. This was a simple task on the semantic level, as a comparative look at Definitions 4.3 and 4.5 reveals. In this case, we were able to derive the syntactic variant from its semantic counterpart by merely replacing the model relation '$\models$' with the provability relation '$\vdash$'. We could proceed in the same way for functions and choose the following formulation based on Definition 4.4:

Let $f : \mathbb{N}^n \to \mathbb{N}$ be a function and φ a formula with $n+1$ free variables. f is represented by φ *syntactically* if the following holds:

$$f(x_1, \ldots, x_n) = y \;\Rightarrow\; \vdash \varphi(\overline{x_1}, \ldots, \overline{x_n}, \overline{y}) \qquad (4.8)$$
$$f(x_1, \ldots, x_n) \neq y \;\Rightarrow\; \vdash \neg\varphi(\overline{x_1}, \ldots, \overline{x_n}, \overline{y}) \qquad (4.9)$$

The concept of arithmetic representability is likewise affected by the Babylonian linguistic confusion that we have to live with in mathematical logic. Particularly in the Anglo-Saxon literature, a variety of different terms are employed. For example, Elliott Mendelson's standard work refers to syntactically representable relations as *expressible* and syntactically representable functions as *representable* [134].
Peter Smith utilizes the term *expressible* in his excellent book on Gödel's incompleteness theorems as a synonym for semantic representability, regardless of whether functions or relations are meant [195]. Contrarily, he refers to syntactically representable functions and relations as *capturable*.

However, this definition would be weak in a crucial point: If we represent a function $f(x_1,\ldots,x_n)$ semantically, $\varphi(\overline{x_1},\ldots,\overline{x_n},\overline{y})$ is a substantively true statement if and only if the function value $f(x_1,\ldots,x_n)$ equals y.

$$\models \varphi(\overline{x_1},\ldots,\overline{x_n},\overline{y}) \Leftrightarrow y = f(x_1,\ldots,x_n)$$

We can also express this fact within Peano arithmetic. If the function f is semantically represented by φ, the following formula instances are all true:

$$\forall y\, (\varphi(\overline{x_1},\ldots,\overline{x_n},y) \leftrightarrow y = \overline{f(x_1,\ldots,x_n)}) \qquad (4.10)$$

However, when syntactically representing function f, there is no guarantee that the formula instances (4.10), albeit true, are also provable. Within a formal system, it is quite possible that we can talk about the relational properties of a function f in the sense of (4.8) and (4.9) but not about the property of f of being a function. In some situations, though, this is relevant. For this reason, we only consider a function f as syntactically represented if the formula φ, in addition to (4.8) and (4.9), satisfies the following:

$$\vdash \forall y\, (\varphi(\overline{x_1},\ldots,\overline{x_n},y) \leftrightarrow y = \overline{f(x_1,\ldots,x_n)})$$

An equivalent formulation is this one:

$$f(x_1,\ldots,x_n) = y \;\Rightarrow\; \vdash \forall y\, (\varphi(\overline{x_1},\ldots,\overline{x_n},y) \leftrightarrow y = \overline{y}) \qquad (4.11)$$

The additional requirement (4.11) turns out strong enough to derive the properties (4.8) and (4.9) with little effort. Therefore, it is sufficient to make (4.11) the sole requirement for the syntactic representability of functions:

Definition 4.6 (Syntactically Representable Function)

> Let $f : \mathbb{N}^n \to \mathbb{N}$ be a function and φ be a formula with $n+1$ free variables. f is *syntactically represented* by φ if the following holds:
>
> $$f(x_1,\ldots,x_n) = y \;\Rightarrow\; \vdash \forall y\, (\varphi(\overline{x_1},\ldots,\overline{x_n},y) \leftrightarrow y = \overline{y})$$

The upcoming theorem states that the property of being syntactically representable within Peano arithmetic applies not only to primitive-recursive relations but also to primitive-recursive functions:

Theorem 4.6

> Every primitive-recursive function $f(x_1,\ldots,x_n)$ is syntactically representable within Peano Arithmetic.

The proof of this theorem, just like the proof of Theorem 4.5, is technically demanding. We allow us to skip the formal details and refer the interested reader to the detailed presentation in [195].

4.2.4 Gödel's Diagonal Argument

Before setting the stage for the grand finale of the proof of the incompleteness theorem, let us briefly summarize the results obtained so far:

- Section 4.2.1 showed how to arithmetize the syntax of a formal language. By assigning each formula φ a Gödel number $\ulcorner \varphi \urcorner$, we succeeded in describing the manipulation of character strings on the arithmetic level.

- Section 4.2.2 introduced the concept of primitive-recursive functions and applied it to numerical relations. Without dwelling on the details, we have hinted that many functions encountered in everyday mathematics are primitive-recursive.

- Section 4.2.3 introduced the notion of arithmetic representability and showed how to utilize Gödel's β function to syntactically represent primitive-recursive relations and functions within Peano arithmetic.

When examined in isolation, these results appear like ordinary mathematical statements; each one sheds light on a particular aspect of mathematical logic, but none seems to threaten the very foundations of mathematics. However, when combined rightly, the three partial results develop a truly destructive power. With meticulous precision, Gödel worked out how the individual pieces of the puzzle fit together, and his work from 1931 occasionally reads like the blueprint for a mathematical explosive device. The devastating power of his work is well known. With the proof of the incompleteness theorems, Gödel shattered the long-cherished hope of a complete formalization of mathematics.

The following pages will retrace the key elements of Gödel's construction. An essential role play specific arithmetic formulas Gödel called *class signs* (*Klassenzeichen*). This term refers to the formulas of the form $\varphi(\xi)$, that is, formulas with one free variable.

For a better understanding of the following considerations, it is helpful to imagine these formulas as entries in an infinitely large table as depicted in Table 4.2. The formulas are arranged such that the formula with the Gödel number i, referred to as $\varphi_i(\xi)$ below, appears in the i-th

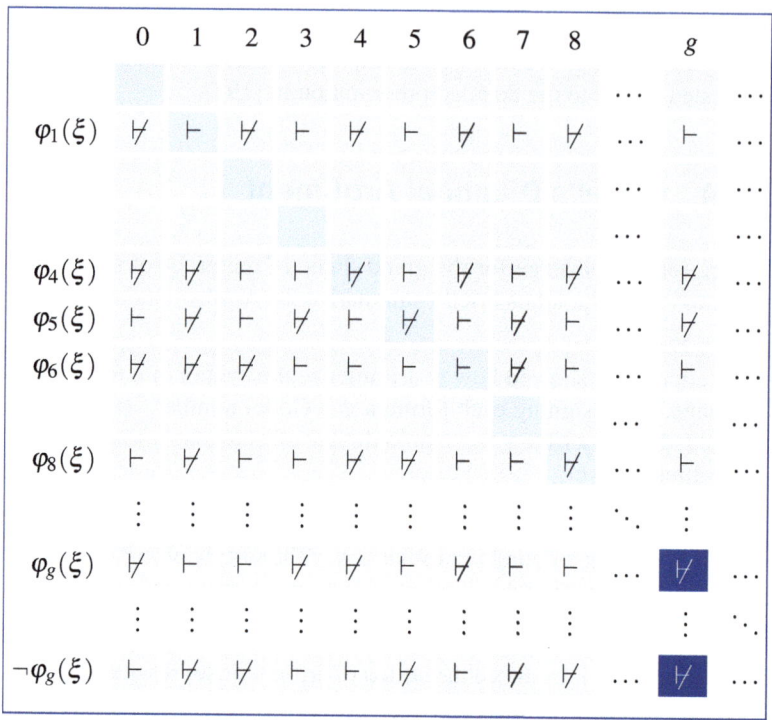

Table 4.2: Shown is a portion of an infinitely large table containing all arithmetic formulas with a single free variable. The table is structured such that the formula with the Gödel number i appears in the i-th row, and all rows whose row number is not the Gödel number of a formula with a single free variable are left empty. The table contains infinitely many columns, each marked with a natural number n. If the formula $\varphi_i(\bar{n})$ is provable within Peano arithmetic, the i-th row in the n-th column is marked with '⊢'. If not, the corresponding field is marked with '⊬'. In his 1931 work, Gödel demonstrated that there must be an undecidable statement on the main diagonal. For this purpose, he constructed a natural number g with the property that neither $\varphi_g(\bar{g})$ nor $\neg\varphi_g(\bar{g})$ is provable within Peano arithmetic.

row. Remember that not every natural number is a Gödel number, and not every Gödel number represents a formula with a single free variable. For that reason, several rows in the table have no entries.

Overall, we are not interested in open but closed formulas. We can obtain these by replacing the free variable in the formula $\varphi_i(\xi)$ with an arithmetic term of the form $\bar{n}$. This way, we acquire a closed formula $\varphi_i(\bar{n})$ for each natural number $n \in \mathbb{N}$. Some of these formulas are provable within Peano arithmetic ($\vdash \varphi(\bar{n})$); others are not ($\nvdash \varphi(\bar{n})$). In order to highlight this property, the table contains a separate column for each $n \in \mathbb{N}$. The formula $\varphi_i(\bar{n})$ is provable if the symbol '⊢' appears in the i-th row and n-th column. Otherwise, the field is marked with '⊬'.

Gödel adopted a clever argument in his proof, similar to Cantor's diagonal argument presented in Section 1.2.2. He succeeded in showing that there must be at least one formula on the table's main diagonal that is undecidable within Peano arithmetic. More specifically, there must be a natural number $g \in \mathbb{N}$ for which neither $\varphi_g(\bar{g})$ nor $\neg\varphi_g(\bar{g})$ is provable.

Let us work out how to calculate the value of g. First of all, consider

4.2 The First Incompleteness Theorem

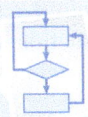

The fact that xBy is a primitive-recursive relation is not self-evident. The proof in Gödel's work extends over six densely written pages filled with formulas. Gödel achieved his result by defining 45 primitive-recursive functions and relations that build on each other and express increasingly complex facts. From a modern perspective, these functions and relations look like the auxiliary routines of a computer program, and this comparison is entirely appropriate. Today, we know that every function a program can calculate without while-loops is primitive-recursive and that every primitive-recursive function is computable by a program of that kind [100]. Despite being obscured by notation, the proof of the first incompleteness theorem is, in fact, one of the first computer programs of the twentieth century. However, Gödel could not have been aware of this; in 1931, programmable computers as we know them today were still a long way off.

1. $x/y \equiv (Ez)\,[z \leq x\,\&\,x = y\,.\,z]^{33})$
   x ist teilbar durch $y^{34})$.

2. $\mathrm{Prim}\,(x) \equiv \overline{(Ez)}\,[z \leq x\,\&\,z \neq 1\,\&\,z \neq x\,\&\,x/z]\,\&\,x > 1$
   x ist Primzahl.

   ...

30. $Sb_0\,(x\,{}^v_y) \equiv x$
    $Sb_{k+1}\,(x\,{}^v_y) \equiv Su\,[Sb_k\,(x\,{}^v_y)]\,(^{k\,St\,v,\,x}_{y})$

31. $Sb\,(x\,{}^v_y) \equiv Sb_{A\,(v,\,x)}\,(x\,{}^v_y)^{36})$
    $Sb\,(x\,{}^v_y)$ ist der oben definierte Begriff $Subst\,a\,(^v_b)^{37})$.

    ...

44. $Bw\,(x) \equiv (n)\,\{0 < n \leq l\,(x) \rightarrow Ax\,(n\,Gl\,x)\,\vee$
    $(Ep, q)\,[0 < p, q < n\,\&\,Fl\,(n\,Gl\,x, p\,Gl\,x, q\,Gl\,x)]\}$
    $\&\,l\,(x) > 0$
    x ist eine $Beweisfigur$ (eine endliche Folge von $Formeln$, deren jede entweder $Axiom$ oder $unmittelbare\,Folge$ aus zwei der vorhergehenden ist).

45. $x\,B\,y \equiv Bw\,(x)\,\&\,[l\,(x)]\,Gl\,x = y$
    x ist ein $Beweis$ für die $Formel\,y$.

    ...

46. $\mathrm{Bew}\,(x) \equiv (Ey)\,y\,B\,x$
    x ist eine $beweisbare\,Formel$. [Bew (x) ist der einzige unter den Begriffen 1—46, von dem nicht behauptet werden kann, er sei rekursiv.]

This excerpt from his original work exposes that Gödel defined 46 rather than 45 functions and relations. The last is the proof relation Bew, which we will encounter in Section 4.2.6. Unlike the first 45 functions and relations, Bew is not primitive recursive.

the function $diag(y)$ and the relation xBy with the following definition:

$$diag(y) := \begin{cases} \ulcorner \varphi_y(\overline{y}) \urcorner & \text{if } y = \ulcorner \varphi_y(\xi) \urcorner \\ 0 & \text{otherwise} \end{cases}$$

$xBy :\Leftrightarrow$ x encodes a proof for the formula with Gödel number y

The function $diag(y)$ maps the Gödel number of the formula $\varphi_y(\xi)$ to the Gödel number of the formula $\varphi_y(\overline{y})$, our table's y-th diagonal element. If y is not a Gödel number or the Gödel number of a formula with no or more than one free variable, the function value is irrelevant to us. For the sake of simplicity, we let those numbers map to 0.

Next, let us figure out the exact meaning of the relation xBy. First, we note that x and y are natural numbers, and we interpret x as the Gödel number of a proof and y as the Gödel number of an arithmetic formula φ. By definition, x and y relate to each other precisely when x encodes a proof of φ. In other words, x encodes a sequence of formulas that derives φ in a finite number of inference steps from the axioms of Peano arithmetic.

In his original work, Gödel proved that *diag* and *B* are primitive recursive. From Theorems 4.5 and 4.6, it then follows that the function *diag* is syntactically represented by a formula $\mathsf{Diag}(y,z)$ and the relation *B* by a formula $\mathsf{B}(x,y)$. In particular, the following applies:

$$diag(y) = z \Rightarrow \vdash \forall z\, (\mathsf{Diag}(\bar{y},z) \leftrightarrow z = \bar{z}) \tag{4.12}$$

$$(x,y) \in B \Rightarrow \vdash \mathsf{B}(\bar{x},\bar{y}) \tag{4.13}$$

$$(x,y) \notin B \Rightarrow \vdash \neg \mathsf{B}(\bar{x},\bar{y}) \tag{4.14}$$

We utilize Diag and B in the following construction:

$$\psi_{\mathrm{Gdl}}(x,y) := \exists z\, (\mathsf{Diag}(y,z) \wedge \mathsf{B}(x,z))$$

The relationships (4.12) to (4.14) let us conclude the following:

$$x \text{ encodes a proof for the formula } \varphi_y(\bar{y}) \Rightarrow \vdash \psi_{\mathrm{Gdl}}(\bar{x},\bar{y})$$
$$x \text{ does not encode a proof for the formula } \varphi_y(\bar{y}) \Rightarrow \vdash \neg \psi_{\mathrm{Gdl}}(\bar{x},\bar{y})$$

This clarifies the meaning of the formula $\psi_{\mathrm{Gdl}}(x,y)$. It is the syntactic representation of the following relation:

$$\mathrm{Gdl}(x,y) :\Leftrightarrow x \text{ is the Gödel number of a proof of } \varphi_y(\bar{y})$$

By definition, x and y are related to each other precisely when x encodes a proof of the formula $\varphi_y(\bar{y})$; that is, x encodes a sequence of formulas that derives the diagonal element $\varphi_y(\bar{y})$ in a finite number of inference steps from the axioms of Peano arithmetic.

From $\psi_{\mathrm{Gdl}}(x,y)$, we construct the following formula:

$$\varphi_g(y) := \forall x\, \neg \psi_{\mathrm{Gdl}}(x,y) \tag{4.15}$$

The result is an arithmetic formula with one free variable. It occurs in the g-th row of our table and is the formula we sought. It bears the fascinating property that its diagonalized statement $\varphi_g(\bar{g})$ is undecidable within PA; neither $\varphi_g(\bar{g})$ nor $\neg \varphi_g(\bar{g})$ is derivable from the axioms. Let us now uncover the reason.

- **Assume** $\vdash \varphi_g(\bar{g})$

    If $\varphi_g(\bar{g})$ were provable, some Gödel number m would encode a proof. $\varphi_g(\bar{g})$ is the diagonal element of the formula $\varphi_g(\xi)$; thus $\mathrm{Gdl}(m,g)$ applies. Since ψ_{Gdl} syntactically represents the relation Gdl, we have:

$$\vdash \psi_{\mathrm{Gdl}}(\bar{m},\bar{g}). \tag{4.16}$$

4.2 The First Incompleteness Theorem

The assumption $\vdash \varphi_g(\overline{g})$ spells out as

$$\vdash \forall x \, \neg \psi_{\text{Gdl}}(x, \overline{g}),$$

and by instantiating x with $\overline{m}$, we obtain

$$\vdash \neg \psi_{\text{Gdl}}(\overline{m}, \overline{g})$$

contradicting (4.16). Hence, the diagonal element $\varphi_g(\overline{g})$ is only provable if Peano arithmetic were contradictory. In that case, however, every formula would be a theorem.

■ **Assume** $\vdash \neg \varphi_g(\overline{g})$

The assumption in full is $\vdash \neg \forall x \, \neg \psi_{\text{Gdl}}(x, \overline{g})$, and from this it follows:

$$\vdash \exists x \, \psi_{\text{Gdl}}(x, \overline{g}). \quad (4.17)$$

If Peano arithmetic is consistent, the formula $\varphi_g(\overline{g})$ cannot be proven simultaneously. Consequently, no natural number can be the Gödel number of a proof for $\varphi_g(\overline{g})$, implying

$$(0, g) \notin \text{Gdl}, (1, g) \notin \text{Gdl}, (2, g) \notin \text{Gdl}, (3, g) \notin \text{Gdl}, \ldots$$

Given that ψ_{Gdl} syntactically represents the relation Gdl, we can draw the following conclusions:

$$\vdash \neg \psi_{\text{Gdl}}(\overline{0}, \overline{g}) \quad (4.18)$$

$$\vdash \neg \psi_{\text{Gdl}}(\overline{1}, \overline{g}) \quad (4.19)$$

$$\vdash \neg \psi_{\text{Gdl}}(\overline{2}, \overline{g}) \quad (4.20)$$

$$\vdash \neg \psi_{\text{Gdl}}(\overline{3}, \overline{g}) \quad (4.21)$$

$$\ldots$$

We find ourselves in a precarious situation. If $\neg \varphi_g(\overline{g})$ were provable within Peano arithmetic, so would the formula (4.17). This formula postulates the existence of a natural number x such that $\psi_{\text{Gdl}}(\overline{x}, \overline{g})$ is substantively true. On the other hand, the formulas (4.18), (4.19), (4.20), etc., seemingly refute this statement: for any natural number x, we can prove $\vdash \neg \psi_{\text{Gdl}}(\overline{x}, \overline{g})$. We succeeded in creating a contradiction, didn't we?

Let us not pass over the result too quickly. We only obtained the contradiction because we interpreted the proven formulas on a semantic level. If we had assumed the correctness of Peano arithmetic, i.e., if we had assumed that only substantively true arithmetic statements are derivable

Earlier, we emphasized that Gödel's proof relies on constructing a statement that postulates its own unprovability. We have already created this formula without explicitly pointing it out. This formula is $\varphi_g(\overline{g})$, and it is easy to see why. First of all, $\varphi_g(\overline{g})$ is the diagonal element of

$$\varphi_g(y) = \forall x \, \neg \psi_{\text{Gdl}}(x, y).$$

Each specific instance $\varphi_g(\overline{y})$ states that no x is the Gödel number of a proof for the diagonal element $\varphi_y(\overline{y})$:

$$\varphi_g(\overline{y}) \mathrel{\hat{=}} \text{``}\varphi_y(\overline{y}) \text{ is unprovable.''}$$

Then the formula $\varphi_g(\overline{g})$ carries the following substantive meaning:

$$\varphi_g(\overline{g}) \mathrel{\hat{=}} \text{``}\varphi_g(\overline{g}) \text{ is unprovable.''}$$

Or, equivalently:

$$\varphi_g(\overline{g}) \mathrel{\hat{=}} \text{``I am unprovable.''}$$

Gödel's proof is often misunderstood as being based on the semantic meaning of $\varphi_g(\overline{g})$. Some critics even see an irregular self-reference in the construction of $\varphi_g(\overline{g})$, reminiscent of Russell's antinomy. However, a quick review of the pages above clarifies that we never interpreted the formula $\varphi_g(\overline{g})$ in substantive terms. The proof that neither $\varphi_g(\overline{g})$ nor $\neg \varphi_g(\overline{g})$ are derivable within the formal system of Peano arithmetic, assuming this system is free of contradictions, was carried out purely on the syntactic level. Nevertheless, the semantic interpretation of $\varphi_g(\overline{g})$ is valuable for understanding Gödel's proof. It sheds a bright light on why undecidable sentences must exist in every sufficiently expressive formal system.

from the axioms, we would have reached our goal: the formulas (4.17), (4.18), (4.19), (4.20), etc. cannot be substantively true at the same time.

However, the syntactic variant of Gödel's first incompleteness theorem only assumes the consistency of Peano arithmetic. To contradict consistency, we must demonstrate that some formula φ is provable alongside its negation $\neg\varphi$. However, this is not achieved by proving (4.17), (4.18), (4.19), (4.20), etc. Even if these formulas cannot be substantively true simultaneously, they do not constitute a contradiction at the syntactic level. Gödel was aware of the situation but failed to devise a solution. At some point, he saw no option but to require ω-consistency rather than consistency as the prerequisite of the first incompleteness theorem. The following definition clarifies the exact meaning of this expression:

Definition 4.7 (ω-Consistency)

> A formal system is ω-*consistent* if the following applies:
>
> - The formal system is consistent.
>
> - If $\vdash \neg\varphi(\bar{n})$ for all $n \in \mathbb{N}$, then $\not\vdash \exists x\, \varphi(x)$.

ω-consistency is a stronger property than consistency. Every ω-consistent formal system is consistent, but not vice versa.

Utilizing the new terminology, we can formulate the variant of the first incompleteness theorem that Gödel proved in his original work. In modern parlance, it reads as follows:

Theorem 4.7 (Gödel, 1931)

> Every ω-consistent formal system that is expressive enough to formalize Peano Arithmetic is negation incomplete.

This result marks a successful ending to our journey along Gödel's historical path, at least for now.

4.2.5 Rosser's Trick

In its original formulation, Gödel's first incompleteness theorem makes a statement about ω-consistent formal systems. The fact that the weaker

4.2 The First Incompleteness Theorem

assumption of consistency can replace Gödel's assumption of ω-consistency was proven by the US mathematician John Barkley Rosser in 1936, around five years after the publication of the incompleteness theorems [173, 196].

It is a remarkable aspect of his work that Rosser could almost entirely retain Gödel's line of reasoning. In order to prove the incompleteness theorem in full generality, it is sufficient to replace Gödel's formula $\varphi_g(y)$ with Rosser's formula

$$\varphi_r(y) := \forall x\, (\psi_{\mathrm{Gdl}}(x,y) \rightarrow \exists (z \leq x)\, \psi_{\mathrm{Gdl}'}(z,y)). \tag{4.22}$$

Replacing $\varphi_g(\bar{g})$ with $\varphi_r(\bar{r})$ is frequently called *Rosser's trick*.

The subformula $\psi_{\mathrm{Gdl}'}(z,y)$ mentioned in (4.22) appears for the first time. It is the syntactic representation of the following primitive-recursive relation:

$$\mathrm{Gdl}'(x,y) :\Leftrightarrow x \text{ is the Gödel number of a proof for } \neg\varphi_y(\bar{y})$$

To demonstrate that formula $\varphi_r(\bar{r})$ is undecidable within Peano arithmetic, let us again distinguish two cases:

■ **Assume** $\vdash \varphi_r(\bar{r})$

If $\varphi_r(\bar{r})$ were provable, some Gödel number m would encode a proof. $\varphi_r(\bar{r})$ is the diagonal element of the formula $\varphi_r(\xi)$; thus $\mathrm{Gdl}(m,r)$ applies. Since ψ_{Gdl} syntactically represents the relation Gdl, we have:

$$\vdash \psi_{\mathrm{Gdl}}(\overline{m},\bar{r}). \tag{4.23}$$

The assumption $\vdash \varphi_r(\bar{r})$ spells out as

$$\vdash \forall x\, (\psi_{\mathrm{Gdl}}(x,\bar{r}) \rightarrow \exists (z \leq x)\, \psi_{\mathrm{Gdl}'}(z,\bar{r})),$$

and by instantiating variable x with $\overline{m}$, we obtain

$$\vdash \psi_{\mathrm{Gdl}}(\overline{m},\bar{r}) \rightarrow \exists (z \leq \overline{m})\, \psi_{\mathrm{Gdl}'}(z,\bar{r}). \tag{4.24}$$

From (4.23) and (4.24), we can deduce

$$\vdash \exists (z \leq \overline{m})\, \psi_{\mathrm{Gdl}'}(z,\bar{r}) \tag{4.25}$$

via the modus ponens. Assuming the consistency of Peano arithmetic, the formula $\neg\varphi_r(\bar{r})$ cannot be proven simultaneously, which

To better understand the essence of Rosser's formula φ_r, let us recapitulate the substantive meaning of Gödel's formula φ_g. We have already uncovered above that every instance $\varphi_g(\bar{y})$ represents the statement:

"$\varphi_y(\bar{y})$ is unprovable."

Hence, the diagonal element $\varphi_g(\bar{g})$ corresponds to:

"I am unprovable."

Let's try a similar analysis of Rosser's formula. After translating the individual formula components into colloquial language, the substantive meaning of $\varphi_r(\bar{y})$ becomes:

"If $\varphi_y(\bar{y})$ is provable, then there exists a shorter proof for $\neg\varphi_y(\bar{y})$."

The diagonal element $\varphi_r(\bar{r})$ thus corresponds to the statement:

"If $\varphi_r(\bar{r})$ is provable, then there exists a shorter proof for $\neg\varphi_r(\bar{r})$."

Or, which is the same:

"If I am provable, then a shorter proof exists for my negation."

Assuming consistency, this statement is equivalent to:

"I am unprovable."

Thus, there is no semantic difference between Gödel's and Rosser's diagonal elements as both postulate their own unprovability. On the other hand, the way both formulas encode their substantive meaning is entirely different, which is precisely where Rosser's proof hides its secret.

John Barkley Rosser was born in Jacksonville on December 6, 1907. After completing his master's degree in physics at the University of Florida, he transferred to the renowned Princeton University, where he received his doctorate in mathematical logic under the supervision of Alonzo Church in 1935 [172]. After shorter stays at Princeton and Harvard, he was appointed professor at Cornell University in 1936, which became his academic home for 30 years.

In addition to refining Gödel's proof (*Rossers Trick*), Rosser is best known for his work in recursion theory. In 1935, he attracted widespread attention when, in collaboration with Stephen Cole Kleene, he found a contradiction in the original formulation of Alonzo Church's λ-calculus (*Kleene-Rosser paradox*). Today, his name is primarily associated with the *Church-Rosser theorem*, which confirms the confluence of certain term substitution systems. (Theorem 2 in [36]). Last but not least, Rosser achieved a high level of recognition through several books that are now part of the standard literature on mathematical logic [174–176].

However, he was not just a theorist. During the Second World War, Rosser was involved in the design of ballistic missiles and later took on important advisory positions in space and military research. In 1963, he was appointed director of the *Army Mathematics Research Center* (AMRC), a US military facility for strategic support of the US invasion of Vietnam. In this role, he was not without controversy and publicly denied any involvement of the AMRC in military projects. Rosser retired in 1973 and died on September 5, 1989, aged 81.

means no natural number can be the Gödel number of a proof for $\neg \varphi_r(\bar{r})$. Therefore, the following applies:

$$\vdash \neg \psi_{\text{Gdl}'}(\overline{0}, \bar{r}) \tag{4.26}$$

$$\vdash \neg \psi_{\text{Gdl}'}(\overline{1}, \bar{r}) \tag{4.27}$$

$$\vdash \neg \psi_{\text{Gdl}'}(\overline{2}, \bar{r}) \tag{4.28}$$

$$\ldots$$

Together, (4.26) and (4.27) imply

$$\vdash \neg \exists (z \leq \overline{1})\, \psi_{\text{Gdl}'}(z, \bar{r}).$$

Adding formula (4.28) lets us derive

$$\vdash \neg \exists (z \leq \overline{2})\, \psi_{\text{Gdl}'}(z, \bar{r}).$$

If we continue along these lines, we eventually obtain

$$\vdash \neg \exists (z \leq \overline{m})\, \psi_{\text{Gdl}'}(z, \bar{r}),$$

contradicting (4.25). Hence, the diagonal element $\varphi_r(\bar{r})$ is only provable if Peano arithmetic were contradictory.

■ **Assume** $\vdash \neg \varphi_r(\bar{r})$

In full, the assumption reads

$$\vdash \neg \forall x\, (\psi_{\text{Gdl}}(x, \bar{r}) \to \exists (z \leq x)\, \psi_{\text{Gdl}'}(z, \bar{r})). \tag{4.29}$$

4.2 The First Incompleteness Theorem

Let m be the Gödel number of a proof for $\neg\varphi_r(\bar{r})$. Then the following holds:
$$\vdash \psi_{\text{Gdl}'}(\bar{m},\bar{r})$$
Peano arithmetic is expressive enough to derive the following, weakened statement:
$$\vdash \forall x\, ((\bar{m} \leq x) \to \exists (z \leq x)\, \psi_{\text{Gdl}'}(z,\bar{r})) \qquad (4.30)$$
As long as Peano arithmetic is consistent, the formula $\varphi_r(\bar{r})$ cannot be proven simultaneously. Consequently, no natural number can be the Gödel number of a proof for $\varphi_r(\bar{r})$. Thus we have:
$$\vdash \neg\psi_{\text{Gdl}}(\bar{0},\bar{r})$$
$$\vdash \neg\psi_{\text{Gdl}}(\bar{1},\bar{r})$$
$$\vdash \neg\psi_{\text{Gdl}}(\bar{2},\bar{r})$$
$$\ldots$$

As in the first case, the following theorems can be derived one after the other:
$$\vdash \neg\exists (z \leq \bar{1})\, \psi_{\text{Gdl}}(z,\bar{r})$$
$$\vdash \neg\exists (z \leq \bar{2})\, \psi_{\text{Gdl}}(z,\bar{r})$$
$$\ldots$$
$$\vdash \neg\exists (z \leq \overline{m-1})\, \psi_{\text{Gdl}}(z,\bar{r})$$

From this, we can draw the subsequent conclusion within PA:
$$\vdash \forall x\, (\psi_{\text{Gdl}}(x,\bar{r}) \to (\bar{m} \leq x)) \qquad (4.31)$$
Transitively combining (4.31) and (4.30) yields the theorem:
$$\vdash \forall x\, (\psi_{\text{Gdl}}(x,\bar{r}) \to \exists (z \leq x)\, \psi_{\text{Gdl}'}(z,\bar{r})) \qquad (4.32)$$
With (4.29) and (4.32), we succeeded in deriving a complementary pair of formulas. If $\neg\varphi_r(\bar{r})$ were provable, we would have identified Peano arithmetic as contradictory.

We are now ready to reap the fruits of our labor. Thanks to Rosser's trick, we can drop the requirement of ω-consistency, safely replacing it with the weaker requirement of consistency:

Theorem 4.8 (Rosser, 1936)

> Any consistent formal system expressive enough to formalize Peano arithmetic is negation incomplete.

The opening quotation of the section about the diagonalization lemma originates from the article *On Undecidable Propositions of Formal Mathematical Systems* [74]. It is a revised transcript of the famous *Princeton lectures* Gödel gave at the Institute for Advanced Study in the spring of 1934. In the original, the quoted passage has a footnote in which Gödel reveals the authorship of the diagonalization principle. It reads:

> "This was first noted by R. Carnap in: Logische Syntax der Sprache, Wien, 1934, page 91."

Indeed. Peeking into Carnap's work, we encounter a construction that follows the classic diagonalization pattern [26]:

> "Let any syntactical property of expressions be chosen [...]. Let $\mathfrak{S}_1$ be that sentence with the free variable 'x' (for which we will take the term-number 3) which expresses this property [...]. Let $\mathfrak{S}_2$ be that sentence which results from $\mathfrak{S}_1$ if for 'x' 'subst[x, 3, str(x)]' is substituted. [...] Thus, if $\mathfrak{S}_2$ is given, the series-number of $\mathfrak{S}_2$ can be calculated; let it be designated by 'b' [...]. Let the sentence 'subst[b, 3, str(b)]' be $\mathfrak{S}_3$; thus $\mathfrak{S}_3$ is the sentence which results from $\mathfrak{S}_2$ when the $\mathfrak{S}t$ with the value b is substituted for 'x'. It is easy to see that, syntactically interpreted, $\mathfrak{S}_3$ mans that $\mathfrak{S}_3$ itself has the chosen syntactical property."

We will employ Carnap's construction scheme one-to-one in the proof of Theorem 4.9. $\mathfrak{S}_1$, $\mathfrak{S}_2$, and $\mathfrak{S}_3$ correspond to our formulas χ, ψ, and γ.

This theorem is often referred to in the literature as the *Gödel-Rosser theorem*. It is identical in wording to Theorem 4.2, which we called the syntactic variant of Gödel's first incompleteness theorem.

4.2.6 The Diagonalization Lemma

> "It is even possible, for any metamathematical property f which can be expressed in the system, to construct a proposition which says of itself that it has this property."
>
> Kurt Gödel [74]

In the previous sections, we have repeatedly emphasized the ingenious construction of Gödel's undecidable formula $\varphi_g(\overline{g})$, which is the diagonal element of $\varphi_g(y)$. For every natural number y, the instance $\varphi_g(\overline{y})$ claims that the diagonal element of the formula with Gödel number y is unprovable within Peano arithmetic. The natural number g is the Gödel number of the formula $\varphi_g(y)$, so $\varphi_g(\overline{g})$ asserts nothing but its own unprovability!

We can talk about the provability of a formula within Peano arithmetic. By setting

$$\mathsf{Bew}(y) := \exists x\, \mathsf{B}(x, y),$$

the formula $\neg\mathsf{Bew}(\overline{y})$ expresses the absence of a proof for the formula with the Gödel number y. We can thus transfer the substantive meaning of $\varphi_g(\overline{g})$ directly into the following formula of Peano arithmetic:

$$\varphi_g(\overline{g}) \leftrightarrow \neg\mathsf{Bew}(\ulcorner \varphi_g(\overline{g}) \urcorner) \tag{4.33}$$

At the end of this section, we will realize that Peano arithmetic can do more than merely allow this formula to be written out. It is also capable of proving it:

$$\vdash \varphi_g(\overline{g}) \leftrightarrow \neg\mathsf{Bew}(\ulcorner \varphi_g(\overline{g}) \urcorner)$$

We will obtain this result as a particular case of a more general principle, usually referred to in the literature as the *diagonalization lemma* or *fixed point theorem*:

 Theorem 4.9 (Diagonalization Lemma)

> For every PA formula $\chi(\xi)$, there exists a PA formula γ satisfying
>
> $$\vdash \gamma \leftrightarrow \chi(\ulcorner \gamma \urcorner)$$

4.2 The First Incompleteness Theorem

To prove this theorem, we start by constructing the following formula from $\chi(\xi)$:

$$\psi(y) := \forall z\,(\mathrm{Diag}(y,z) \to \chi(z))$$

We have already learned about the formula Diag on Page 212. It syntactically represents the diagonalization function $diag(y)$, which maps a natural number y to the Gödel number of the formula $\varphi_y(\bar{y})$. The substantive meaning of ψ is thus apparent: For any chosen natural number y, the formula instance $\psi(\bar{y})$ expresses that the diagonal element $\varphi_y(\bar{y})$ has the property χ.

One of these formula instances is the formula γ, which we will use to prove the diagonalization lemma. It is created by diagonalizing $\psi(y)$ itself, i.e., by replacing y with the Gödel number of ψ:

$$\begin{aligned}\gamma &:= \psi(\ulcorner\psi\urcorner) \\ &= \forall z\,(\mathrm{Diag}(\ulcorner\psi\urcorner, z) \to \chi(z))\end{aligned}$$

We will now show that this formula fulfills the postulated property

$$\vdash \gamma \leftrightarrow \chi(\ulcorner\gamma\urcorner).$$

Before revealing the derivation, let's remember that the formula $\mathrm{diag}(y,z)$ represents the function $diag(y)$ syntactically, which means the following:

$$diag(y) = z \;\Rightarrow\; \vdash \forall z\,(\mathrm{Diag}(\bar{y},z) \leftrightarrow z = \bar{z}) \qquad (4.35)$$

By definition, γ is the diagonal element of ψ, so we have:

$$diag(\ulcorner\psi\urcorner) = \ulcorner\gamma\urcorner$$

From (4.35), it follows that

$$\vdash \forall z\,(\mathrm{Diag}(\ulcorner\psi\urcorner, z) \leftrightarrow z = \ulcorner\gamma\urcorner), \qquad (4.36)$$

which implies:

$$\vdash \mathrm{Diag}(\ulcorner\psi\urcorner, \ulcorner\gamma\urcorner) \qquad (4.37)$$

$$\vdash \mathrm{Diag}(\ulcorner\psi\urcorner, z) \to z = \ulcorner\gamma\urcorner \qquad (4.38)$$

After the groundwork has been laid, we can move on to deriving the formula $\chi(\ulcorner\gamma\urcorner) \leftrightarrow \gamma$. For this purpose, we first prove both directions of the equivalence separately:

 The proof of the diagonalization lemma relies on the formula $\chi(\ulcorner\gamma\urcorner)$ being deducible from the formula $z = \ulcorner\gamma\urcorner \to \chi(z)$. In fact, the derivation turns out more complex than expected: It makes use of the *substitution property of equality*, which can be formally captured as follows: If x and y are equal and E is a property, then x posesses the property E if and only if y does:

$$x = y \Rightarrow (E(x) \Leftrightarrow E(y))$$

Due to the symmetry of equality, we are free to choose the following equivalent characterization:

$$x = y \Rightarrow (E(y) \Rightarrow E(x))$$

Peano arithmetic is sufficiently expressive to prove the substitution property; every formula of the form

$$\xi = \zeta \to (\varphi(\zeta) \to \varphi(\xi)) \qquad (4.34)$$

is a theorem. From here, the derivation is simple. From

$$\vdash z = y \to (\chi(y) \to \chi(z)),$$

we can conclude

$$\vdash z = \ulcorner\gamma\urcorner \to (\chi(\ulcorner\gamma\urcorner) \to \chi(z)),$$

which implies

$$\vdash \chi(\ulcorner\gamma\urcorner) \to (z = \ulcorner\gamma\urcorner \to \chi(z))$$

Together with $\vdash \chi(\ulcorner\gamma\urcorner)$, the modus ponens yields the desired result:

$$\vdash z = \ulcorner\gamma\urcorner \to \chi(z)$$

It remains to show that all formulas of the type (4.34) are theorems, and this is where the real effort hides. It requires an inductive proof over the formula structure, as described in detail in [134].

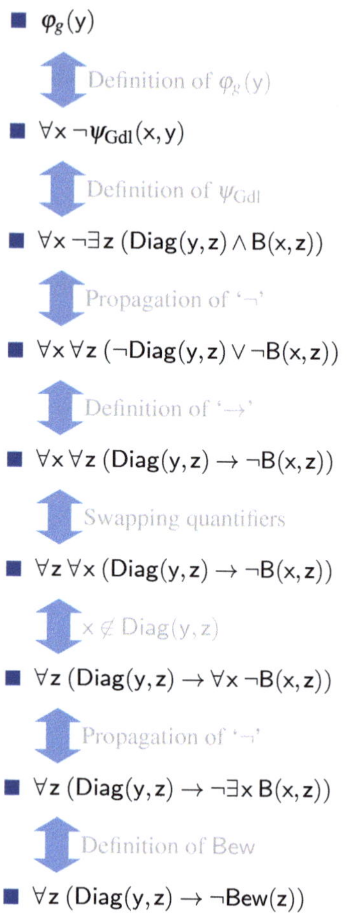

- $\varphi_g(y)$

  ↕ Definition of $\varphi_g(y)$

- $\forall x \, \neg \psi_{Gdl}(x,y)$

  ↕ Definition of ψ_{Gdl}

- $\forall x \, \neg \exists z \, (Diag(y,z) \wedge B(x,z))$

  ↕ Propagation of '$\neg$'

- $\forall x \, \forall z \, (\neg Diag(y,z) \vee \neg B(x,z))$

  ↕ Definition of '$\rightarrow$'

- $\forall x \, \forall z \, (Diag(y,z) \rightarrow \neg B(x,z))$

  ↕ Swapping quantifiers

- $\forall z \, \forall x \, (Diag(y,z) \rightarrow \neg B(x,z))$

  ↕ $x \notin Diag(y,z)$

- $\forall z \, (Diag(y,z) \rightarrow \forall x \, \neg B(x,z))$

  ↕ Propagation of '$\neg$'

- $\forall z \, (Diag(y,z) \rightarrow \neg \exists x \, B(x,z))$

  ↕ Definition of Bew

- $\forall z \, (Diag(y,z) \rightarrow \neg Bew(z))$

Figure 4.12: All depicted transformations can be replicated within Peano Arithmetic.

- **Direction from left to right:** $\vdash \gamma \rightarrow \chi(\overline{\ulcorner \gamma \urcorner})$

  1. $\{\gamma\} \vdash \gamma$ (Theorem 2.4)
  2. $\{\gamma\} \vdash \forall z \, (Diag(\overline{\ulcorner \psi \urcorner}, z) \rightarrow \chi(z))$ (Def)
  3. $\vdash \forall z \, (Diag(\overline{\ulcorner \psi \urcorner}, z) \rightarrow \chi(z)) \rightarrow$
     $\qquad (Diag(\overline{\ulcorner \psi \urcorner}, \overline{\ulcorner \gamma \urcorner}) \rightarrow \chi(\overline{\ulcorner \gamma \urcorner}))$ (A4)
  4. $\{\gamma\} \vdash Diag(\overline{\ulcorner \psi \urcorner}, \overline{\ulcorner \gamma \urcorner}) \rightarrow \chi(\overline{\ulcorner \gamma \urcorner})$ (MP, 2,3)
  5. $\vdash Diag(\overline{\ulcorner \psi \urcorner}, \overline{\ulcorner \gamma \urcorner})$ (4.37)
  6. $\{\gamma\} \vdash \chi(\overline{\ulcorner \gamma \urcorner})$ (MP, 4,5)
  7. $\vdash \gamma \rightarrow \chi(\overline{\ulcorner \gamma \urcorner})$ (DT)

- **Direction from righ to left:** $\vdash \chi(\overline{\ulcorner \gamma \urcorner}) \rightarrow \gamma$

  8. $\{\chi(\overline{\ulcorner \gamma \urcorner})\} \vdash \chi(\overline{\ulcorner \gamma \urcorner})$ (Theorem 2.4)
     $\ldots$
  9. $\{\chi(\overline{\ulcorner \gamma \urcorner})\} \vdash z = \overline{\ulcorner \gamma \urcorner} \rightarrow \chi(z)$ (from 8)
  10. $\vdash Diag(\overline{\ulcorner \psi \urcorner}, z) \rightarrow z = \overline{\ulcorner \gamma \urcorner}$ (4.38)
  11. $\{\chi(\overline{\ulcorner \gamma \urcorner})\} \vdash Diag(\overline{\ulcorner \psi \urcorner}, z) \rightarrow \chi(z)$ (MB, 10,9)
  12. $\{\chi(\overline{\ulcorner \gamma \urcorner})\} \vdash \forall z \, (Diag(\overline{\ulcorner \psi \urcorner}, z) \rightarrow \chi(z))$ (G, 11)
  13. $\{\chi(\overline{\ulcorner \gamma \urcorner})\} \vdash \gamma$ (Def)
  14. $\vdash \chi(\overline{\ulcorner \gamma \urcorner}) \rightarrow \gamma$ (DT)

- **Both parts together result in:**

     $\ldots$
  15. $\vdash \gamma \leftrightarrow \chi(\overline{\ulcorner \gamma \urcorner})$ (from 7 and 14)

This concludes the proof of the diagonalization lemma. $\square$

It is time to resume our original plan, the proof of

$$\varphi_g(\overline{g}) \leftrightarrow \neg Bew(\overline{\ulcorner \varphi_g(\overline{g}) \urcorner}). \qquad (4.39)$$

The diagonalization lemma only gets us part of the way here. Although it guarantees the existence of a formula γ satisfying

$$\gamma \leftrightarrow \neg Bew(\overline{\ulcorner \gamma \urcorner}),$$

it does not reveal what the formula γ looks like. It would be a mere coincidence if the formula γ matched Gödel's formula $\varphi_g(\overline{g})$.

Nevertheless, we are almost there: We can prove (4.39) in the same way as the diagonalization lemma itself. A slight modification of the derivation sequence is all we need.

4.2 The First Incompleteness Theorem

We repeat the proof of the diagonalization lemma but use Gödel's formula $\varphi_g(y)$ instead of $\psi(y)$. Then, $\gamma = \varphi_g(\bar{g})$ and

$$diag(g) = diag(\ulcorner \varphi_g(y) \urcorner) = \ulcorner \gamma \urcorner.$$

Analogously to (4.36), we can conclude

$$\vdash \forall z \, (\mathrm{Diag}(\bar{g}, z) \leftrightarrow z = \ulcorner \varphi_g(\bar{g}) \urcorner) \qquad (4.40)$$

which, in turn, implies

$$\vdash \mathrm{Diag}(\bar{g}, \ulcorner \varphi_g(\bar{g}) \urcorner) \qquad (4.41)$$
$$\vdash \mathrm{Diag}(\bar{g}, z) \to z = \ulcorner \varphi_g(\bar{g}) \urcorner \qquad (4.42)$$

Figure 4.12 demonstrates that the formula $\varphi_g(y)$ is logically equivalent to the formula

$$\forall z \, (\mathrm{Diag}(y, z) \to \neg \mathrm{Bew}(z)).$$

The equivalence was shown purely by elementary predicate-logical transformations, all reproducible within Peano arithmetic. Consequently, we can derive the theorem

$$\vdash \varphi_g(y) \leftrightarrow \forall z \, (\mathrm{Diag}(y, z) \to \neg \mathrm{Bew}(z)),$$

and thus the following applies a fortiori:

$$\vdash \varphi_g(\bar{g}) \leftrightarrow \forall z \, (\mathrm{Diag}(\bar{g}, z) \to \neg \mathrm{Bew}(z))$$

This, in turn, can be split into:

$$\vdash \varphi_g(\bar{g}) \to \forall z \, (\mathrm{Diag}(\bar{g}, z) \to \neg \mathrm{Bew}(z)) \qquad (4.43)$$
$$\vdash \forall z \, (\mathrm{Diag}(\bar{g}, z) \to \neg \mathrm{Bew}(z)) \to \varphi_g(\bar{g}) \qquad (4.44)$$

Now we are in a position to translate the derivation sequence of the diagonalization lemma into a derivation sequence for the formula (4.39):

■ **Direction from left to right:** $\vdash \varphi_g(\bar{g}) \to \neg \mathrm{Bew}(\ulcorner \varphi_g(\bar{g}) \urcorner)$

1. $\{\varphi_g(\bar{g})\} \vdash \varphi_g(\bar{g})$ (Theorem 2.4)
2. $\vdash \varphi_g(\bar{g}) \to \forall z \, (\mathrm{Diag}(\bar{g}, z) \to \neg \mathrm{Bew}(z))$ (4.43)
3. $\{\varphi_g(\bar{g})\} \vdash \forall z \, (\mathrm{Diag}(\bar{g}, z) \to \neg \mathrm{Bew}(z))$ (MP, 1,2)
4. $\vdash \forall z \, (\mathrm{Diag}(\bar{g}, z) \to \neg \mathrm{Bew}(z)) \to$
   $\qquad (\mathrm{Diag}(\bar{g}, \ulcorner \varphi_g(\bar{g}) \urcorner) \to \neg \mathrm{Bew}(\ulcorner \varphi_g(\bar{g}) \urcorner))$ (A4)
5. $\{\varphi_g(\bar{g})\} \vdash \mathrm{Diag}(\bar{g}, \ulcorner \varphi_g(\bar{g}) \urcorner) \to \neg \mathrm{Bew}(\ulcorner \varphi_g(\bar{g}) \urcorner)$ (MP, 3,4)
6. $\vdash \mathrm{Diag}(\bar{g}, \ulcorner \varphi_g(\bar{g}) \urcorner)$ (4.41)
7. $\{\varphi_g(\bar{g})\} \vdash \neg \mathrm{Bew}(\ulcorner \varphi_g(\bar{g}) \urcorner)$ (MP, 5,6)
8. $\vdash \varphi_g(\bar{g}) \to \neg \mathrm{Bew}(\ulcorner \varphi_g(\bar{g}) \urcorner)$ (DT)

- **Direction from right to left:** $\vdash \neg\mathsf{Bew}(\ulcorner\varphi_g(\overline{g})\urcorner) \to \varphi_g(\overline{g})$

9. $\{\neg\mathsf{Bew}(\ulcorner\varphi_g(\overline{g})\urcorner)\} \vdash \neg\mathsf{Bew}(\ulcorner\varphi_g(\overline{g})\urcorner)$ (Theorem 2.4)

  ...

10. $\{\neg\mathsf{Bew}(\ulcorner\varphi_g(\overline{g})\urcorner)\} \vdash z = \ulcorner\varphi_g(\overline{g})\urcorner \to \neg\mathsf{Bew}(z)$ (from 9)
11. $\vdash \mathsf{Diag}(\overline{g}, z) \to z = \ulcorner\varphi_g(\overline{g})\urcorner$ (4.42)
12. $\{\neg\mathsf{Bew}(\ulcorner\varphi_g(\overline{g})\urcorner)\} \vdash \mathsf{Diag}(\overline{g}, z) \to \neg\mathsf{Bew}(z)$ (MB, 11, 10)
13. $\{\neg\mathsf{Bew}(\ulcorner\varphi_g(\overline{g})\urcorner)\} \vdash \forall z\, (\mathsf{Diag}(\overline{g}, z) \to \neg\mathsf{Bew}(z))$ (G, 12)
14. $\vdash \forall z\, (\mathsf{Diag}(\overline{g}, z) \to \neg\mathsf{Bew}(z)) \to \varphi_g(\overline{g})$ (4.44)
15. $\{\neg\mathsf{Bew}(\ulcorner\varphi_g(\overline{g})\urcorner)\} \vdash \varphi_g(\overline{g})$ (MP, 13, 14)
16. $\vdash \neg\mathsf{Bew}(\ulcorner\varphi_g(\overline{g})\urcorner) \to \varphi_g(\overline{g})$ (DT)

- **Both parts together yield:**

  ...

17. $\vdash \varphi_g(\overline{g}) \leftrightarrow \neg\mathsf{Bew}(\ulcorner\varphi_g(\overline{g})\urcorner)$ (from 8 and 16)

The finish line has been crossed. We have found a proof for

 Theorem 4.10

In Peano arithmetic, the following holds:
$$\vdash \varphi_g(\overline{g}) \leftrightarrow \neg\mathsf{Bew}(\ulcorner\varphi_g(\overline{g})\urcorner)$$

4.2.7 Tarski's Truth Predicate

> "THEOREM I. (α) In whatever way the symbol 'Tr', denoting a class of expressions, is defined in the metatheory, it will be possible to derive from it the negation of one of the sentences which were described in the condition (α) of the convention T;
> (β) assuming that the class of all provable sentences of the metatheory is consistent, it is impossible to construct an adequate definition of truth in the sense of convention T on the basis of the metatheory."
>
> Alfred Tarski [207]

4.2 The First Incompleteness Theorem

The previous sections have repeatedly emphasized how important it is to strictly distinguish between a formula's provability and its truth. Today, we distinguish sharply between a syntactic and a semantic level, but this is by no means self-evident; it results from a long-term process with many renowned mathematicians involved. A particular contribution came from a man who undoubtedly ranks among the great logicians of the twentieth century: Alfred Tarski (Figure 4.13).

Tarski developed his first semantic arguments in a logic seminar he held at the University of Warsaw between 1927 and 1929 [79]. In 1931, he channeled his ideas and published his work *"Sur les ensembles définissables de nombres réels,"* which delves deeply into the semantic aspects of logic [203, 208]. Tarski defined a formal language capable of formulating statements about real numbers and addressed in depth what it means to *"define"* such a number within logic.

His work from 1931 was the precursor to a publication written in Polish, which, in hindsight, is one of Tarski's most important contributions to logic [204]. This work is mainly concerned with a rigorous definition of truth and the question of to what extent it can be captured within a formal language. Tarski constructed specific *truth predicates* and investigated the requirements for defining them within a formal system.

Using Peano arithmetic as an example, we aim to clarify the precise meaning of a truth predicate. Our starting point is the following definition:

$$T := \{x \mid x \text{ is the Gödel number of a true formula}\} \qquad (4.45)$$

The set T captures the concept of truth at the level of sets; it reduces whether a formula φ is true or false to whether the Gödel number of φ, the natural number $\ulcorner \varphi \urcorner$, belongs to T.

Next, we will attempt to grasp the concept of truth at the formula level, which is where Tarski's truth predicate comes into play. We call a formula $\mathcal{T}(\xi)$ with the free variable ξ a truth predicate if it applies precisely to the Gödel numbers of the true formulas, i.e., if the following applies:

$$\models \mathcal{T}(\bar{x}) \Leftrightarrow x \in T \qquad (4.46)$$

Let us examine the consequences of this definition by distinguishing two cases:

- ■ **φ is a formula satisfying $\models \mathcal{T}(\ulcorner \varphi \urcorner)$**

  Because of (4.46), $\ulcorner \varphi \urcorner \in T$, and from (4.45) it follows that φ is substantively true. Hence, $\models \varphi$ holds.

"[...] it is evident that all these results only receive a clear content and can only then be exactly proved, if a concrete and precisely formulated definition of [true] sentence is accepted as a basis for the investigation" [206]

Alfred Tarski [8]
(1901 – 1983)

VIII

THE CONCEPT OF TRUTH IN FORMALIZED LANGUAGES†

INTRODUCTION

THE present article is almost wholly devoted to a single problem—*the definition of truth*. Its task is to construct—with reference to a given language—*a materially adequate and formally correct definition of the term 'true sentence'*. This problem, which belongs to the classical questions of philosophy, raises considerable difficulties. For although the meaning of the term 'true sentence' in colloquial language seems to be quite clear and intelligible, all attempts to define this meaning more precisely have hitherto been fruitless, and many investigations in which this term has been used and which started with apparently evident premisses have often led to paradoxes and antinomies (for which, however, a more or less satisfactory solution has been found). The concept of truth shares in this respect the fate of other analogous concepts in the domain of the semantics of language.

Figure 4.13: The solid foundation of the concept of truth is one of Tarski's most profound contributions to mathematical logic. It opened the door to model theory, which draws almost exclusively on semantic arguments.

Alfred Tarski was born on January 14, 1901, in Warsaw as Alfred Tajtelbaum. The exceptionally talented boy spent his childhood in wealthy circumstances. After graduating from school with brilliant grades, he enrolled as a student at Warsaw University in 1918. He initially developed a strong interest in the natural sciences but then found his real love in mathematics and logic. Tarski enjoyed a top-class academic environment in Warsaw and counted famous mathematicians such as Jan Łukasiewicz and Wacław Sierpiński among his teachers.

In 1923, Alfred took the surname Tarski and converted from the Jewish religion to Catholicism. This was primarily a strategic move, as he would have an easier time with a Polish identity back then. The same year, Tarski submitted his doctoral thesis. After that, he failed several times to obtain a professorship, forcing him to earn his living by lecturing at schools and the university. During this phase of his life, Tarski wrote some of his most notable publications, including his groundbreaking work on the formal definition of the concept of truth [204, 205].

On September 1, 1939, the German Wehrmacht invaded Poland. At the time, Tarski was on a trip abroad to Harvard, thus escaping the Nazis by chance. Returning to his home country was out of the question, as it would have meant the death sentence for the mathematician of Jewish origin. Tarski failed several times to obtain an emigration permit for his family and had to wait until the war's end before he could bring his wife and two children to the USA. He never saw his parents and brother again, as neither survived the disastrous time.

Tarski held several temporary positions at renowned universities and research institutions in the US and accepted a permanent professorship at the University of California in Berkeley in 1945. He finally held the position he had deserved for so long due to his brilliant scientific achievements. Tarski completed several visiting positions at other universities but always returned to Berkeley. Even in old age, the restless mathematician did not abandon his work and continued researching and teaching for many years. He invested the last twelve years of his life in the book *"A Formalization of Set Theory without Variables"* [209]. It was his last major work. Shortly after completion, Alfred Tarski died at the age of 82.

The introductory quote of this section is the original wording of what is now known as Tarski's theorem on the indefinability of truth. You may have noticed that Tarski's formulation refers to a *"metatheory,"* but such a term does not appear anywhere in our own presentation. We can think of Tarski's metatheory as a metalanguage that enriches a formal language L with the necessary means of expression to talk about the syntactic structure of its formulas. In this section, however, we relate Tarski's results exclusively to Peano arithmetic, which, as we already know, is expressive enough to talk about its own syntactic structure. In this sense, we can equate Peano arithmetic with its own metalanguage and refrain from making a distinction in this respect.

- **φ is a formula satisfying $\models \varphi$**

  Due to (4.45), $\ulcorner \varphi \urcorner \in T$, and from (4.46), it follows that $\mathcal{T}(\ulcorner \varphi \urcorner)$ is substantively true. Hence, $\models \mathcal{T}(\ulcorner \varphi \urcorner)$.

In summary, we obtain the following relationship:

$$\text{For all formulas } \varphi: \models \varphi \Leftrightarrow\, \models \mathcal{T}(\ulcorner \varphi \urcorner)$$

Or, equivalently:

$$\text{For all formulas } \varphi: \models \varphi \leftrightarrow \mathcal{T}(\ulcorner \varphi \urcorner) \tag{4.47}$$

Tarski suspected that formal systems are inherently incapable of capturing the concept of truth in this manner. In the context of Peano arithmetic, this would imply that a formula $\mathcal{T}$, as defined above, cannot exist. However, Tarski had no proof to support his assumption back then.

Shortly after, events came thick and fast. The same year Tarski published his work, Gödel published his two incompleteness theorems. From the first incompleteness theorem, it does indeed follow that there can be no truth predicates in systems such as Peano arithmetic. Tarski

4.2 The First Incompleteness Theorem

found this result very convenient. He planned to publish a German version of his Polish paper and could suddenly back up his conjecture with a formal proof. It explains why the translation published in 1935 with the title *"Der Wahrheitsbegriff in den formalisierten Sprachen"* differs from the original Polish treatise in both form and content. Tarski explicitly mentions this fact in a footnote. We quote from the English translation:

> *"We owe the method used here to Gödel, who has employed it for other purposes in his recently published work [...] I take this opportunity of mentioning that Th. I and the sketch of its proof was only added to the present work after it had already gone to press. At the time the work was presented at the Warsaw Society of Sciences (21 March 1931), Gödels article – so far as I know – had not yet appeared. In this place therefore I had originally expressed, instead of positive results, only certain suppositions in the same direction."*
>
> Alfred Tarski [207]

Tarski's work also appeared in English under *"The Concept of Truth in Formalized Languages"* when the German language lost its status as the academic lingua franca after World War II. This version is the most frequently cited today, and we have also quoted from it above.

With the knowledge acquired so far, it is easy to derive Tarski's brilliant negative result. By choosing

$$\chi(y) := \neg \mathcal{T}(y),$$

the diagonalization lemma guarantees the existence of a formula γ with the following property:

$$\vdash \gamma \leftrightarrow \neg \mathcal{T}(\ulcorner \gamma \urcorner) \qquad (4.48)$$

If Peano arithmetic is correct, as we assume, then every provable statement is substantively true. Hence, we can draw the following conclusion from (4.48):

$$\models \gamma \leftrightarrow \neg \mathcal{T}(\ulcorner \gamma \urcorner)$$

From (4.47), however, we conclude

$$\models \gamma \leftrightarrow \mathcal{T}(\ulcorner \gamma \urcorner),$$

which results in a contradiction:

$$\models \neg \mathcal{T}(\ulcorner \gamma \urcorner) \leftrightarrow \mathcal{T}(\ulcorner \gamma \urcorner)$$

Hence, we must drop the assumption about the existence of a truth predicate within Peano arithmetic and may add the following theorem to our stock of knowledge:

Theorem 4.11 (Tarski, 1935)

> The concept of truth is not definable within Peano arithmetic; there is no truth predicate.

At this juncture, we aim to more closely explore the relationship between Tarski's theorem and Gödel's first incompleteness theorem. We know that Peano arithmetic is expressive enough to talk about the provability of a formula. To this end, in Section 4.2.6, we constructed a formula Bew(y) with the following substantive meaning:

$$\models \mathsf{Bew}(\bar{y}) \Leftrightarrow y \text{ is the Gödel number of a provable formula}$$

Then the following applies:

- If Peano Arithmetic is correct, every formula φ satisfies

$$\models \mathsf{Bew}(\ulcorner \varphi \urcorner) \to \varphi \qquad (4.49)$$

- If Peano Arithmetic is complete, every formula φ satisfies

$$\models \varphi \to \mathsf{Bew}(\ulcorner \varphi \urcorner) \qquad (4.50)$$

If Peano arithmetic were both correct and complete, we could therefore draw the following conclusion from (4.49) and (4.50):

$$\models \varphi \leftrightarrow \mathsf{Bew}(\ulcorner \varphi \urcorner)$$

Consequently, Bew would be a truth predicate, which cannot exist according to Tarski's theorem. The contradiction resolves itself only by dropping the assumption that Peano arithmetic is simultaneously correct and complete. Finally, we are facing a marvelous result: the semantic variant of the first incompleteness theorem follows directly from Tarski's theorem.

The fact that Peano arithmetic can formalize the concept of provability, but not the concept of truth, was seen by Gödel as a key property for the existence of undecidable propositions. We read in a letter to the US mathematician Arthur Burks:

4.2 The First Incompleteness Theorem

> "It is this theorem which is the true reason for the existence of undecidable propositions in the formal systems containing arithmetic. I did not formulate it explicitly in my paper of 1931, but only in my Princeton lectures of 1934. The same theorem was proved by Tarski in his paper on the concept of truth published in 1933."
>
> Kurt Gödel [54, 145]

The considerations in the previous two sections nourish the suspicion that the first incompleteness theorem and the principle of diagonalization are inextricably interwoven. The following section will show that this impression is at least partially deceptive. We will derive the semantic variant of the first incompleteness theorem with a clever argument that does not directly reference the principle of diagonalization.

4.2.8 Berry Paradox

> "What strikes the author as of interest in the proof via Berry's paradox is not its brevity but that it provides a different sort of reason for [...] incompleteness [...]."
>
> George Boolos [14]

The *Berry paradox* is a colloquially formulated paradox first described in the introduction to the *Principia Mathematica*. It is named after its originator, the librarian G. G. Berry from the Bodleian Library in Oxford. In a footnote, Russell and Whitehead explicitly pointed out that Berry brought them the example.

Figure 4.14 shows the original wording of the Berry paradox and illustrates how to paraphrase it in other languages by slightly adapting the substantive wording. It is precisely the concise formulation that leads to its explosive power. In its original version, the paradox names a number that cannot be defined with fewer than nineteen syllables but manages to do so with eighteen. An untenable situation!

In the late 1980s, the American logician George Boolos noticed that the Berry paradox could be recreated within Peano arithmetic if it was correct and complete at the same time. His efforts resulted in a remarkable proof of the semantic variant of the first incompleteness theorem. For one thing, it appears extremely simple compared to classical proofs. For another, it manages without a direct reference to the principle of diagonalization [14].

Original wording

"The least integer not nameable in fewer than nineteen syllables."

Paraphrased in German

"Die kleinste natürliche Zahl, die nicht mit weniger als vierzehn Wörtern definierbar ist."

Figure 4.14: The Berry Paradox, above in the original wording and below in a German adaption.

In the following, we will take a closer look at how Boolos achieved his feat. We start with a simple definition:

Definition 4.8

A formula $\psi(\xi)$ *names* the number n if the following applies:

$$\vdash \forall x\, (\psi(x) \leftrightarrow x = \bar{n})$$

In colloquial terms, the definition states that a formula $\psi(\xi)$ names the number n if and only if $\forall x\, (\psi(x) \leftrightarrow x = \bar{n})$ is derivable from the axioms of Peano arithmetic. Keep in mind that the property of naming a number is linked to the provability and not to the truth of a formula!

Figure 4.15 lists several formulas, all naming the natural number 1. The examples emphasize that the choice of $\psi(\xi)$ is not unique; each natural number is namable in infinitely many ways.

In his proof, Boolos utilizes a formula $\varphi_C(x,y)$ with the following substantive meaning:

$$\models \varphi_C(\bar{x},\bar{y}) \Leftrightarrow x \text{ is named by a formula } \varphi \text{ with } |\varphi| = y$$

Herein, $|\varphi|$ denotes the number of symbols in φ. The fact that a formula with this property exists comes as no surprise after all we have learned. Its existence follows from Peano arithmetic being expressive enough to talk about the syntactic structure and the provability of formulas.

Boolos makes use of the formula $\varphi_C(x,y)$ in the following definition:

$$\varphi_B(x,y) := \exists z\, (z < y \wedge \varphi_C(x,z))$$

The meaning of this formula is evident. If x and y are two natural numbers, the instance $\varphi_B(\bar{x},\bar{y})$ is substantively true if and only if the number x is named by a formula that contains fewer than y symbols:

$$\models \varphi_B(\bar{x},\bar{y}) \Leftrightarrow x \text{ is named by a formula } \varphi \text{ with } |\varphi| < y$$

The meaning of the next formula is just as easy to understand:

$$\varphi_A(x,y) := \neg \varphi_B(x,y) \wedge \forall (a < x)\, \varphi_B(a,y)$$

If x and y are natural numbers, the instance $\varphi_A(\bar{x},\bar{y})$ is substantively true if and only if

---

The following formulas are provable in PA:

$$\forall x\, (x = s(0) \leftrightarrow x = \bar{1})$$
$$\forall x\, (x + x = s(s(0)) \leftrightarrow x = \bar{1})$$
$$\forall x\, (x + x + x = s(s(s(0))) \leftrightarrow x = \bar{1})$$
$$\ldots$$

Definition 4.8

All formulas below name the natural number 1:

$$x = s(0)$$
$$x + x = s(s(0))$$
$$x + x + x = s(s(s(0)))$$
$$\ldots$$

Figure 4.15: The colloquial formulation of the Berry paradox remains fuzzy about what it means to name a number. In the formal variant, this term is defined with mathematical precision.

4.2 The First Incompleteness Theorem

- no formula φ with $|\varphi| < y$ names the number x ☞ $\neg \varphi_B(\bar{x}, \bar{y})$

- and every number smaller than x ☞ $\forall (a < \bar{x})$

- is named by a formula φ with $|\varphi| < y$.   ☞ $\varphi_B(a, \bar{y})$

In other words:

$$\models \varphi_A(\bar{x}, \bar{y}) \Leftrightarrow \text{x is the smallest number that is not named by a formula with fewer than y symbols.}$$

There is a simple reason why for any given number y, there must be a smallest number not named by a formula with less than y symbols: It holds because

- there are only finitely many formulas with less than y symbols, and

- each formula names at most one number if PA is correct.

The first property is obvious, and the second is easy to verify. If there were a formula ψ that simultaneously denotes several numbers, the following would apply to two different natural numbers, n and m:

$$\vdash \forall x\, (\psi(x) \leftrightarrow x = \bar{n})$$
$$\vdash \forall x\, (\psi(x) \leftrightarrow x = \bar{m})$$

Then, the substantively false formula

$$x = \bar{m} \leftrightarrow x = \bar{n}$$

would be a theorem, contradicting the assumed correctness of Peano arithmetic. Consequently, only finitely many numbers are nameable by formulas with fewer than y symbols, which means there must be a largest number. The successor to this number is the smallest number that cannot be named by a formula with fewer than y symbols.

Here comes the final step: We choose $k := |\varphi_A(x,y)|$ and make a last definition:

$$\varphi_F(x) := \exists y\, ((y = \overline{50} \times \bar{k}) \wedge (\varphi_A(x,y)))$$

After what we have said so far, the meaning of this formula is evident:

$$\models \varphi_F(\bar{x}) \Leftrightarrow \text{x is the smallest number that is not named by a formula with fewer than $50 \cdot k$ symbols.}$$

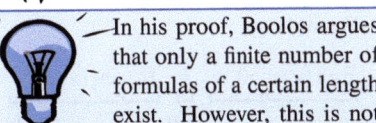

In his proof, Boolos argues that only a finite number of formulas of a certain length exist. However, this is not correct for Peano arithmetic according to Definition 3.1. We have allowed ourselves to draw from an infinitely large pool of variables and can thus form an infinite number of formulas with the same length, e.g., the formulas shown here:

$$\exists x\, (x = \bar{1}),$$
$$\exists y\, (y = \bar{1}),$$
$$\exists z\, (z = \bar{1}),$$
$$\ldots$$

However, Boolos does not make a mistake here, as his syntax definition differs slightly from ours. In his variant of PA, there is only a finite supply of elementary symbols, and variables are represented by compound strings of the form

$$x', x'', x''', \ldots$$

The change in syntax ensures the existence of only finitely many formulas of a certain length while simultaneously allowing the generation of an unlimited number of identifiers.

There is no need to worry about our different syntax definitions, though. We can follow Boolos' line of reasoning step by step by considering a variable such as y or z as syntactic abbreviations for x' or x''. Then, the above examples are merely abbreviations for the following formulas, each having a different length:

$$\exists x\, (x = \bar{1}),$$
$$\exists x'\, (x' = \bar{1}),$$
$$\exists x''\, (x'' = \bar{1}),$$
$$\ldots$$

- Formula φ_F

$$\exists y \, ((\underbrace{y = \overline{50}}_{6 \quad 151 \quad 1} \times \underbrace{\overline{k}}_{1+3k}) \underbrace{\wedge}_{3} (\underbrace{\varphi_A(x,y)}_{k} \underbrace{)}_{2})$$

$\overline{0} = 0$ ☞ 1 symbol)
$\overline{1} = s(0)$ ☞ 4 symbols)
$\overline{2} = s(s(0))$ ☞ 7 symbols)
$\overline{3} = s(s(s(0)))$ ☞ 10 symbols)
$\ldots$
$\overline{50} = \underbrace{s(\ldots s\,(0)\ldots)}_{50 \text{ times}}$ ☞ 151 symbols)
$\ldots$
$\overline{k} = \underbrace{s(\ldots s\,(0)\ldots)}_{k \text{ times}}$ ☞ 1+3k symbols)

- Length of φ_F

$$\begin{array}{r} 6 \\ + \quad 151 \\ + \quad 1+3k \\ + \quad 3 \\ + \quad k \\ + \quad 2 \\ \hline 163+4k \end{array}$$

$k > 3$ implies $|\varphi_F| < 50 \cdot k$

Figure 4.16: According to of our definition, n is the smallest number that cannot be named by a formula that contains less than $50 \cdot k$ symbols. The formula φ_F consists of less than $50 \cdot k$ symbols. Consequently, the number n is not named by φ_F.

Denoting the smallest number not named by a formula with less than $50 \cdot k$ symbols by n, we can express the fact as follows:

$$\models \varphi_F(\bar{x}) \Leftrightarrow x = n$$

Or, equivalently:

$$\models \forall x \, (\varphi_F(x) \leftrightarrow x = \bar{n}) \quad (4.51)$$

Next, we want to learn more about the number of symbols in φ_F. The calculation in Figure 4.16 reveals that the number is less than $50 \cdot k$, meaning that φ_F cannot denote n. By definition, this implies that the formula

$$\forall x \, (\varphi_F(x) \leftrightarrow x = \bar{n}) \quad (4.52)$$

is unprovable in PA:

$$\not\vdash \forall x \, (\varphi_F(x) \leftrightarrow x = \bar{n}) \quad (4.53)$$

At this point, we have reached our goal. We have shown that Peano arithmetic includes a formula, φ_F, that is substantively true according to (4.51) but formally unprovable according to (4.53). The existence of such a formula is precisely what the semantic variant of Gödel's first incompleteness theorem asserts.

Above all, the presented proof impresses with its brevity. Boolos seemingly derived the first incompleteness theorem in only a few strokes of the pen, which raises the question of whether it was worth investing the large amount of work earlier in this book. To settle this question, we must keep two aspects in mind:

- Unlike Gödel, Boolos does not prove the syntactic variant of the first incompleteness theorem, but only the semantic one. This variant is much easier to derive, and similar short proofs already existed in 1987. In Section 5.4.2, we will present such a proof and derive the semantic variant of Gödel's first incompleteness via the arithmetization of Turing machines. This much in advance: we will encounter an argument as brief and elegant as the one presented here.

- In reality, Boolos' proof is not as short as our presentation may suggest. The reason is that we assumed the existence of formula φ_C without providing any formal backup. However, ensuring the existence of φ_C with mathematical precision requires the entire Gödelization apparatus and the anything but trivial property that Peano arithmetic is expressive enough to talk about the concept of provability. In full detail, Boolos' proof covers many pages and includes large parts of Gödel's original proof.

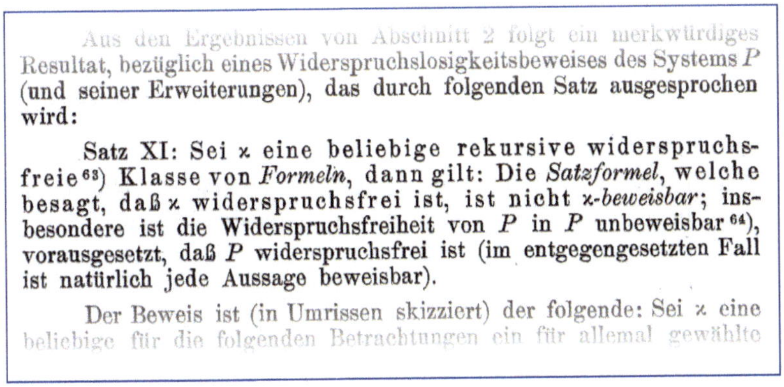

> Aus den Ergebnissen von Abschnitt 2 folgt ein merkwürdiges Resultat, bezüglich eines Widerspruchslosigkeitsbeweises des Systems P (und seiner Erweiterungen), das durch folgenden Satz ausgesprochen wird:
>
> Satz XI: Sei $\varkappa$ eine beliebige rekursive widerspruchsfreie[63]) Klasse von *Formeln*, dann gilt: Die *Satzformel*, welche besagt, daß $\varkappa$ widerspruchsfrei ist, ist nicht $\varkappa$-*beweisbar*; insbesondere ist die Widerspruchsfreiheit von P in P unbeweisbar[64]), vorausgesetzt, daß P widerspruchsfrei ist (im entgegengesetzten Fall ist natürlich jede Aussage beweisbar).
>
> Der Beweis ist (in Umrissen skizziert) der folgende: Sei $\varkappa$ eine beliebige für die folgenden Betrachtungen ein für allemal gewählte

Figure 4.17: Gödel's original work concludes with Theorem XI, today known as Gödel's second incompleteness theorem.

Thus, the distinguishing aspect of Boolos' proof is not its brevity. Instead, unlike most other proofs, the chain of reasoning does not rely on direct reference to the principle of diagonalization. This property is extraordinary and adds a unique charm. Boolos also saw it this way, as the opening quotation to this section points out.

4.3 The Second Incompleteness Theorem

In his 1931 paper, Gödel proved far more than the incompleteness of arithmetic. In the second part, he explored the consequences of the first incompleteness theorem and made a far-reaching discovery. He encountered what we now call *Gödel's second incompleteness theorem* (Figure 4.17). It states that no formal system expressive enough to talk about the additive and multiplicative properties of the natural numbers can prove its consistency. In this section, we will clarify the exact meaning of this assertion and investigate its consequences for the entire discipline of mathematics.

When we say that a formal system can prove its consistency, we mean the following:

- There is a formula Con, being true precisely if the formal system is consistent.
- The formula Con is provable within the formal system ($\vdash$ Con).

Remember that we call a formal system consistent if, for no formula φ, both φ and its negation $\neg\varphi$ are derivable from the axioms. Using the formula Bew(y) from Page 218, we can formalize the consistency within Peano arithmetic as follows:

$$\text{Con} := \neg \exists x \, (\text{Bew}(x) \wedge \text{Bew}(\neg x))$$

Substantively, this formula states what we are looking for: No formula is provable together with its negation within Peano arithmetic:

$$\models \text{Con} \Leftrightarrow \text{PA is consistent}$$

There is an even more straightforward way to characterize consistency. It suffices to pick any provable formula, e.g., the formula $0 \neq \overline{1}$, and to demand the unprovability of its negation:

$$\text{Con} := \neg \text{Bew}(\ulcorner 0 = \overline{1} \urcorner) \tag{4.54}$$

We will use this definition of Con from now on.

At this point, the first incompleteness theorem comes into play, which states that every consistent formal system expressive enough to formalize Peano arithmetic must be negation incomplete. For proving the second incompleteness theorem, the following two facts are crucial:

- Peano arithmetic is sufficiently expressive to formulate the first incompleteness theorem. For our purposes, however, we do not need the entire theorem but only the assertion that the consistency of PA implies the unprovability of Gödel's diagonal element $\varphi_g(\overline{g})$. This assertion was the easier direction in the proof in Section 4.2.4. Within Peano arithmetic, we can describe the statement by the following formula:

$$\text{Con} \to \neg \text{Bew}(\ulcorner \varphi_g(\overline{g}) \urcorner) \tag{4.55}$$

- We can not only carry over the incompleteness theorem into Peano arithmetic, but also its proof. Section 3.2.1.3 has shown how to accomplish a task like this by formally reproducing the proof of the theorem on the equality of ordered pairs within Zermelo-Fraenkel set theory. In the context of the first incompleteness theorem, reproducing the proof within PA means to construct a sequence of formulas that begins with PA's axioms and ends as follows:

$$\vdash \text{Con} \to \neg \text{Bew}(\ulcorner \varphi_g(\overline{g}) \urcorner) \tag{4.56}$$

The fact that Con, as defined in (4.54), actually describes the consistency of Peano arithmetic is due to a property that we discussed in the marginal note on Page 89. There, we explained that every formula is derivable from the axioms in a contradictory formal system that includes the usual propositional reasoning apparatus. Therefore, if φ and $\neg \varphi$ are provable, so are $0 = \overline{1}$ and $0 \neq \overline{1}$. Conversely, the unprovability of $0 = \overline{1}$ implies that no formula φ is provable alongside its negation $\neg \varphi$.

In his 1931 paper, Gödel described consistency slightly differently but still followed the same fundamental idea. He exploited that a formal system is consistent if at least one unprovable formula exists. Gödel translated this formulation one-to-one into a formula that he called Wid [73]:

$$\text{Wid} := \exists x \, (\text{Form}(x) \wedge \neg \text{Bew}(x))$$

Herein, Form(x) is a formula that is provable if and only if x is the Gödel number of a syntactically well-formed arithmetic expression.

4.3 The Second Incompleteness Theorem

	$\vdash \ldots$	$(\ldots)$
PA can prove the formalized first incompleteness theorem. ☞	$\vdash \text{Con} \to \neg\text{Bew}(\ulcorner \varphi_g(\overline{g}) \urcorner)$	
If the consistency of PA were provable within PA, ... ☞	$\vdash \text{Con}$	(Hypothesis)
	$\vdash \neg\text{Bew}(\ulcorner \varphi_g(\overline{g}) \urcorner)$	(MP)
	$\vdash \neg\text{Bew}(\ulcorner \varphi_g(\overline{g}) \urcorner) \to \varphi_g(\overline{g})$	(4.57)
...Gödel's diagonal statement would also be a theorem. However, this statement is unprovable in PA! ☞	$\vdash \varphi_g(\overline{g})$	(MP)
	We know from the first incompleteness theorem: $\not\vdash \varphi_g(\overline{g})$	

Figure 4.18: The final step in the proof of Gödel's second incompleteness theorem. If Con were provable within Peano arithmetic, we could also prove $\varphi_g(\overline{g})$, contradicting the first incompleteness theorem.

Figure 4.18 depicts the crucial step: If the formula Con were derivable within Peano arithmetic, so would the formula

$$\neg\text{Bew}(\ulcorner \varphi_g(\overline{g}) \urcorner).$$

According to Theorem 4.10, formula

$$\varphi_g(\overline{g}) \leftrightarrow \neg\text{Bew}(\ulcorner \varphi_g(\overline{g}) \urcorner)$$

is a theorem and thus even more so the weakened variant

$$\neg\text{Bew}(\ulcorner \varphi_g(\overline{g}) \urcorner) \to \varphi_g(\overline{g}). \tag{4.57}$$

Applying the modus ponens yields a proof for $\varphi_g(\overline{g})$. However, this formula is unprovable according to the first incompleteness theorem. At this point, we have crossed the finishing line, ready to proudly proclaim Gödel's second incompleteness theorem:

 Theorem 4.12 (Gödel, 1931)

> In every consistent formal system expressive enough to formalize Peano arithmetic, we have $\not\vdash \text{Con}$.

4.3.1 Hilbert–Bernays–Löb Provability Conditions

The line of argument that led us to the second incompleteness theorem in the previous section is lacking at a crucial point. We trusted that the

> The detailed derivation by David Hilbert and Paul Bernays from 1939 highlights the complexity of reproducing the proof of the first incompleteness theorem in Peano arithmetic [97]. Still, Gödel's proof sketch of the second incompleteness theorem from 1931 was never seriously doubted. The proof sketch was so well received that Gödel refrained from filling the gaps in the announced second publication. How could that be?
>
> To clarify the situation, let us recall that Gödel did not prove his result for Peano arithmetic but for a formal system called P, which built upon the logical substructure of *Principia Mathematica*. The fact that a colloquially formulated proof is reproducible within the logical framework of the *Principia* was not spectacular news in 1931. In their three-volume opus, Russell and Whitehead impressively demonstrated that all the conclusions of ordinary mathematics are reproducible in their system.
>
> As such, Hilbert and Bernays' work is far more than just the complementation of Gödel's proof sketch. It was the first formal demonstration that the proof of Gödel's first incompleteness theorem is reproducible within theories significantly less complex than the logical framework of the *Principia Mathematica*.

proof of the first incompleteness theorem is formalizable within Peano arithmetic, which is anything but self-evident.

In his 1931 paper, Gödel did not provide the details either. He outlined the derivation in a form similar to ours and concluded his work with the following words:

> *"The results will be expressed and proved in full generality in a sequal to appear shortly. Also in that paper, the proof of Theorem XI, which has only been sketched here, will be presented in detail."*
>
> Kurt Gödel [77]

The announced sequel never appeared; most mathematicians found the proof sketch so convincing that hardly anyone doubted its correctness. Besides, there was no question that the complete elaboration of the proof would be lengthy and technically demanding.

The first to take on this task were David Hilbert and Paul Bernays. In 1939, they carried out the proof for Z and Z_μ, two specialized variants of Peano arithmetic [97]. After completing their work, a natural question arose: Can the proof be applied to other formal systems, too, and if so, what criteria must these systems fulfill?

Hilbert and Bernays quickly found what they were looking for. They discovered that their chain of reasoning applies to all formal systems whose proof predicate Bew fulfills certain *derivability conditions*. These were simplified in 1955 by Martin Löb to what we now refer to as the *Hilbert-Bernays-Löb criteria*. Specifically, this term refers to the following three properties:

$$\vdash \varphi \Rightarrow \vdash \mathsf{Bew}(\ulcorner \varphi \urcorner) \tag{4.58}$$

$$\vdash \mathsf{Bew}(\ulcorner \varphi \to \psi \urcorner) \to (\mathsf{Bew}(\ulcorner \varphi \urcorner) \to \mathsf{Bew}(\ulcorner \psi \urcorner)) \tag{4.59}$$

$$\vdash \mathsf{Bew}(\ulcorner \varphi \urcorner) \to \mathsf{Bew}(\ulcorner \mathsf{Bew}(\ulcorner \varphi \urcorner) \urcorner) \tag{4.60}$$

In connection with these criteria, a simplified notation materialized that we are happy to adopt. It pursues the idea of symbolizing the proof predicate using the *modal operator* '$\Box$':

$$\Box \varphi := \mathsf{Bew}(\ulcorner \varphi \urcorner)$$

4.3 The Second Incompleteness Theorem

In the new notation, the Hilbert-Bernays-Löb criteria read as follows:

$$\vdash \varphi \Rightarrow \vdash \Box\varphi \tag{DC1}$$
$$\vdash \Box(\varphi \to \psi) \to (\Box\varphi \to \Box\psi) \tag{DC2}$$
$$\vdash \Box\varphi \to \Box\Box\varphi \tag{DC3}$$

The formalized first incomplete theorem then takes on the form

$$\vdash \text{Con} \to \neg\Box\varphi_g(\bar{g}) \tag{4.61}$$

with

$$\text{Con} = \neg\Box(0 = \bar{1}).$$

We will now show how (4.61) can be formally derived from the Hilbert-Bernays-Löb criteria within PA.

First, let us recall Theorem 4.10, which implies that the formulas

$$\varphi_g(\bar{g}) \to \neg\Box\varphi_g(\bar{g}) \tag{4.62}$$
$$\neg\Box\varphi_g(\bar{g}) \to \varphi_g(\bar{g}) \tag{4.63}$$

are theorems of PA. Our proof commences with the first of the two:

1. $\vdash \varphi_g(\bar{g}) \to \neg\Box\varphi_g(\bar{g})$ \hfill (4.62)
2. $\vdash \Box(\varphi_g(\bar{g}) \to \neg\Box\varphi_g(\bar{g}))$ \hfill (DC1, 1)
3. $\vdash \Box(\varphi_g(\bar{g}) \to \neg\Box\varphi_g(\bar{g})) \to (\Box\varphi_g(\bar{g}) \to \Box\neg\Box\varphi_g(\bar{g}))$ \hfill (DC2)
4. $\vdash \Box\varphi_g(\bar{g}) \to \Box\neg\Box\varphi_g(\bar{g})$ \hfill (MP, 2,3)
5. $\vdash \neg\Box\varphi_g(\bar{g}) \to (\Box\varphi_g(\bar{g}) \to 0 = \bar{1})$ \hfill (T8)
6. $\vdash \Box(\neg\Box\varphi_g(\bar{g}) \to (\Box\varphi_g(\bar{g}) \to 0 = \bar{1}))$ \hfill (DC1, 5)
7. $\vdash \Box(\neg\Box\varphi_g(\bar{g}) \to (\Box\varphi_g(\bar{g}) \to 0 = \bar{1})) \to$
   $\quad (\Box\neg\Box\varphi_g(\bar{g}) \to \Box(\Box\varphi_g(\bar{g}) \to 0 = \bar{1}))$ \hfill (DC2)
8. $\vdash \Box\neg\Box\varphi_g(\bar{g}) \to \Box(\Box\varphi_g(\bar{g}) \to 0 = \bar{1})$ \hfill (MP, 6,7)
9. $\vdash \Box\varphi_g(\bar{g}) \to \Box(\Box\varphi_g(\bar{g}) \to 0 = \bar{1})$ \hfill (MB, 4,8)
10. $\vdash \Box(\Box\varphi_g(\bar{g}) \to 0 = \bar{1}) \to (\Box\Box\varphi_g(\bar{g}) \to \Box(0 = \bar{1}))$ \hfill (DC2)
15. $\vdash \Box\varphi_g(\bar{g}) \to (\Box\Box\varphi_g(\bar{g}) \to \Box(0 = \bar{1}))$ \hfill (MB, 9,10)
16. $\vdash (\Box\varphi_g(\bar{g}) \to (\Box\Box\varphi_g(\bar{g}) \to \Box(0 = \bar{1}))) \to$
    $\quad ((\Box\varphi_g(\bar{g}) \to \Box\Box\varphi_g(\bar{g})) \to (\Box\varphi_g(\bar{g}) \to \Box(0 = \bar{1})))$ \hfill (A2)
17. $\vdash (\Box\varphi_g(\bar{g}) \to \Box\Box\varphi_g(\bar{g})) \to (\Box\varphi_g(\bar{g}) \to \Box(0 = \bar{1}))$ \hfill (MP, 15,16)
18. $\vdash \Box\varphi_g(\bar{g}) \to \Box\Box\varphi_g(\bar{g})$ \hfill (DC3)
19. $\vdash \Box\varphi_g(\bar{g}) \to \Box(0 = \bar{1})$ \hfill (MP, 18,17)
20. $\vdash (\Box\varphi_g(\bar{g}) \to \Box(0 = \bar{1})) \to (\neg\Box(0 = \bar{1}) \to \neg\Box\varphi_g(\bar{g}))$ \hfill (T6)

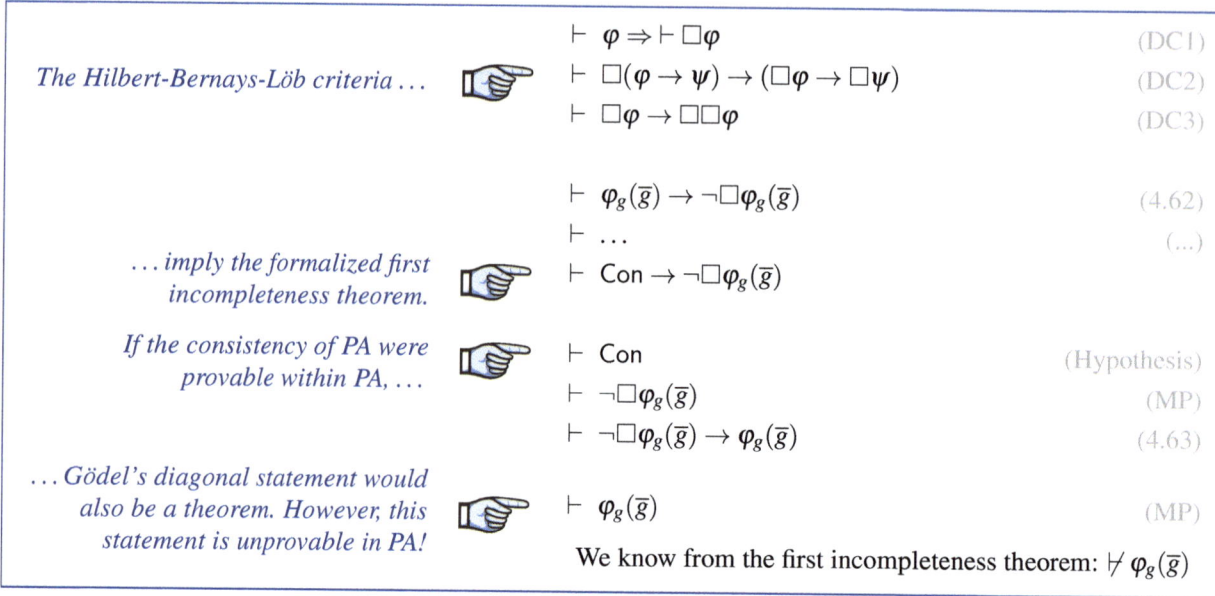

Figure 4.19: Gödel's second incompleteness theorem, derived from the Hilbert-Bernays-Löb criteria.

$$21. \vdash \neg\Box(0 = \overline{1}) \to \neg\Box\varphi_g(\overline{g}) \quad \text{(MP, 19,20)}$$
$$22. \vdash \text{Con} \to \neg\Box\varphi_g(\overline{g}) \quad \text{(Def)}$$

We are ready to complement the proof sketch of the second incompleteness theorem in Figure 4.18. The result is shown in Figure 4.19.

Even now, however, not all gaps in the proof have been filled. To do so, we had to demonstrate that Peano arithmetic fulfills all three Hilbert-Bernays-Löb criteria. This part is technically demanding, and we will deliberately skip the proof. It is elaborated in detail, for example, in [195].

4.3.2 Löb's Theorem

The previous section has outlined how to derive the formalized variant of Gödel's first incompleteness theorem from the Hilbert-Bernays-Löb criteria. Historically, the criteria originate from the work

"Solution of a problem of Leon Henkin"

by the German mathematician Martin Löb, published in 1955 [125]. The title arouses curiosity! Which problem did Leon Henkin express back then, and how was it solved? Peeking into Löb's work provides us with the answer. It is about a question published in the June 1955 issue of the renowned *Journal of Symbolic Logic* under the heading *Problems*. It reads:

> "*A problem concerning provability. If Σ is any standard formal system adequate for recursive number theory, a formula (having a certain integer q as its Gödel number) can be constructed which expresses the proposition that the formula with Gödel number q is provable in Σ. Is this formula provable or independent in Σ?*"
>
> Leon Henkin [87]

Henkin uses Σ to denote a formal system that is expressive enough to formalize Peano arithmetic. We assume Σ to be PA to keep our considerations simple.

Henkin's reasoning is easy to understand when recalling the construction of Gödel's formula $\varphi_g(\bar{g})$. With $\varphi_g(\bar{g})$, Gödel set up a formula that claims to be unprovable. Likewise, he could have constructed a formula that claims its provability. We can straightforwardly obtain such a formula from the diagonalization lemma, proven in Section 4.2.6. By choosing $\chi := \mathrm{Bew}$, the diagonalization lemma guarantees the existence of a formula $\mathcal{H}$ with the following property:

$$\vdash \mathcal{H} \leftrightarrow \mathrm{Bew}(\ulcorner \mathcal{H} \urcorner)$$

The modal operator '□' allows us to rewrite this property as such:

$$\vdash \mathcal{H} \leftrightarrow \Box \mathcal{H} \qquad (4.64)$$

We call $\mathcal{H}$ a *Henkin formula* because it claims itself to be provable. It does this in the same way as Gödel's formula $\varphi_g(\bar{g})$ claims itself to be unprovable.

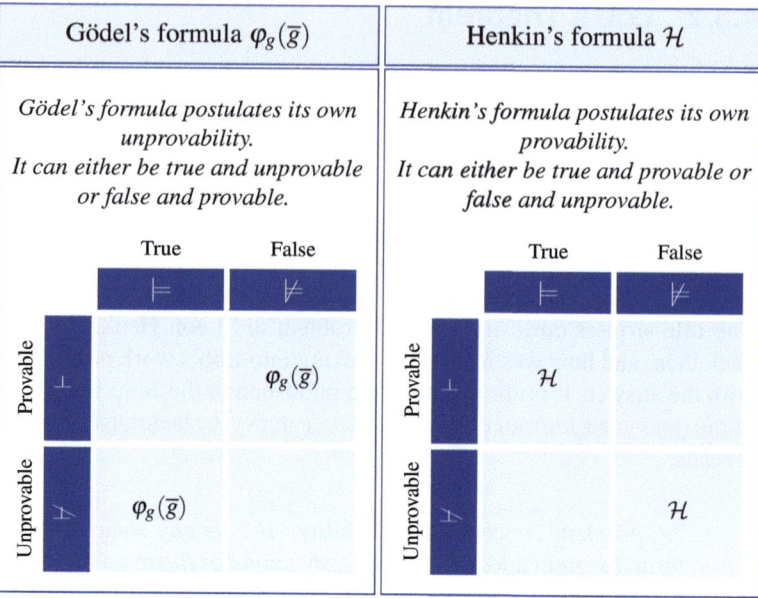

Figure 4.20: A comparison of Gödel's formula and Henkin's formula

At the time, Henkin asked himself whether $\mathcal{H}$ was provable within Peano arithmetic but found no answer. In contrast to Gödel's formula $\varphi_g(\overline{g})$, which cannot hide its proximity to semantic paradoxes, Henkin's variant is well-behaved. Neither the assumption of its provability nor its unprovability appears contradictory (Figure 4.20). Let us take some time and play through both variants:

- **Suppose** $\vdash \mathcal{H}$

  Since the formula claims its provability, it would be substantively true. It would be simultaneously true and provable.

- **Suppose** $\nvdash \mathcal{H}$

  Since the formula claims its provability, it would be substantively false. It would be simultaneously false and unprovable.

While neither of the two possibilities leads to an immediate contradiction, one thing is for sure: only one of them is true.

Henkin's question remained unanswered for three years until Martin Löb ultimately found an answer. We will take a closer look at how Löb has solved the riddle and assume that φ is a formula satisfying

$$\vdash \Box \varphi \to \varphi. \tag{4.65}$$

4.3 The Second Incompleteness Theorem

Below, we will use the Henkin formula $\mathcal{H}$ for φ, which fulfills this property due to (4.64).

By choosing

$$\chi(y) := \text{Bew}(y) \to \varphi,$$

the diagonalization lemma guarantees the existence of a formula γ with the following property:

$$\vdash \gamma \leftrightarrow (\text{Bew}(\ulcorner \gamma \urcorner) \to \varphi)$$

Or, in the new modal spelling:

$$\vdash \gamma \leftrightarrow (\Box \gamma \to \varphi) \qquad (4.66)$$

Figure 4.21 depicts a compelling derivation sequence involving this formula. It demonstrates that the provability of $\Box \varphi \to \varphi$ entails the provability of φ, regardless of the concrete choice of φ. This relationship is the content of Löb's theorem, which reads as follows in the case of Peano arithmetic:

Theorem 4.13 (Löb, 1955)

> In Peano Arithmetic, $\vdash \Box \varphi \to \varphi$ implies $\vdash \varphi$.

With Löb's theorem at our disposal, we can finally answer Henkin's question. We know from (4.64) that $\mathcal{H}$ satisfies

$$\vdash \Box \mathcal{H} \to \mathcal{H}$$

and thus fulfills the condition of Theorem 4.13. From this, it follows that Henkin's formula $\mathcal{H}$ is a theorem; it is provable within PA:

$$\vdash \mathcal{H}$$

As noteworthy as it is that we could solve Henkin's problem by applying Theorem 4.13, it is not the main reason for mentioning it in this book. A more intensive study of Löb's theorem shows that its substantive meaning is much deeper than initially suggested. Figure 4.22 illustrates its far-reaching power. It reveals that Löb's theorem resides in the direct vicinity of a theorem that is undoubtedly one of the paramount results of mathematical logic: Gödel's second incompleteness theorem. This is the real significance of Löb's theorem.

1. $\vdash \Box \varphi \to \varphi$		(Hypothesis)
...		(...)
2. $\vdash \gamma \to (\Box \gamma \to \varphi)$		(from 4.66)
3. $\vdash (\Box \gamma \to \varphi) \to \gamma$		(from 4.66)
4. $\vdash \Box(\gamma \to (\Box \gamma \to \varphi))$		(DC1, 2)
5. $\vdash \Box(\gamma \to (\Box \gamma \to \varphi)) \to$		
$\quad (\Box \gamma \to \Box(\Box \gamma \to \varphi))$		(DC2)
6. $\vdash \Box \gamma \to \Box(\Box \gamma \to \varphi)$		(MP, 4,5)
7. $\vdash \Box(\Box \gamma \to \varphi) \to$		
$\quad (\Box\Box \gamma \to \Box \varphi)$		(DC2)
8. $\vdash \Box \gamma \to (\Box\Box \gamma \to \Box \varphi)$		(MB, 6,7)
9. $\{\Box \gamma\} \vdash \Box\Box \gamma \to \Box \varphi$		(DT)
10. $\vdash \Box \gamma \to \Box\Box \gamma$		(DC3)
11. $\{\Box \gamma\} \vdash \Box\Box \gamma$		(DT)
12. $\{\Box \gamma\} \vdash \Box \varphi$		(MP, 9,11)
13. $\vdash \Box \gamma \to \Box \varphi$		(DT)
14. $\vdash \Box \gamma \to \varphi$		(MB, 13,1)
15. $\vdash \gamma$		(MP, 14,3)
16. $\vdash \Box \gamma$		(DC1, 15)
17. $\vdash \varphi$		(MP, 16,14)

Figure 4.21: If $\Box \varphi \to \varphi$ is provable within Peano arithmetic, so is φ [15].

If the consistency of PA were provable within PA, ...	☞	$\vdash \text{Con}$ (Hypothesis)
		$\vdash \neg\Box(0 = \overline{1})$ (Def)
		$\vdash \neg\Box(0 = \overline{1}) \to (\Box(0 = \overline{1}) \to (0 = \overline{1}))$ (T8)
		$\vdash \Box(0 = \overline{1}) \to (0 = \overline{1})$ (MP. 2,3)
... Löb's theorem allowed us to derive a formula,	☞	$\vdash (0 = \overline{1})$ (Theorem 4.13)
... whose negation is also a theorem.	☞	$\vdash (0 \neq \overline{1})$

Conclusion: If Con is provable, PA is inconsistent.

Figure 4.22: Deriving Gödel's second incompleteness theorem from Löb's theorem.

4.4 Common Misconceptions

Few mathematical discoveries were received as controversially as Gödel's incompleteness theorems – and misunderstood as often. In fact, there are good reasons for this. Some critics studied the theorems with little precision; others ignored the premises or overinterpreted their substantive meaning. Still others, consciously or unconsciously, pulled the incompleteness theorems out of their mathematical context, abusing them for legitimizing this or that obscurity.

This section will outline some common misunderstandings and try to dispel them with a solid explanation (cf. [56]). Not everything in this section is new. If you have read the previous text carefully, you should be able to catch quickly where most of the misunderstandings come from. In this case, you have already developed a good understanding of what Gödel's theorems say and, more importantly, what they do not say.

Misconception 1

"Gödel has shown that there are true propositions in mathematics that cannot be proven."

Some interpret Gödel's first incompleteness theorem as postulating the existence of mathematical propositions that are unprovable in an absolute sense. The misunderstanding disappears after recalling what it means to prove a statement. In a formal sense, a formula φ is provable if it can be derived from the axioms of a formal system by repeated application of inference rules. As a result, provability is always tied to a specific formal system. It is easy to see that for every formula φ, there is a formal system in which φ is provable. Hence, provability is always a relative property and never absolute.

4.4 Common Misconceptions

Suppose the formula φ stands for Goldbach's conjecture, for which we do not know up to this day whether it is provable in Zermelo-Fraenkel set theory. If it turns out that φ is unprovable in ZF, we could add φ to the axioms and receive with $ZF \cup \{\varphi\}$ a formal system in which Goldbach's conjecture is provable. Whether it makes sense to build the edifice of mathematics on this system is a different story.

Ordinary mathematics also ties the concept of provability to a formal system. However, this system is usually not explicitly mentioned, and hardly any proof is written down in the formal precision we encounter in mathematical logic. Here, we use *provable* to mean that a statement is deducible via the ordinary reasoning apparatus of mathematics. The formal counterpart to this reasoning apparatus is Zermelo-Fraenkel set theory, represented by the formal systems ZF and ZFC.

Keep in mind that not all formal systems are subject to incompleteness, but only those with a certain expressiveness. In particular, all systems capable of formalizing the additive and multiplicative properties of natural numbers are affected. This includes Peano arithmetic and Zermelo-Fraenkel set theory, in which natural numbers can be represented by sets and addition and multiplication by set operations.

Misconception 2

"Gödel has shown that undecidable statements exist in every formal system."

We already know from the discussion in Section 2.1 that not every formal system is incomplete. There, we defined a correct and complete calculus capable of deriving a limited number of primitive statements about natural numbers. Needless to say, this calculus is far from having any practical application in mathematics.

Let us summarize: Not every formal system is incomplete. This observation inevitably raises the question of when the incompleteness phenomenon starts to manifest. How expressive must a formal system be to be subject to Gödel's first incompleteness theorem? The two formal systems in Table 4.3 get us closer to the answer.

Listed on the left are the axioms of *Presburger arithmetic*, which are identical to Peano arithmetic except for the missing multiplication axioms. In 1929, the Polish mathematician Mojżesz Presburger showed that Presburger arithmetic is correct and complete. As a result, all arithmetic formulas not containing the multiplication sign can be derived from the axioms. In order to succumb to the phenomenon of incompleteness, we need more than the mere capability of talking about the additive capabilities of natural numbers. We must also be able to talk about multiplication.

Presburger arithmetic		Robinson arithmetic	
$\sigma = \tau \to (\sigma = \rho \to \tau = \rho)$	(P1)	$\sigma = \tau \to (\sigma = \rho \to \tau = \rho)$	(R1)
$\sigma = \tau \to s(\sigma) = s(\tau)$	(P2)	$\sigma = \tau \to s(\sigma) = s(\tau)$	(R2)
$\neg(0 = s(\sigma))$	(P3)	$\neg(0 = s(\sigma))$	(R3)
$s(\sigma) = s(\tau) \to \sigma = \tau$	(P4)	$s(\sigma) = s(\tau) \to \sigma = \tau$	(R4)
$\sigma + 0 = \sigma$	(P5)	$\sigma + 0 = \sigma$	(R5)
$\sigma + s(\tau) = s(\sigma + \tau)$	(P6)	$\sigma + s(\tau) = s(\sigma + \tau)$	(R6)
$\varphi(0) \to (\forall x\, (\varphi(x) \to \varphi(s(x)))) \to \forall x\, \varphi(x)$	(P7)	$\sigma \times 0 = 0$	(R7)
		$\sigma \times s(\tau) = (\sigma \times \tau) + \sigma$	(R8)
		$\sigma = 0 \vee \exists \xi\, \sigma = s(\xi)$	(R9)

Table 4.3: Non-logical axioms of Presburger arithmetic (left) and Robinson arithmetic (right)

However, Presburger's result does not imply that a formal system must have the full expressive power of Peano arithmetic to fall victim to the incompleteness theorem. A close analysis of Gödel's proof reveals that mathematical induction plays virtually no role in this context. By replacing the induction axiom in PA with the much weaker axiom (R9) from Table 4.3, we reach *Robinson arithmetic*, whose expressive power resides strictly between that of Presburger's and Peano's arithmetics. Everything needed to carry out Gödel's proof is present in Robinson arithmetic, which means we can weaken the requirements of the incompleteness theorems even further. Gödel's theorems apply to all formal systems that are expressive enough to formalize Robinson arithmetic.

Misconception 3

"Gödel's first incompleteness theorem contradicts Gödel's completeness theorem."

When the first incompleteness theorem is seen to contradict Gödel's completeness theorem, this is almost always due to the careless use of concepts and terminology. On the surface, Gödel's completeness theorem guarantees the connection $\vdash \varphi \Leftrightarrow \models \varphi$, while the first incompleteness theorem attests that the relations '$\vdash$' and '$\models$' do not align.

The contradiction disappears as soon as we remember how the symbol '$\models$' is to be read. In the context of the completeness theorem, $\models \varphi$

states that φ is *universally valid*; that is, the formula is true under all possible interpretations. However, in the context of the incompleteness theorem, $\models \varphi$ expresses that φ is substantively true under a particular interpretation. In the case of Peano arithmetic, under the interpretation that associates the symbols '$+$', '$\times$', and 's' with addition, multiplication, and the successor function over the domain of the natural numbers. Also, note that Gödel's completeness theorem only makes a statement about first-order predicate logic and loses validity in predicate logics of second or higher order.

On the contrary, Gödel's first incompleteness theorem is valid in all formal systems sufficiently expressive to formalize Peano arithmetic, thereby applying to predicate logics of higher order, too. It becomes evident by now that the substantive meanings of the two theorems are fundamentally different. Both have little in common apart from their similar-sounding names.

A well-known result of Euclidean geometry states that neither the *parallel postulate* (Figure 4.23) nor its negation is deducible from the other axioms. The parallel postulate is undecidable because the other Euclidean axioms allow for several consistent interpretations. Among them is the geometry of the plane; it is, in a sense, the standard interpretation of Euclidean geometry, and the parallel postulate is a true statement here. However, other interpretations, such as *elliptical* or *hyperbolic geometry*, are also consistent with the other Euclidean axioms. In these non-Euclidean geometries, the parallel postulate is a false statement.

If several consistent, non-isomorphic interpretations exist for the axioms of a correct formal system, undecidable propositions must inevitably exist. It is important not to confuse or even equate the incompleteness phenomenon discovered by Gödel with this type of incompleteness. In particular, the undecidability of the parallel postulate is due to the axioms' deficiency in not uniquely characterizing the geometric objects we have in mind. Where the axioms leave room for interpretation, undecidable propositions unavoidably pop up.

Gödel's incompleteness phenomenon goes much deeper. We also encounter it in formal systems with a single consistent interpretation. An example of such a system is Peano arithmetic, formulated in second-order predicate logic. Although Dedekind's isomorphism theorem guarantees that all models of this formal system are isomorphic to the standard model, undecidable theorems still exist.

Misconception 4

"The existence of undecidable propositions is caused by the inability of the axioms to characterize the properties of the described objects uniquely. Undecidable propositions only arise because the axioms allow more than one consistent interpretation."

"That, if a straight line falling on two straight lines make the interior angles on the same side less than two right angles, the two straight lines, if produced indefinitely, meet on that side on which are the angles less than two right angles."

Euclid
(ca. 365 BC – ca. 300 BC)

Figure 4.23: Euclid's parallel postulate

Misconception 5

"Incomplete formal systems can be completed by iterating through all undecidable formulas and adding either the formula itself or its negation as an axiom."

Gödel has demonstrated that for every formal system that meets the requirements of the first incompleteness theorem, an undecidable formula $\varphi_g(\overline{g})$ can be constructed. Neither this formula nor its negation is derivable from the axioms so we can add one of them to the axioms without contradiction. It feels as if we could systematically close the leaks and eventually arrive at a complete formal system.

At a second glance, it becomes evident that this approach is futile. As soon as we have broadened the set of axioms, we can apply Gödel's construction again and obtain a formula $\varphi_{g'}(\overline{g'})$ that differs from $\varphi_g(\overline{g})$ and is undecidable within the new system. But how can a different formula be obtained the second time? The reason is simple: Since we have added a new axiom, the Gödel numbers of all formulas involved in constructing Gödel's diagonal element alter, and so does the diagonal element itself. There is no escape: Formal systems expressive enough to formalize Peano arithmetic cannot be completed.

Misconception 6

"The second incompleteness theorem implies the unprovability of the consistency of Peano arithmetic."

The second incompleteness theorem states that a formal system expressive enough to formalize Peano arithmetic cannot prove its own consistency. It follows that such a system is even more incapable of proving the consistency of a more expressive system. In short, the consistency of PA is just as impossible to prove within PA as is the consistency of ZF or ZFC.

It is incorrect to reverse the conclusion described above. The second incompleteness theorem in no way excludes the possibility of proving the consistency of a formal system in a more expressive system. In particular, the second incompleteness theorem does not rule out the consistency of PA being provable within ZF or ZFC. In 1936, the German mathematician Gerhard Gentzen demonstrated that this is indeed possible [70]. By cleverly coding the proofs of PA as ordinal numbers, he succeeded in proving consistency via the principle of transfinite induction. Gentzen's result does not contradict the second incompleteness theorem in any way. With transfinite induction, he has resorted to a set-theoretical technique that is unavailable in PA. As a result, Gentzen's proof can be formalized within ZF or ZFC but not in PA.

Misconception 7

"The formula Con formalizes consistency. It follows that a formal system that proves Con is consistent."

Gödel's second incompleteness theorem is often incorrectly understood to suggest the provability of Con would imply the consistency of the formal system. However, this is by no means true. If a formal system that includes the propositional reasoning apparatus is contradictory, all formulas would be derivable from the axioms, including Con.

The reverse conclusion is correct: if we succeed in proving the consistency of a sufficiently expressive formal system, it must necessarily be inconsistent. As a result, we can only utilize the second incompleteness theorem to prove the inconsistency, but not the consistency, of a formal system.

The true meaning of the second incompleteness theorem is something else: If a formal system cannot prove its own consistency, the proof cannot succeed in any less expressive system either. From this, it follows that the consistency of ordinary mathematics is not provable by any standard methods of ordinary mathematics. However, this was the plan that Hilbert had pursued so vehemently for many years: proving the consistency of classical mathematics by finite means.

Does this result suggest that we should distrust Peano arithmetic, to name just one example? The answer is no. Even if the second incompleteness theorem destroys the hope of securing PA with arguments that are more primitive and thus more credible than Peano arithmetic itself, there is no reason for distrust. Hardly anyone questions the consistency of PA. To seriously raise any doubt, the natural numbers are too familiar, and the axioms too simple.

And what about set theory? Is the foundation axiom sufficient to banish all antinomies from set theory? Again, the prevailing opinion is that mathematics can be consistently built on ZF or ZFC, but formal proof is lacking. The second incompleteness theorem unmistakably clarifies that a consistency proof can only be conducted in a formal system more complex than ZF or ZFC. Hence, we would merely defer the question to another system. In fact, the second incompleteness theorem shatters any hope of ever confirming the consistency of ZF and ZFC.

4.5 Goodstein's Theorem

In 1944, the English mathematician Reuben Louis Goodstein proved a theorem that pushes Gödel's incompleteness phenomenon into bright light. At first glance, Goodstein's theorem looks like an ordinary proposition of number theory; it makes a statement about the progression of particular number sequences, called *Goodstein sequences* today. Moreover, the statement can be proven comparatively easily using the means of ordinal number theory, as described in Section 3.2.2.

Goodstein's theorem is exceptional because, just as Gödel's formula $\varphi_g(\overline{g})$ or Rosser's formula $\varphi_r(\overline{r})$, it is undecidable within Peano arith-

 Reuben Louis Goodstein was an English mathematician born in London on December 15, 1912. In 1931, he transferred from St. Paul's School in London to the renowned University of Cambridge, where he graduated with flying colors in mathematics.
His next stop was a lecturer position at the University of Reading. Besides his teaching duties, Goodstein continued to research and was awarded a doctorate by the University of London in 1947. In 1948, he accepted an appointment from the University of Leicester, where he worked as a professor until his retirement in 1977.

Goodstein dedicated his life to teaching and was highly regarded as an exceptional educator. Today, he is best known for *Goodstein's theorem*, a prominent example of what is referred to as a *natural independence phenomenon* in mathematical logic. Broadly speaking, this category includes ordinary mathematical theorems that, much like the artificially constructed formulas of Gödel and Rosser, are unprovable. Less well known is that Goodstein developed a widely used conceptual framework for naming operations beyond exponentiation. Goodstein coined the term *tetration* for iterated exponentiation, followed, in order of the Greek prefixes, by *pentation, hexation, heptation, octation*, and so on. Reuben Goodstein died in Leicester on March 8, 1985, aged 72.

metic; that is, neither the theorem itself nor its negation is derivable from the axioms unless Peano arithmetic is inconsistent. This is the astonishing result of a 1982 paper by Laurie Kirby and Jeff Paris [116]. In contrast to the artificially constructed formulas of Gödel and Rosser, Goodstein's theorem is anything but an artificial product: it is an ordinary theorem of number theory, and, unlike $\varphi_g(\overline{g})$ and $\varphi_r(\overline{r})$, it is not self-referential in any way.

To understand Goodstein's theorem, we need to recall some basic knowledge about the representation of natural numbers. First, we note that every natural number x can be written in the form

$$x = a_n \cdot b^n + a_{n-1} \cdot b^{n-1} + a_{n-2} \cdot b^{n-2} + \ldots + a_1 b + a_0 \quad (4.67)$$

with $a_0, \ldots, a_n \geq 0$. b is called the *base* and may be any natural number greater than 1. Requiring $a_i < b$ for all i, the digits $a_0, \ldots, a_n$ are unique, and (4.67) becomes the so-called *b-adic representation* of x.

For instance, choosing $b = 2$ and $x = 36$, we get:

$$36 = 1 \cdot 2^5 + 1 \cdot 2^2$$

The construction of the Goodstein sequence requires a specific representation of x, called the *expanded b-adic representation*. It is created by recursively replacing the exponents in (4.67) with their own *b*-adic representation. For the number 36, the expanded representation in base 2 reads as follows:

$$36 = 1 \cdot 2^{2^2+1} + 1 \cdot 2^2$$

Furthermore, we need a particular substitution function, which Goodstein calls $S_c^b(x)$ in his original paper. For a natural number x, the value

4.5 Goodstein's Theorem

$S_c^b(x)$ is calculated by first converting x into the expanded b-adic representation and then replacing all bases with c. For example, $S_3^2(36)$ is the number

$$S_3^2(36) = S_3^2(2^{2^2+1} + 2^2) = 3^{3^3+1} + 3^3 = 22876792454988 \quad (4.68)$$

We are now ready to describe Goodstein sequences formally. First, we note that for each natural number x, there is a separate Goodstein sequence, which we denote by

$$g_0(x), g_1(x), g_2(x), g_3(x), \ldots$$

x is the start value of the sequence, i.e., $g_0(x) = x$. To calculate g_1, we first translate g_0 into its b-adic representation with $b = 2$. We then replace the base b, as shown in Figure 4.24, with $b+1$ and reduce the result by 1. Calculating the other sequence members is carried out the same way. In each step, we first increase the base by 1 (*base bumping*). After that, we decrease the result by 1:

$$g_{n+1}(x) = \begin{cases} S_{n+3}^{n+2}(g_n(x)) - 1 & \text{if } g_n(x) > 0 \\ 0 & \text{if } g_n(x) = 0 \end{cases}$$

How likely is a Goodstein sequence to reach 0? Figure 4.24 shows that the sequence members are rapidly growing due to continuously increasing the base, and we can barely pen them down in decimal notation after

$$g_0(36) = 36$$
$$= 2^{2^2+1} + 2^2$$

$g_1(36) := S_3^2(g_0(36)) - 1$

$$g_1(36) = \left(3^{3^3+1} + 3^3\right) - 1$$
$$= 3^{3^3+1} + 2 \cdot 3^2 + 2 \cdot 3 + 2$$

$g_2(36) := S_4^3(g_1(36)) - 1$

$$g_2(36) = \left(4^{4^4+1} + 2 \cdot 4^2 + 2 \cdot 4 + 2\right) - 1$$
$$= 4^{4^4+1} + 2 \cdot 4^2 + 2 \cdot 4 + 1$$

$g_3(36) := S_5^4(g_2(36)) - 1$

$$g_3(36) = \left(5^{5^5+1} + 2 \cdot 5^2 + 2 \cdot 5 + 1\right) - 1$$
$$= 5^{5^5+1} + 2 \cdot 5^2 + 2 \cdot 5$$

Figure 4.24: The initial four elements of the Goodstein sequence with the start value 36

Start value 1	Start value 2	Start value 3	Start value 4	Start value 5	Start value 6
$g_0(1) = 1$	$g_0(2) = 2$	$g_0(3) = 3$	$g_0(4) = 4$	$g_0(5) = 5$	$g_0(6) = 6$
$g_1(1) = 0$	$g_1(2) = 2$	$g_1(3) = 3$	$g_1(4) = 26$	$g_1(5) = 27$	$g_1(6) = 29$
$g_2(1) = 0$	$g_2(2) = 1$	$g_2(3) = 3$	$g_2(4) = 41$	$g_2(5) = 255$	$g_2(6) = 257$
$g_3(1) = 0$	$g_3(2) = 0$	$g_3(3) = 2$	$g_3(4) = 60$	$g_3(5) = 467$	$g_3(6) = 3125$
$g_4(1) = 0$	$g_4(2) = 0$	$g_4(3) = 1$	$g_4(4) = 83$	$g_4(5) = 775$	$g_4(6) = 46655$
$g_5(1) = 0$	$g_5(2) = 0$	$g_5(3) = 0$	$g_5(4) = 109$	$g_5(5) = 1197$	$g_5(6) = 98039$
$g_6(1) = 0$	$g_6(2) = 0$	$g_6(3) = 0$	$g_6(4) = 139$	$g_6(5) = 1751$	$g_6(6) = 187243$
$g_7(1) = 0$	$g_7(2) = 0$	$g_7(3) = 0$	$g_7(4) = 173$	$g_7(5) = 2454$	$g_7(6) = 332147$
$g_8(1) = 0$	$g_8(2) = 0$	$g_8(3) = 0$	$g_8(4) = 211$	$g_8(5) = 3325$	$g_8(6) = 555551$
$g_9(1) = 0$	$g_9(2) = 0$	$g_9(3) = 0$	$g_9(4) = 253$	$g_9(5) = 4382$	$g_9(6) = 885775$
$g_{10}(1) = 0$	$g_{10}(2) = 0$	$g_{10}(3) = 0$	$g_{10}(4) = 299$	$g_{10}(5) = 5643$	$g_{10}(6) = 1357259$
$g_{11}(1) = 0$	$g_{11}(2) = 0$	$g_{11}(3) = 0$	$g_{11}(4) = 348$	$g_{11}(5) = 7126$	$g_{11}(6) = 2011162$
...	...	...	...	...	...

Figure 4.25: Development of the Goodstein sequences with the start values 1 to 6

- Goodstein sequence starting at 1

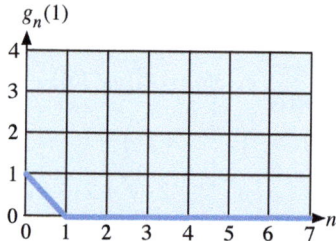

- Goodstein sequence starting at 2

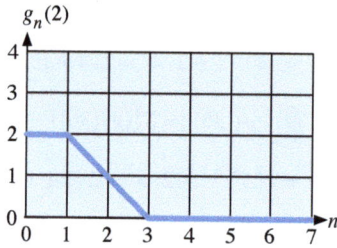

- Goodstein sequence starting at 3

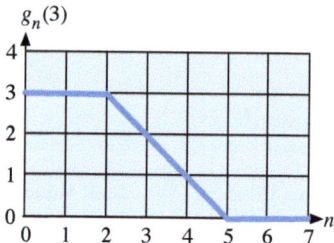

- Goodstein sequence starting at 4

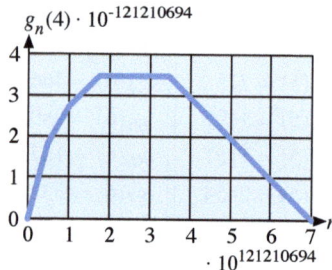

Figure 4.26: Course of values of the first four Goodstein sequences

only a few steps. The examples in Figure 4.25 further illustrate that this phenomenon already applies to small start values. From $x = 4$, Goodstein sequences seem to head toward infinity, and increasing the start value fuels their growth.

The time has come to articulate Goodstein's theorem. In the light of the studied examples, it reveals an astonishing phenomenon:

 Theorem 4.14 (Goodstein, 1944)

> Every Goodstein sequence eventually reaches 0.

Figure 4.26 depicts when the first four Goodstein sequences reach 0 [6, 169]. The first three sequences quickly do so. When starting at 4, however, the sequence rises continuously for a long time, reaching its maximum at

$$i = \frac{1}{4} 24 \cdot 2^{24} 2^{24 \cdot 2^{24}} - 3 \approx 1{,}72 \cdot 10^{121210694}.$$

After remaining stable for a while, it starts a continuous descent and reaches 0 at

$$i = 24 \cdot 2^{24} 2^{24 \cdot 2^{24}} - 3 \approx 6{,}89 \cdot 10^{121210694}.$$

This number is far beyond our imagination; it corresponds to a decimal number with more than 121 million digits!

That every Goodstein sequence reaches 0 is a fascinating result on its own. Even more astonishing, though, is that we can prove this property with little effort using the ordinal number theory introduced in Section 3.2.2. In essence, the proof builds upon the idea of translating a Goodstein sequence into a parallel sequence of ordinal numbers according to the following scheme:

$$g'_n(x) = S_\omega^{n+2}(g_n(x))$$

Figure 4.27 illustrates the construction. We begin by writing down the elements of a Goodstein sequence in their expanded b-adic representation. After that, we replace all bases with the ordinal number ω.

The monotonicity of the substitution S_ω^k is essential for our subsequent argumentation. It means that the ordering of two numbers is preserved; that is:

$$x < y \;\Rightarrow\; S_\omega^k(x) < S_\omega^k(y) \quad \text{for all } k \in \mathbb{N} \qquad (4.69)$$

4.5 Goodstein's Theorem

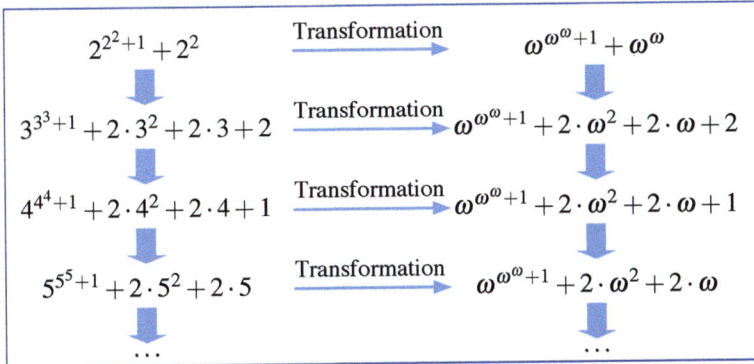

Figure 4.27: Each Goodstein sequence can be translated into a strictly monotonically decreasing sequence of ordinal numbers until the Goodstein sequence reaches the value 0. If Theorem 4.14 were false, there would be a Goodstein sequence whose elements all differ from 0, resulting in an infinitely descending sequence of ordinal numbers. However, we already know that such a sequence cannot exist.

For all n with $g_n(x) > 0$, we can now estimate $g'_{n+1}(x)$ as follows:

$$\begin{aligned}
g'_{n+1}(x) &= S_\omega^{n+3}(g_{n+1}(x)) \\
&= S_\omega^{n+3}(S_{n+3}^{n+2}(g_n(x)) - 1) \\
&< S_\omega^{n+3}(S_{n+3}^{n+2}(g_n(x))) \qquad \text{(due to (4.69))} \\
&= S_\omega^{n+2}(g_n(x)) \\
&= g'_n(x)
\end{aligned}$$

Voilà: The parallel sequence is strictly decreasing. Consequently, we can construct an infinitely descending sequence of ordinal numbers from any Goodstein sequence whose elements all differ from 0. However, we already know from Chapter 3 that such a sequence cannot exist. Conversely, this means that every Goodstein sequence must inevitably reach 0 at some point.

With the help of ordinal number theory, it took little effort to back up Goodstein's theorem. However, the more exciting question is different: Why can Goodstein's theorem be formulated but not proved within Peano arithmetic? How can it happen that we have to resort to proofs that are *outside* the theory in which the theorem is formulated?

First, let us convince ourselves that Peano arithmetic is expressive enough to talk about Goodstein's theorem. To this end, we introduce the *Goodstein function* $\mathfrak{G} : \mathbb{N} \to \mathbb{N}$, which is defined as follows:

$$\mathfrak{G}(x) := \min\{n \mid g_n(x) = 0\}$$

Expressed colloquially, the function value $\mathfrak{G}(x)$ tells us when the Goodstein sequence with the start value x reaches 0. Figure 4.28 illustrates the calculation of $\mathfrak{G}(x)$ for the first four Goodstein sequences.

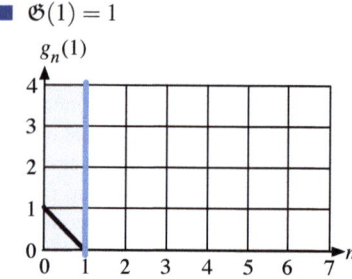

- $\mathfrak{G}(1) = 1$

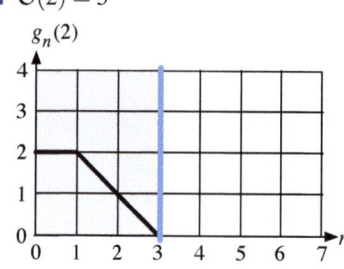

- $\mathfrak{G}(2) = 3$

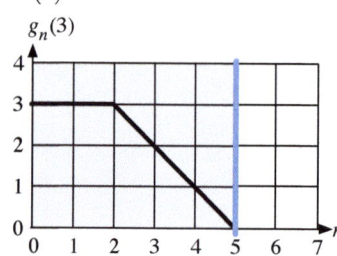

- $\mathfrak{G}(3) = 5$

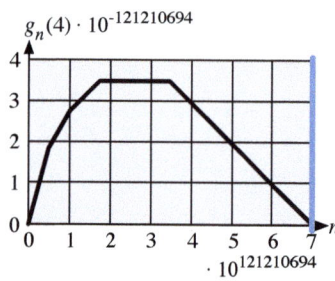

- $\mathfrak{G}(4) = 24 \cdot 2^{24} 2^{24 \cdot 2^{24}} - 3$

Figure 4.28: The first four function values of the *Goodstein function* $\mathfrak{G}$

Next, we encode the Goodstein function using an arithmetic formula $\varphi_\mathfrak{G}$ with two free variables, x and y. This formula is supposed to fulfill the following relationship:

$$\models \varphi_\mathfrak{G}(\overline{x}, \overline{y}) \Leftrightarrow \mathfrak{G}(x) = y$$

The existence of a formula with this property is a result that we will work out in Chapter 5. There, we will introduce the arithmetization of Turing machines and implicitly prove that Peano arithmetic is sufficiently expressive to represent any function a systematic procedure can compute. Though we can only have a vague idea of what a systematic procedure makes up at this point, we can already identify the Goodstein function as computable. As every Goodstein sequence eventually reaches 0, we can systematically determine the function value $\mathfrak{G}(x)$ by calculating the sequence elements one after another until 0 comes up.

Next, we will establish a relationship between the Goodstein function $\mathfrak{G}(x)$ and Goodstein's theorem. Substantively, the theorem is equivalent to the statement that the function $\mathfrak{G}(x)$ is *defined* for all n, which, in turn, is equivalent to saying that the Goodstein function is a *total function*. When we talk about proving the totality of $\mathfrak{G}(x)$ in PA, we mean to derive the following theorem:

$$\vdash \forall x \exists y \, \varphi_\mathfrak{G}(x, y) \qquad (4.70)$$

For formulas of this type, a strong result by Georg Kreisel applies:

Theorem 4.15 (Kreisel, 1952)

> If a computable function $f : \mathbb{N} \to \mathbb{N}$ can be proved to be total within Peano arithmetic, then there exists a function f_α with $\alpha < \varepsilon_0$ that dominates f.

At the center of this theorem is a *dominance statement*. Kreisel refers to a hierarchy of fast-growing functions that goes back to Stanley Wainer and Martin Löb [126, 216]. Even if the hierarchy is complex in detail, it follows the simple basic idea of ordering functions according to their growth rate. This is a common approach in mathematics and not all that difficult as long as we do not leave the realm of ordinary functions. For instance, the sequence

$$n, 2 \cdot n, 3 \cdot n, \ldots, n^2, n^3, \ldots, 2^n, 3^n, \ldots, 2^{2^n}, 2^{2^{2^n}}, \ldots \qquad (4.71)$$

forms a natural hierarchy of ever-faster-growing functions. Things become trickier once we attempt to integrate functions that grow even

4.5 Goodstein's Theorem

faster. Consider the *Ackermann function* $\mathfrak{A}(m,n)$, which we have already encountered on Page 201. Its definition appears innocuous at first, yet it is practically impossible to calculate the function value $\mathfrak{A}(m,n)$ even for small values of m.

The reason for this becomes evident if we keep the parameter m constant for different values. In this way, $\mathfrak{A}(m,n)$ becomes a separate function $\mathfrak{A}_m(n)$ for each number $m \in \mathbb{N}$, whose growth characteristics are depicted in Figure 4.29. The function $\mathfrak{A}_0(n)$ grows linearly, $\mathfrak{A}_1(n)$ exponentially, $\mathfrak{A}_2(n)$ hyper-exponentially, and so on. The Ackermann function thus turns out to be a universal function uniting an infinite number of ever-faster-growing functions.

Even faster grows the *diagonalized Ackermann function*

$$\mathfrak{A}_\omega(n) := \mathfrak{A}_n(n).$$

For each value of m, $\mathfrak{A}_\omega(n)$ will exceed $\mathfrak{A}_m(n)$ beyond a certain n. We say $\mathfrak{A}_\omega$ dominates $\mathfrak{A}_m$. Indexing this function with the ordinal number ω was a natural choice; after all, we can regard $\mathfrak{A}_\omega$ as a limit function in the same way as we have defined ω as the limit ordinal for the natural numbers.

Wainer and Löb proceeded similarly. The Löb-Wainer hierarchy is formed by a sequence of functions f_α with an ordinal index α. The functions f_0 and f_1 grow linearly, f_2 grows exponentially, and for transcribing f_3, we would have to resort to exponential towers as utilized in Section 3.2.2.2 in the construction of ordinal numbers.

The Löb-Wainer hierarchy allows us to capture the growth rate of the diagonalized Ackermann function quantitatively: $\mathfrak{A}_\omega$ grows at the same rate as the function f_ω.

At this point, it becomes evident how to read Kreisel's theorem. Suppose an arithmetic formula φ_f describes a computable function f. In that case, we can only hope to prove

$$\forall x \exists y\, \varphi_f(x,y)$$

within PA if some function f_α dominates f with $\alpha < \varepsilon_0$.

The time has come to take a closer look at the growth rate of the Goodstein function. The striking jump from $\mathfrak{G}(3)$ to $\mathfrak{G}(4)$ already raises the suspicion that we are dealing with a growth rate that significantly exceeds that of the diagonalized Ackermann function. Thanks to Laurie Kirby and Jeff Paris, we now have a precise understanding of the growth

■ Ackermann function

$$\mathfrak{A}(0,n) := 2 \cdot n + 1$$
$$\mathfrak{A}(m+1,0) := \mathfrak{A}(m,1)$$
$$\mathfrak{A}(m+1,n+1) := \mathfrak{A}(m,\mathfrak{A}(m+1,n))$$

■ Function hierarchy

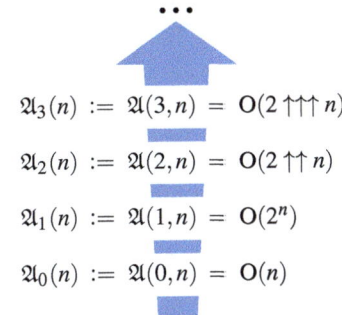

$$\mathfrak{A}_3(n) := \mathfrak{A}(3,n) = O(2 \uparrow\uparrow\uparrow n)$$
$$\mathfrak{A}_2(n) := \mathfrak{A}(2,n) = O(2 \uparrow\uparrow n)$$
$$\mathfrak{A}_1(n) := \mathfrak{A}(1,n) = O(2^n)$$
$$\mathfrak{A}_0(n) := \mathfrak{A}(0,n) = O(n)$$

Figure 4.29: Operator hierarchy derived from of the Ackermann function

The notation $2 \uparrow^k n$ used in Figure 4.29 goes back to the American computer scientist Donald E. Knuth [120]. In 1976, he proposed the *arrow notation* to solve the long-standing problem in classical mathematics of lacking a separate symbolism for functions beyond exponentiation. Formally, $m \uparrow^k n$ is defined as follows:

$$m \uparrow^k n := \begin{cases} a^b & \text{if } k=1 \\ 1 & \text{if } n=0 \\ m \uparrow^{k-1} (m \uparrow^k (n-1)) & \text{else} \end{cases}$$

His novel notation paved the way for naming functions of different growth rates according to a uniform scheme. Exponentiation is the slowest growth rate the arrow notation can represent ($m \uparrow n$). Then comes tetration ($m \uparrow\uparrow n$), etc.

behavior of the Goodstein function. In 1982, they were the first to prove that $\mathfrak{G}(x)$ grows as fast as the function f_{ε_0} from the Löb-Wainer hierarchy [116]. Conversely, this means that $\mathfrak{G}$ dominates every function f_α with $\alpha < \varepsilon_0$, which, by Kreisel's theorem, implies that it is impossible to prove formula (4.70) within PA. From the equivalence of Goodstein's theorem and the totality of $\mathfrak{G}(x)$, we can immediately conclude the result we were looking for:

Theorem 4.16 (Kirby, Paris, 1982)

> The Goodstein theorem is unprovable within PA.

We do not want to leave an interesting aspect unmentioned at this point. Even if it is impossible to derive formula (4.70) within PA, we can prove function $\mathfrak{G}$ to be defined for each given value of x. Consequently, for all x, we have:

$$\vdash \exists y\, \varphi_\mathfrak{G}(\overline{x}, y) \tag{4.72}$$

At the same time, Theorem 4.16 attests:

$$\nvdash \forall x\, \exists y\, \varphi_\mathfrak{G}(x, y) \tag{4.73}$$

Both variants only differ in the quantification over x. In (4.72), the quantification happens outside the formal system, whereas in (4.73), it happens inside. Once again, it becomes apparent how meticulously we need to distinguish between the meta-level (for all x) and the object level ($\forall x$).

Do functions exist that grow even faster than the Goodstein function? The answer is yes! A well-known example is the beaver function $\mathfrak{B}(n)$ (*busy beaver function*), formulated by the Hungarian mathematician Tibor Radó in 1962 as part of a competition [17, 166]. The competition aimed to construct a Turing machine (*busy beaver*) with as few states as possible, writing as many ones as possible onto an empty tape [17, 166]. The function value $\mathfrak{B}(n)$ is the maximum possible number for a beaver with n states. Since there are only a finite number of beavers for every n, the value of the beaver function is well-defined for all natural numbers. Nevertheless, the function values are only known exactly up to $n = 4$:

$\mathfrak{B}(1)$	$\mathfrak{B}(2)$	$\mathfrak{B}(3)$	$\mathfrak{B}(4)$	$\mathfrak{B}(5)$
1	4	6	13	≥ 4098

Although only rough estimates exist for $n \geq 5$, impressive statements can be made about the growth rate of $\mathfrak{B}(n)$. It can be proven that the beaver function must grow faster than any computable function. Consequently, $\mathfrak{B}(n)$ dominates both the Ackermann and the Goodstein function. The beaver function itself is uncomputable; that is, it is impossible to determine the function value $\mathfrak{B}(n)$ with a systematic method for all n.

4.6 Exercises

Exercise 4.1

On page 198, you have learned how to map formulas of Peano arithmetic to natural numbers. This exercise will give you an idea about the size of the numbers we deal with. Find the formulas that correspond to the following Gödel numbers:

a) 27945122556290792802283166332500000000000

b) 920783852754905293279042680914408826637119384453120000

Note: Factorizing such large numbers by hand is almost impossible. Revert to software tools such as Mathematica, Maple, or WolframAlpha for this task.

Exercise 4.2

Use the method from Page 198 to calculate the Gödel number of the formula $\exists x\, s(x) = x$. Repeat the calculation for the formula $\bar{1} + 0 = \bar{1}$.

Note: It is sufficient to write down the Gödel numbers in factorized notation. Entirely written out, both numbers have well over a hundred decimal digits.

Exercise 4.3

In [197], Raymond Smullyan has proposed a clever type of Gödelization. His coding initially assigns one of the following numeric constants to each formula character:

0	'	(	)	f	,	v	∼	⊃	∀	=	≤	#
↕	↕	↕	↕	↕	↕	↕	↕	↕	↕	↕	↕	↕
1	0	2	3	4	5	6	7	8	9	10	11	12

Smullyan then interprets the assigned constants as digits of a number in base 13. In particular, if a formula φ consists of n characters and c_i denotes the i-th character, the following Gödel number is assigned:

$$\ulcorner \varphi \urcorner := \sum_{i=1}^{n} c_i \cdot 13^{n-i}.$$

a) Determine the Gödel number for the formula $\forall v\, v = v$.

b) Smullyan's logic employs the apostrophe as a symbol for the successor function: $0'$ represents the number 1, $0''$ represents the number 2, and so on. Which Gödel number corresponds to the expression representing the natural number n?

Exercise 4.4

Let K be a formal system capable of arithmetically representing relations and functions like we are used to from Peano arithmetic. Further, let A_1 and A_2 be the following propositions:

A_1: If φ represents a relation R semantically, it also represents R syntactically.
A_2: If φ represents a relation R syntactically, it also represents R semantically.

Mark all correct statements:
If the formal system K is …

	propositions A_1 is		propositions A_2 is	
	true	false	true	false
■ correct and complete, then …	○	○	○	○
■ correct and negation complete, then …	○	○	○	○
■ consistent and complete, then …	○	○	○	○
■ consistent und negation complete, then …	○	○	○	○

Exercise 4.5

Gödel's paramount paper from 1931 contains a technical part that defines 45 primitive-recursive functions and relations. It starts with two relations and three functions:

$$P_1(x,y) :\Leftrightarrow \exists z\, (z \leq x \wedge x = y \cdot z)$$

$$P_2(x) :\Leftrightarrow \neg \exists (z \leq x)\, (z \neq 1 \wedge z \neq x \wedge P_1(x,z)) \wedge x > 1$$

$$f_3(0,x) := 0$$
$$f_3(n+1,x) := \min\{y \leq x \mid P_2(y) \wedge P_1(x,y) \wedge y > f_3(n,x)\}$$

$$f_4(0) := 1$$
$$f_4(n+1) := (n+1) \cdot f_4(n)$$

$$f_5(0) := 0$$
$$f_5(n+1) := \min\{y \leq f_4(f_5(n)) + 1 \mid P_2(y) \wedge y > f_5(n)\}$$

Find out the substantive meaning of the defined relations and functions.

4.6 Exercises

Exercise 4.6

Consider the following line of reasoning:

a) Every true statement in number theory is a logical consequence of the Peano axioms.

b) Peano arithmetic formalizes the Peano axioms. As a first-order theory, it is subject to Gödel's completeness theorem, which states that first-order theories can prove all logical conclusions.

c) It follows from a) and b) that every true statement of number theory is provable within Peano arithmetic.

The result contradicts Gödel's first incompleteness theorem. Where does the error hide?

Exercise 4.7

In Section 4.2, you became familiar with the semantic and syntactic variants of Gödel's first incompleteness theorem. Another variant is this one:

"Any correct formal system expressive enough to formalize Peano arithmetic is negation-incomplete."

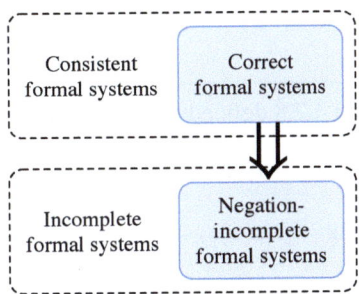

The formulation is stronger than the semantic variant but weaker than the syntactic variant. Can it still be derived from the semantic variant with little effort?

Exercise 4.8

On page 212, we claimed that it easy to derive from

$$z = diag(y) \Rightarrow \vdash \forall z \, (\mathsf{Diag}(\bar{y}, z) \leftrightarrow z = \bar{z}) \quad (4.74)$$
$$(x, y) \in B \Rightarrow \vdash \mathsf{B}(\bar{x}, \bar{y}) \quad (4.75)$$
$$(x, y) \notin B \Rightarrow \vdash \neg \mathsf{B}(\bar{x}, \bar{y}) \quad (4.76)$$

the following relationships:

x encodes a proof for the formula $\varphi_y(\bar{y}) \Rightarrow \vdash \psi_{\mathrm{Gdl}}(\bar{x}, \bar{y})$

x does not encode a proof for the formula $\varphi_y(\bar{y}) \Rightarrow \vdash \neg \psi_{\mathrm{Gdl}}(\bar{x}, \bar{y})$

Outline the missing proof.

Exercise 4.9

Section 2.6 introduced second-order predicate logic and highlighted that the expressive power of PL1 increases with the addition of predicate and function variables. We also mentioned the considerable price to pay. Unlike first-order predicate logic, PL2 is no longer complete if this term relates to the standard semantics; no calculus can derive precisely those formulas of PL2 that are universally valid.

a) Construct a PL2 formula φ that is true under exactly those interpretations whose individual domains are isomorphic to the natural numbers:

$$(U,I) \models \varphi \Leftrightarrow U \cong \mathbb{N} \tag{4.77}$$

b) Show that the incompleteness of PL2 is a direct consequence of Gödel's first incompleteness theorem.

c) Explain why no PL1 formula satisfies (4.77).

Exercise 4.10

Section 4.2.8 defined what it means to *name* a natural number within Peano arithmetic.

Which numbers are named by the following formulas?

a) $x + x = (s(s(0)))$

b) $x + x = s(s(s(0)))$

c) $x \times x = s(s(0))$

d) $x \times x = s(s(s(0)))$

Which of the following statements about the formulas of Peano arithmetic are correct?

e) "Each formula names a natural number."

f) "Each natural number is named by a formula."

g) "There are an infinite number of ways to name a natural number."

5 Computability Theory

> *"If it should turn out that the basic logics of a machine designed for the numerical solution of differential equations coincide with the logics of a machine intended to make bills for a department store, I would regard this as the most amazing coincidence that I have ever encountered."*
>
> Howard Aiken [47]

Alongside proof theory, computability theory is the second cornerstone of mathematical logic. Under its umbrella, it unites all insights and methods concerned with the possibilities and limits of the *algorithmic method*. Two questions are of primary importance in this context:

- **How can the concept of computability be formally defined?**

   We all intuitively know what it means to perform a calculation. On closer inspection, however, our mental models quickly become too vague to draw tangible conclusions. Computability theory solidifies our intuitive ideas by defining precise *models of computation*. Some of these models are thoroughly mathematical in character, while others orient themselves closely to the hardware architecture of real computers.

- **Where are the limits of computability?**

   One of the core results of computability theory is the disclosure of *undecidable problems*, problems with existing solutions that cannot be algorithmically determined. The consequences are far-reaching and extend well beyond algorithms and computers. Today, we know that computability theory and proof theory are closely intertwined and that many negative results from one area carry over to the other.

   In this chapter, we will exploit this connection in two ways. First, we will reveal the existence of amazingly simple proofs for several well-known proof-theoretical results. Second, we will utilize the results of computability theory to answer previously unanswered questions.

> Always keep in mind that computability theory deals with the existence of algorithmic procedures rather than their efficiency. The first studies in this field date back to when the computer in its modern form did not yet exist, so questions about the resource consumption of an algorithmic procedure were irrelevant. Almost by chance, computability theory took on practical significance in constructing the first calculating machines. Subsequently, *complexity theory* emerged as an independent branch of research that deals with the runtime and space complexity of algorithms. Computability and complexity theory have become an integral part of computer science curricula, and all today's graduates are familiar with the basics of both theories. However, few people realize that computability theory, in particular, does not have its roots in computer science. It was developed to answer questions of mathematical logic and is older than the first computer ever built.

5.1 Models of Computation

In the following two subsections, we will look closer at two of the most important models of computation: the Turing machine and the register machine. We will continue with a discussion of Church's thesis and establish that the chosen model to investigate the concept of computability is irrelevant.

5.1.1 Turing Machines

In Section 1.2.8, we have explained the basic functionality of Turing machines and have already seen one in action. We will now proceed to define the concept formally:

 Definition 5.1 (Turing Machine)

A Turing machine is a triple (Q, S, I), consisting of

- the finite *set of states* $Q = \{q_1, \ldots, q_N\}$,
- the *tape alphabet* $S = \{S_0, \ldots, S_M\}$, and
- the *instruction set* $I = \{I_1, \ldots, I_K\}$.

Each instruction from the set I has the form

$$(q_i, S_j, S_k, \mathrm{L}, q_l) \text{ or } (q_i, S_j, S_k, \mathrm{R}, q_l) \text{ or } (q_i, S_j, S_k, \mathrm{N}, q_l)$$

with $q_i, q_l \in Q$ and $S_j, S_k \in S$.

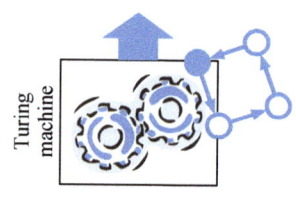

Configuration		Behaviour	
m-config.	symbol	operations	final m-config.
$\mathfrak{b}$	None	$P0, R$	$\mathfrak{c}$
$\mathfrak{c}$	None	R	$\mathfrak{e}$
$\mathfrak{e}$	None	$P1, R$	$\mathfrak{f}$
$\mathfrak{f}$	None	R	$\mathfrak{b}$

$\mathfrak{b} \to q_1$
$\mathfrak{c} \to q_2$
$\mathfrak{e} \to q_3$
$\mathfrak{f} \to q_4$

'None' $\to S_0$
$0 \to S_1$
$1 \to S_2$

$Q = \{q_1, q_2, q_3, q_4\}$
$S = \{S_0, S_1, S_2\}$
$I = \{(q_1, S_0, S_1, \mathrm{R}, q_2),$
$(q_2, S_0, S_0, \mathrm{R}, q_3),$
$(q_3, S_0, S_2, \mathrm{R}, q_4),$
$(q_4, S_0, S_0, \mathrm{R}, q_1)\}$

Figure 5.1: Formal description of the first machine from Turing's celebrated paper

Figure 5.1 illustrates how to represent the first example machine from Turing's original paper in the agreed nomenclature. We had already used the same machine as a demonstration object in Section 1.2.8.

The state q_1 and the tape symbol S_0 have a special meaning in our model. q_1 is called the *initial state* or *starting state*, where every Turing machine starts by definition. The symbol S_0 marks an empty cell on the tape. To distinguish it visually from the others, we frequently use the symbol $\square$ instead of S_0. Turing always started his machines on an empty tape with all cells pre-labeled with the symbol $\square$.

After launch, a Turing machine executes a series of elementary computation steps (Figure 5.2). Each step commences with reading the character underneath the read-write head. For example, if the machine detects

5.1 Models of Computation

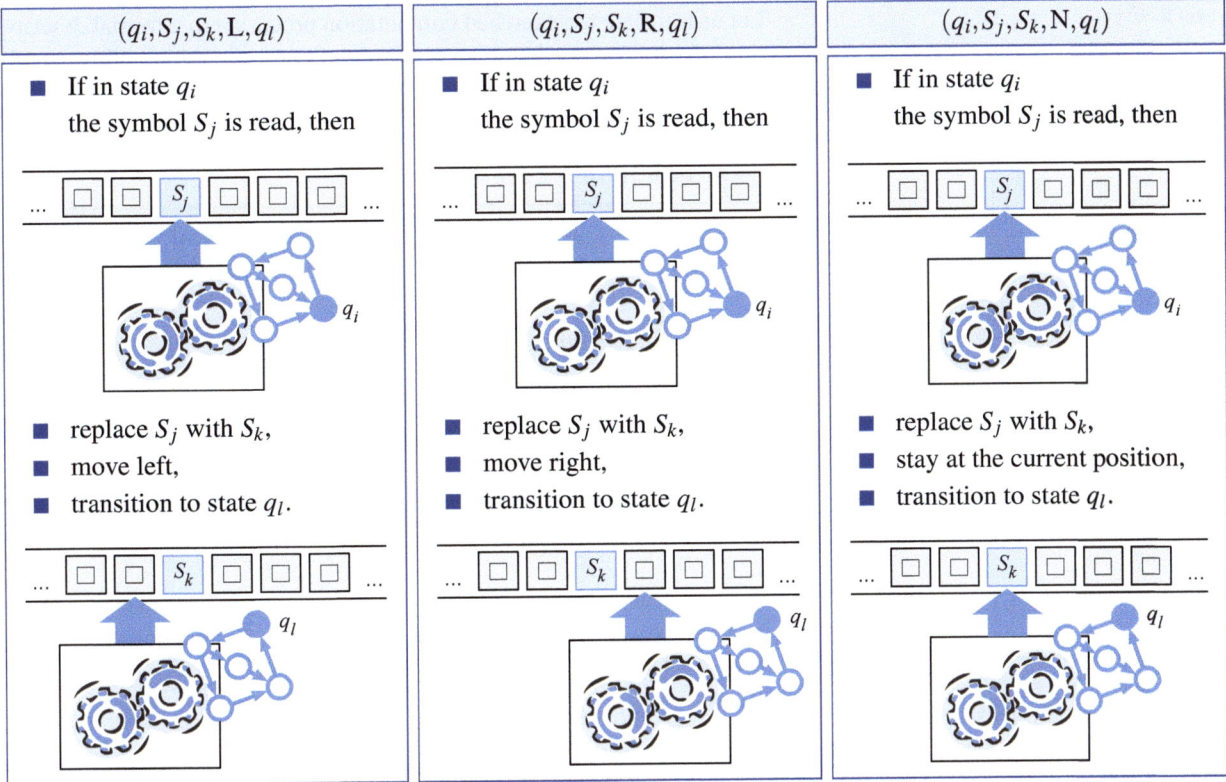

Figure 5.2: Interpretation of the instructions of a Turing machine

the symbol S_j, it searches, depending on the current state q_i, for a suitable instruction of the following form:

$$(q_i, S_j, \_, \_, \_). \tag{5.1}$$

After executing the located instruction, the entire process repeats.

Two particular cases need our attention. Our definition does not exclude that more than one instruction of the form (5.1) exists for a tape symbol S_j and a state q_i. Such machines are called *indeterministic* and play a crucial role in complexity theory (see e.g. [100]). For our considerations, however, it is safe to assume that Turing machines are deterministic, which means that no more than one instruction of the form (5.1) exists for any tape symbol S_j and state q_i. It is still possible, though, that no suitable rule exists. In this case, the machine stops and performs no further calculations; we say the machine *terminates*.

$(q_1, 0, \square)$

⬇ $(q_1, \square, S_1, R, q_2)$

$(q_2, 1, S_1, \square)$

⬇ $(q_2, \square, \square, R, q_3)$

$(q_3, 2, S_1, \square, \square)$

⬇ $(q_3, \square, S_2, R, q_4)$

$(q_4, 3, S_1, \square, S_2, \square)$

⬇ $(q_4, \square, \square, R, q_1)$

$(q_1, 4, S_1, \square, S_2, \square, \square)$

⬇ $(q_1, \square, S_1, R, q_2)$

$(q_2, 5, S_1, \square, S_2, \square, S_1, \square)$

⬇ $(q_2, \square, \square, R, q_3)$

$(q_3, 6, S_1, \square, S_2, \square, S_1, \square, \square)$

⬇ $(q_3, \square, S_2, R, q_4)$

$(q_4, 7, S_1, \square, S_2, \square, S_1, \square, S_2, \square)$

⬇ $(q_4, \square, \square, R, q_1)$

$(q_1, 8, S_1, \square, S_2, \square, S_1, \square, S_2, \square, \square)$

⬇ $(q_1, \square, S_1, R, q_2)$

$(q_2, 9, S_1, \square, S_2, \square, S_1, \square, S_2, \square, S_1, \square)$

⬇ $(q_2, \square, \square, R, q_3)$

$(q_3, 10, S_1, \square, S_2, \square, S_1, \square, S_2, \square, S_1, \square, \square)$

⬇ $(q_3, \square, S_2, R, q_4)$

...

Figure 5.3: Every computation sequence translates to a sequence of configurations.

Let us translate the sketched computation process into a formal description. At its core is the concept of a *configuration*, which allows us to capture the current state of a Turing machine in the sense of a snapshot.

 Definition 5.2 (Configuration)

Let $M = (Q, S, I)$ be a Turing machine. Any vector of the form

$$x = (q, i, s_0, \ldots, s_n)$$

is called a *configuration* of M.

- q is the current state of the machine,
- i is the position of the read-write head ($0 \leq i \leq n$), and
- $s_0, \ldots, s_n$ is the portion of the tape used so far.

We stipulated above that a Turing machine begins in state q_1 on an empty tape; all cells contain the tape symbol $\square$. Accordingly, every Turing machine starts in the initial configuration

$$x_{\text{Start}} := (q_1, 0, \square).$$

Starting from the initial configuration, we can translate the calculation sequence of a Turing machine into a sequence of configurations. Figure 5.3 illustrates the translation for our example machine.

Bear in mind that Turing originally designed his machines to generate real numbers. Such a machine writes the digits of a real number one after the other on an initially empty tape and typically never stops.

At this juncture, we will leave Turing's historical route and reinterpret the behavior of his machines in a more modern sense; we will use them to compute functions of the following form:

$$f : S^* \to S^*$$

To do so, we write an *input word* $\omega \in S^*$ at an arbitrary spot on the tape and position the read-write head on the first character. Subsequently, we perform the calculation steps described above. If the machine terminates, we interpret the tape content as the function value $f(\omega)$. Otherwise, we consider the function f undefined at ω. Turing machines are thus able to calculate partial functions naturally. In the following, we refer to any function computable in this manner as *Turing-computable*.

5.1 Models of Computation

Let us summarize: Turing machines receive their input as character sequences, not numbers. To perform arithmetic operations, i.e., to evaluate functions of the form

$$f : \mathbb{N} \to \mathbb{N},$$

we must fall back to a suitable encoding that maps numerical values to character sequences from the set S^*. Two encodings suggest themselves at this point:

- **Unary encoding**

  The input and output values are represented by sequences of corresponding length (Figure 5.4 center). Unary coding has the advantage that many algorithms easily translate into a Turing machine. However, it is unsuitable for complexity considerations, as simple tasks such as writing a number on the tape have linear complexity.

- **Binary encoding**

  The input and output values are written onto the tape as binary numbers (Figure 5.4 below). The coding is the same as the one used in actual computer systems. In complexity theory, binary coding is the preferred representation, too, as in that case, many results translate directly to real-life computer architectures.

A Turing machine can translate the unary coding of a natural number comparatively easily into binary coding and vice versa. Therefore, for problems in computability theory, the chosen encoding is irrelevant, as the primary question in this field is *whether* and not *how efficiently* a solution can be found.

5.1.1.1 Extensions to the Basic Model

Over time, various adaptations of the original Turing machine have emerged. We will briefly examine the three popular variants depicted in Figure 5.5.

- **Partially tape-restricted Turing machines**

  Partially tape-restricted Turing machines employ a tape that extends infinitely in one direction only. Without loss of generality, we may assume the tape is limited to the left and enumerate the cells with natural numbers. The first cell is indexed 0 and stores the input sequence's first character. As the read-write head of a tape-restricted

- General computation scheme

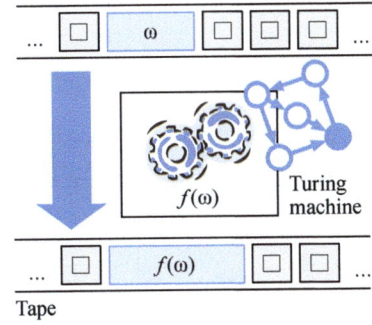

- Unary encoding

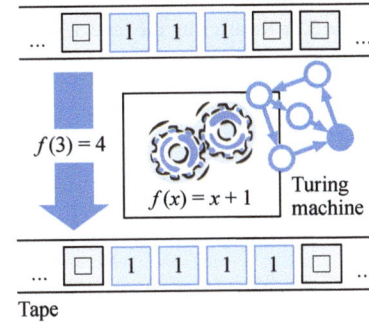

- Binary encoding

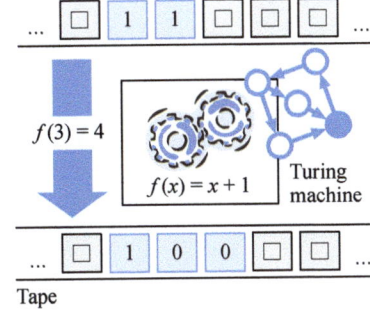

Figure 5.4: A comparison of unary and binary encoding

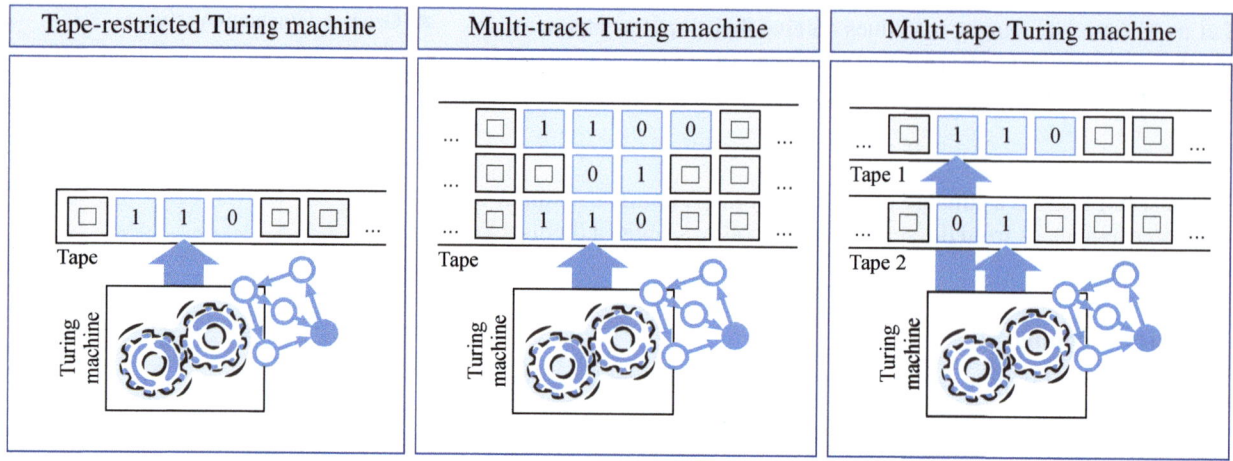

Figure 5.5: Restricting the tape at one side or adding new tracks or tapes does not affect the computational power of the Turing machine.

Turing machine can never move beyond the end of the tape, the machine ignores any request to cross the boundary and keeps the read-write head in its current position instead.

- **Multi-track Turing machines**

  The tape of a *k-track Turing machine* is divided into k separate tracks. The individual tracks are addressed by read-write heads that are tightly connected. Similar to the principle of computer hard drives, the heads may all move to the left or right at the same time but not independently of each other.

- **Multi-tape Turing machines**

  A *k-tape Turing machine* consists of k tapes addressed by k separate read-write heads. In contrast to multi-track Turing machines, all heads can move independently of each other.

A well-known result of computability theory states that the presented machine models are convertible into each other, thus having the same computational power as the base model [100]. More specifically, if a function is computable with a Turing machine, it can also be computed, for instance, with a partially bounded Turing machine and vice versa. The same applies to the other machine types. Computability theory often exploits this property. To prove a property about the Turing machine, it usually suffices to prove the property for one of the simpler machine models. The equivalent computational power ensures that the result carries over to the other machine models.

5.1 Models of Computation

Using the example of partially bounded Turing machines, we want to demonstrate how the different machine models can simulate each other. Partially bounded Turing machines can be simulated by the basic model by marking the end of the tape with a special symbol '◊' and adding an instruction

$$(q_i, \Diamond, \Diamond, \mathrm{R}, q_i)$$

for all $q_i \in Q$ that moves the read-write head one cell back.

The reverse is also true; we can simulate any Turing machine on a partially bounded tape. Again, we start by marking the beginning of the tape with the special symbol '◊'. After that, we move the read-write head to the right of the first character of the input and start the machine. The calculation proceeds as usual as long as the head is to the right of the first input character. However, if the machine moves the read-write head beyond the left boundary, i.e., if we encounter the previously inserted symbol '◊', we must spend some extra work. We first create space for a new symbol by moving the entire tape content one place to the right (Figure 5.6). Then, we continue the original calculation.

The construction indicates that we can translate any Turing machine into an equivalent machine that will never attempt to move the read-write head left of the starting position.

5.1.1.2 Alternative Representations

The previous description of the Turing machine closely resembled Turing's original work. This section will discuss a different type of representation invented by the British mathematician Stephen Wolfram. It builds upon *linear automata*, a special kind of *cellular automata*.

 Definition 5.3 (Cellular Automaton)

A *cellular automaton* (CA) is a 4-tuple (Z, Q, ν, δ). It consists of

- the *set of cells* Z,
- the finite *set of states* Q,
- the *neighborhood function* $\nu : Z \to Z^n$,
- the *state transition function* $\delta : Q \times Q^n \to Q$.

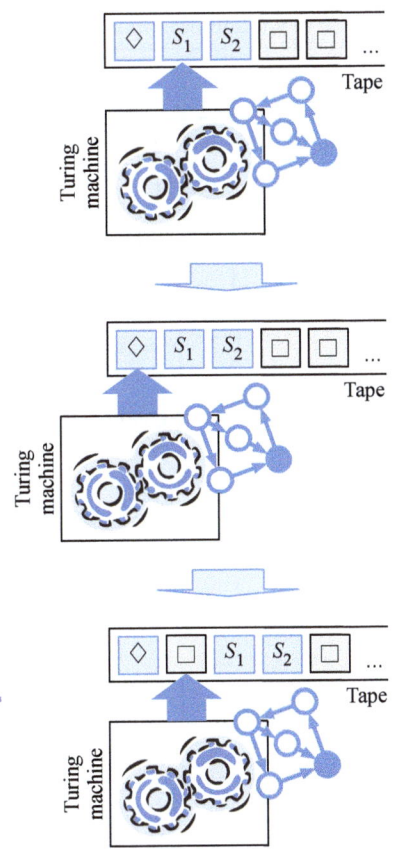

Figure 5.6: Each Turing machine can be simulated by a machine with a partially bounded tape. If the read-write head is on the far left, as in this example, the head movement is simulated by copying the entire tape content one position to the right. The symbol '◊' is utilized to detect the tape boundary.

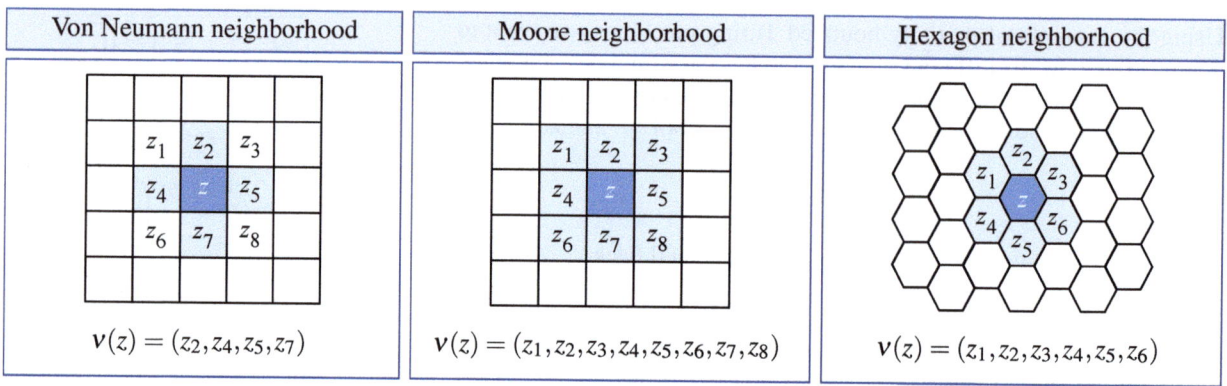

Figure 5.7: Different neighborhood relations of cellular automata

The set Z of a cellular automaton consists of several elementary automata called *cells*. At any given time, each cell is in one of a finite number of states from the set Q. A frequently used representation of this automata type uses colors to symbolize the current state. In this case, Q plays the role of the color palette.

The individual cells of a cellular automaton interact continuously. The exact behavior depends on a cell's current state as well as the state of its neighboring cells. Function v defines the neighborhood relation; it maps a cell z to an n-elementary vector comprising the neighbors of z. Figure 5.7 shows a selection of neighborhood topologies v can model.

The state transition function δ determines the switching behavior of a cellular automaton. If the neighboring cells $z_1, \ldots, z_n$ of z are in the states

$$q_{z_1}, \ldots, q_{z_n},$$

respectively, the successor state of z is given by

$$q'_z = \delta(q_z, q_{z_1}, \ldots, q_{z_n}). \tag{5.2}$$

Each evaluation of δ reflects the state change of a single cell.

Linear automata are cellular automata with a one-dimensional topology. The cells are arranged next to each other; they extend infinitely to the left and right, thereby reproducing the infinite tape required to model Turing machines. The coloring of the cells represents the tape content, making the set of available colors correspond to the tape alphabet of a Turing machine.

5.1 Models of Computation

Linear automata rely on distributed calculation, which means all cells change state in parallel. In contrast, Turing machines operate with a single read-write head in a well-defined position. To simulate a Turing machine, we augment the linear automaton with a distinguished head cell, acting as the read-write head. This extended linear automaton mimics the inner workings of a Turing machine, with the head cell recolored and, if necessary, shifted one position to the left or right in each calculation step. We also introduce an additional state to the *head cell*, representing the state of the modeled Turing machine.

Figure 5.8 describes the example machine from Figure 5.1 in the notation of the modified linear automaton. The arrow directions correspond to the states q_1, q_2, q_3, and q_4, while the different cell colors represent the tape symbols S_0, S_1, and S_2. Two color squares and two arrow icons describe each of the four instructions. The upper color square indicates the current tape symbol, and the lower square defines the replacement symbol. The direction of the upper arrow defines which state the machine must be in for the corresponding rule to apply. The lower arrow defines the follow-up state and the head movement. If the arrow appears left of the successor state, the read-write head moves to the left; otherwise, it moves to the right. In our example machine, all instructions move the head to the right.

According to Wolfram's terminology, we have constructed a *4,3-machine*, as it distinguishes four arrow directions and three colors.

■ Schema

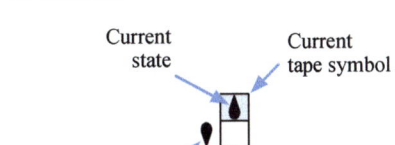

■ Rule set

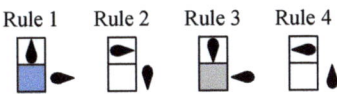

■ The automaton in action

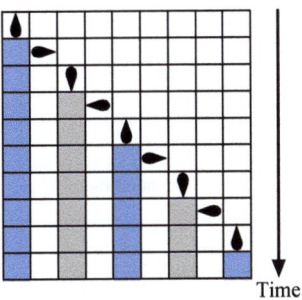

Figure 5.8: A slight modification of the basic model allows linear cellular automata to simulate Turing machines.

5.1.1.3 The Universal Turing Machine

In his seminal 1936 paper, Turing demonstrated that his machines could do much more than perform simple calculations. In particular, he constructed a universal machine capable of simulating other machines. In §6, Turing introduced the universal machine with the following words [212]:

> "It is possible to invent a single machine which can be used to compute any computable sequence. If this machine U is supplied with a tape on the beginning of which is written the S.D of some computing machine M, then U will compute the same sequence as M."

Figure 5.9 illustrates how the universal Turing machine operates. It receives a coded description of another machine, M, as input, which Turing calls the *standard description*, S.D for short. After launch, the universal Turing machine begins to simulate the behavior of M step by step. Once the calculation is complete, the tape contains the same content that M would have produced; there is no way to tell whether M created the tape or its simulation. Overall, the universal Turing machine acts pretty much like a modern computer; it works as an interpreter that simulates the behavior of another machine stoically. The standard description is the *program* the universal machine executes.

To define the standard description, Turing exploited that a machine is uniquely determined by its instruction set. Writing down all instructions one after the other produces a string with enough information to reconstruct the Turing machine's functionality. For the example machine from Figure 5.1, this string could look like this:

$$;q_1 S_0 S_1 R q_2 ; q_2 S_0 S_0 R q_3 ; q_3 S_0 S_2 R q_4 ; q_4 S_0 S_0 R q_1$$

Turing introduced the semicolon as a new symbol acting as an alignment mark. His universal machine specifically searches for this symbol to quickly navigate to the beginning or end of an instruction. This symbol is not strictly necessary; if no semicolon is present, it is still possible to reconstruct the instructions with some extra effort.

Figure 5.10 illustrates the next step. After forming the instruction chain, the universal machine replaces the symbols q_i and S_i with the tape symbols D, A, and C according to the following scheme:

$$q_i := D\underbrace{AAA\ldots A}_{i \text{ times}} \qquad S_i := D\underbrace{CCC\ldots C}_{i \text{ times}}$$

■ Turing machine M

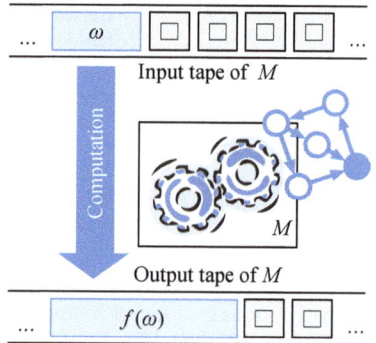

■ Universal Turing machine U

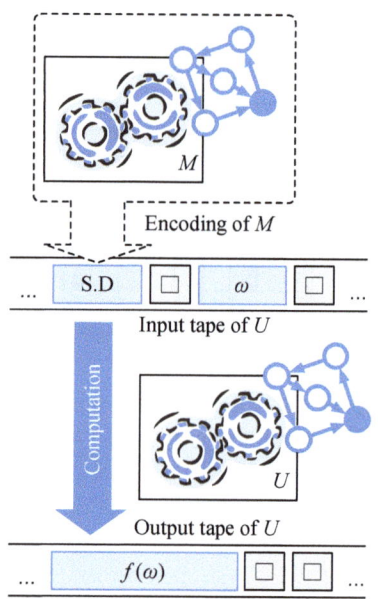

Figure 5.9: The inner workings of the universal Turing machine. Another machine, M, is simulated by encoding M as a character string and writing it onto the tape of U together with the input word ω. After launch, U will analyze the tape content and simulate the behavior of M step by step.

5.1 Models of Computation

The result is what Turing calls the *standard description*, or S.D for short. It is a sequence of characters that uniquely describes the behavior of the encoded machine. In addition, Turing introduced the term *description number*, D.N. We can obtain it directly from the standard description by replacing each character with a fixed digit.

The influence that Gödel's work from 1931 must have had on Turing becomes nowhere as apparent as here. Gödel had shown how to translate the formulas of a formal system into natural numbers. Turing's encoding serves the same purpose; it shows that we can fully encode the behavior of a Turing machine in a single natural number. We also want to express the similarity in linguistic terms and will refer to the *description number* of a Turing machine as its *Gödel number*.

Turing was aware of the immense significance of his universal machine and thus described it in meticulous detail. In addition to a multi-page explanation of the basic functionality, he included the complete instruction table in his paper (see Figures 5.11 and 5.12). The universal machine has a modular structure. It consists of more than 50 individual machines referencing each other. To achieve a compact representation, he let the subsequent state of one machine refer to the start state of another while adding one or more parameters to the reference. In this way, Turing had created a jump instruction, thus using a concept that would become the standard repertoire of imperative programming languages many years later.

In [15], J. P. Burgess describes the universal machine as *"one of the intellectual landmarks of the last century,"* and there is no doubt that Turing wrote an inspiring chapter in the history of science in 1936. However, this chapter also includes minor flaws that should not go unmentioned:

- Strictly speaking, Turing's universal machine does not work exactly as Figure 5.9 claims. The problem is as follows: When machine U simulates machine M, it does indeed generate the same sequence of digits as M, but the digits do not appear at the same tape positions. Turing had designed his machine in such a way that it writes extra symbols on the tape that serve as control characters. The different tape layouts did not matter for the purpose he developed his machine back then. However, once the complete tape content is interpreted as output, as is common today, the universal machine U produces a different output than the simulated machine M.

- Turing's original instruction table is flawed in several places. Occasionally, he swapped symbols, used incorrect indices, or forgot an

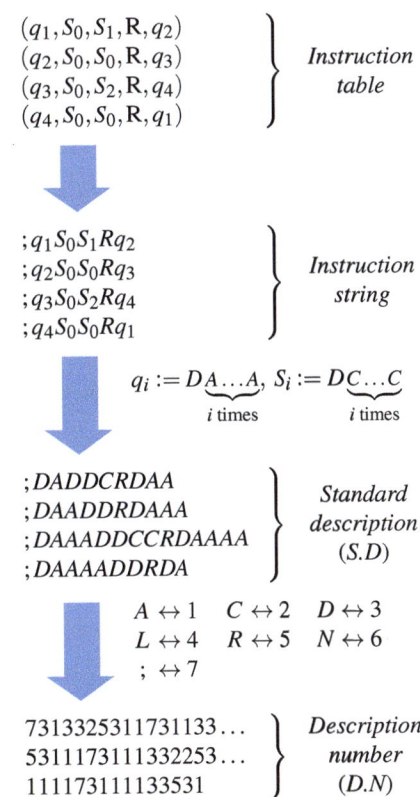

Figure 5.10: Gödelization of Turing machines

A glance at Turing's original paper reveals that it contains a different Gödel number than the one we calculated in Figure 5.10. The reason is the position of the semicolon. In his explanations, Turing used the symbol to mark the end of an instruction, just as it is customary in modern programming languages. In reality, however, his universal machine works differently; it expects the semicolon at the beginning and not at the end of an instruction [162].

$\mathfrak{f}(\mathfrak{C}, \mathfrak{B}, \alpha)$	ə	L	$\mathfrak{f}_1(\mathfrak{C}, \mathfrak{B}, \alpha)$	$\mathfrak{cp}_1(\mathfrak{C}, \mathfrak{U}, \beta)$	γ		$\mathfrak{f}'(\mathfrak{cp}_2(\mathfrak{C}, \mathfrak{U}, \gamma), \mathfrak{U}, \beta)$
	Not ə	L	$\mathfrak{f}(\mathfrak{C}, \mathfrak{B}, \alpha)$	$\mathfrak{cp}_2(\mathfrak{C}, \mathfrak{U}, \gamma)$	γ		$\mathfrak{C}$
	None	L	$\mathfrak{f}(\mathfrak{C}, \mathfrak{B}, \alpha)$		Not γ		$\mathfrak{U}$
$\mathfrak{f}_1(\mathfrak{C}, \mathfrak{B}, \alpha)$	α		$\mathfrak{C}$	$\mathfrak{cpe}(\mathfrak{C}, \mathfrak{U}, \mathfrak{F}, \alpha, \beta)$			$\mathfrak{cp}(\mathfrak{e}(\mathfrak{e}(\mathfrak{C}, \mathfrak{C}, \beta), \mathfrak{C}, \alpha), \mathfrak{U}, \mathfrak{F}, \alpha, \beta)$
	Not α	R	$\mathfrak{f}_1(\mathfrak{C}, \mathfrak{B}, \alpha)$	$\mathfrak{cpe}(\mathfrak{U}, \mathfrak{F}, \alpha, \beta)$			$\mathfrak{cpe}(\mathfrak{cpe}(\mathfrak{U}, \mathfrak{F}, \alpha, \beta), \mathfrak{U}, \mathfrak{F}, \beta)$
	None	R	$\mathfrak{f}_2(\mathfrak{C}, \mathfrak{B}, \alpha)$	$\mathfrak{q}(\mathfrak{C})$	Any	R	$\mathfrak{q}(\mathfrak{C})$
$\mathfrak{f}_2(\mathfrak{C}, \mathfrak{B}, \alpha)$	α		$\mathfrak{C}$		None	R	$\mathfrak{q}_1(\mathfrak{C})$
	Not α	R	$\mathfrak{f}_1(\mathfrak{C}, \mathfrak{B}, \alpha)$	$\mathfrak{q}_1(\mathfrak{C})$	Any	R	$\mathfrak{q}(\mathfrak{C})$
	None	R	$\mathfrak{B}$		None		$\mathfrak{C}$
$\mathfrak{pe}(\mathfrak{C}, \beta)$			$\mathfrak{f}(\mathfrak{pe}_1(\mathfrak{C}, \beta), \mathfrak{C}, \text{ə})$	$\mathfrak{q}(\mathfrak{C}, \alpha)$			$\mathfrak{q}(\mathfrak{q}_1(\mathfrak{C}, \alpha))$
$\mathfrak{pe}_1(\mathfrak{C}, \beta)$	Any	R,R	$\mathfrak{pe}_1(\mathfrak{C}, \beta)$	$\mathfrak{q}_1(\mathfrak{C}, \alpha)$	α		$\mathfrak{C}$
	None	Pβ	$\mathfrak{C}$		Not α	L	$\mathfrak{q}_1(\mathfrak{C}, \alpha)$
$\mathfrak{l}(\mathfrak{C})$		L	$\mathfrak{C}$				
$\mathfrak{r}(\mathfrak{C})$		R	$\mathfrak{C}$	$\mathfrak{pe}_2(\mathfrak{C}, \alpha, \beta)$			$\mathfrak{pe}(\mathfrak{pe}(\mathfrak{C}, \beta), \alpha)$
$\mathfrak{f}'(\mathfrak{C}, \mathfrak{B}, \alpha)$			$\mathfrak{f}(\mathfrak{l}(\mathfrak{C}), \mathfrak{B}, \alpha)$	$\mathfrak{ce}_2(\mathfrak{B}, \alpha, \beta)$			$\mathfrak{ce}(\mathfrak{ce}(\mathfrak{B}, \beta), \alpha)$
$\mathfrak{f}''(\mathfrak{C}, \mathfrak{B}, \alpha)$			$\mathfrak{f}(\mathfrak{r}(\mathfrak{C}), \mathfrak{B}, \alpha)$	$\mathfrak{ce}_3(\mathfrak{B}, \alpha, \beta, \gamma)$			$\mathfrak{ce}(\mathfrak{ce}_2(\mathfrak{B}, \beta, \gamma), \alpha)$
$\mathfrak{c}(\mathfrak{C}, \mathfrak{B}, \alpha)$			$\mathfrak{f}'(\mathfrak{c}_1(\mathfrak{C}), \mathfrak{B}, \alpha)$	$\mathfrak{ce}_4(\mathfrak{B}, \alpha, \beta, \gamma, \delta)$			$\mathfrak{ce}(\mathfrak{ce}_3(\mathfrak{B}, \beta, \gamma, \delta), \alpha)$
$\mathfrak{c}_1(\mathfrak{C})$	β		$\mathfrak{pe}(\mathfrak{C}, \beta)$	$\mathfrak{ce}_5(\mathfrak{B}, \alpha, \beta, \gamma, \delta, \varepsilon)$			$\mathfrak{ce}(\mathfrak{ce}_4(\mathfrak{B}, \beta, \gamma, \delta, \varepsilon), \alpha)$
$\mathfrak{ce}(\mathfrak{C}, \mathfrak{B}, \alpha)$			$\mathfrak{c}(\mathfrak{e}(\mathfrak{C}, \mathfrak{B}, \alpha), \mathfrak{B}, \alpha)$	$\mathfrak{e}(\mathfrak{C})$	ə	R	$\mathfrak{e}_1(\mathfrak{C})$
$\mathfrak{ce}(\mathfrak{B}, \alpha)$			$\mathfrak{ce}(\mathfrak{ce}(\mathfrak{B}, \alpha), \mathfrak{B}, \alpha)$		Not ə	L	$\mathfrak{e}(\mathfrak{C})$
$\mathfrak{cp}(\mathfrak{C}, \mathfrak{U}, \mathfrak{F}, \alpha, \beta)$			$\mathfrak{f}'(\mathfrak{cp}_1(\mathfrak{C}_1, \mathfrak{U}, \beta), \mathfrak{f}(\mathfrak{U}, \mathfrak{C}, \beta), \alpha)$	$\mathfrak{e}_1(\mathfrak{C})$	Any	R,E,R	$\mathfrak{e}_1(\mathfrak{C})$
					None		$\mathfrak{C}$

Figure 5.11: Auxiliary routines for constructing Turing's universal machine from 1936 [212]. The passages highlighted in color are corrections made by Emil Post and Donald Davies [43, 162].

instruction. Many of these errors were later discovered by Emil Post and Donald Davies [43, 162]. Most of them are relatively easy to rectify; they are what we call implementation errors today.

- More seriously, though, Turing also introduced architectural mistakes that are almost impossible to rectify. For example, the simulation fails with machines that repeatedly move the read-write head to the same position and replace a previously written symbol with another. Presumably, Turing did not consider this case, as he only had machines in mind that wrote sequences of numbers, digit by digit, from left to right. Nevertheless, his machine model explicitly allows the described behavior.

5.1 Models of Computation

m-config	symbol	operations	final m-config		m-config	symbol	operations	final m-config
b			$f(b_1, b_1, ::)$		$m\mathfrak{k}_1$	Not A	R, R	$m\mathfrak{k}_1$
b_1		$R, R, P:, R, R, PD, R, R, PA$	anf			A	L, L, L, L	$m\mathfrak{k}_2$
anf			$q(anf_1, :)$		$m\mathfrak{k}_2$	C	R, Px, L, L, L	$m\mathfrak{k}_2$
anf_1			$con(\mathfrak{k}om, y)$			:		$m\mathfrak{k}_4$
$con(\mathfrak{C}, \alpha)$	Not A	R, R	$con(\mathfrak{C}, \alpha)$			D	R, Px, L, L, L	$m\mathfrak{k}_3$
	A	$L, P\alpha, R$	$con_1(\mathfrak{C}, \alpha)$		$m\mathfrak{k}_3$	Not :	R, Pv, L, L, L	$m\mathfrak{k}_3$
$con_1(\mathfrak{C}, \alpha)$	A	$R, P\alpha, R$	$con_1(\mathfrak{C}, \alpha)$			:		$m\mathfrak{k}_4$
	D	$R, P\alpha, R$	$con_2(\mathfrak{C}, \alpha)$		$m\mathfrak{k}_4$			$con(l(l(m\mathfrak{k}_5)),)$
	None	$PD, R, P\alpha, R, R, R$	$\mathfrak{C}$		$m\mathfrak{k}_5$	Any	$R, P\omega, R$	$m\mathfrak{k}_5$
$con_2(\mathfrak{C}, \alpha)$	C	$R, P\alpha, R$	$con_2(\mathfrak{C}, \alpha)$			None	$P:$	$\mathfrak{sh}$
	Not C	R, R	$\mathfrak{C}$		$\mathfrak{sh}$			$f(\mathfrak{sh}_1, inst, u)$
$\mathfrak{k}om$	;	R, Pz, L	$con(\mathfrak{k}mp, x)$		$\mathfrak{sh}_2$	D	R, R, R, R	$\mathfrak{sh}_3$
	z	L, L	$\mathfrak{k}om$			Not D		$inst$
	Not z nor ;	L	$\mathfrak{k}om$		$\mathfrak{sh}_3$	C	R, R	$\mathfrak{sh}_4$
$\mathfrak{k}mp$			$cpe(e(\mathfrak{k}om, x, y), sim, x, y)$			Not C		$inst$
sim			$f'(sim_1, sim_1, z)$		$\mathfrak{sh}_4$	C	R, R	$\mathfrak{sh}_5$
sim_1			$con(sim_2,)$			Not C		$pe_2(inst, 0, :)$
sim_2	A		sim_3		$\mathfrak{sh}_5$	C		$inst$
	Not A	L, Pu, R, R, R	sim_2			Not C		$pe_2(inst, 1, :)$
sim_3	Not A	L, Py	$e(m\mathfrak{k}, z)$		$inst$			$q(l(inst_1), u)$
	A	L, Py, R, R, R	sim_3		$inst_1$	L	R, E	$ce_5(ov, v, y, x, u, w)$
$m\mathfrak{k}$			$q(m\mathfrak{k}_1, :)$			R	R, E	$ce_5(ov, v, x, u, y, w)$
						N	R, E	$ce_5(ov, v, x, y, u, w)$

Figure 5.12: A landmark in the history of science: Alan Turing's *universal machine* from 1936 [212]. The passages highlighted in color are corrections made by Emil Post and Donald Davies [43, 162].

However, the flaws of his machine in its original form do not diminish Turing's accomplishments. The defects were not fundamental, so it was only a matter of time before the first machines appeared that simulated *any* other Turing machine flawlessly, thus precisely functioning as described in Figure 5.9. A milestone along these lines was the universal Turing machine by Marvin Minsky in 1962. It eliminated all errors and limitations of Turing's original design and was less complex in structure; the Minsky machine only distinguishes seven states and four tape symbols. In Wolfram's terminology, it is termed a 7,4 machine.

In 2002, Wolfram presented a refined machine that is also universal but manages with even fewer states (Figure 5.13 above). While Minsky's model still required seven states, the new 2,5 machine managed with only two states. Yet, Wolfram had to raise the number of tape characters from four to five. More significant, though, was his simultaneously published 2,3 machine (Figure 5.13 below). Wolfram proposed the machine as a potential candidate for the smallest possible universal Turing machine [225]. Although he could not prove his conjecture, he achieved a remarkable intermediate result: he proved that two tape characters and two states do not suffice to achieve universality. Therefore, if the 2,3 machine were universal, it would also be the smallest.

In 2007, the search for the smallest possible universal machine ended. That year, 20-year-old Briton Alex Smith succeeded in proving the universality of the 2,3 candidate. Unfortunately, *"programming"* this machine is cumbersome. A specialized compiler must translate the input machine into a suitable input string, quickly reaching a gigantic size.

5.1.2 Register Machines

This section explores the *register machine*, a computational model similar in structure and function to real-life computers [109, 138, 139]. Unlike the Turing machine, there is no tape; instead, several registers hold natural numbers of any size. Every register is accessible directly through an individual memory address, eliminating the need for the time-consuming back-and-forth movement of a read-write head (Figure 5.14). The register machine operates under the control of a program composed of a numbered list of instructions.

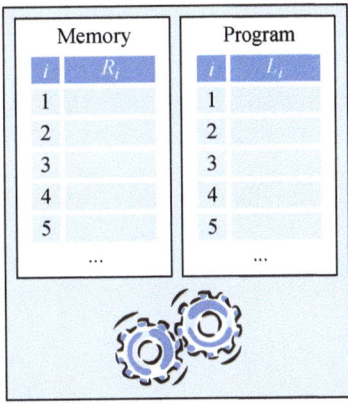

Figure 5.14: General layout of a register machine

 Definition 5.4 (Register Machine)

A *register machine* is a tuple (R, I). It consists of

- the finite set of *registers* $R = \{R_1, \ldots, R_r\}$ and
- the finite set of *instructions* $I = \{L_1, \ldots, L_l\}$.

Each instruction takes one of the following forms:

- $L_i : R_j \leftarrow R_j + 1$
- $L_i : R_j \leftarrow R_j - 1$
- $L_i : \text{stop}$
- $L_i : \text{goto } L_n$ $(n \neq i+1)$
- $L_i : \text{if } R_j = 0 \text{ goto } L_n$ $(n \neq i+1)$
- $L_i : \text{if } R_j \neq 0 \text{ goto } L_n$ $(n \neq i+1)$

After launch, all registers reset to 0, and the execution of the instruction L_1 begins. The selection of the subsequent instruction works similarly to what we are used to from imperative programming languages. The instruction L_i is followed by L_{i+1} unless an unconditional jump (goto) or a conditional jump (if goto) redirects the control flow or a stop command terminates the computation.

The arithmetic capabilities of register machines are spartan compared to real-life computers; apart from decrementing or incrementing the content of a register by one, they support no other operations. There is a special rule for subtraction because register machines, as defined in this book, cannot process negative numbers. Subtraction is always *saturated*, which means, for instance, that the calculation $0 - 1$ yields 0 instead of the expected result -1.

The literature confronts the reader with a broad range of register machine definitions, one differing from the other. Some authors equip the machines with an infinite number of registers, either capable of storing natural numbers of any size or only numbers from a limited range. Even more diverse are the instruction sets. The machine type presented here utilizes a language often referred to as *Goto language* [100, 184, 198]. Other machine models use instructions reminiscent of the assembler languages of early microprocessors. Likewise, they frequently handle the input and output behavior differently. Some machine types do not transfer the input and output values via registers, as described here, but via dedicated memory tapes. It is a fundamental result of computability theory that these differences do not affect the machine's computational power. Therefore, it does not matter which of these models we use to investigate the concept of computability.

Just like Turing machines, register machines can be utilized in two different ways:

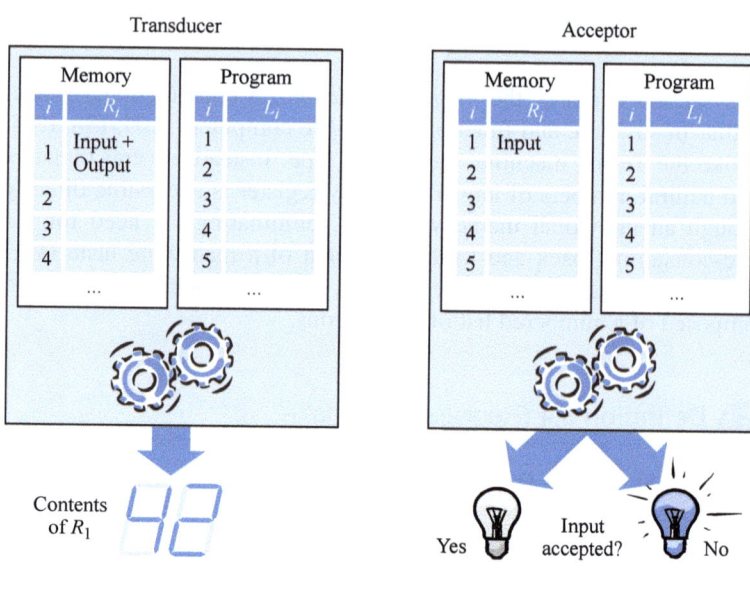

Figure 5.15: Comparison of transducers and acceptors

- **As a calculating machine** (Figure 5.15 left)

  In this scenario, the machine processes the input in register R_1 and stores the final result in this register, too [110]. The remaining registers' purpose is to hold intermediate values. Such a calculating machine is also known as a *transducer*.

- **As accepting machine** (Figure 5.15 right)

  Instead of calculating a specific value, the machine merely supplies a yes-no answer. We say that a register machine *accepts* the input in R_1 if it terminates after a finite number of steps, with all registers holding the value 0 [109]. Otherwise, the input is *not accepted*. We also say the input is *rejected* or *discarded*.

As an example, Figure 5.16 depicts the register machine program from the renowned 1991 paper *Proof of Recursive Unsolvability of Hilbert's Tenth Problem* by James P. Jones and Yuri Matiyasevich [110]. The program is supposed to run on a transductor. Started with the input $R_1 = 2$, it executes the computations shown in Figure 5.17. After 23 operations, the machine stops with the result 1 left in R_1.

You may have noticed that Jones and Matiyasevich deviated slightly from what we have agreed in Definition 5.14; they have combined sev-

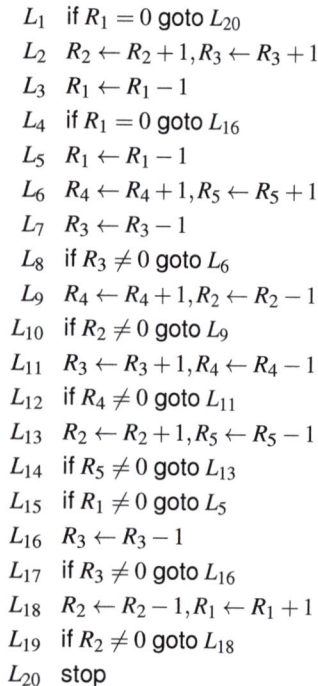

L_1 if $R_1 = 0$ goto L_{20}
L_2 $R_2 \leftarrow R_2 + 1, R_3 \leftarrow R_3 + 1$
L_3 $R_1 \leftarrow R_1 - 1$
L_4 if $R_1 = 0$ goto L_{16}
L_5 $R_1 \leftarrow R_1 - 1$
L_6 $R_4 \leftarrow R_4 + 1, R_5 \leftarrow R_5 + 1$
L_7 $R_3 \leftarrow R_3 - 1$
L_8 if $R_3 \neq 0$ goto L_6
L_9 $R_4 \leftarrow R_4 + 1, R_2 \leftarrow R_2 - 1$
L_{10} if $R_2 \neq 0$ goto L_9
L_{11} $R_3 \leftarrow R_3 + 1, R_4 \leftarrow R_4 - 1$
L_{12} if $R_4 \neq 0$ goto L_{11}
L_{13} $R_2 \leftarrow R_2 + 1, R_5 \leftarrow R_5 - 1$
L_{14} if $R_5 \neq 0$ goto L_{13}
L_{15} if $R_1 \neq 0$ goto L_5
L_{16} $R_3 \leftarrow R_3 - 1$
L_{17} if $R_3 \neq 0$ goto L_{16}
L_{18} $R_2 \leftarrow R_2 - 1, R_1 \leftarrow R_1 + 1$
L_{19} if $R_2 \neq 0$ goto L_{18}
L_{20} stop

Figure 5.16: Register machine program from [110]

	R_1	R_2	R_3	R_4	R_5		Instruction
0	2	0	0	0	0	L_1	if $R_1 = 0$ goto L_{20}
1	2	0	0	0	0	L_2	$R_2 \leftarrow R_2 + 1$
							$R_3 \leftarrow R_3 + 1$
2	2	1	1	0	0	L_3	$R_1 \leftarrow R_1 - 1$
3	1	1	1	0	0	L_4	if $R_1 = 0$ goto L_{16}
4	1	1	1	0	0	L_5	$R_1 \leftarrow R_1 - 1$
5	0	1	1	0	0	L_6	$R_4 \leftarrow R_4 + 1$
							$R_5 \leftarrow R_5 + 1$
6	0	1	1	1	1	L_7	$R_3 \leftarrow R_3 - 1$
7	0	1	0	1	1	L_8	if $R_3 \neq 0$ goto L_5
8	0	1	0	1	1	L_9	$R_4 \leftarrow R_4 + 1$
							$R_2 \leftarrow R_2 - 1$
9	0	0	0	2	1	L_{10}	if $R_2 \neq 0$ goto L_9
10	0	0	0	2	1	L_{11}	$R_3 \leftarrow R_3 + 1$
							$R_4 \leftarrow R_4 - 1$
11	0	0	1	1	1	L_{12}	if $R_4 \neq 0$ goto L_{11}
12	0	0	1	1	1	L_{11}	$R_3 \leftarrow R_3 + 1$
							$R_4 \leftarrow R_4 - 1$
13	0	0	2	0	1	L_{12}	if $R_4 \neq 0$ goto L_{11}
14	0	0	2	0	1	L_{13}	$R_2 \leftarrow R_2 + 1$
							$R_5 \leftarrow R_5 - 1$
15	0	1	2	0	0	L_{14}	if $R_5 \neq 0$ goto L_{13}
16	0	1	2	0	0	L_{15}	if $R_1 \neq 0$ goto L_5
17	0	1	2	0	0	L_{16}	$R_3 \leftarrow R_3 - 1$
18	0	1	1	0	0	L_{17}	if $R_3 \neq 0$ goto L_{16}
19	0	1	1	0	0	L_{16}	$R_3 \leftarrow R_3 - 1$
20	0	1	0	0	0	L_{17}	if $R_3 \neq 0$ goto L_{16}
21	0	1	0	0	0	L_{18}	$R_2 \leftarrow R_2 - 1$
							$R_1 \leftarrow R_1 + 1$
22	1	0	0	0	0	L_{19}	if $R_2 \neq 0$ goto L_{18}
23	1	0	0	0	0	L_{20}	stop

Figure 5.17: Instruction trace for the input $R_1 = 2$

eral individual instructions into a single instruction in lines L_2, L_6, L_9, L_{11}, L_{13}, and L_{18}. However, this generalization presents us with little difficulties, as we can always split combined instructions into several lines.

5.2 Church's Thesis

The Turing machine and the register machine are two calculation models that appear very different from the outset. From a distance, the register machine seems to be the more capable model, as all registers are freely addressable, allowing us to translate many algorithms into a register machine program without substantial detours. On the surface, its memory also appears more spacious than a Turing machine's. Instead of a single tape, an arbitrary number of registers can store natural numbers of any size. Only at a second glance does it become apparent that the machine's generous design does not increase its computational power.

Many achievements in computability theory are associated with the name Alonzo Church. The American logician was born on June 14, 1903, in Washington, D.C., and attended school in Ridgefield, Connecticut. After studying and completing his doctorate at Princeton University, he spent time in Chicago, Harvard, Göttingen, and Amsterdam. After his return to the USA, he moved to Princeton, where he was appointed Assistant Professor in 1929, Associate Professor in 1939, and Full Professor in 1947. Church remained loyal to his alma mater for a long time. Only after his retirement in 1967 did he move to the University of California, Los Angeles, where he taught and researched for another 23 years. Three years after his second retirement, on August 11, 1995, Alonzo Church died in Hudson, Ohio, aged 92.

One of his most outstanding achievements was the invention of the λ-calculus in 1930, intended as a formal foundation for mathematics, free of paradoxes but less cumbersome than the competing type theory of Russell and Whitehead. At that time, however, it was not foreseeable that its future would lie in computer science instead. Over time, the λ-calculus became a valuable tool for formally analyzing programming languages and the operational core of the functional programming language Lisp.

In 1936, Church derived the same result from the λ calculus that Turing achieved a few months later using the Turing machine: the undecidability of first-order predicate logic [34]. Although he anticipated the main result from Turing's famous publication, his proof was nowhere near as clear and elegant. *Church's thesis* [35], discussed in Section 5.2, also dates from 1936.

In hindsight, we can regard Church as the mentor of a new generation of logicians. Among his 31 doctoral students were the well-known logicians Martin Davis, Leon Henkin, Stephen Kleene, Michael Oser Rabin, Barkley Rosser, Dana Scott, Raymond Smullyan and Alan Turing, most of whom we have already met or will meet elsewhere in this book.

Every function a register machine can calculate is also calculable by a Turing machine [139].

Can we perhaps generalize this observation? To get closer to the answer, let us discover yet other models of computation and find out whether one of them surpasses the computational power of the Turing machine.

- **While programs** (Figure 5.18)

   The While language is a fictitious computer language based on the imperative programming paradigm. A While program draws from an infinite supply of variables $x_i, i \in \mathbb{N}$, where $x_1, \ldots, x_n$ are designated for receiving the input values. x_0 will store the result, and the purpose of the remaining variables is to keep intermediate values.

   Visually, the While language is reminiscent of classic imperative programming languages such as C or Pascal, with the vocabulary kept to a minimum. It only includes the operators succ and pred, the assignment operator ':=', the composition operator ';', and the loop construct while do end.

- **μ-recursive functions**

   The set of *μ-recursive functions* is the smallest set, which contains all primitive-recursive functions and is also closed under the μ *operator*. This operator reduces a function $f : \mathbb{N}^{n+1} \to \mathbb{N}$ of arity $n+1$

5.2 Church's Thesis

to a function of arity n according to the following scheme:

$$(\mu f)(x_1,\ldots,x_n) := \min\left\{m \;\middle|\; \begin{array}{l} f(m,x_1,\ldots,x_n) = 0 \\ \text{and for all } k < m, \\ f(k,x_1,\ldots,x_n) \neq \bot \end{array}\right\} \quad (5.3)$$

The symbol '$\bot$' signifies an undefined function value. If the right-hand side of equation (5.3) is the empty set, then there is no minimal element and the function value is undefined $((\mu f)(x_1,\ldots,x_n) = \bot)$.

■ **Lambda calculus** (Figure 5.19)

The *lambda calculus* (short λ-*calculus*) builds upon the idea of defining complex mathematical functions by combining general arithmetic rules. The basic operation is the application of a function f to an argument x, written as $(f x)$. For example, if add is a function for adding two numbers, $((\text{add } x) y)$ calculates the sum $x + y$. The λ operator allows us to *bind* variables, thus creating new functions from existing ones. For example, $(\lambda x.((\text{add } x) x))$ is a function with the parameter x that calculates the value $2 \cdot x$. λ expressions can be combined arbitrarily. It is thus possible for a function to receive λ expressions as arguments and, in particular, to be applied to functions. As the expression $((\lambda x.x)(\lambda x.x))$ shows, a function can even take itself as an argument.

■ **Term rewriting systems** (Figure 5.20)

A *term rewriting system* consists of a set of rewriting rules of the form $l \to r$, the so-called *productions*. The left and right-hand sides contain terms that may also include variables in addition to the symbols of a set Σ. A production transforms a given input word $\omega \in \Sigma^*$. To see if a production is applicable, we need to check whether ω and the left-hand side of a production $l \to r$ match for some substitution $\mathfrak{S}$ ($l\mathfrak{S} = r\mathfrak{S}$). In this case, $r\mathfrak{S}$ is the result. Rewriting systems exist in many variations. Important representatives are the *phrase structure grammars* (type-0 grammars) [100] or the *semi-Thue systems*, invented by the Norwegian mathematician Axel Thue in 1914.

We know today that it is irrelevant whether we base the notion of computability on one model of computation or another; despite their noticeable visual differences, all models mentioned above share the same expressive power. Consequently, the concept of computability remains the same regardless of whether we define it through the Turing machine, the register machine, the While language, the set of μ-recursive functions, the λ calculus, or with the help of term rewriting systems.

```
while x₁ ≠ 0 do
    x₃ := x₂;
    while x₃ ≠ 0 do
        x₀ := succ(x₀);
        x₃ := pred(x₃)
    end;
    x₁ := pred(x₁)
end
```

	x_0	x_1	x_2	x_3	Instruction
1	0	2	2	0	while $x_1 \neq 0$ do
2	0	2	2	0	$x_3 := x_2$
3	0	2	2	2	while $x_3 \neq 0$ do
4	0	2	2	2	$x_0 := \text{succ}(x_0)$
5	1	2	2	2	$x_3 := \text{pred}(x_3)$
6	1	2	2	1	while $x_3 \neq 0$ do
7	1	2	2	1	$x_0 := \text{succ}(x_0)$
8	2	2	2	1	$x_3 := \text{pred}(x_3)$
9	2	2	2	0	while $x_3 \neq 0$ do
10	2	2	2	0	$x_1 := \text{pred}(x_1)$
11	2	1	2	0	$x_3 := x_2$
12	2	1	2	2	while $x_3 \neq 0$ do
13	2	1	2	2	$x_0 := \text{succ}(x_0)$
14	3	1	2	2	$x_3 := \text{pred}(x_3)$
15	3	1	2	1	while $x_3 \neq 0$ do
16	3	1	2	1	$x_0 := \text{succ}(x_0)$
17	4	1	2	1	$x_3 := \text{pred}(x_3)$
18	4	1	2	0	while $x_3 \neq 0$ do
19	4	1	2	0	$x_1 := \text{pred}(x_1)$
20	4	0	2	0	while $x_1 \neq 0$ do

Figure 5.18: While program for multiplying two natural numbers x_1 and x_2. The depicted trace demonstrates the program execution for $x_1 = 2$ and $x_2 = 2$. At the end of the calculation, the register x_0 contains the result 4.

- Reduction rules

    (α) $\quad \lambda\xi.\varphi \stackrel{\mu\notin\varphi}{\Rightarrow} \lambda\mu.\varphi[\xi \leftarrow \mu]$

    (β) $\quad ((\lambda\xi.\varphi)\psi) \rightarrow \varphi[\xi \leftarrow \psi]$

    (η) $\quad \lambda\xi.\varphi\xi \rightarrow \varphi$

- Derivation

    $((((\lambda y.(\lambda z.(\lambda x.((yz)x))))$
    $\quad (\lambda w.(\lambda x.(wx))))P)v)$
    $\stackrel{\beta}{\Rightarrow} (((\lambda z.(\lambda x.(((\lambda w.(\lambda x.(wx)))z)x)))P)v)$
    $\stackrel{\beta}{\Rightarrow} ((\lambda x.(((\lambda w.(\lambda x.(wx)))P)x))v)$
    $\stackrel{\beta}{\Rightarrow} (((\lambda w.(\lambda x.(wx)))P)v)$
    $\stackrel{\beta}{\Rightarrow} ((\lambda x.(Px))v)$
    $\stackrel{\beta}{\Rightarrow} (Pv)$

Figure 5.19: The λ calculus in action

- Productions

$x\mathrm{I}$	$\rightarrow \; x\mathrm{IU}$	(Rule 1)
$\mathrm{M}x$	$\rightarrow \; \mathrm{M}xx$	(Rule 2)
$x\mathrm{III}y$	$\rightarrow \; x\mathrm{U}y$	(Rule 3)
$x\mathrm{UU}y$	$\rightarrow \; xy$	(Rule 4)

- Derivation of MUIIU from MI

MI	$\Rightarrow$ MII	(Rule 2)
	$\Rightarrow$ MIIII	(Rule 2)
	$\Rightarrow$ MUI	(Rule 3)
	$\Rightarrow$ MUIU	(Rule 1)
	$\Rightarrow$ MUIUUIU	(Rule 2)
	$\Rightarrow$ MUIIU	(Rule 4)

Figure 5.20: MIU system by Douglas Hofstadter [106], formulated as a term rewriting system

The American logician Alonzo Church considered this the empirical confirmation of the thesis that the intuitive concept of computability coincides with the notion of Turing computability.

 Theorem 5.1 (Church's Thesis)

> The class of Turing-computable functions coincides with the class of intuitively computable functions.

The expression *intuitively calculable function* requires our special attention in this context. It denotes a function that can be calculated by a person – in whatever form. Thus, Church's thesis states nothing less than that every function calculable in one way or another is also calculable by a Turing machine.

Church's thesis is not a theorem in a precise mathematical sense since the concept of an intuitively computable function has no formal definition. If there were, we would have already committed ourselves – consciously or unconsciously – to a particular model of computation and reduced the actual meaning of this term to absurdity. Consequently, it is not possible to ever prove Church's thesis. We can only gather evidence for its correctness, which is what researchers have succeeded in many times in the past. All previous attempts to broaden the range of calculable functions by specifying a more expressive model of computation have been in vain. Even unusual concepts such as the *quantum computer* [148] or *DNA computing* [5] have not been able to push the limits of what is computable by machines.

Church's thesis legitimizes the following definition, which finally provides a formal foundation for the concept of computability, a concept we have already been using several times:

 Definition 5.5 (Computability)

> A partial function $f : \Sigma^* \rightarrow \Sigma^*$ is computable if there exists a Turing machine M with the following properties for every input ω:
>
> - If $f(\omega) \neq \bot$, then M writes $f(\omega)$ onto the tape and halts.
> - If $f(\omega) = \bot$, then M does not halt.

Herein, Σ is an arbitrary set eligible as a Turing machine's tape alphabet. From here, it is only a tiny step to formally capture the concept of *decidability*:

5.2 Church's Thesis

Definition 5.6 (Decidability, Semi-Decidability)

A set $N \subseteq \Sigma^*$ is *decidable* if the *characteristic function* $\chi_N : \Sigma^* \to \{0,1\}$ is computable, defined by:

$$\chi_N(\omega) := \begin{cases} 1 & \text{if } \omega \in N \\ 0 & \text{if } \omega \notin N \end{cases}$$

A set $N \subseteq \Sigma^*$ is *semi-decidable* if the *partial characteristic function* $\chi'_N : \Sigma^* \to \{1\}$ is computable, defined by:

$$\chi'_N(\omega) := \begin{cases} 1 & \text{if } \omega \in N \\ \bot & \text{if } \omega \notin N \end{cases}$$

■ Decidability

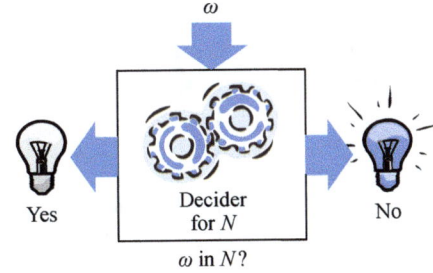

■ Semi-Decidability

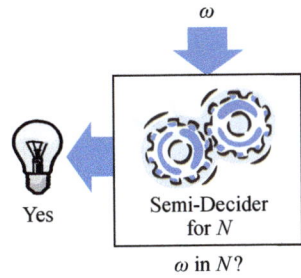

Figure 5.21: Visual representation of the two decidability concepts

At the heart of this definition is the *characteristic function*. It is the formal link between the notion of computability formulated for functions and the decidability criterion formulated for sets.

Figure 5.21 visualizes the two concepts of decidability. If a set N is decidable, there exists an algorithmic machine capable of determining whether ω belongs to N or not. In the diagram, two separate light bulbs represent the two possible outcomes. One of these bulbs will light up after some time, indicating the result. However, the exact timing of when a bulb will glow is unknown; we only know that it will happen eventually, regardless of whether $\omega \in N$ or $\omega \notin N$. Therefore, to decide on the set membership, all we need to do is be patient.

In contrast, a machine that implements semi-decidability only has a single light bulb. When starting with an element $\omega \in N$, the lamp lights up eventually. However, no definite statement is made if $\omega \notin N$. In this case, we can never say with certainty whether the machine has entered an infinite loop or whether it will provide a positive answer later. Semi-decidability is, therefore, tantamount to a partial statement. The machine only answers the question "Is $\omega \in N$?" in the positive case after a finite amount of time. However, if the answer is negative, the machine will never respond.

If a set N is decidable, so is the complement $\overline{N}$, and it follows that N and $\overline{N}$ are a fortiori semi-decidable. Interestingly, the converse holds, too: If, alongside N, the complement $\overline{N}$ is also semi-decidable, N becomes decidable. Figure 5.22 illustrates how to combine the semi-deciders for N and $\overline{N}$ to a decider for N. The two semi-deciders are supplied with the input word ω and simulated in parallel. If ω is in N, the first semi-decider will react eventually; if ω is not in N, the second semi-decider will respond. In the form of a theorem, our result reads as follows:

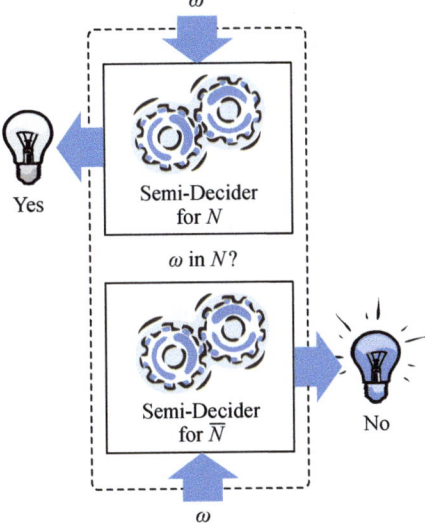

Figure 5.22: A set N is decidable if N and its complement $\overline{N}$ are both semi-decidable.

- Decidability

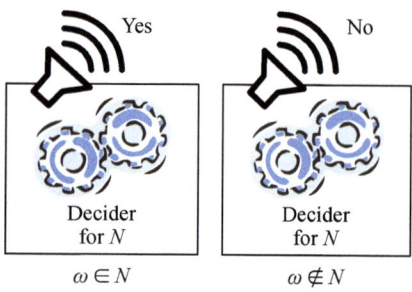

- Semi-decidability

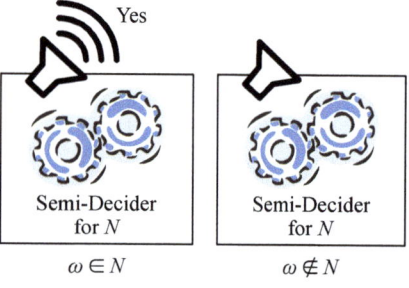

- Enumerability

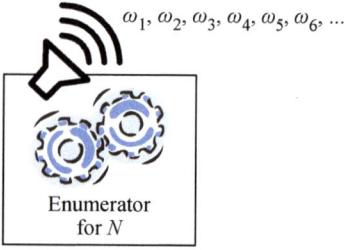

Figure 5.23: The concepts of *decidability*, *semi-decidability*, and *enumerability*

 Theorem 5.2

For every set N, the following holds:

N is decidable $\Leftrightarrow$ N and $\overline{N}$ are semi-decidable

Along the same lines, let us formally consolidate another concept that we have used informally on several occasions: the *enumerability* of sets. Just as in the case of decidability, we can reduce this concept to the computability of a function:

Definition 5.7 (Enumerability)

A set N is called enumerable if it is the empty set or a surjective computable function $f : \mathbb{N} \to N$ exists.

Only upon closer inspection does it become apparent that the concept described in this definition matches our intuitive understanding of the phrase enumerability. So, let us take a closer look! If a set N is enumerable, then, by definition, a Turing machine M calculates the function value $f(x)$ for any given natural number x. Hence, M enables us to calculate the values

$$f(0), f(1), f(2), f(3), \ldots$$

one after the other. In this way, an ever-growing list of elements from N comes into being. However, we know more than that. Since f is surjective by definition, every element $\omega \in N$ will eventually appear in our enumeration. As a result, we obtain an algorithmic machine that sequentially lists all elements of N (Figure 5.23). Observe that we have only required the surjectivity of f and not its bijectivity. Consequently, we explicitly permit elements to occur more than once in the enumeration.

Let us try to establish a connection between enumerability, countability, and semi-decidability. On the one hand, every enumerable set is also countable since we can relate each element to at least one natural number. On the other hand, for numerous countable sets N, there is no function $f : \mathbb{N} \to N$ being both surjective and computable. We will soon learn about a prominent example: the set of all true arithmetic formulas.

There is also a close relationship between the enumerability and semi-decidability of a set. First, every enumerable set N is also semi-decidable; after all, we can recite all elements one after the other and

5.2 Church's Thesis

stop when the element ω we are seeking shows up. If $\omega \in N$, we will find the element after a finite number of steps. If $\omega \notin N$, we will continue forever.

The reverse is also true: every semi-decidable set N is enumerable. However, realizing this remarkable fact is anything but trivial. To enumerate the elements of N, we must prevent the semi-decider from entering an infinite loop. But how is this possible? Once again, Cantor's pairing function helps us to succeed. Let us recall: Cantor's pairing function π establishes a mapping between the set of all tuples $(i,j) \in \mathbb{N}^2$ and the natural numbers. Figure 5.24 reveals how the clever usage of this function allows us to recite the elements of a semi-decidable set one after the other.

- In an infinite loop, we calculate the elements

$$\pi^{-1}(0), \pi^{-1}(1), \pi^{-1}(2), \pi^{-1}(3), \ldots \quad (5.4)$$

The result is a sequence in which every tuple $(i,j) \in \mathbb{N}^2$ appears eventually.

- For each tuple (i,j), we start the semi-decider with the i-th element ω of Σ^*. If it determines the set membership within j steps, it outputs ω. If the semi-decider has not stopped after j steps, we abort the calculation and continue with the next tuple. Since for each $\omega \in N$, the semi-decider answers the set membership positively in j steps for some $j \in \mathbb{N}$, all elements of N eventually show up.

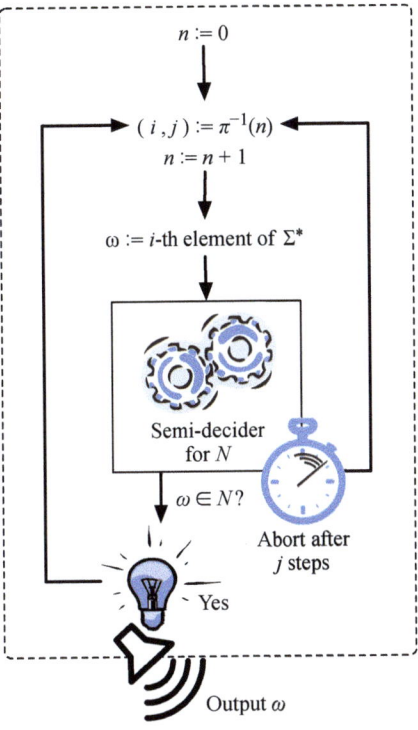

Figure 5.24: The clever use of Cantor's pairing function allows us to enumerate the elements of a semi-decidable set.

We have thus succeeded in proving the following theorem:

Theorem 5.3 (Enumerability and Semi-Decidability)

For every set $N \neq \emptyset$, the following holds:

N is enumerable $\Leftrightarrow$ N is semi-decidable

By combining the statements of Theorems 5.2 and 5.3, we get the following result without further ado:

Corallary 5.1

For every set $N \neq \emptyset$, the following holds:

N is decidable $\Leftrightarrow$ N and $\overline{N}$ are enumerable

5.3 Limits of Computability

In this section, we will push the algorithmic method to its limits. We will start our discussion with different variants of the *halting problem* and conclude by generalizing our findings with *Rice's theorem*.

5.3.1 Halting Problem

The term *halting problem* encompasses various questions related to the termination behavior of Turing machines. Specifically, it addresses whether it is possible to algorithmically determine if a given Turing machine will halt on a particular input or continue computing indefinitely. We begin with the definition of the *general halting problem*:

 Definition 5.8 (General Halting Problem)

The *general halting problem* is defined as follows:

- Given: Turing machine M and input word ω
- Asked: Does M halt with ω as input?

✋ : Machine terminates
💫 : Machine runs forever

Table 5.1: A simple diagonalization argument reveals the undecidability of the halting problem. The adjacent table lists all inputs on the horizontal axis and all Turing machines on the vertical axis. The table entry in the i-th row and the j-th column indicates whether the machine M_i halts on input ω_j. If the halting problem were decidable, a Turing machine would exist that determines the diagonal entry (i, i) and terminates exactly when M_i does not halt on input ω_i. This machine cannot appear anywhere in the table, contradicting its construction.

5.3 Limits of Computability

Let us assume the halting problem was decidable. We can quickly turn this assumption into a contradiction with a diagonalization argument similar to the one we have been using in Section 1.2.2.

To do so, we start constructing a matrix, as shown in Table ??. The matrix lists all Turing machines on the vertical and all input words on the horizontal axis. The actual order is irrelevant; it only matters that every Turing machine and every input word appear in some row and some column, respectively. The individual matrix cells tell us more about the termination behavior of the machines. In particular, the i-th row and the j-th column indicate whether the Turing machine M_i terminates on input ω_j.

If the halting problem were decidable, some Turing machine H would exist that receives an input word ω and a Turing machine M in encoded form and always correctly determines whether M terminates on input ω. The fictitious Turing machine H is a machine for calculating the just constructed matrix. From H, as sketched in Figure 5.25, we build another machine, H', that calculates the matrix element (i,i) for the input word ω_i and behaves reciprocally to the received answer. More precisely, H' terminates with input ω_i if and only if M_i would continue to run forever.

Since H' is also a Turing machine, it must be located in some row of the constructed matrix, as all machines are listed. However, no matter which row we check, the diagonal construction invariably leads to a contradiction. For all $i \in \mathbb{N}$, $H' \neq M_i$ applies, as M_i accepts input ω_i precisely when H' rejects it. The contradiction forces us to drop the assumption about the existence of H; no Turing machine can decide the halting problem.

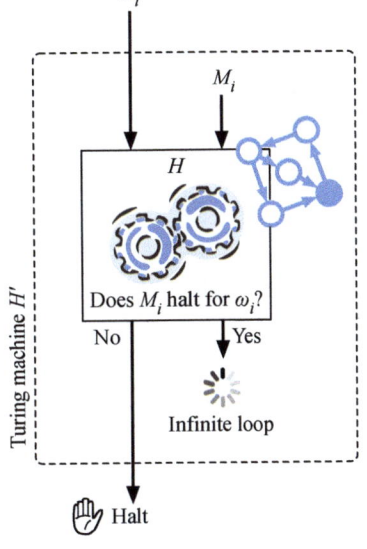

Figure 5.25: If a Turing machine H decides the halting problem, we could transform it into a machine H' that terminates for the input word ω_i precisely when the Turing machine M_i runs infinitely long on the input ω_i. The method of diagonalization reveals the construction as contradictory. Consequently, such a machine cannot exist.

 Theorem 5.4

> The general halting problem is undecidable.

 Remember that the assertion of Theorem 5.4 applies to any model of computation with the same expressive power as the Turing machine, which includes the register machine from Section 5.1.2 and all discussed variants of the Turing machine from Section 5.1.1.1. There, we introduced three popular extensions: the partially tape-restricted machine, the multi-track machine, and the multi-tape machine, all exhibiting the same expressive power.

A simple diagonalization argument was sufficient to establish a theorem that ranks among the most important results in computability theory. The following analysis will reveal why. It will progressively uncover the theorem's enormous power.

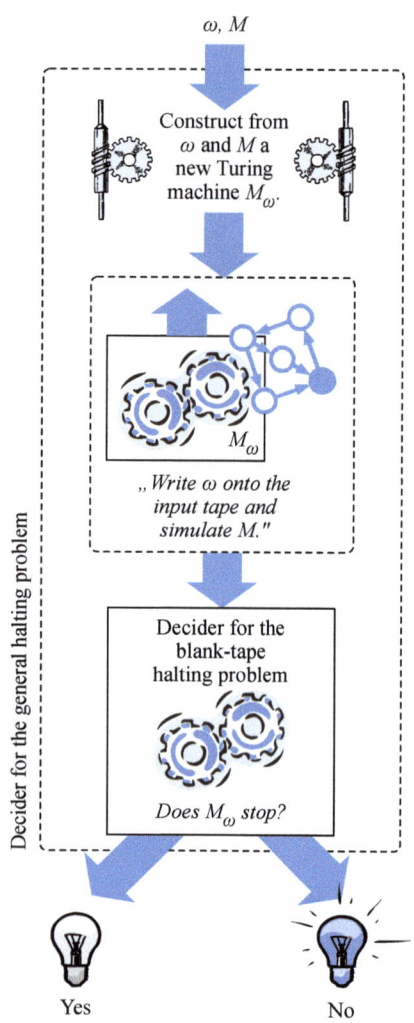

Figure 5.26: Reduction of the general halting problem to the blank-tape halting problem. If the blank-tape halting problem were decidable, so would the general halting problem. Consequently, since the general halting problem is undecidable, the blank-tape halting problem cannot be decidable either.

The Blank-Tape Halting Problem

Let us consider a weakened variant of the halting problem:

 Definition 5.9 (Blank-tape halting problem)

> The *blank-tape halting problem* is defined as follows:
> - Given: Turing machine M
> - Asked: Does M halt with the empty word ε as input?

While the general halting problem has to decide the termination property for arbitrary Turing machines and arbitrary inputs, the blank-tape halting problem restricts itself to computation runs starting on a blank tape. In this case, the input is the empty word, denoted by ε.

At first glance, the blank-tape halting problem seems more straightforward than the general halting problem. Nevertheless, a simple *reduction proof* exposes this problem as undecidable, too. A reduction proof shows that if the blank-tape halting problem were decidable, then the general halting problem would be as well. In other words, the general halting problem is reducible to the blank-tape halting problem. Figure 5.26 illustrates how to carry out the reduction.

To decide whether a Turing machine halts for an input word ω, we first construct a Turing machine M_ω that writes the symbols of ω onto the tape and simulates M afterward. Starting on a blank tape, M_ω terminates precisely when the original machine M terminates with input ω. Consequently, we could decide the general halting problem with a Turing machine capable of only deciding the blank-tape halting problem. It immediately follows from Theorem 5.4 that the blank-tape halting problem must be undecidable, too.

 Theorem 5.5

> The blank-tape halting problem is undecidable.

5.3.2 Rice's Theorem

The undecidability of the halting problem has demonstrated the existence of statements about Turing machines that defy mechanical prov-

5.3 Limits of Computability

ability. No algorithmic method can correctly predict the termination for all Turing machines and all inputs. Moreover, we identified the blank-tape halting problem as undecidable through a suitable reduction. This section will explore whether there are more undecidable Turing machine properties. This much in advance: We will receive an astonishing answer.

Let M denote an arbitrary Turing machine, f_M the function computed by M, and E a functional property of M, that is, a property of f_M. We require E to be non-trivial, meaning there must be at least one machine that possesses the property and at least one that does not. The following list offers some initial suggestions, though the possibilities are endless at this point:

- There is an output of M that contains the symbol 0.
- All outputs of M are at least n characters long.
- M calculates a total function.

Let us explore the consequences of the existence of a decision procedure for E. For this purpose, we first introduce the Turing machine $M_\perp$, calculating the everywhere undefined function $f(\omega) = \perp$. $M_\perp$ is straightforward to construct, as it does not terminate on any input.

For the moment, let us assume that $M_\perp$ possesses the chosen property E. Since E is non-trivial, at least one other machine $M_{\overline{E}}$ does not fulfill E. We summarize:

$$M_\perp \text{ possesses the property } E \tag{5.5}$$
$$M_{\overline{E}} \text{ does not possess the property } E \tag{5.6}$$

We now join M and $M_{\overline{E}}$ into a combined machine H. As the upper part of Figure 5.27 illustrates, H first starts machine M with the empty word. If M halts after a finite number of steps, H applies the machine $M_{\overline{E}}$ to the input word ω.

To understand the behavior of H, let us distinguish two cases:

- **M does not terminate**

  In this case, H is functionally identical to $M_\perp$. Hence, it possesses the property E.

- **M does terminate**

  In this case, H is functionally identical to $M_{\overline{E}}$. Hence, it does not possess the property E.

- Case 1: $M_\perp$ satisfies E

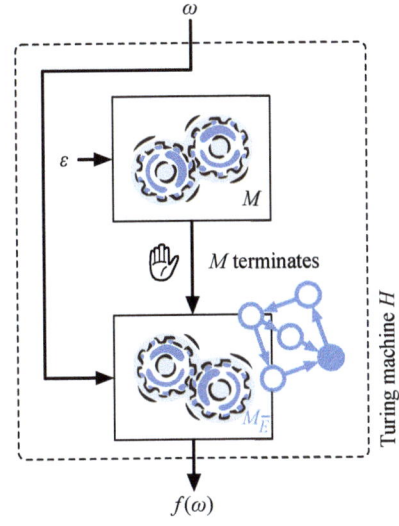

- Case 2: $M_\perp$ does not satisfy E

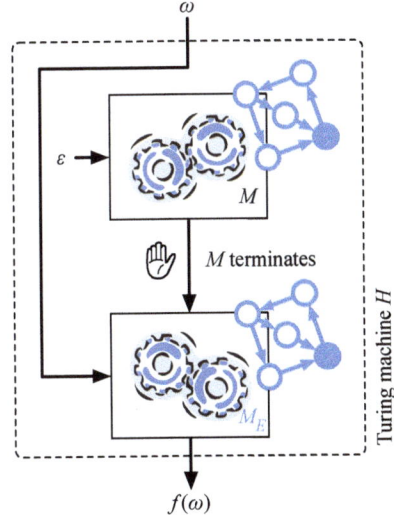

Figure 5.27: The centerpiece of the proof of Rice's theorem. The illustration shows how to establish a direct relationship between the investigated machine property E and the halting problem.

>
> The implications of Rice's theorem are vast; it destroys the hope of algorithmically deciding any nontrivial functional property about Turing machines in a sweeping blow. The limitations imposed by this theorem stretch deep into the real world of software development. It follows that no algorithm can automatically verify, for any given program, whether the program meets its functional specification. Even problems as simple as the existence of infinite loops elude an algorithmic solution. Its striking generality makes Rice's theorem one of the most valuable results in computability theory.

Voilà: Our construction has established a direct relationship between E and the termination of M. Hence, if some procedure decided the property E, we could solve the halting problem for any machine M. In other words, we had found a decision procedure for the halting problem.

Bear in mind that the above considerations assumed that the selected property E applies to $M_\perp$. If this is not the case, we can modify the machine H as shown in the lower part of Figure 5.27. Instead of $M_{\overline{E}}$, we start an arbitrary machine M_E that fulfills E. The case distinction then reads as follows:

- **M does not terminate**

  In this case, H is functionally identical to $M_\perp$. Hence, it does not possess the property E.

- **M does terminate**

  In this case, H is functionally identical to $M_{\overline{E}}$. Hence, it possesses the property E.

Again, we have successfully established a one-to-one relationship between E and the termination of M. If there were a decision procedure for property E, we could also solve the halting problem.

The halting problem's undecidability thus inevitably leads to the realization that a decision procedure for E cannot exist. This is precisely the statement of Henry Gordon Rice's famous theorem from 1953.

Theorem 5.6 (Rice's Theorem)

> Let E be a nontrivial functional property of Turing machines. Then the following problem is undecidable:
>
> - Given: Turing machine M
> - Asked: Does M possess the property E?

5.4 Impact on Mathematics

At the beginning of this chapter, we have justified the great significance of computability theory, at least in part, by its profound impact on mathematics. In this section, we will demonstrate how closely interwoven

5.4 Impact on Mathematics

computability theory, on the one hand, and proof theory, on the other, really are. In particular, we will derive three prominent negative results, namely

- the unsolvability of Hilbert's decision problem,
- the incompleteness of arithmetic, and
- the unsolvability of Hilbert's tenth problem.

We will prove all three results by reducing the halting problem. Specifically, we will show that the solution to one of the three problems puts us in a position to decide the termination of Turing machines or register machines (Figures 5.28 and 5.29). Hence, if only one of them were solvable, so would the halting problem. However, we already know from Section 5.3.1 that the halting problem is undecidable.

We will commence with a rough outline of all three reductions to maintain a clear view of the big picture. After that, we will provide the technical details in separate subsections.

- **Unsolvability of Hilbert's decision problem**

  Section 5.4.1 will demonstrate that the blank-tape halting problem is solvable with a decision procedure for first-order predicate logic. The core of the proof builds upon the idea of translating a tape-restricted Turing machine M into a predicate-logic formula φ_M with the following property:

  $$M \text{ terminates} \Leftrightarrow \varphi_M \text{ is universally valid}$$

  We can then exploit the established relationship to decide the termination of a tape-restricted Turing machine with a decision procedure for predicate logic. From the halting problem's undecidability, we can conclude that predicate logic has no decision procedure.

- **Incompleteness of arithmetic**

  Section 5.4.2 will show that the blank-tape halting problem is also solvable using a decision procedure for Peano arithmetic. The proof idea remains the same. We will show that a Turing machine M can be translated into an arithmetic formula φ_M satisfying:

  $$\varphi_M \text{ is true} \Leftrightarrow M \text{ terminates}$$

  This relationship allows us to decide the termination of a Turing machine with a decision procedure for PA, thus concluding that such

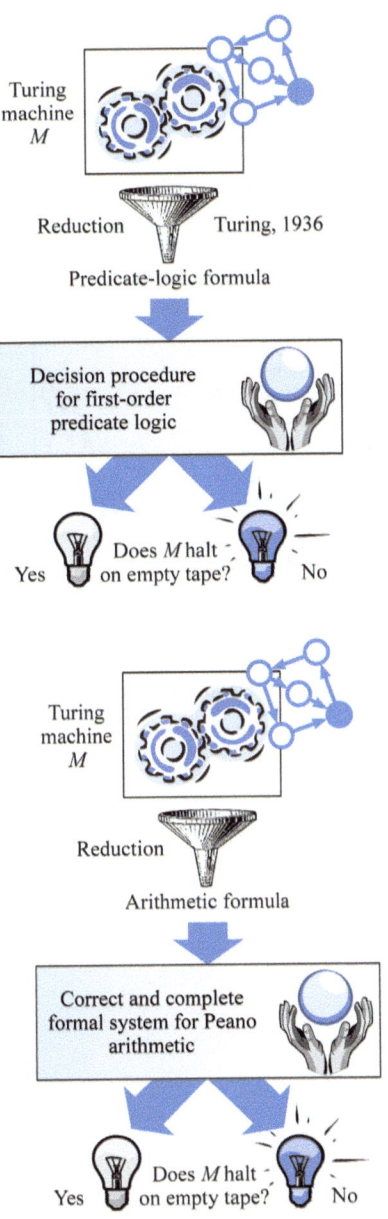

Figure 5.28: If there were a decision procedure for first-order predicate logic (above) or a correct and, at the same time, complete formal system for Peano arithmetic (below), the halting problem would be decidable.

a procedure cannot exist. Combined with Theorem 2.5, we obtain an astonishing result as a byproduct: The semantic variant of Gödel's first incompleteness theorem.

■ **Unsolvability of Hilbert's tenth problem**

Section 5.4.3 will demonstrate that the halting problem is also solvable with a decision procedure for Diophantine equations. This time, however, we will refrain from proving the result via Turing machines. Instead, we will show how to translate a register machine into a Diophantine equation $\varphi_R = 0$ with the following property:

$$\varphi_R = 0 \text{ has a solution} \Leftrightarrow R \text{ terminates}$$

Based on this relationship, we can decide the termination of a register machine using a solution procedure for Diophantine equations. Again, it immediately follows from the undecidability of the halting problem that no general solution procedure for Diophantine equations can exist.

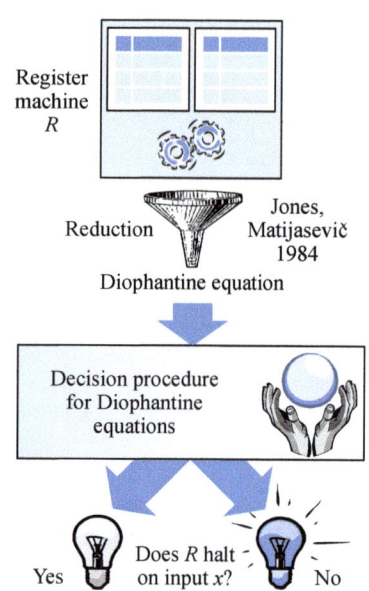

Figure 5.29: A solution procedure for Diophantine equations can also decide the halting problem for register machines.

Now that the roadmap is settled, let us delve into the details. It is safe to skip them on the first reading, as they are not essential for understanding the topics covered in the rest of the book.

5.4.1 Undecidability of PL1

In this section, we will work out how to translate a tape-restricted Turing machine M into a predicate-logic formula that is universally valid if and only if M terminates on a blank tape. The construction of this formula is a cornerstone of Turing's historical proof and is discussed in detail in §11 of his paper [212]. On the following pages, we will elucidate the core of his thought process.

Turing commences with the definition of several predicates describing the configurations of tape-restricted Turing machines (Figure 5.30):

$$\begin{aligned}
\mathsf{I}(t,y) \;&:\; \text{At time } t, \text{ the head is positioned over cell } y \\
\mathsf{R}_{S_j}(t,y) \;&:\; \text{At time } t, \text{ cell } y \text{ contains symbol } S_j \\
\mathsf{K}_{q_i}(t) \;&:\; \text{At time } t, \text{ the machine is in state } q_i \\
\mathsf{F}(x,y) \;&:\; x \text{ and } y \text{ are natural numbers with } y = x+1
\end{aligned}$$

The aforementioned meaning of the predicates is their *intended* meaning. For the following construction to succeed, we must ensure that the

5.4 Impact on Mathematics

symbols I, R_{S_j}, K_{q_i}, and F are indeed interpreted in the desired sense. Below, we will discuss in depth how this is achievable. In the meantime, we will content ourselves with the assumption that the predicates carry their desired meaning.

According to our assumption, we can translate each instruction into a predicate-logic formula 'Inst' describing the transition from one configuration to the next. First, we consider an instruction of the form (q_i, S_j, S_k, L, q_l). The transition triggered by this instruction can be characterized as follows:

■ If in the configuration at time t...

- cell y contains symbol S_j, ☞ $R_{S_j}(t,y)$
- and the read-write head is positioned over cell y, ☞ $I(t,y)$
- and the machine is in state q_i, ☞ $K_{q_i}(t)$

■ then in the configuration at time $t+1$...

- the read-write head has moved left, ☞ $I(t+1, y-1)$
- and symbol S_j has been replaced by S_k, ☞ $R_{S_k}(t+1, y)$
- and state q_l is entered. ☞ $K_{q_l}(t+1)$

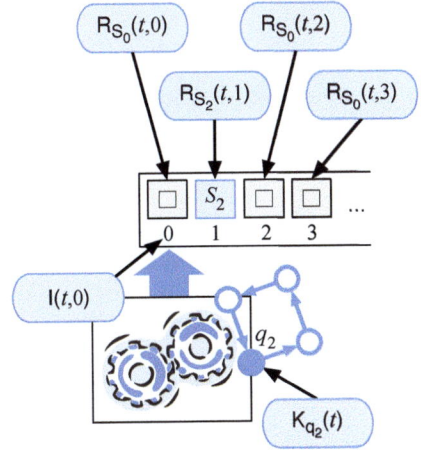

Figure 5.30: In his 1936 paper, Turing introduced several predicates for describing the configurations of tape-restricted Turing machines.

Combining the partial results, we obtain the following intermediate formula:

$$\forall t \, \forall y \, ((R_{S_j}(t,y) \wedge I(t,y) \wedge K_{q_i}(t)) \\ \to (I(t+1, y-1) \wedge R_{S_k}(t+1, y) \wedge K_{q_l}(t+1))) \quad (5.7)$$

There is one more thing. We must ensure that the tape content remains unchanged anywhere the read-write head is *not* located:

$$\forall z \, (z \neq y \to \bigwedge_{i=0}^{M} (R_{S_i}(t,z) \to R_{S_i}(t+1,z))) \quad (5.8)$$

Conjunctively joining (5.7) and (5.8) gives us:

$$\forall t \, \forall y \, ((R_{S_j}(t,y) \wedge I(t,y) \wedge K_{q_i}(t)) \\ \to (I(t+1, y-1) \wedge R_{S_k}(t+1, y) \wedge K_{q_l}(t+1) \\ \wedge \forall z \, (z \neq y \to \bigwedge_{i=0}^{M} (R_{S_i}(t,z) \to R_{S_i}(t+1,z))))) \quad (5.9)$$

This formula is not a genuine predicate-logic formula yet, as it utilizes a function character with '+' and a predicate character with '≠', neither of which is available in PL1. However, we can quickly eliminate the expressions $t+1$, $y-1$, and $z \neq y$ using the predicate $F(x,y)$:

$$\text{Inst}(q_i, S_j, S_k, L, q_l) :=$$
$$\forall t \, \forall y \, \forall t' \, \forall y' \, ((R_{S_j}(t,y) \wedge I(t,y) \wedge K_{q_i}(t) \wedge F(t,t') \wedge F(y',y))$$
$$\to (I(t',y') \wedge R_{S_k}(t',y) \wedge K_{q_l}(t')$$
$$\wedge \forall z \, (F(y',z) \vee \bigwedge_{i=0}^{M} (R_{S_i}(t,z) \to R_{S_i}(t',z))))) \tag{5.10}$$

At this point, it is evident how to translate the other instruction types. All we need to do is slightly rewrite (5.10):

$$\text{Inst}(q_i, S_j, S_k, R, q_l) :=$$
$$\forall t \, \forall y \, \forall t' \, \forall y' \, ((R_{S_j}(t,y) \wedge I(t,y) \wedge K_{q_i}(t) \wedge F(t,t') \wedge F(y,y'))$$
$$\to (I(t',y') \wedge R_{S_k}(t',y) \wedge K_{q_l}(t')$$
$$\wedge \forall z \, (F(z,y') \vee \bigwedge_{i=0}^{M} (R_{S_i}(t,z) \to R_{S_i}(t',z))))) \tag{5.11}$$

$$\text{Inst}(q_i, S_j, S_k, N, q_l) :=$$
$$\forall t \, \forall y \, \forall t' \, \forall y' \, ((R_{S_j}(t,y) \wedge I(t,y) \wedge K_{q_i}(t) \wedge F(t,t') \wedge F(y',y))$$
$$\to (I(t',y) \wedge R_{S_k}(t',y) \wedge K_{q_l}(t')$$
$$\wedge \forall z \, (F(y',z) \vee \bigwedge_{i=0}^{M} (R_{S_i}(t,z) \to R_{S_i}(t',z))))) \tag{5.12}$$

By translating all instructions of a Turing machine into formulas and joining them conjunctively, we can encode the entire instruction set into a single formula. For the example machine from Figure 5.1, this formula reads as follows:

$$\text{Des}(M) = \text{Inst}(q_1, S_0, S_1, R, q_2) \wedge \text{Inst}(q_2, S_0, S_0, R, q_3) \wedge$$
$$\text{Inst}(q_3, S_0, S_2, R, q_4) \wedge \text{Inst}(q_4, S_0, S_0, R, q_1)$$

The term $\text{Des}(M)$ was coined by Turing and is the abbreviation for *"de-*

Formula (5.10) differs from Turing's formulation in two respects. For one thing, we have renamed the original paper's variable x to t to emphasize its meaning. t denotes a point in time, the t-th calculation step, and y is a cell number. For another, Turing made a severe mistake in the last line of the formula. In the original, it reads as follows:

$$\forall z \, (F(y',z) \vee (R_{S_j}(x,z) \to R_{S_k}(x',z)))$$

This formula did not make sense and was quickly rectified by Turing himself. One year after the release of the original paper, Turing published several corrections in an article entitled *"On Computable Numbers, with an Application to the Entscheidungsproblem. A Correction"* [213]. In this article, he brought his original formula into the correct form except for a forgotten universal quantifier.

5.4 Impact on Mathematics

scription of M". Written out, the formula is a true monstrosity:

$$\forall t \forall y \forall t' \forall y' ((R_{S_0}(t,y) \wedge I(t,y) \wedge K_{q_1}(t) \wedge F(t,t') \wedge F(y,y'))$$
$$\rightarrow (I(t',y') \wedge R_{S_1}(t',y) \wedge K_{q_2}(t')$$
$$\wedge \forall z (F(z,y') \vee ((R_{S_0}(t,z) \rightarrow R_{S_0}(t',z)) \wedge$$
$$(R_{S_1}(t,z) \rightarrow R_{S_1}(t',z)) \wedge (R_{S_2}(t,z) \rightarrow R_{S_2}(t',z)))))) \wedge$$
$$\forall t \forall y \forall t' \forall y' ((R_{S_0}(t,y) \wedge I(t,y) \wedge K_{q_2}(t) \wedge F(t,t') \wedge F(y,y'))$$
$$\rightarrow (I(t',y') \wedge R_{S_0}(t',y) \wedge K_{q_3}(t')$$
$$\wedge \forall z (F(z,y') \vee ((R_{S_0}(t,z) \rightarrow R_{S_0}(t',z)) \wedge$$
$$(R_{S_1}(t,z) \rightarrow R_{S_1}(t',z)) \wedge (R_{S_2}(t,z) \rightarrow R_{S_2}(t',z)))))) \wedge$$
$$\forall t \forall y \forall t' \forall y' ((R_{S_0}(t,y) \wedge I(t,y) \wedge K_{q_3}(t) \wedge F(t,t') \wedge F(y,y'))$$
$$\rightarrow (I(t',y') \wedge R_{S_2}(t',y) \wedge K_{q_4}(t')$$
$$\wedge \forall z (F(z,y') \vee ((R_{S_0}(t,z) \rightarrow R_{S_0}(t',z)) \wedge$$
$$(R_{S_1}(t,z) \rightarrow R_{S_1}(t',z)) \wedge (R_{S_2}(t,z) \rightarrow R_{S_2}(t',z)))))) \wedge$$
$$\forall t \forall y \forall t' \forall y' ((R_{S_0}(t,y) \wedge I(t,y) \wedge K_{q_4}(t) \wedge F(t,t') \wedge F(y,y'))$$
$$\rightarrow (I(t',y') \wedge R_{S_0}(t',y) \wedge K_{q_1}(t')$$
$$\wedge \forall z (F(z,y') \vee ((R_{S_0}(t,z) \rightarrow R_{S_0}(t',z)) \wedge$$
$$(R_{S_1}(t,z) \rightarrow R_{S_1}(t',z)) \wedge (R_{S_2}(t,z) \rightarrow R_{S_2}(t',z))))))$$

Des(M) describes the transition from one configuration to the next. However, it conveys nothing about the starting configuration of the machine. By definition, we agreed that at time 0,

- the read-write head is positioned over cell 0,   ☞ $I(0,0)$
- the machine resides in the start state q_1, and   ☞ $K_{q_1}(0)$
- a blank tape is present.   ☞ $\forall y\, R_{S_0}(0,y)$

Consequently, we can describe the overall behavior of M as follows:

$$I(0,0) \wedge K_{q_1}(0) \wedge \forall y\, R_{S_0}(0,y) \wedge \text{Des}(M) \qquad (5.13)$$

To turn this formula into a genuine predicate-logic formula, the constant 0 has to vanish. For doing so, Turing exploited that it is irrelevant whether the machine is in the start configuration at time 0 or at a later time u, allowing him to solve the problem by replacing 0 with an existentially quantified constant symbol u. Formula (5.13) then becomes:

$$A(M) := \exists u\, (I(u,u) \wedge K_{q_1}(u) \wedge \forall y\, R_{S_0}(u,y)) \wedge \text{Des}(M)$$

> In Turing's historical 1936 paper, the definition of the formula Halt(M) is nowhere present, nor is the word *halting problem*. In his original proof, Turing did not reduce the halting problem as we did. Instead, he showed that a decision procedure for first-order predicate logic can be employed to determine whether a machine will eventually write the symbol S_1 onto the tape. Following Turing's path, we obtain a formula that is even simpler than the one constructed here. Instead of the lengthy formula Halt(M), it contains the much shorter expression:
>
> $$\exists t\, \exists y\, R_{S_1}(t, y)$$
>
> This formula states that the machine will reach a configuration where cell y contains the tape symbol S_1. In other words: the machine eventually writes S_1.
> Just like the halting problem, the problem of whether a Turing machine writes a particular symbol onto the tape is undecidable, so both formulas serve their purpose.

Last, we will characterize the termination of M by a formula Halt(M). To understand its structure, remember that a Turing machine continues to compute whenever an instruction is applicable. If we consider an instruction of the form $(q_i, S_j, \_, \_)$, it is applicable if and only if

- the machine is in state q_i, and   ☞ $K_{q_i}(t)$
- the currently addressed cell is y, and   ☞ $I(t, y)$
- the symbol S_j is stored in cell y.   ☞ $R_{S_j}(t, y)$

This allows us to describe the continuation of a machine with a formula Cont(M, t) by forming the partial formula

$$\exists y\, (K_{q_i}(t) \wedge I(t, y) \wedge R_{S_j}(t, y))$$

for each instruction $(q_i, S_j, \_, \_)$ and then joining them disjunctively. For our example machine, we get:

$$\begin{aligned}
\text{Cont}(M, t) = &\underbrace{\exists y\, (K_{q_1}(t) \wedge I(t, y) \wedge R_{S_0}(t, y))}_{\text{Instruction } (q_1, S_0, S_1, R, q_2)} \vee \\
&\underbrace{\exists y\, (K_{q_2}(t) \wedge I(t, y) \wedge R_{S_0}(t, y))}_{\text{Instruction } (q_2, S_0, S_0, R, q_3)} \vee \\
&\underbrace{\exists y\, (K_{q_3}(t) \wedge I(t, y) \wedge R_{S_0}(t, y))}_{\text{Instruction } (q_3, S_0, S_2, R, q_4)} \vee \\
&\underbrace{\exists y\, (K_{q_4}(t) \wedge I(t, y) \wedge R_{S_0}(t, y))}_{\text{Instruction } (q_4, S_0, S_0, R, q_1)}
\end{aligned}$$

With the help of the formula just generated, we can describe the termination of a Turing machine with ease:

$$\text{Halt}(M) := \exists t\, \neg \text{Cont}(M, t)$$

In plain language, the formula states that the machine will eventually reach a configuration with no further progress possible.

Combining the partial formulas $A(M)$ and Halt(M) into

$$\text{Un}(M) := A(M) \rightarrow \text{Halt}(M)$$

gives us a formula with the following substantive meaning:

"*M terminates on a blank tape.*"

5.4 Impact on Mathematics

We must pay attention to a crucial aspect that we already mentioned above. The construction of the formula $\mathrm{Un}(M)$ relies on the assumption that the involved predicates convey their intended meaning. Denoting this interpretation by (U,I), we can specify the relationship between $\mathrm{Un}(M)$ and the termination of M as follows:

$$M \text{ terminates} \Leftrightarrow (U,I) \models \mathrm{Un}(M)$$

To answer Hilbert's decision problem negatively, however, we need a formula that is *universally valid* precisely when the machine M terminates:

$$M \text{ terminates} \Leftrightarrow \;\models \mathrm{Un}(M) \tag{5.14}$$

Let us examine whether our formula $\mathrm{Un}(M)$ fulfills this property. The direction from right to left is almost trivial. If $\mathrm{Un}(M)$ is universally valid, the formula is true under *any* interpretation and thus a fortiori under the interpretation assigning all predicate symbols their intended meaning. In this case, the formula $\mathrm{Un}(M)$ says the desired: M will eventually reach a configuration where no further progress is possible. In short, M terminates.

Does the direction from left to right also hold? The answer is no! Since we are considering the formula's universal validity, we must also consider those interpretations under which the applied predicates lose their intended meaning. For example, we have assumed without justification that $F(x,y)$ expresses the relationship $y = x+1$. However, when the symbols have a different meaning, $\mathrm{Un}(M)$ may be wrong, albeit the Turing machine M still terminates. To ensure the relationship (5.14) in both directions, we must add a subformula to $\mathrm{Un}(M)$ that compels F to its intended meaning. In [213], Turing introduced a formula precisely for that purpose, the formula Q:

$$Q = \forall x\, \exists w\, \forall y\, \forall z\; (F(x,w) \wedge (F(x,y) \rightarrow G(x,y)) \wedge$$
$$(F(x,z) \wedge G(z,y) \rightarrow G(x,y)) \wedge$$
$$(G(z,x) \vee (G(x,y) \wedge F(y,z)) \vee (F(x,y) \wedge F(z,y)) \rightarrow \neg F(x,z)))$$

Herein, G denotes a new predicate symbol with the following intended meaning:

$G(x,y)$: x and y are integers, with x being less than y.

The individual components of the formula Q then state the following:

- Every number has a successor
  ☞ $F(x,w)$

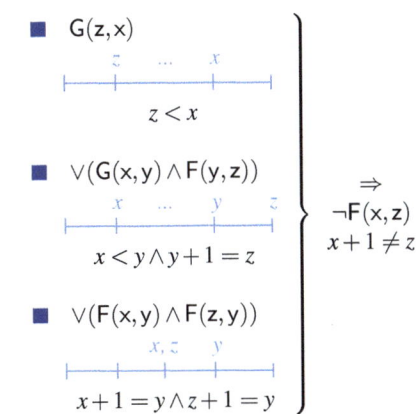

Figure 5.31: The formula Q ensures, among others, that the predicates F and G comply with the depicted ordering properties.

> The formula Q is missing entirely in Turing's 1936 original paper, as he was trying to describe the intended meaning of the predicate sign F as follows:
>
> $N(u) \wedge \forall x\, (N(x) \rightarrow \exists x'\, F(x,x')) \wedge$
> $\forall y\, \forall z\, (F(y,z) \rightarrow N(y) \wedge N(z))$
>
> In Turing's formula, $N(\xi)$ expresses that ξ corresponds to a natural number. Under this interpretation, the first part of the formula states that $u \in \mathbb{N}$, the middle part postulates that every number has a successor, and the last part expresses that $y+1 = z$ implies $y,z \in \mathbb{N}$.
> Turing missed out on the fact that his formula, as presented in 1936, does not guarantee the uniqueness of a number's successor. His proof, however, explicitly relies on this property. In the corrections published in 1937, Turing rectified this error by introducing the formula Q as a replacement for the above expression.

- $x+1$ is greater than x
  ☞ $F(x,y) \to G(x,y)$

- If $x+1 < y$, then $x < y$ holds as well
  ☞ $F(x,z) \land G(z,y) \to G(x,y)$

- F and G fulfill the ordering properties depicted in Figure 5.31
  ☞ $G(z,x) \lor (G(x,y) \land F(y,z)) \lor (F(x,y) \land F(z,y)) \to \neg F(x,z)$

In his 1937 corrections, Turing showed that the formula Q adequately describes our predicates' intended meaning. With

$$(A(M) \land Q) \to \text{Halt}(M), \qquad (5.15)$$

we thus get the formula we sought; it is universally valid if the Turing machine M terminates on a blank tape.

Above, we have already anticipated the final step in Turing's proof. If there were a decision procedure for first-order predicate logic, it could decide the halting problem. To determine whether a tape-restricted Turing machine M terminates on a blank tape, it would suffice to construct formula (5.15) and check whether it is universally valid or not. In the first case, we knew M would terminate; in the second, it continued calculating indefinitely. However, we already know that the halting problem is undecidable, which implies the unsolvability of Hilbert's decision problem:

Theorem 5.7 (Turing, 1936)

> There is no decision procedure for predicate logic.

5.4.2 Incompleteness of Arithmetic

This section will discuss an alternative encoding of Turing machines. In contrast to the previous section, the result will no longer be an ordinary formula of PL1 but a formula of Peano arithmetic. Specifically, we will show how a Turing machine M translates into an arithmetic formula being true precisely when M terminates. The notion of universal validity will no longer play any role since we are only interested in whether the constructed formulas are true or false under the standard interpretation of Peano arithmetic. In this interpretation, the domain is the set of natural numbers, '$+$' corresponds to addition, '$\times$' to multiplication, and 's' to the successor operation.

5.4 Impact on Mathematics

The construction relies on encoding a sequence of transitions into a single natural number and claiming that this number describes a terminating computation sequence.

We will approach our goal in several steps. First, let us consider how to code a single configuration as a natural number. A configuration has the form

$$(q, i, s_0, s_1, \ldots, s_n) \tag{5.16}$$

with q representing the current state, i the position of the read-write head, and $s_0, s_1, \ldots, s_n$ the portion of the tape that has been used so far. We represent the current state and the tape symbols with natural numbers and utilize the number n to represent the state q_n or the tape symbol S_n. In this way, we can regard all elements of a configuration as natural numbers.

To talk about the configuration elements within a formula, we introduce several new function symbols, I, R_j, K, and L, as representatives of the following functions:

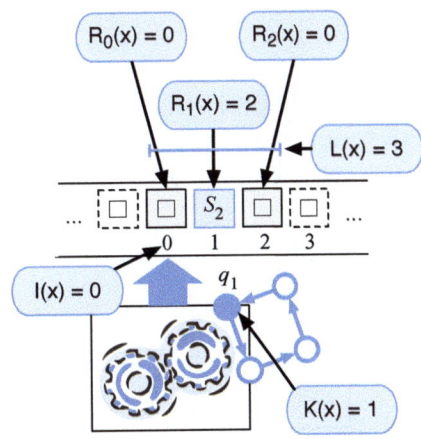

Figure 5.32: Describing a Turing machine configuration with I, R_j, K, and L

$$\text{I} : (q, i, s_0, \ldots, s_n) \mapsto i \quad \text{(head position)}$$
$$R_j : (q, i, s_0, \ldots, s_n) \mapsto s_j \quad \text{(symbol in cell } j\text{)}$$
$$\text{K} : (q, i, s_0, \ldots, s_n) \mapsto q \quad \text{(machine state)}$$
$$\text{L} : (q, i, s_0, \ldots, s_n) \mapsto n+1 \quad \text{(number of used tape cells)}$$

Figure 5.32 illustrates how to describe a Turing machine's configuration with the functions just defined.

Remember that we are looking for a formula of Peano arithmetic. In particular, such a formula must not contain any function symbols besides '0', 's', '+', and '×' and no predicate symbols besides '='. Therefore, we will eventually have to replace the newly introduced symbols with equivalent arithmetic expressions. For the moment, however, we will refrain from dwelling on this shortcoming and postpone the details.

Next, we will incorporate the introduced functions into two important sub-formulas. The first is the formula $\varphi_{\text{Start}}(x)$, which is supposed to be true precisely if the variable x describes the start configuration of a Turing machine. Above, we specified that in the start configuration,

- the read-write head is above cell 0,   ☞ $\text{I}(x) = 0$
- the current state is the start state q_1,   ☞ $\text{K}(x) = \overline{1}$
- and the tape is blank.   ☞ $\text{L}(x) = \overline{1} \wedge R_0(x) = 0$

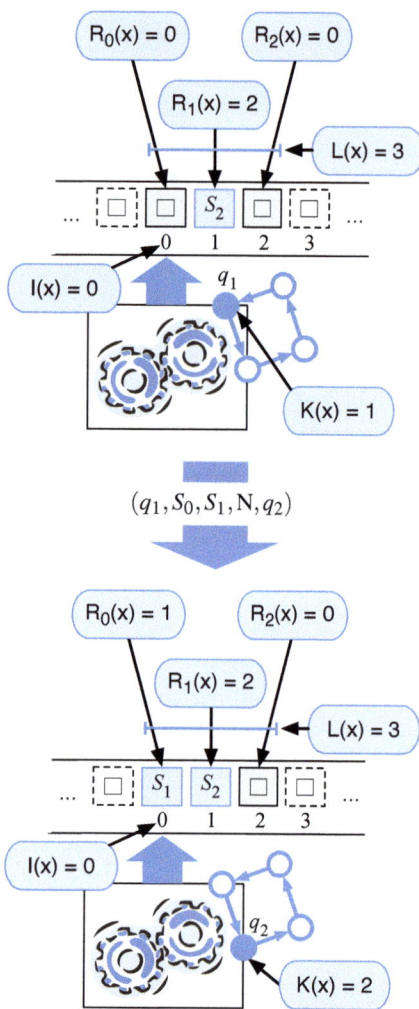

Figure 5.33: Executing an instruction of the form (q_i, S_j, S_k, N, q_l)

We obtain the desired formula $\varphi_{\text{Start}}(x)$ by conjunctive linking all partial expressions:

$$\varphi_{\text{Start}}(x) := (\mathsf{I}(x) = 0 \land \mathsf{K}(x) = \overline{1} \land \mathsf{L}(x) = \overline{1} \land \mathsf{R}_0(x) = 0)$$

The second is the formula $\varphi_{\text{Cont}}(x)$. It is supposed to be true if the machine does not terminate in the configuration x, i.e., if some instruction translates x into a subsequent configuration. We can construct the formula similarly to the formula $\text{Cont}(M)$ from Section 5.4.1. For the machine considered there, it reads as such:

$$\varphi_{\text{Cont}}(x) := \underbrace{\exists (y < \mathsf{L}(x)) \; \bigl(\mathsf{I}(x) = y \land \mathsf{K}(x) = \overline{1} \land \mathsf{R}_y(x) = 0\bigr)}_{\text{Instruction } (q_1, S_0, S_1, R, q_2)} \lor$$

$$\underbrace{\exists (y < \mathsf{L}(x)) \; \bigl(\mathsf{I}(x) = y \land \mathsf{K}(x) = \overline{2} \land \mathsf{R}_y(x) = 0\bigr)}_{\text{Instruction } (q_2, S_0, S_0, R, q_3)} \lor$$

$$\underbrace{\exists (y < \mathsf{L}(x)) \; \bigl(\mathsf{I}(x) = y \land \mathsf{K}(x) = \overline{3} \land \mathsf{R}_y(x) = 0\bigr)}_{\text{Instruction } (q_3, S_0, S_2, R, q_4)} \lor$$

$$\underbrace{\exists (y < \mathsf{L}(x)) \; \bigl(\mathsf{I}(x) = y \land \mathsf{K}(x) = \overline{4} \land \mathsf{R}_y(x) = 0\bigr)}_{\text{Instruction } (q_4, S_0, S_0, R, q_1)}$$

In addition, we define a formula $\varphi_I(x, y)$ for every instruction I, describing the transition from the configuration x to the configuration y. First, we consider an instruction (q_i, S_j, S_k, N, q_l), which keeps the read-write head in place (Figure 5.33). When executed, the Turing machine switches from configuration x to configuration y,

- if the initial configuration x ...

  - encodes n written cells,   ☞ $\mathsf{L}(x) = \overline{n}$
  - state q_i is entered,   ☞ $\mathsf{K}(x) = \overline{i}$
  - the read-write head is positioned over cell h,   ☞ $\mathsf{I}(x) = \overline{h}$
  - cell h contains symbol S_j,   ☞ $\mathsf{R}_h(x) = \overline{j}$

- and the subsequent configuration y ...

  - encodes n written cells,   ☞ $\mathsf{L}(y) = \overline{n}$
  - state q_l is entered,   ☞ $\mathsf{K}(y) = \overline{l}$
  - the read-write head is positioned over cell h,   ☞ $\mathsf{I}(y) = \overline{h}$
  - cell h contains symbol S_k.   ☞ $\mathsf{R}_h(y) = \overline{k}$

5.4 Impact on Mathematics

Further, we must keep in mind that the tape content remains unchanged wherever the read-write head is not positioned:

$$\forall (h' < n)\, (h' \neq h \rightarrow \exists s\, (R_{h'}(x) = s \wedge R_{h'}(y) = s))$$

Joining all formulas together yields the following result:

- Instruction type (q_i, S_j, S_k, N, q_l)

$$\varphi_I(x,y) := \exists h\, \exists n\, ($$
$$L(x) = n \wedge K(x) = \bar{i} \wedge I(x) = h \wedge R_h(x) = \bar{j}\, \wedge$$
$$L(y) = n \wedge K(y) = \bar{l} \wedge I(y) = h \wedge R_h(y) = \bar{k}\, \wedge$$
$$\forall (h' < n)\, (h' \neq h \rightarrow \exists s\, (R_{h'}(x) = s \wedge R_{h'}(y) = s)))$$

We can proceed similarly for the other instruction types. However, we must spend some extra work when the read-write head addresses the first or last cell of the encoded tape sequence. The more straightforward case is the right-hand movement (Figure 5.34). If n_1 and n_2 denote the lengths of the tape segments encoded in x and y, respectively, and h_1 and h_2 the positions of the read-write head, the following applies:

- If the read-write head is not on the far right,   ☞ $h_1 + \bar{1} \neq n_1$
  - then y encodes as many cells as x.   ☞ $n_1 = n_2$
- Otherwise,   ☞ $h_1 + \bar{1} = n_1$
  - y has one more encoded cell than x,   ☞ $n_1 + \bar{1} = n_2$
  - and this cell is empty.   ☞ $R_{n_1}(y) = 0$
- In both cases, the head moves to the right.   ☞ $h_1 + \bar{1} = h_2$

The formula $\varphi_I(x, y)$ thus appears as follows:

- Instruction type (q_i, S_j, S_k, R, q_l)

$$\varphi_I(x,y) := \exists h_1\, \exists h_2\, \exists n_1\, \exists n_2\, ($$
$$L(x) = n_1 \wedge K(x) = \bar{i} \wedge I(x) = h_1 \wedge R_{h_1}(x) = \bar{j}\, \wedge$$
$$L(y) = n_2 \wedge K(y) = \bar{l} \wedge I(y) = h_2 \wedge R_{h_1}(y) = \bar{k}\, \wedge$$
$$(h_1 + \bar{1} \neq n_1 \rightarrow n_1 = n_2)\, \wedge$$
$$(h_1 + \bar{1} = n_1 \rightarrow (n_1 + \bar{1} = n_2 \wedge R_{n_1}(y) = 0))\, \wedge$$
$$h_1 + \bar{1} = h_2\, \wedge$$
$$\forall (h < n_1)\, (h \neq h_1 \rightarrow \exists s\, (R_h(x) = s \wedge R_h(y) = s)))$$

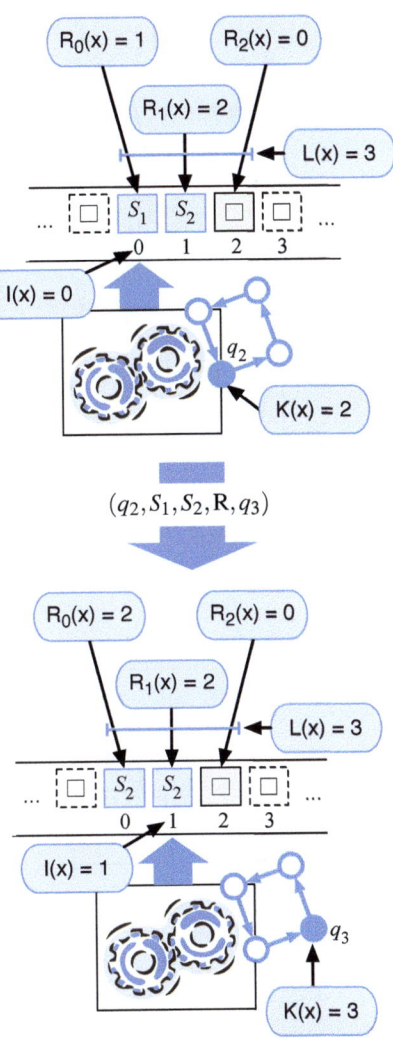

Figure 5.34: Execution of an instruction of the form (q_i, S_j, S_k, R, q_l)

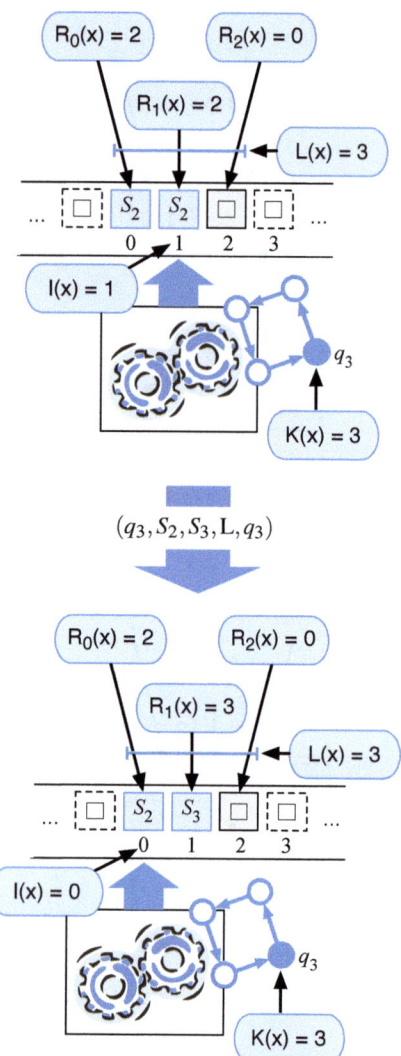

Figure 5.35: Execution of an instruction of the form (q_i, S_j, S_k, L, q_l)

In case of a leftward movement, we must be extra vigilant (Figures 5.35 and 5.36). If the read-write head is already on the far left ($h_1 = 0$), we simulate the head movement by copying the entire tape content one cell to the right. To summarize:

- If the read-write head is not at the far left, ☞ $h_1 \neq 0$
  - then y encodes the same number of cells as x, and ☞ $n_1 = n_2$
  - the read-write head moves left. ☞ $h_1 = h_2 + \overline{1}$
- If the read-write head is at the far left, ☞ $h_1 = 0$
  - then y encodes one more cell than x, ☞ $n_1 + 1 = n_2$
  - this additional cell is empty, and ☞ $R_0(y) = 0$
  - the read-write head remains in place. ☞ $h_1 = h_2$

The following formula describes the copy process:

$$\forall (h < n_1)\, (h \neq 0 \to \exists s\, (R_h(x) = s \wedge R_{h+1}(y) = s))$$

Putting all the pieces together yields the following formula:

- Instruction type (q_i, S_j, S_k, L, q_l)

$$\varphi_I(x, y) := \exists h_1\, \exists h_2\, \exists n_1\, \exists n_2\, ($$
$$L(x) = n_1 \wedge K(x) = \overline{i} \wedge I(x) = h_1 \wedge R_{h_1}(x) = \overline{j} \wedge$$
$$L(y) = n_2 \wedge K(y) = \overline{l} \wedge I(y) = h_2 \wedge$$
$$(h_1 \neq 0 \to ($$
$$n_1 = n_2 \wedge h_1 = h_2 + \overline{1} \wedge R_{h_1}(y) = \overline{k} \wedge$$
$$\forall (h < n_1)\, (h \neq h_1 \to \exists s\, (R_h(x) = s \wedge R_h(y) = s)))) \wedge$$
$$(h_1 = 0 \to ($$
$$n_1 + \overline{1} = n_2 \wedge R_0(y) = 0 \wedge h_1 = h_2 \wedge R_1(y) = \overline{k} \wedge$$
$$\forall (h < n_1)\, (h \neq 0 \to \exists s\, (R_h(x) = s \wedge R_{h+1}(y) = s)))))$$

By joining all instruction formulas disjunctively, we obtain a formula that is true precisely when the machine can transition from configuration x to configuration y:

$$\varphi_{\text{Trans}}(x, y) := \varphi_{I_1}(x, y) \vee \varphi_{I_2}(x, y) \vee \varphi_{I_3}(x, y) \vee \ldots \vee \varphi_{I_K}(x, y)$$

Next, we will encode an arbitrary initial segment of a calculation sequence, i.e., a finite sequence of configurations $\kappa_0, \kappa_1, \kappa_2, \ldots, \kappa_m$, as a

5.4 Impact on Mathematics

natural number. For this purpose, we introduce a unary function symbol, M, and a binary function symbol, C, with the following intended meaning:

$$M : (\kappa_0, \kappa_1, \kappa_2, \ldots, \kappa_m) \mapsto m \qquad \text{(number of transitions)}$$
$$C : (\kappa_0, \kappa_1, \kappa_2, \ldots, \kappa_m), k \mapsto \kappa_k \qquad \text{(k-th configuration)}$$

The two new functions allow us to describe a terminating calculation sequence straight away. All we need is to recap that a Turing machine M terminates after n calculation steps if it

- starts in the initial configuration ☞ $\varphi_{\text{Start}}(C(z,0))$
- and computes ☞ $n = M(z) \wedge \forall (u < n) \; \varphi_{\text{Trans}}(C(z,u), C(z,s(u)))$
- until it can't proceed further. ☞ $\neg \varphi_{\text{Cont}}(C(z,n))$

To summarize,

$$\varphi_M := \exists z \, \exists n \, (\varphi_{\text{Start}}(C(z,0)) \wedge \tag{5.17}$$
$$n = M(z) \wedge \forall (u < n) \; \varphi_{\text{Trans}}(C(z,u), C(z,s(u))) \wedge$$
$$\neg \varphi_{\text{Cont}}(C(z,n)))$$

gives us the formula we seek. φ_M is true precisely when the Turing machine M terminates. Its only flaw: it is not a formula of Peano arithmetic since it contains several function symbols that are not genuinely available in PA. Furthermore, the variable y in φ_{Cont} also occurs as the index of a function symbol. Let us take care of this shortcoming by eliminating the function symbols one after the other. Once again, the key to the solution is Gödel's β function, thoroughly discussed in Section 4.2.3:

$$\beta(x,y,z) := x \bmod (1 + y \cdot (z+1))$$

We proved that for every finite sequence of natural numbers $a_0, \ldots, a_n$, there exist two numbers, b and c, satisfying

$$\beta(b,c,0) = a_0, \; \beta(b,c,1) = a_1, \; \beta(b,c,2) = a_2, \ldots$$

This enables us to represent each configuration

$$x = (q, i, s_0, s_1, \ldots, s_n)$$

with two numbers, b and c. We can choose these numbers to satisfy the

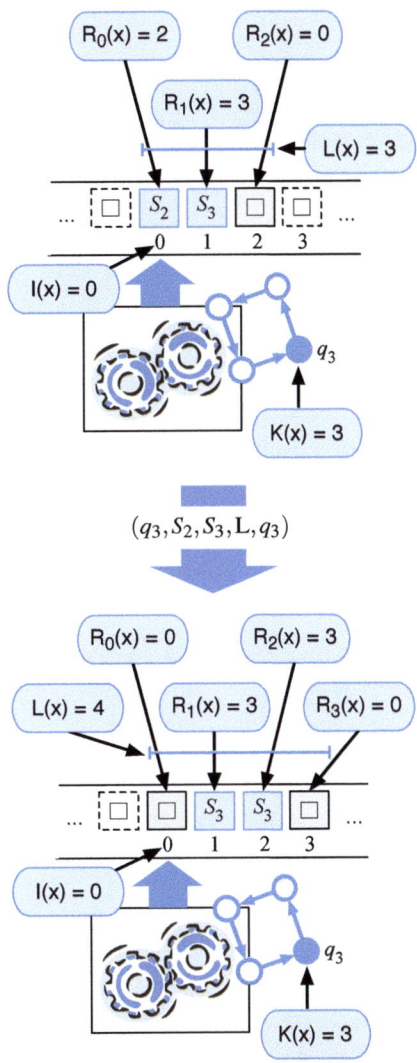

Figure 5.36: Execution of an instruction of the form (q_i, S_j, S_k, L, q_l). As the read-write head is already on the far left in the depicted case, the head movement is simulated by copying the entire tape content one position to the right.

following properties:

$$\beta(b,c,0) = n+1 \quad \text{(number of encoded cells, L(x))}$$
$$\beta(b,c,1) = q \quad \text{(machine state, K(x))}$$
$$\beta(b,c,2) = i \quad \text{(head position, I(x))}$$
$$\beta(b,c,3) = s_0 \quad \text{(content of the first cell, } R_0(x)\text{)}$$
$$\ldots$$
$$\beta(b,c,n+3) = s_n \quad \text{(content of last cell, } R_n(x)\text{)}$$

From here, it is just a mere technicality to translate formulas $\varphi_{\text{Start}}(x)$, $\varphi_{\text{Trans}}(x)$, and $\varphi_{\text{Cont}}(x)$ into genuine formulas of Peano arithmetic. For example, we can rewrite

$$\varphi_{\text{Start}}(x) = (I(x) = 0 \wedge K(x) = \overline{1} \wedge L(x) = \overline{1} \wedge R_0(x) = 0)$$

as follows:

$$\varphi_{\text{Start}}(b,c) := (\beta(b,c,\overline{2}) = 0) \wedge (\beta(b,c,\overline{1}) = \overline{1}) \wedge$$
$$(\beta(b,c,0) = \overline{1}) \wedge (\beta(b,c,\overline{3}) = 0)$$

By replacing the placeholder $\beta(b,c,\overline{n})$ with the equivalent arithmetic expression (4.7) from Section 4.2.3, we obtain a proper PA formula:

$$\varphi_{\text{Start}}(b,c) := \exists d \; b = s(c \times s(\overline{2})) \times d + 0 \wedge 0 < s(c \times s(\overline{2})) \wedge$$
$$\exists d \; b = s(c \times s(\overline{1})) \times d + \overline{1} \wedge \overline{1} < s(c \times s(\overline{1})) \wedge$$
$$\exists d \; b = s(c \times s(0)) \times d + \overline{1} \wedge \overline{1} < s(c \times s(0)) \wedge$$
$$\exists d \; b = s(c \times s(\overline{3})) \times d + 0 \wedge 0 < s(c \times s(\overline{3}))$$

In the same way, we can translate formulas $\varphi_{\text{Trans}}(x)$ and $\varphi_{\text{Cont}}(x)$ into the corresponding formulas $\varphi_{\text{Trans}}(b_1,c_1,b_2,c_2)$ and $\varphi_{\text{Cont}}(b,c)$, respectively. These formulas are difficult to grasp intuitively, so we will avoid writing them out in full. From now on, we also refrain from eliminating the placeholder $\beta(b,c,\overline{n})$.

After having managed to code individual configurations arithmetically, we want to convince ourselves that finite calculation sequences are encodable, too. For this purpose, we employ Gödel's β function for a second time. According to what was said above, we can understand a finite calculation sequence

$$\kappa_0, \kappa_1, \kappa_2, \ldots, \kappa_m$$

as a sequence of two natural numbers:

$$b_0, c_0, b_1, c_1, \ldots, b_m, c_m$$

5.4 Impact on Mathematics

Employing Gödel's β function, we can condense this sequence into two natural numbers, b and c, in an accustomed way. We can choose b and c to satisfy the following:

$\beta(b,c,0) = m$ (number of computation steps, $M(z)$)
$\beta(b,c,2 \cdot i + 1) = b_i$ (first component of $C(z,i)$)
$\beta(b,c,2 \cdot i + 2) = c_i$ (second component of $C(z,i)$)

We can now rewrite formula (5.17) as follows:

$$\begin{aligned}\varphi_M = \exists b\, \exists c\, \exists n\, (&\varphi_{\text{Start}}(\beta(b,c,\overline{1}),\beta(b,c,\overline{2})) \wedge \\ &n = \beta(b,c,0) \wedge \\ &\forall (u<n)\, \varphi_{\text{Trans}}(\beta(b,c,\overline{2}\times u+\overline{1}),\beta(b,c,\overline{2}\times u+\overline{2}), \\ &\qquad\qquad\qquad \beta(b,c,\overline{2}\times u+\overline{3}),\beta(b,c,\overline{2}\times u+\overline{4})) \wedge \\ &\neg\varphi_{\text{Cont}}(\beta(b,c,\overline{2}\times n+\overline{1}),\beta(b,c,\overline{2}\times n+\overline{2})))\end{aligned} \qquad (5.18)$$

At this point, we have crossed the finish line. φ_M is a genuine PA formula that is true precisely when the encoded Turing machine M terminates. The rest of our argument aligns with the previous section. If there were a decision procedure for Peano arithmetic, it could solve the halting problem. All we need to do is to construct formula (5.18) and check whether it is true or false. Hence, from the undecidability of the halting, we immediately get

Theorem 5.8

The semantic variant of the decision problem is unsolvable for Peano arithmetic.

The theorem asserts that no systematic procedure can reliably determine for every given arithmetic formula whether it is true or false. As a result, no formal system for Peano arithmetic can be both correct and complete. If such a system did exist, it would allow us to solve the semantic variant of the decision problem, as suggested by Theorem 2.6. However, Theorem 5.8 demonstrates that this is impossible. Hence, we obtain one of the most important results of proof theory as a corollary: the semantic variant of Gödel's first incompleteness theorem.

Corallary 5.2 (First Incompleteness Theorem, Semantical)

Any correct formal system expressive enough to formalize Peano arithmetic is incomplete.

The discussion in this section has shown that we can derive the semantic variant of Gödel's first incompleteness theorem in an amazingly straightforward way. All we need is the knowledge of the halting problem's undecidability and the realization that we can encode any Turing machine as an arithmetic formula. What a marvelous result.

5.4.3 Hilbert's Tenth Problem

The two preceding sections have presented elegant proofs for the unsolvability of predicate logic's decision problem and the incompleteness of arithmetic, both based on reducing the halting problem. In this section, we will derive another negative result of great significance in the same way: the unsolvability of Hilbert's tenth Problem. At the 2nd International Congress of Mathematicians in Paris, David Hilbert formulated the problem as such:

> "Given a Diophantine equation with any number of unknown quantities and with rational integral numerical coefficients: To devise a process according to which it can be determined in a finite number of operations whether the equation is solvable in rational integers."
>
> David Hilbert, 1900 [98]

In Chapter 1, we anticipated that Hilbert's tenth Problem is unsolvable; it is impossible to devise a systematic procedure that always correctly decides whether or not a given Diophantine equation is solvable. Historically, this fantastic result was achieved in several stages, summarized in Figure 5.37:

- In 1953, Martin Davis published a paper titled *Arithmetical Problems and Recursively Enumerable Predicates* [44]. Davis showed that every semi-decidable relation $R(a_1, \ldots, a_n)$ is representable in the following form:

$$\exists y \, \forall (z \leq y) \, \exists x_1 \ldots \exists x_m \, p(a_1, \ldots, a_n, y, z, x_1, \ldots, x_m) = 0 \quad (5.19)$$

In this equation, p is a multivariable polynomial with $n + m + 2$ unknowns. This equation put Davis on the edge of the finish line; only the quantifier $\forall (z \leq y)$ prevented him from crossing. If Davis had succeeded in eliminating the universal quantifier, he would have solved Hilbert's tenth problem. In plain words, equation (5.19)

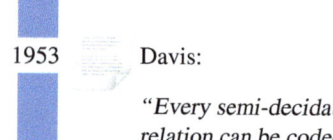

1953 Davis:

"Every semi-decidable relation can be coded as a Diophantine equation with a restrictedly universally quantified variable." [44]

1961 Davis, Putnam, Robinson:

"Every semi-decidable relation can be coded as an exponential Diophantine equation." [48]

1970 Matiyasevich:

"Every exponential Diophantine equation can be translated into an ordinary Diophantine equation." [132]

⇒ Hilbert's tenth problem has no solution!

1984 Jones, Matiyasevich:

"Every register machine can be represented exponentially diophantically." [109]

Figure 5.37: Landmark moments in mathematics. Since 1970, we know that Hilbert's tenth problem has no solution, and since 1984, we possess an elegant proof.

5.4 Impact on Mathematics

would then state that for every semi-decidable relation R, a Diophantine equation existed that is solvable for $(a_1,\ldots,a_n) \in R$ and unsolvable for $(a_1,\ldots,a_n) \notin R$. The semi-decidability of the halting problem would then lead to the unsolvability of Hilbert's tenth problem. In 1953, the solution seemed within reach. Nevertheless, all attempts to eliminate the universal quantifier from equation (5.19) bore no fruit at first.

- In 1961, Martin Davis, Hilary Putnam, and Julia Robinson published a seminal paper titled *The Decision Problem for Exponential Diophantine Equations* [48]. In this paper, the authors proved a theorem known today as the *bounded quantifier theorem*. The theorem permitted them to eliminate Davis' universal quantifier, but at tremendous cost. In particular, the resulting equation had the following form:

$$\exists x_1 \ldots \exists x_m \, r(a_1,\ldots,a_n,x_1,\ldots,x_m) = s(a_1,\ldots,a_n,x_1,\ldots,x_m)$$

What is crucial here is that $r = s$ was no longer an ordinary Diophantine equation but an exponential one. In contrast to ordinary Diophantine equations, variables may also occur as exponents. Even if the result did not link directly to Hilbert's tenth problem, it constituted an important milestone: Davis, Putnam, and Robinson had proven that no decision procedure exists for *exponential* Diophantine equations.

- The certainty that no decision procedure exists even for ordinary Diophantine equations has been with us since 1970. That year, Yuri Matiyasevich proved that every exponential Diophantine equation translates into an ordinary Diophantine equation (Figure 5.38 below) [132]. The proof is not very descriptive, and we will not go into detail here. Nevertheless, it plays a vital role in our line of argument. Only through this theorem does the unsolvability of the decision problem for exponential Diophantine equations imply the unsolvability of Hilbert's tenth problem.

- Although the puzzle was solved in 1970, interest remained strong. Despite being correct, the proof was tremendously technical and complicated. A new proof for the undecidability of exponential Diophantine equations was finally published in 1984 by James Jones and Yuri Matiyasevich (Figure 5.38 above) [109]. In terms of content, they proved the same result as Davis, Putnam, and Robinson several years earlier, but their approach was remarkably elegant. They discovered how to code register machines exponentially diophantically. The undecidability of the halting problem then implied the

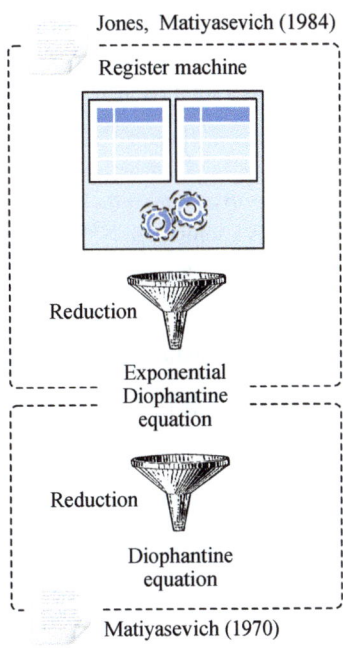

Figure 5.38: In 1984, Jones and Matiyasevich showed how to represent register machines exponentially diophantically. Combined with Matiyasevich's result from 1970, an elegant proof of the unsolvability of Hilbert's tenth problem emerges.

- Reduction from $\mathbb{Z}$ to $\mathbb{N}$

$$p(x_1,\ldots,x_m) = 0$$

has a solution in $\mathbb{Z}^m$

$$\Downarrow\Uparrow$$

$$p(p_1-q_1,\ldots,p_m-q_m) = 0$$

has a solution in $\mathbb{N}^{2m}$

- Reduction from $\mathbb{N}$ to $\mathbb{Z}$

$$p(x_1,\ldots,x_m) = 0$$

has a solution in $\mathbb{N}^m$

Lagrange[1] $\Downarrow\Uparrow$

$$p(w_1^2+x_1^2+y_1^2+z_1^2,\ldots,w_m^2+x_m^2+y_m^2+z_m^2) = 0$$

has a solution in $\mathbb{Z}^{4m}$

Figure 5.39: Mutual reduction of the decision problems of Diophantine equations

[1] *"Every natural number is the sum of four square numbers."*

Joseph Louis de Lagrange (1736 – 1813)

Figure 5.40: In 1770, the French mathematician Joseph Louis de Lagrange proved a conjecture made by Bachet de Méziriac in 1612. It states that every natural number is the sum of four squares.

non-existence of a decision procedure for exponential Diophantine equations.

Jones and Matiyasevich's proof is so striking in clarity and elegance that it virtually demands a closer examination in this book. In order to fully grasp its details, however, we need to do some preliminary work. First, we need to clarify when we consider a Diophantine equation solvable. Contrary to Hilbert's demand, we will not seek the solutions in the integers but in the natural numbers. At first glance, we are about to make a huge mistake here. For example, the equation

$$(x+1)^3 + (y+1)^3 = (z+1)^3 \tag{5.20}$$

has no solutions in the natural numbers but infinitely many in the integers. The fact that we can nonetheless change the domain without hesitation is due to our interest not in the solvability of specific equations but in the existence of decision procedures. In this case, we can use any method that seeks solutions in the integers to decide the solvability in the natural numbers, and vice versa. Figure 5.39 summarizes the respective reductions. One of the four is not immediately apparent as its legitimation requires Lagrange's famous *four-square theorem*, already encountered in Section 3.1.2. It stipulates that every natural number is representable as the sum of four squares (Figure 5.40).

Concerning the example equation (5.20), the reduction means the following: The question of being solvable in the integers is equivalent to the question of whether the equation

$$(p_1-q_1+1)^3 + (p_2-q_2+1)^3 = (p_3-q_3+1)^3$$

has a solution in the natural numbers. Similarly, (5.20) is solvable in the natural numbers if and only if equation

$$(w_1^2+x_1^2+y_1^2+z_1^2+1)^3 + (w_2^2+x_2^2+y_2^2+z_2^2+1)^3 = \\ (w_3^2+x_3^2+y_3^2+z_3^2+1)^3$$

has a solution in the integers. Therefore, the existence of a decision procedure is independent of the chosen domain. If a procedure correctly decides for any given Diophantine equation whether the equation is solvable in the natural numbers, a decision procedure for the solvability in the integers would likewise exist and vice versa.

5.4.3.1 Diophantine Representability

This section explains how to represent relations *diophantically*. The idea we are pursuing here is similar to the one in Section 4.2.1, where we showed how to represent relations *arithmetically*.

 Definition 5.10 (Diophantine Representable Relations)

Let $R \subseteq \mathbb{N}^n$. The (exponential) Diophantine equation

$$p(a_1, \ldots, a_n, x_1, \ldots, x_m) = 0$$

represents R if it satisfies the following condition:

$(a_1, \ldots, a_n) \in R \Leftrightarrow p(a_1, \ldots, a_n, x_1, \ldots, x_m) = 0$ has a solution

In colloquial terms, the definition states that a relation R is represented diophantically by the equation $p = 0$ if we can find natural numbers $x_1, \ldots, x_m$ for every combination $(a_1, \ldots, a_n) \in R$, with

$$p(a_1, \ldots, a_n, x_1, \ldots, x_m) = 0.$$

Conversely, if $(a_1, \ldots, a_n) \notin R$ applies, the equation must not be solvable for any combination of $x_1, \ldots, x_m$. In mathematical notation, we can express the situation as such:

$$R(a_1, \ldots, a_n) \Leftrightarrow \exists x_1 \ldots \exists x_m \, p(a_1, \ldots, a_n, x_1, \ldots, x_m) = 0$$

At this point, it is evident how to read equation (5.19). The form used by Davis is identical to the one presented here except for the additional universal quantifier.

Figure 5.41 lists several diophantically representable relations. Although most are relatively basic, the last example illustrates that more complex relations have a corresponding representation, too. The depicted formula is an equation with 26 unknowns, solvable in the positive natural numbers if and only if $k+2$ is prime [111, 170, 219].

Let us spend a little extra time on the prime number example. You may have noticed that the equation comprises 14 sub-expressions connected conjunctively. Therefore, it is no longer a genuine Diophantine equation in the strict sense, as it contains logical connectives besides the arithmetic operations. Fortunately, it is easy to eliminate the operators '∧' and '∨', so we can use them within Diophantine equations without

A selection of Diophantine representable relations	
▪ "x is an even natural number" $x - 2 \cdot y = 0$	▪ "$k+2$ is a prime number" $\begin{aligned} wz + h + j - q &= 0 \;\wedge \\ (gk + 2g + k + 1)(h + j) + h - z &= 0 \;\wedge \\ 16(k+1)^3(k+2)(n+1)^2 + 1 - f^2 &= 0 \;\wedge \\ 2n + p + q + z - e &= 0 \;\wedge \\ e^3(e+2)(a+1)^2 + 1 - o^2 &= 0 \;\wedge \\ (a^2 - 1)y^2 + 1 - x^2 &= 0 \;\wedge \\ 16r^2 y^4 (a^2 - 1) + 1 - u^2 &= 0 \;\wedge \\ n + l + v - y &= 0 \;\wedge \\ (a^2 - 1)l^2 + 1 - m^2 &= 0 \;\wedge \\ ai + k + 1 - l - i &= 0 \;\wedge \\ ((a + u^2(u^2 - a))^2 - 1)(n + 4dy)^2 + 1 - (x + cu)^2 &= 0 \;\wedge \\ p + l(a - n - 1) + b(2an + 2a - n^2 - 2n - 2) - m &= 0 \;\wedge \\ q + y(a - p - 1) + s(2ap + 2a - p^2 - 2p - 2) - x &= 0 \;\wedge \\ z + pl(a - p) + t(2ap - p^2 - 1) - pm &= 0 \end{aligned}$
▪ "x is a square number" $x - y^2 = 0$	
▪ "x divides y" $x \cdot z - y = 0$	
▪ "x is greater or equal than y" $y + z - x = 0$	
▪ "x is greater than y" $y + z + 1 - x = 0$	

Figure 5.41: Selection of Diophantine representable relations. The set of all prime numbers also has a Diophantine representation [219]. The number $k+2$ is prime if and only if the equation on the right is solvable in the positive natural numbers.

hesitation. In particular, the following relationships apply:

$$p = 0 \vee q = 0 \quad \Leftrightarrow \quad p \cdot q = 0 \qquad (5.21)$$

$$p = 0 \wedge q = 0 \quad \Leftrightarrow \quad p^2 + q^2 = 0 \qquad (5.22)$$

Next, we define the *masking relation* '$\preccurlyeq$', which plays a central role in the proof of Jones and Matiyasevich:

 Definition 5.11 (Masking Relation '$\preccurlyeq$')

> Let r and s be natural numbers with the binary representations
>
> $$r = \sum_{i=0}^{n} r_i 2^i \quad (0 \leq r_i \leq 1), \qquad s = \sum_{i=0}^{n} s_i 2^i \quad (0 \leq s_i \leq 1).$$
>
> The *masking relation* '$\preccurlyeq$' is defined as follows:
>
> $$r \preccurlyeq s \;:\Leftrightarrow\; r_i \leq s_i \text{ for all } i \text{ with } 0 \leq i \leq n$$

In plain words, $r \preccurlyeq s$ expresses that the i-th binary digit of r never exceeds the i-th binary digit of s. For the moment, let us assume the

5.4 Impact on Mathematics

masking relation has a Diophantine representation. Later on, we will convince ourselves that this is indeed the case.

We will frequently utilize the masking relation in connection with a particular constant, which Jones and Matiyasevich denoted by I. This constant signifies the following number:

$$I := \sum_{t=0}^{s} Q^t = \underbrace{111\ldots111}_{s+1 \text{ digits}}{}_Q \tag{5.23}$$

Q is the base of I and will be chosen appropriately below.

At first glance, the meaning of I is hard to recognize because its exact value plays no role in our analysis. Its relevance is due to its unique digit distribution: represented in base Q, the constant I corresponds to a sequence of $s+1$ ones.

5.4.3.2 Encoding of Register Machines

At this point, we are ready to encode the calculation sequence of a register machine with an exponential Diophantine equation. The following variables capture the machine's behavior:

$$r_{j,t} = \text{The content of register } R_j \text{ at time } t$$

$$l_{j,t} = \begin{cases} 1 & \text{At time } t, \text{ instruction } L_j \text{ executes} \\ 0 & \text{Another instruction executes} \end{cases}$$

The computation of a register machine can be tracked with a two-dimensional matrix, as shown in Figure 5.42. The time axis extends from right to left. Vertically, we have two sub-matrices. The *data-flow matrix* (top) has a separate row for each register and describes how their contents change over time. The *control-flow matrix* (bottom) has a separate row for each instruction, and the values 1 and 0 indicate whether the instruction is currently executing or not.

Figure 5.43 (above) features the data and control flow matrix for the register machine program from Figure 5.16. The machine starts with the input $R_1 = 2$ and terminates after 23 steps. We already know the resulting calculation sequence; it is the same one we used in Section 5.1.2 to explain the behavior of register machines. Only the type of presentation is different. In Figure 5.17, we analyzed the behavior using an ordinary execution log.

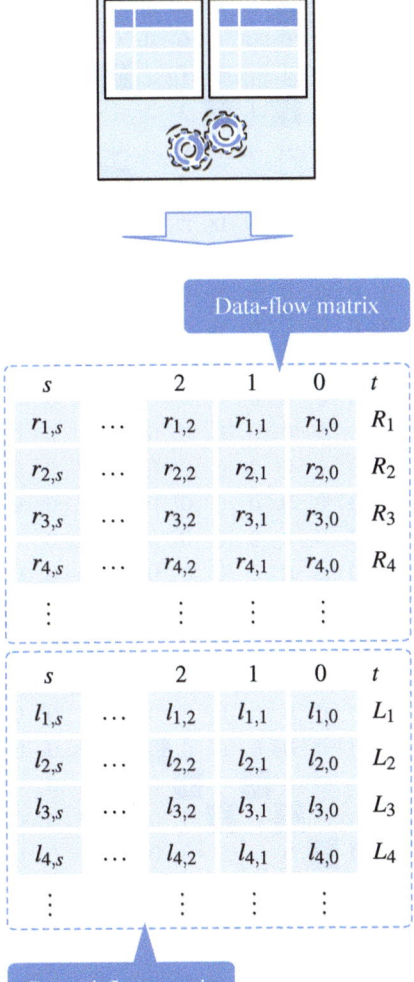

Figure 5.42: Tabular representation of a register machine's computation sequence. The horizontal axis includes a separate column for each computation step, while the vertical axis provides a distinct row for each register and instruction. The data flow matrix captures changes in register contents, while the control flow matrix tracks the program execution. A value of 1 in the control flow matrix indicates that a specific instruction executes.

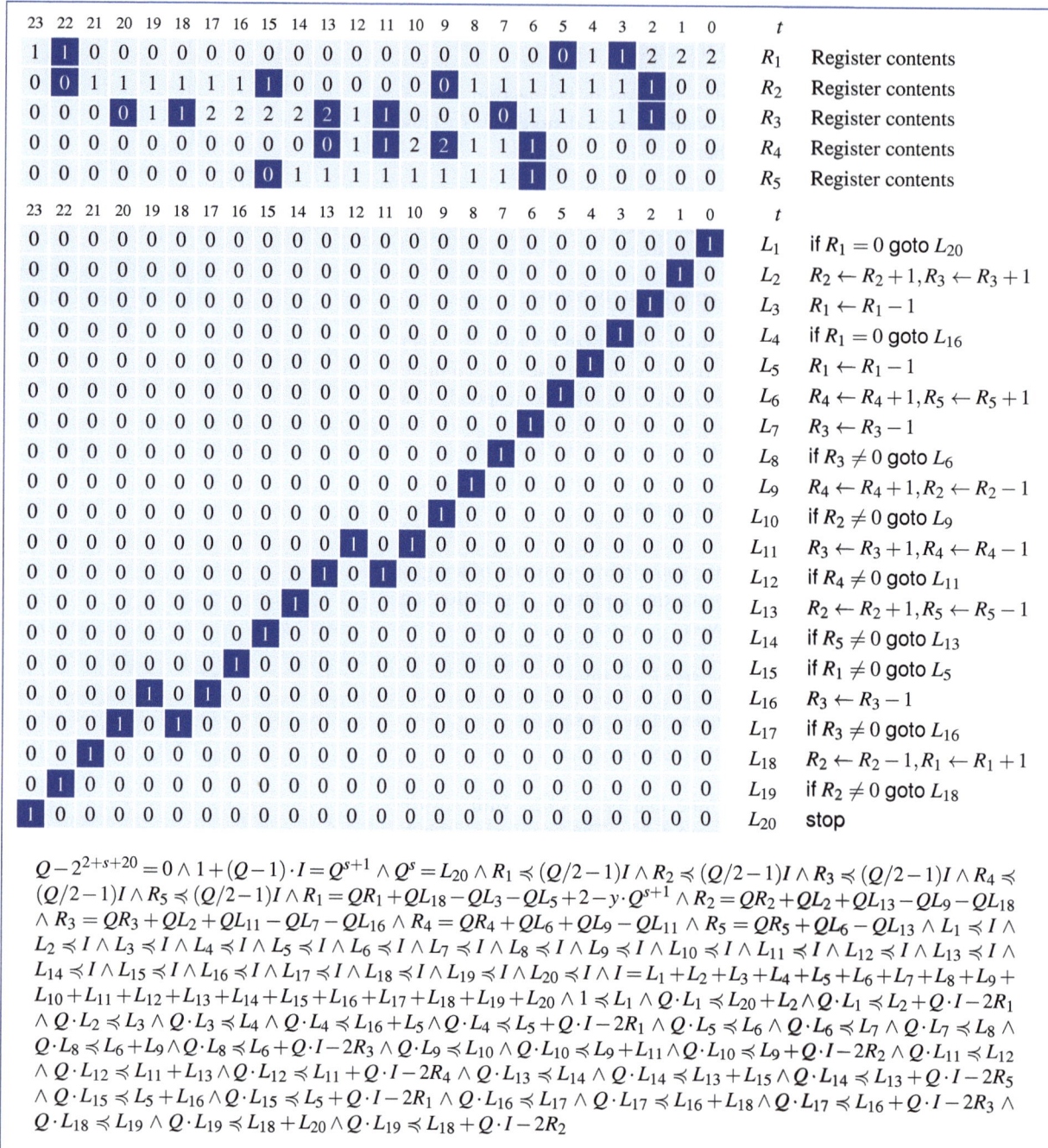

Figure 5.43: Encoded calculation sequence of our example machine for the input $R_1 = 2$

5.4 Impact on Mathematics

To create a Diophantine encoding of a register machine's computation sequence, we start by representing each row of the data flow matrix and the control flow matrix by a number with base Q. We pursue the idea of coding the entry in the t-th column into the t-th digit:

$$R_j = \sum_{t=0}^{s} r_{j,t} Q^t, \quad 0 \leq r_{j,t} < \frac{Q}{2} \qquad (5.24)$$

$$L_j = \sum_{t=0}^{s} l_{j,t} Q^t, \quad 0 \leq l_{j,t} \leq 1 \qquad (5.25)$$

In these equations, s is the number of computation steps performed by a machine with the registers $R_1, \ldots, R_r$, and the code lines $L_1, \ldots, L_l$. For the coding to work, Q must be a power of two and chosen large enough to hold all possible values in a single digit. In particular, we must ensure that no overflows occur between adjacent digits. To be on the safe side, we choose the value of Q as follows for a register machine with r registers, l code lines, and input x [109]:

$$Q = 2^{x+s+l} \qquad (5.26)$$

The fact that we require the digits $r_{j,t}$ to exhaust only the smaller range 0 to $\frac{Q}{2} - 1$ rather than the entire range 0 to $Q - 1$, will become relevant below. It will ensure that the modeling of the conditional branching commands works as expected.

Let us now proceed to set up the Diophantine equation we seek. As shown in Figure 5.44, the equation contains the variables

$$Q, I, s, x, y, R_1, \ldots, R_r, L_1, \ldots, L_l$$

and is solvable for a given value x precisely if the encoded register machine terminates on input x. Even more applies, though: Every solution allows us to reconstruct the entire calculation sequence. In this case, the following holds:

- I and Q equal the values given in (5.23) and (5.26),
- the register machine terminates after s steps,
- output register R_1 contains y at the end of the computation,
- R_j encodes the data flow as stated in (5.24),
- L_j encodes the control flow as stated in (5.25).

We will now formulate a series of Diophantine equations to enforce the above properties:

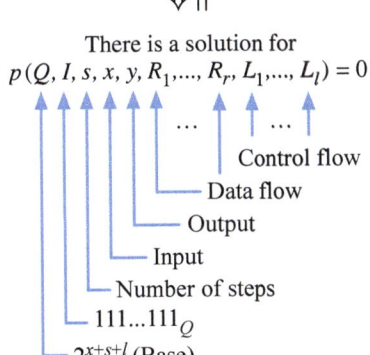

Figure 5.44: Encoding of register machines using exponential Diophantine equations.

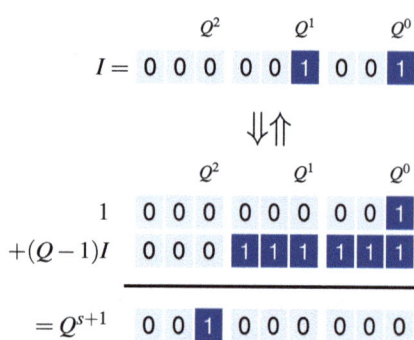

Figure 5.45: Visualization of equation (5.27) for $Q = 8$ and $s = 1$. The individual digits are shown in binary. The numerical values range from 0 to 7, thus being 3 bits wide.

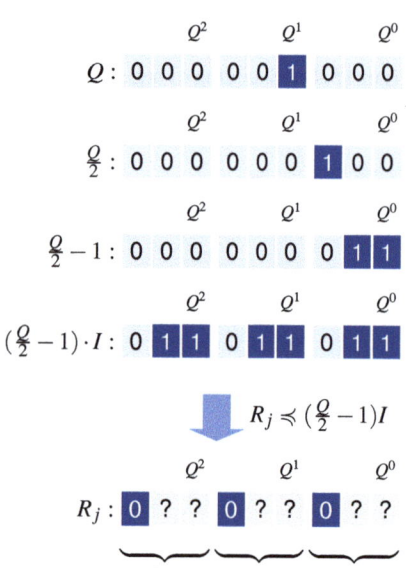

Figure 5.46: $R_j \preccurlyeq (\frac{Q}{2} - 1)I$ ensures that the most significant bit in each digit $r_{j,t}$ is 0.

- The following constraint ensures that Q and I take on their intended values:
$$Q - 2^{x+s+l} = 0 \;\land\; 1 + (Q-1) \cdot I = Q^{s+1} \quad (5.27)$$
Figure 5.45 shows an example.

- To describe the register machine's termination, we require that the machine executes the stop instruction after s computation steps. For this purpose, we use an equation of the form
$$Q^s = \sum_k L_k,$$
with the sum iterating over all lines of the form L_k : stop.

- To adequately describe the data flow, we first require the variables R_j to have the form agreed in (5.24). As Figure 5.46 demonstrates, the masking relation becomes handy here:
$$R_j \preccurlyeq (\frac{Q}{2} - 1)I$$

Furthermore, we add a separate *register equation* for each R_j, describing how the content changes over time. These equations are as follows:
$$R_1 = QR_1 + \sum_k QL_k - \sum_i QL_i + x - y \cdot Q^{s+1} \quad (5.28)$$
$$R_j = QR_j + \sum_k QL_k - \sum_i QL_i \quad (2 \le j \le r) \quad (5.29)$$

The first sum iterates over all rows of the form
$$L_k : R_j \leftarrow R_j + 1,$$
whereas the second iterates over all rows of the form
$$L_i : R_j \leftarrow R_j - 1.$$
The former increases, and the latter decreases the j-th register. The equations also encode the start and end configuration. They are designed such that the rightmost digit equals x in R_1 and 0 in $R_2, \ldots, R_r$. Furthermore, the digit at position $s+1$ must be equal to y in R_1 and equal to 0 in $R_2, \ldots, R_r$.

- To correctly describe the program flow, we require the elements of the control flow matrix to exclusively take on the values 0 and 1, and the program execution to start with instruction L_1:
$$L_1 \preccurlyeq I \;\land\; \ldots \;\land\; L_l \preccurlyeq I \;\land\; 1 \preccurlyeq L_1$$

5.4 Impact on Mathematics

Furthermore, we must ensure that exactly one instruction executes at any given time, which is the case if each column of the control flow matrix contains a single 1:

$$I = \sum_{i=1}^{l} L_i$$

In addition, we need to formulate several equations defining the transition from one configuration to the next. For non-branching instructions, they flow from the pen without further ado (Figure 5.47):

$$L_i : R_j \leftarrow R_j + 1 : \quad Q \cdot L_i \preccurlyeq L_{i+1}$$
$$L_i : R_j \leftarrow R_j - 1 : \quad Q \cdot L_i \preccurlyeq L_{i+1}$$
$$L_i : \text{goto } L_n : \quad Q \cdot L_i \preccurlyeq L_n$$

Encoding the branch instructions

$$L_i : \text{if } R_j = 0 \text{ goto } L_n$$
$$L_i : \text{if } R_j \neq 0 \text{ goto } L_n$$

is more demanding. It is comparatively easy to formulate that the program execution continues either in line n or in line $i+1$:

$$Q \cdot L_i \preccurlyeq L_n + L_{i+1}$$

However, we must also ensure that the proper successor instruction executes. To understand how this is possible, let us look at the expression

$$Q \cdot I - 2 \cdot R_j.$$

Figure 5.48 shows the resulting bit pattern when carrying out the subtraction in binary. It is crucial to note that we required the digits $r_{j,t}$ in equation (5.24) to meet the constraint

$$r_{j,t} < \frac{Q}{2}.$$

Consequently, we can multiply R_j by 2 without generating any overflows between adjacent digits. As a result, all digits of $2 \cdot R_j$ are even, or to put it another way: In all digits of $2 \cdot R_j$, the rightmost bit equals 0. This, in turn, means that the subtraction

$$Q \cdot I - 2 \cdot R_j$$

generates a carry bit from one binary packet to the next if register R_j contains a value $\neq 0$. Thus, we can tell from the bit pattern of

■ $L_i : R_j \leftarrow R_j + 1, L_i : R_j \leftarrow R_j - 1$

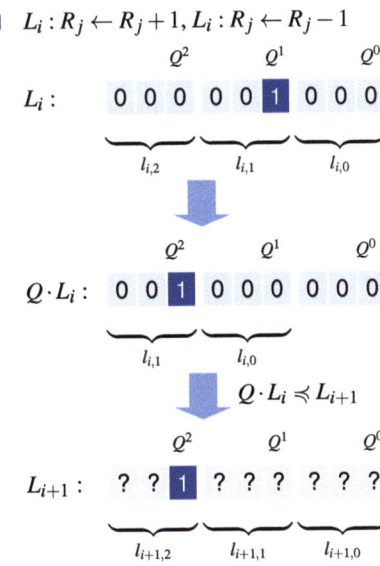

■ $L_i : \text{goto } L_n$

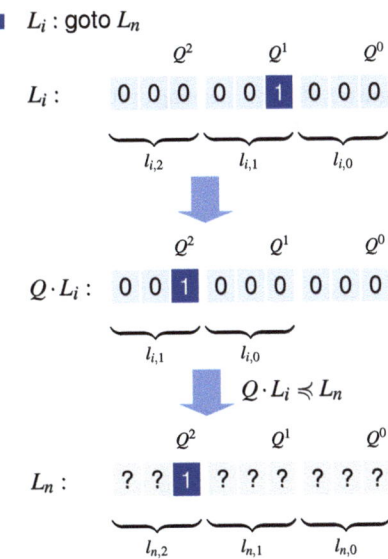

Figure 5.47: Modeling the control flow with the masking relation '$\preccurlyeq$'

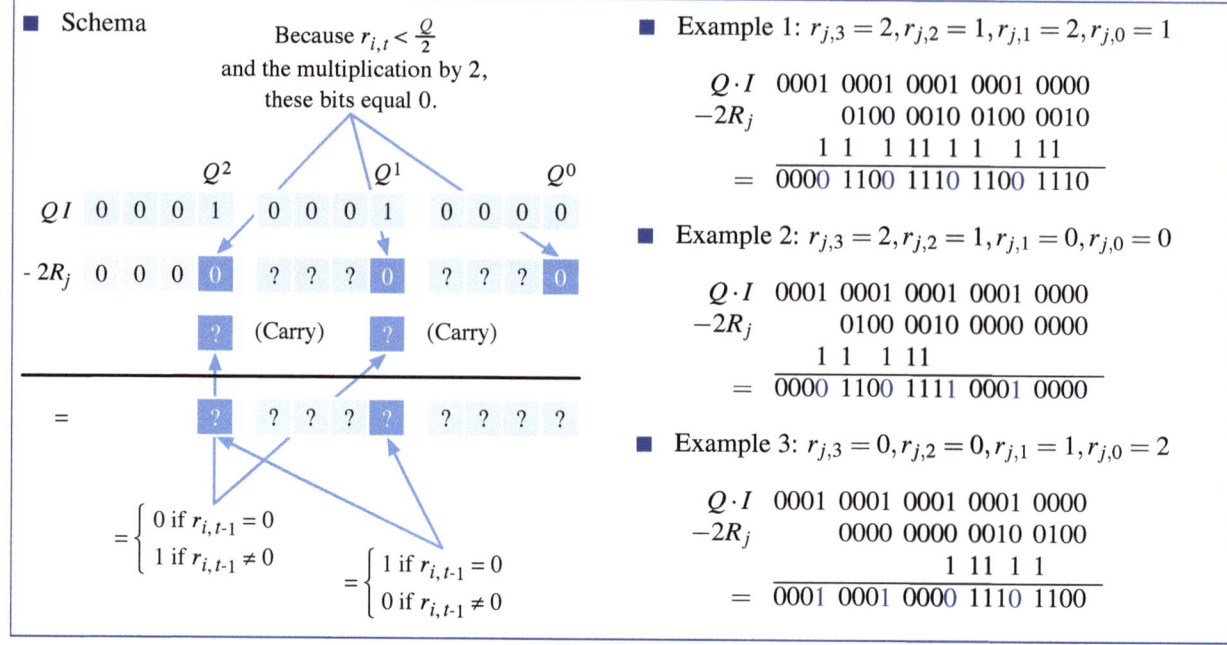

Figure 5.48: The meaning of the formula $Q \cdot I - 2 \cdot R_j$. The rightmost bit of the t-th digit indicates whether the register content R_j is zero or non-zero at time $t-1$.

$Q \cdot I - 2 \cdot R_j$ whether the content of register R_j equals 0 at time t. All we have to do is examine the digit with the significance Q^{t+1}. If the rightmost bit of this digit equals 1, then $r_{j,t} = 0$, otherwise $r_{j,t} \neq 0$.

Summed up, we can characterize the behavior of the conditional branch instructions as follows:

$$L_i : \text{if } R_j = 0 \text{ goto } L_n : \quad Q \cdot L_i \preccurlyeq L_n + L_{i+1} \wedge$$
$$Q \cdot L_i \preccurlyeq L_{i+1} + Q \cdot I - 2R_j$$

$$L_i : \text{if } R_j \neq 0 \text{ goto } L_n : \quad Q \cdot L_i \preccurlyeq L_n + L_{i+1} \wedge$$
$$Q \cdot L_i \preccurlyeq L_n + Q \cdot I - 2R_j$$

Putting all pieces together, we have almost hit the finish line. We obtain a Diophantine equation solvable in the natural numbers precisely when the encoded register machine terminates for the given input. The lower part of Figure 5.43 shows what the formula looks like for our specific example machine.

5.4 Impact on Mathematics

There is just one piece of the puzzle we need to add. So far, we have assumed that the masking relation '$\preccurlyeq$' has a Diophantine representation, but we have not provided any justification yet. The time has come to make up for this shortcoming.

First, let us clarify the specific values of r and s that fulfill the relationship $r \preccurlyeq s$. We distinguish two cases:

- **Case 1:** $r > s$

  In at least one bit position, r equals 1 and s equals 0. As such, $r \not\preccurlyeq s$ always applies.

- **Case 2:** $r \leq s$

  For some combinations $r \preccurlyeq s$ applies, but not for others. Figure 5.50 shows the actual distribution.

The structure in Figure 5.50 is the famous *Sierpiński triangle*, named after the Polish mathematician Wacław Sierpiński (Figure 5.49). The triangle is a classic example in fractal geometry to visualize the concept of self-similarity. Roughly speaking, an object is self-similar if it exhibits the same structure at different scales. The Sierpiński triangle beautifully illustrates this concept – removing one of its three sub-triangles results in another Sierpiński triangle with an identical structure.

The Sierpiński triangle is closely related to *Pascal's triangle*, shown in the upper half of Figure 5.51. To quickly calculate the entries of Pascal's triangle, we first fill the boundary cells with 1. The value of an inner cell then equates to the sum of the two values above.

In fact, Pascal's triangle has a very practical meaning. The cell in row s and column r contains the value of the *binomial coefficient*

$$\binom{s}{r}.$$

This property follows immediately from the above construction rule and the well-known equation

$$\binom{s+1}{r+1} = \binom{s}{r} + \binom{s}{r+1}.$$

For our purposes, Pascal's triangle becomes valuable when considering its entries modulo 2. Every even number then becomes 0, and every odd number becomes 1. As the bottom half of Figure 5.51 reveals,

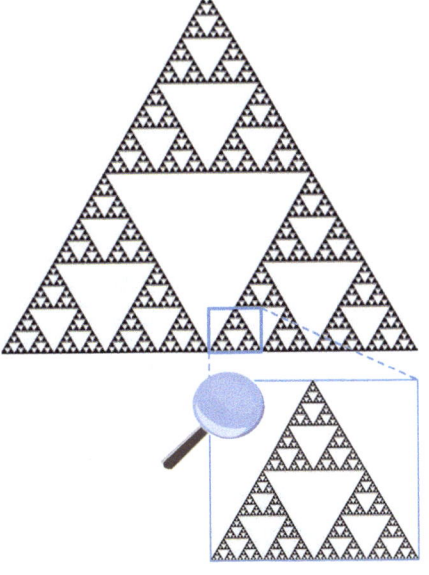

Wacław Sierpiński (1882 – 1969)

Figure 5.49: In In 1915, the Polish mathematician Wacław Sierpiński discovered the fractal that we now refer to as the Sierpiński triangle [186]. It is frequently used in fractal geometry to illustrate the concept of self-similarity.

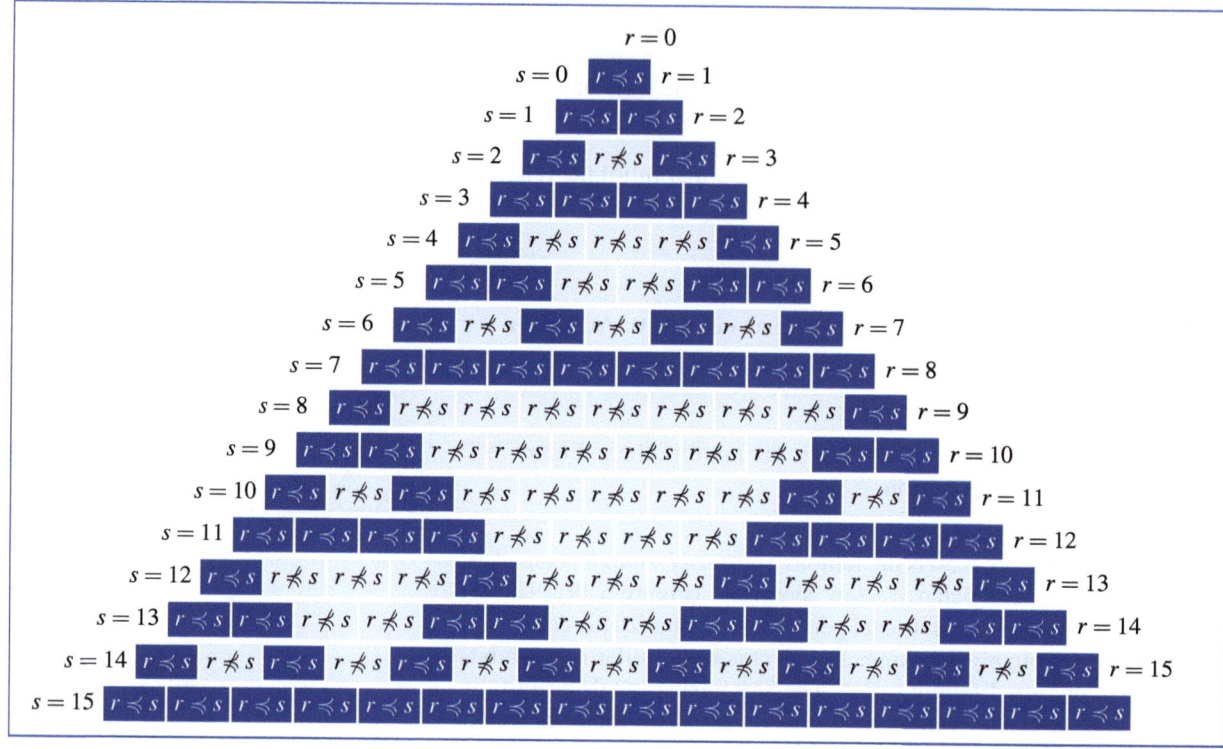

Figure 5.50: Arranging the combinations of r and s with $r \preccurlyeq s$ in two dimensions produces the fractal structure of the Sierpiński triangle. The values of s are plotted line by line. The values of r reside on the diagonals running to the bottom left.

we obtain the exact structure we are looking for, the Sierpiński triangle from Figure 5.50. We have thus managed to establish a connection between the property $r \preccurlyeq s$ and the elements of Pascal's triangle: $r \preccurlyeq s$ holds precisely when Pascal's triangle contains an odd number in row s and column r:

$$r \preccurlyeq s \iff \binom{s}{r} \text{ is an odd number} \qquad (5.30)$$

Now, the following applies according to the binomial theorem:

$$(u+1)^s = \sum_{r=0}^{s} \binom{s}{r} u^r$$

In other words, we can consider each row of Pascal's triangle as the digit sequence of the number $(u+1)^s$ if the base u is sufficiently large (Figure 5.52).

5.4 Impact on Mathematics

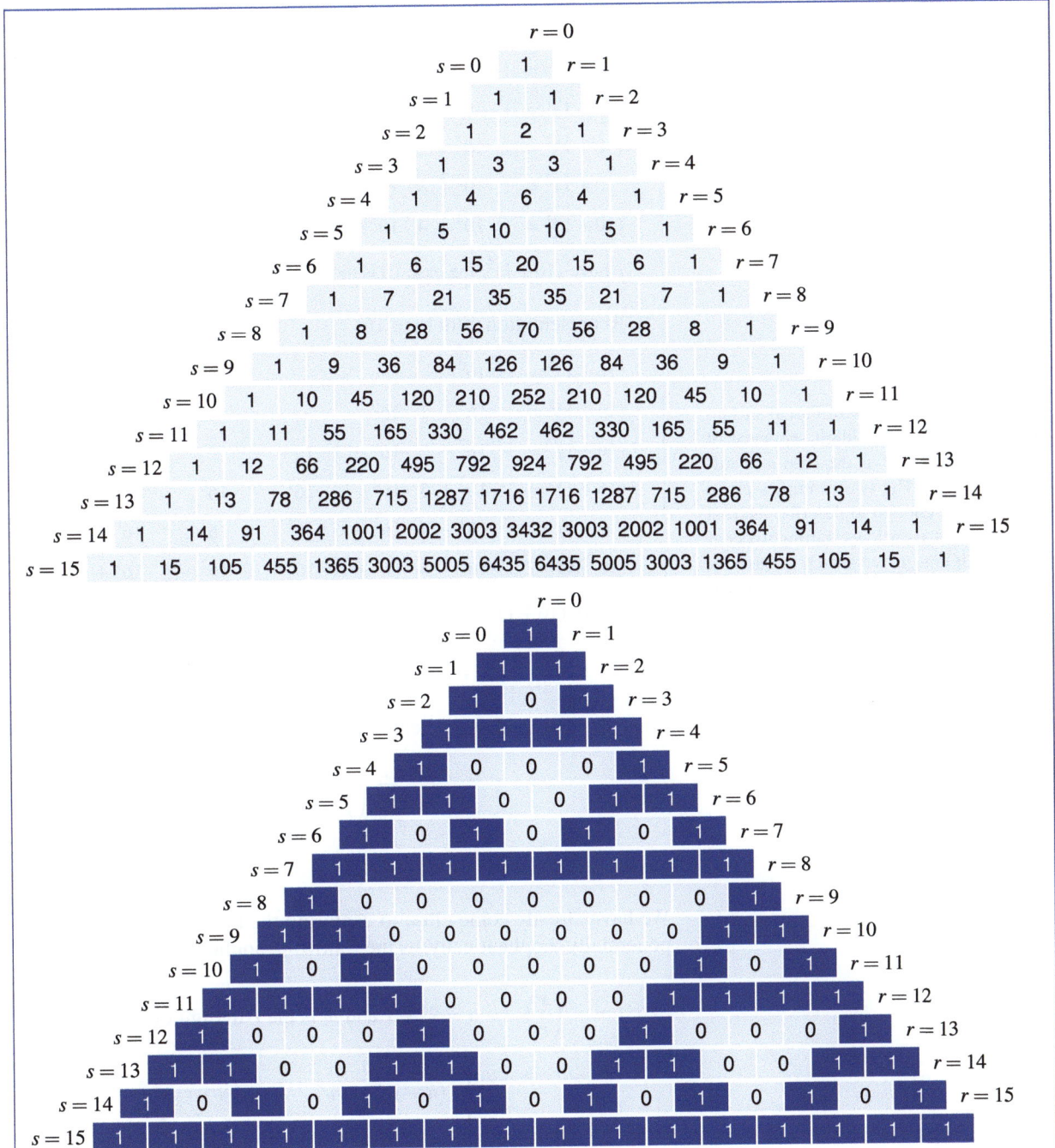

Figure 5.51: Reducing all coefficients modulo 2 transforms Pascal's triangle into the Sierpiński triangle.

- **Representation with base 10**

$$(10+1)^0 = 1 \cdot 10^0$$
$$(10+1)^1 = 1 \cdot 10^1 + 1 \cdot 10^0$$
$$(10+1)^2 = 1 \cdot 10^2 + 2 \cdot 10^1 + 1 \cdot 10^0$$
$$(10+1)^3 = 1 \cdot 10^3 + 3 \cdot 10^2 + 3 \cdot 10^1 + 1 \cdot 10^0$$
$$(10+1)^4 = 1 \cdot 10^4 + 4 \cdot 10^3 + 6 \cdot 10^2 + 4 \cdot 10^1 + 1 \cdot 10^0$$

Above $(10+1)^5$, the base 10 is too small. **Overflows occur.**

- **Representation with base 16**

$$(16+1)^0 = 1 \cdot 16^0$$
$$(16+1)^1 = 1 \cdot 16^1 + 1 \cdot 16^0$$
$$(16+1)^2 = 1 \cdot 16^2 + 2 \cdot 16^1 + 1 \cdot 16^0$$
$$(16+1)^3 = 1 \cdot 16^3 + 3 \cdot 16^2 + 3 \cdot 16^1 + 1 \cdot 16^0$$
$$(16+1)^4 = 1 \cdot 16^4 + 4 \cdot 16^3 + 6 \cdot 16^2 + 4 \cdot 16^1 + 1 \cdot 16^0$$
$$(16+1)^5 = 1 \cdot 16^5 + 5 \cdot 16^4 + 10 \cdot 16^3 + 10 \cdot 16^2 + 5 \cdot 16^1 + 1 \cdot 16^0$$

Above $(16+1)^6$, the base 16 is too small. **Overflows occur.**

Figure 5.52: Every row of Pascal's triangle corresponds to the digit sequence of the number $(u+1)^s$ if the base u is sufficiently large. The two examples prove base 10 large enough to describe the first five lines, with overflows starting to occur in the sixth line. If we switch to base 16, the sixth line will also display correctly. However, we must raise the base again to include the seventh line.

Since the binomial coefficients fulfill the well-known relationship

$$\sum_{r=0}^{s} \binom{s}{r} = 2^s$$

we are safe by choosing $u = 2^s + 1$.

We have already come close to capturing the binomial coefficients exponentially diophantically, as the following applies:

$$m = \binom{s}{r} \Leftrightarrow u = 2^s + 1 \text{ and } m \text{ is the } r\text{-th digit of } (u+1)^s$$

$$\Leftrightarrow \begin{cases} \exists u \, \exists w \, \exists v \, u = 2^s + 1 & \wedge \\ (u+1)^s = vu^{r+1} + mu^r + w & \wedge \\ w < u^r & \wedge \\ m < u & \end{cases} \quad (5.31)$$

Now, we can transcribe the relationship (5.30) with no difficulty. All we need to do is to supplement (5.31) slightly:

$$r \preccurlyeq s \Leftrightarrow \begin{cases} \exists m \, \exists z \, \exists u \, \exists w \, \exists v \, u = 2^s + 1 & \wedge \\ (u+1)^s = vu^{r+1} + mu^r + w & \wedge \\ w < u^r & \wedge \\ m < u & \wedge \\ m = 2z + 1 & \end{cases} \quad (5.32)$$

Voilà: The masking relation '$\preccurlyeq$' has an exponential Diophantine representation, which closes the last gap in our proof; we now know that every computation sequence of a register machine is encodable as an exponential Diophantine equation. From here, the undecidability of the halting problem leads directly to

 Theorem 5.9 (Jones, Matiyasevich, 1984)

> There is no decision procedure for exponential Diophantine equations.

Combined with Matiyasevich's result from 1970, we finally get what we were after in this section: The answer to Hilbert's tenth problem.

 Corallary 5.3

> Hilbert's tenth problem has no solution.

5.5 Exercises

Exercise 5.1

The instruction sets of two Turing machines, M_1 and M_2, are given as follows:

$$I_1 := \{(q_1, S_0, S_1, R, q_2), (q_2, S_0, S_0, R, q_3), (q_3, S_0, S_2, R, q_1)\}$$
$$I_2 := \{(q_1, S_0, S_1, R, q_2), (q_2, S_0, S_0, R, q_3), (q_3, S_0, S_2, R, q_1), (q_4, S_0, S_0, R, q_1)\}$$

a) Create the standard descriptions of M_1 and M_2.

b) Analyze the behavior of the machines. How do both differ?

Exercise 5.2

Figure 5.12 contains the instruction table of Turing's historical *universal machine*. Among others, Turing defined the following submachine:

$$\mathfrak{con}(\mathfrak{C}, \alpha) \begin{cases} \text{Not } A & R, R & \mathfrak{con}(\mathfrak{C}, \alpha) \\ A & L, P\alpha, R & \mathfrak{con}_1(\mathfrak{C}, \alpha) \end{cases}$$

$$\mathfrak{con}_1(\mathfrak{C}, \alpha) \begin{cases} A & R, P\alpha, R & \mathfrak{con}_1(\mathfrak{C}, \alpha) \\ D & R, P\alpha, R & \mathfrak{con}_2(\mathfrak{C}, \alpha) \\ \text{None} & PD, R, P\alpha, R, R, R & \mathfrak{C} \end{cases}$$

$$\mathfrak{con}_2(\mathfrak{C}, \alpha) \begin{cases} C & R, P\alpha, R & \mathfrak{con}_2(\mathfrak{C}, \alpha) \\ \text{Not } C & R, R & \mathfrak{C} \end{cases}$$

When started in the configuration

the machine terminates after 10 computation steps with the read-write head located at the following tape position:

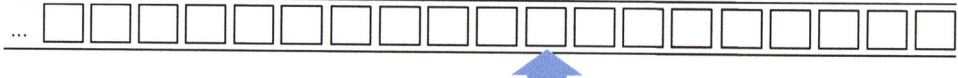

a) Add the tape content in the illustration above.

b) Which subtask might the machine fulfill in Turing's overall design?

5.5 Exercises

Exercise 5.3

A universal Turing machine takes another Turing machine as input in encoded form. This exercise focuses on the encoding proposed in Turing's 1936 paper.

Rate the following statements:

	true	false
Every natural number is the Gödel number of a Turing machine.	○	○
The encoding is injective.	○	○
The encoding is surjective.	○	○
The encoding is bijective.	○	○

Exercise 5.4

A Turing machine or a register machine may operate in two distinct ways: as a transducer or an acceptor.. For a given function $f : \mathbb{N} \to \mathbb{N}$, this means the following: As a transducer, the machine takes x as input and produces y as output. As an acceptor, it receives the tuple (x, y) and accepts the input precisely if x and y satisfy the relationship $f(x) = y$.

This exercise aims at establishing a connection between the two concepts.

a) Can a transducer simulate an acceptor?

b) Can an acceptor simulate a transducer?

Exercise 5.5

Section 5.1.2 has introduced a register machine that James P. Jones and Yuri Matiyasevich invented in 1991 [110]. The machine, which works as a transducer, utilizes register R_1 to receive the input x and store the output $f(x)$. The execution log in Figure 5.17 provided insight into how the machine acts for the input $R_1 = 2$. It stopped after 23 steps, leaving the result 1 in R_1.

a) Complete the list below by simulating the machine for additional inputs:

$f(0) = $ ▯ , $f(1) = $ ▯ , $f(2) = $ 1 , $f(3) = $ ▯ , $f(4) = $ ▯ , $f(5) = $ ▯

b) Which famous number sequence does the machine compute?

Exercise 5.6

All discussed register machines offered three branch instructions:

- L_i : goto L_n
- L_i : if $R_j = 0$ goto L_n
- L_i : if $R_j \neq 0$ goto L_n

a) Do these instructions suffice to implement the following extended branch instruction?

$$L_i : \text{if } R_j = 0 \text{ goto } L_n \text{ else goto } L_m$$

b) Does the expressiveness decrease if we restrict ourselves to the instruction

$$L_i : \text{if } R_j = 0 \text{ goto } L_n ?$$

Exercise 5.7

This exercise is about the cellular automaton defined by the following rules:

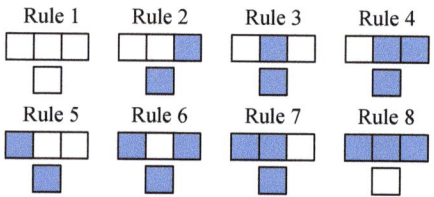
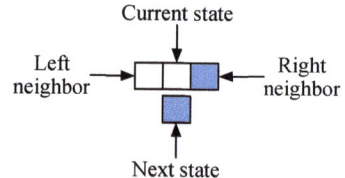

Which familiar structure is generated by this automaton? To answer the question, complement the following diagram.

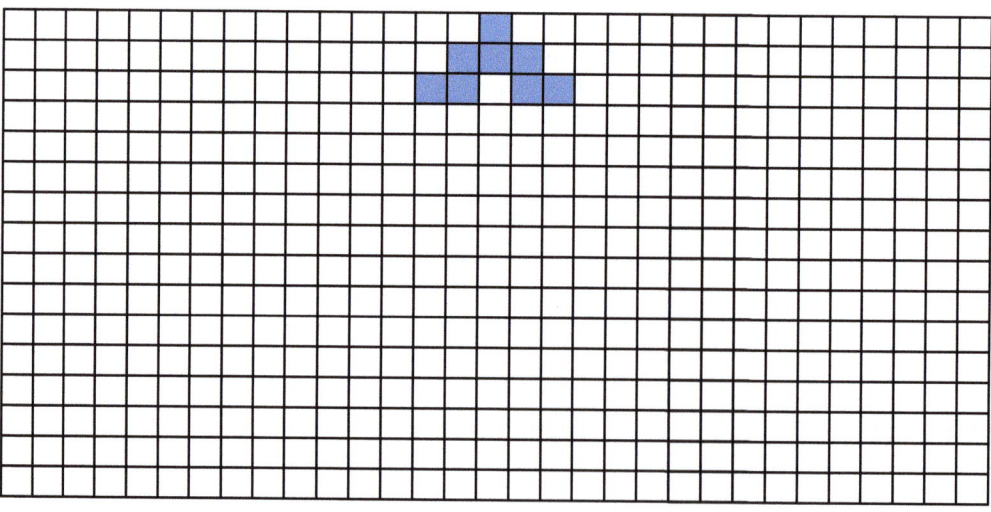

5.5 Exercises

Exercise 5.8

Consider the following Turing machine:

$Q = \{q_1, q_2, q_3, q_4, q_5, q_6\}$
$S = \{0, 1\}$
$I = \{(q_1, 0, 1, R, q_2), (q_2, 0, 1, R, q_3), (q_3, 0, 1, R, q_4), (q_4, 0, 1, L, q_1), (q_5, 0, 1, R, q_6),$
$(q_1, 1, 1, L, q_3), (q_2, 1, 1, R, q_2), (q_3, 1, 0, L, q_5), (q_4, 1, 1, L, q_4), (q_5, 1, 0, L, q_1)\}$

a) Complete the simulation run printed below.

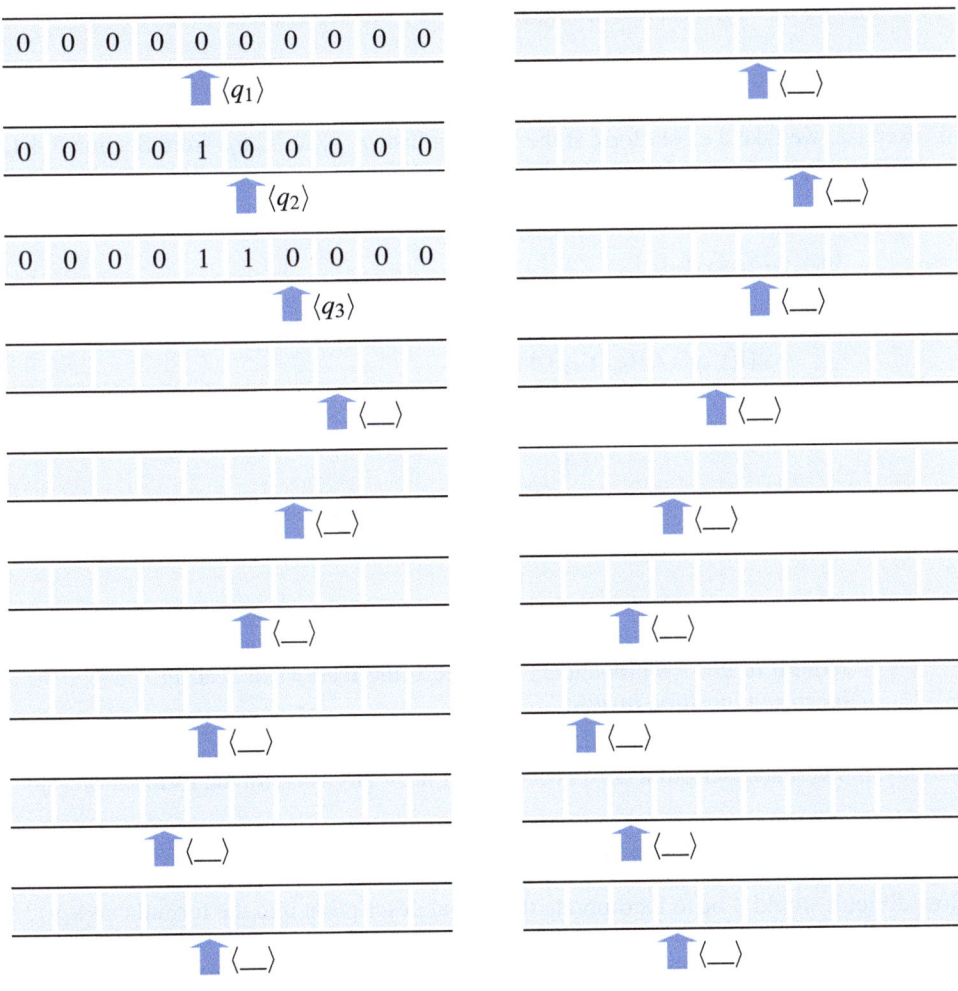

b) Find out which well-known Turing machine we are looking at here. Will the machine terminate?

Exercise 5.9

In Definition 5.7, we agreed that a set N is *enumerable* if and only if a surjective and computable function $f : \mathbb{N} \to N$ exists. What are the consequences of replacing the requirement of surjectivity with the requirement of bijectivity?

Exercise 5.10

Section 5.3.2 presented Rice's theorem. In a sweeping blow, it dashed any hope of algorithmically deciding any non-trivial property of Turing machines. However, is this really so? For example, consider the property of a machine to have precisely five states. For any machine, we can easily decide on this property by briefly inspecting the instruction table. However, this should be impossible according to Rice's theorem, shouldn't it?

Exercise 5.11

In this exercise, we take a closer look at the formula $\mathrm{Inst}(q_i, S_j, S_k, \mathrm{L}, q_l)$ from Section 5.4.1. In his 1936 paper, Turing employed it to describe the leftward movement of a tape-restricted Turing machine:

$$\mathrm{Inst}(q_i, S_j, S_k, \mathrm{L}, q_l) :=$$

$$\forall t\, \forall y\, \forall t'\, \forall y'\, ((R_{S_j}(t,y) \wedge I(t,y) \wedge K_{q_i}(t) \wedge F(t,t') \wedge F(y',y))$$

$$\to (I(t',y') \wedge R_{S_k}(t',y) \wedge K_{q_l}(t')$$

$$\wedge \forall z\, (F(y',z) \vee \bigwedge_{i=0}^{M} (R_{S_i}(t,z) \to R_{S_i}(t',z)))))$$

On page 261, we have defined how such a machine behaves when the read-write head is on the far left:

> "As the read-write head of a tape-restricted Turing machine can never move beyond the end of the tape, the machine ignores any request to cross the boundary and keeps the read-write head in its current position instead."

Apparently, the verbal description does not manifest itself anywhere in the formula $\varphi_I(x,y)$.

a) Is Turing's proof perhaps incomplete?

b) How difficult would it be to incorporate the verbal description into the formula $\varphi_I(x,y)$?

5.5 Exercises

Exercise 5.12

Section 5.4.2 dealt with the arithmetization of Turing machines. In this context, we introduced the PA formula $\varphi_l(x,y)$, describing the transition from one configuration x to another configuration y. For the left movement, it reads as follows:

$$\varphi_l(x,y) := \exists h_1 \exists h_2 \exists n_1 \exists n_2 ($$
$$L(x) = n_1 \wedge K(x) = \bar{i} \wedge I(x) = h_1 \wedge R_{h_1}(x) = \bar{j} \wedge$$
$$L(y) = n_2 \wedge K(y) = \bar{l} \wedge I(y) = h_2 \wedge$$
$$(h_1 \neq 0 \rightarrow ($$
$$n_1 = n_2 \wedge h_1 = h_2 + \bar{1} \wedge R_{h_1}(y) = \bar{k} \wedge$$
$$\forall (h < n_1)(h \neq h_1 \rightarrow \exists s (R_h(x) = s \wedge R_h(y) = s)))) \wedge$$
$$(h_1 = 0 \rightarrow ($$
$$n_1 + \bar{1} = n_2 \wedge R_0(y) = 0 \wedge h_1 = h_2 \wedge R_1(y) = \bar{k} \wedge$$
$$\forall (h < n_1)(h \neq 0 \rightarrow \exists s (R_h(x) = s \wedge R_{h+1}(y) = s)))))$$

We could have simplified the proof by restricting ourselves to tape-restricted Turing machines. How would the formula read in that case?

Exercise 5.13

By considering the example

$$(x+1)^3 + (y+1)^3 = (z+1)^3,$$

we argued in Section 5.4.3 that it makes a difference whether we seek the solutions of a Diophantine equation in the integers or the natural numbers.

a) Demonstrate that the equation has an infinite number of solutions in the integers.

b) Why is the equation unsolvable in the natural numbers?

Exercise 5.14

Section 5.4.3 worked out how to combine Diophantine equations conjunctively or disjunctively.

a) Which relations do the following two equations represent?

$$a+x+1-b = 0 \vee b+y+1-a = 0$$
$$a+x-b = 0 \wedge b+y-a = 0$$

b) Translate the expressions into ordinary Diophantine equations.

Exercise 5.15 Let R and S be two diophantically representable relations.

a) Does the relation $R \cup S$ have a Diophantine representation?

b) Does the relation $R \cap S$ have a Diophantine representation?

Exercise 5.16 Figure 5.41 features a Diophantine equation with 26 unknowns, having a solution in the positive natural numbers if and only if $k+2$ is a prime number.

a) Can this equation be utilized to enumerate all prime numbers?

b) Does the set of all prime twins also have a Diophantine representation?

Exercise 5.17 The register machine program below originates from the repeatedly cited work by Jones and Matiyasevich from 1984 [109]:

L_0 $R_2 \leftarrow R_2 + 1$	$R_2 \leftarrow R_2 - 1$	L_{11} if $R_2 < R_1$ goto L_{10}
L_1 $R_2 \leftarrow R_2 + 1$	L_6 if $0 < R_2$ goto L_5	L_{12} $R_1 \leftarrow R_1 - 1$
L_2 if $R_3 = 0$ goto L_5	L_7 $R_2 \leftarrow R_2 + 1$	$R_2 \leftarrow R_2 - 1$
L_3 $R_3 \leftarrow R_3 - 1$	$R_4 \leftarrow R_4 - 1$	$R_3 \leftarrow R_3 - 1$
L_4 goto L_2	L_8 if $0 < R_4$ goto L_7	L_{13} if $0 < R_1$ goto L_{12}
L_5 $R_3 \leftarrow R_3 + 1$	L_9 if $R_3 < R_1$ goto L_5	L_{14} stop
$R_4 \leftarrow R_4 + 1$	L_{10} if $R_1 < R_3$ goto L_1	

a) With $R_i < R_j$, the machine utilizes an operation unsupported by the original register machine model. Show that the native language elements are sufficiently expressive to simulate the operation.

b) Manually execute the program with input $R_1 = 2$ and create an execution log similar to the one in Figure 5.17. Note that the program starts in line L_0 and not, as before, in line L_1.

c) Set up the data and control flow matrix for this computation sequence. You already know what such a matrix looks like from Figure 5.43.

d) By starting the register machine with different input values, you will notice that it halts in some cases and keeps calculating forever in others. Try to establish a connection between the input value and the termination property.

Exercise 5.18

Section 5.4.3.2 demonstrated how to encode register machines diophantically. Among others, the constructed equation contained the subexpression

$$L_1 \preccurlyeq I \wedge \ldots \wedge L_l \preccurlyeq I \tag{5.33}$$

to ensure that $L_1, \ldots, L_l$ only consists of the digits 0 and 1. In addition, we utilized the subexpression

$$I = \sum_{i=1}^{l} L_i \tag{5.34}$$

to guarantee that there is at most one '1' in each column of the control flow matrix. At first glance, we can do without (5.33), as it seems to follow from (5.34). Prove or disprove this assertion.

6 Algorithmic Information Theory

> *"Any one who considers arithmetical methods of producing random digits is, of course, in a state of sin."*
>
> John von Neumann [144]

In Chapter 5, we have explored the functionality of Turing machines and made two fundamental observations: On the one hand, we can utilize Turing machines to generate symbol sequences. On the other hand, we can treat them like programs and thus regard them as mere symbol sequences, too.

We will now generalize this relationship by relating a sequence of symbols s to the shortest *program* generating s. This way, we will arrive at an exact measure of the *information content* or *information complexity* of a finitely or infinitely long string. The first investigations of this kind were conducted towards the end of the 1960s by Ray Solomonoff, Andrej Kolmogorov, and Gregory Chaitin. Their research has developed into a new theory of information called *algorithmic information theory*.

This chapter aims to outline the fundamental concepts of algorithmic information theory. To this end, we will first clarify in Section 6.1 how to formally capture the *complexity* of a string by linking it to the existence of certain algorithms. Among other things, we will learn to distinguish precisely between random and non-random strings. In Section 6.2, we will revisit Turing's halting problem. The concept of a *halting probability* will lead us straight to *Chaitin's constant*, a truly miraculous number whose discovery is among the shining moments of modern mathematical logic. Finally, in Section 6.3, we will establish an insightful connection between formal systems and programs, uncovering the true essence of algorithmic information theory. This exploration will lead us to *Chaitin's incompleteness theorem*, which again demonstrates the limits of mathematics in razor-sharp clarity.

6.1 Algorithmic Complexity

We open our discussion with two questions:

- How much information does the structure in Figure 6.1 contain?
- What is the difference between the binary sequences in Figure 6.2?

We will address the second question first. In both examples, the sequence of zeros and ones seems to follow no apparent law; both sequences appear as being generated by a series of unbiased coin tosses. However, can we trust our intuition at this point? Are the zeros and ones truly random, or might their arrangements be governed by hidden laws that we fail to recognize with the naked eye?

To get to the bottom of this question, let us try to shorten both number sequences using classic compression software such as Zip. The underlying idea is simple: Compression tools scan for regularities that allow them to reconstruct the given input from a shortened bit sequence. If compression succeeds, this would prove that the presented number sequences are not random. However, if the sequences were genuinely produced by a series of coin tosses, the first n digits would provide no information about the subsequent ones. The binary sequence would thus lack redundancy, forcing the compression tool to output a file approximately the same length as the original input.

Figure 6.3 presents the result, which, at first sight, backs up our assumption that the two binary sequences are random; in both cases, Zip

Figure 6.1: A three-dimensional Mandelbrot set, visualized by Daniel White and Paul Nylander [160, 202]

```
10111001110111101000000110101110000011110111111101000110100100001010000000011110010110101001001001111101001100011000110110010
1010101000000000011000001000110011100110010000011100010000101    1110101111011111101000010101110001001011100011111000111110100
100000111010001110111100010000111101000010010111010101011110000  0101001110111110011101111001110000101111001010010010010100010100111101001110001010011100010111011001100100110111010100100011100001110011100010010111100110101100010111111111101010010111010010011101111
0001001101101011100000010110011110001000110000111001011010100  111100111111011011010001100111110000100010010100010011111011
011011010001010001001001010000011110010010110110111111001010110100  101110010011100000110101000001001011010101001011011101111
100000011110011000110010000100101111100100101011011110  1011110110111001100010101011101101110110100011111001010010
110100000000101100110111011111001011101010110111111101010    011010001010001110001011110110110010100000000100010010101
110111001111111011110100010001111100010110110110110101110001100111  1100011011011101001011110101011110001110110101010000010010
010101101011010100000010000100000011010011100011110101111  011011010110100111110011010000001001001010110100101011011000
000101111001011111100111011010010001000010010011101110100   100001000000001011000110100011111100100110100110011110010
010001100001011111001110101011101110011000010010010010010111010101   01111001111000000001101001010011000111101110101010011011001
010110111011001011100010101011010111011100100101111001011101101010110010010111001110111010100110010010101001001000010010111    01110110001101011110010010010011111010001110110010010
01100111011110100110101011101111101101100011110111011110100   11100011001011110110010110101111100011010010110110011101111
1011011110000111110001110110110001100011110110011101111    0101111000010000001110100010001100110011110101000010101001100100101011101101010001111110001111100000010010111101111111111100111011110010011011110000000110011
00001111111100000100011111111111010111001101101110000000110011   10001110100010011110011101111100111011110001010110101110011
01010011111010100001101...                                        01111110011000001000101...
```

Figure 6.2: Are these binary sequences random? (Cf. [169])

6.1 Algorithmic Complexity

produces an output slightly longer than the input. Nevertheless, how conclusive is our experiment? While the result suggests randomness, it does not rule out the possibility that the seemingly random sequences conceal a regularity the compression program fails to recognize.

Figure 6.4 reveals that the sequence on the left is indeed subject to such regularity; it comprises the first 1000 decimal digits of the number π, calculated modulo 2. The digit sequence is thus anything but random and can be generated systematically with a simple program.

The other sequence does not exhibit such regularity; it originates from the digits of a random number taken from the book *A Million Random Digits with 100,000 Normal Deviates* published in 1955 [40]. On more than 600 densely written pages, this extraordinary publication contains 1,000,000 digits of an experimentally determined number sequence. As the succession of zeros and ones is not subject to any laws, it is impossible to calculate the digits systematically.

In order to output a finite random sequence programmatically, we had to encode all digits in plain text. Such a program would look similar to this one:

PRINT "000000011111001011010100100100111111010011 0..."

Apparently, the length of such a program roughly matches the length of the output. This is not a significant issue for a sequence of, let's say, 1,000 digits, as the program is still likely shorter than a program calculating π. However, the situation changes dramatically with longer sequences. If we wanted to output not just 1,000 digits but 1,000,000, the program for generating the random sequence would grow in length by a factor of 1,000. In contrast, the program for calculating π would only grow slightly[1]. We say the first sequence has a lower *algorithmic complexity*.

 Definition 6.1 (Algorithmic Complexity)

> The *algorithmic complexity* $\kappa(s)$ of a finite binary sequence s is the length of the shortest program that outputs s.

To avoid misunderstandings from the outset, let us elaborate on two aspects of the definition:

---
[1] The program length increases because we must encode the length of the generated output into the program as a constant.

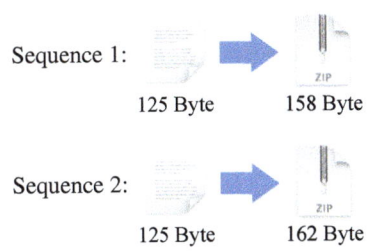

Figure 6.3: None of the two sequences can be reduced in size using the Zip compression tool.

■ Sequence 1

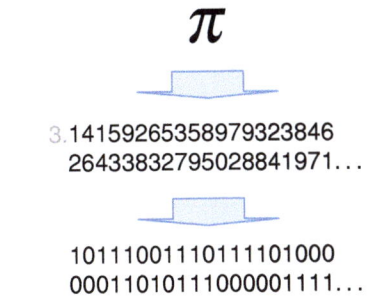

■ Sequence 2

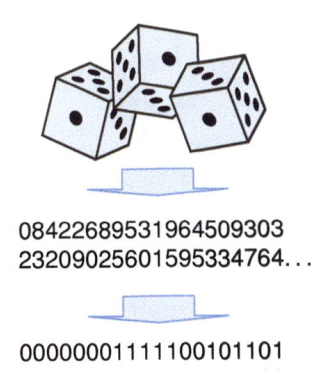

Figure 6.4: Only the second sequence is random. It was generated from the first 1,000 digits of a number determined in a random experiment.

Algorithmic complexity goes by many names. It is commonly referred to as *Kolmogorov complexity*, sometimes also as *Kolmogorov-Chaitin complexity*. Kolmogorov coined the term in 1965 [121] (Figure 6.5). Chaitin took it up in 1969 [32] and brought it to the attention of a broad public, not least through the publication of several popular science books [29–31]. However, the true father of algorithmic complexity is someone else. The idea of relating the complexity of character strings to the concept of algorithms was published as early as 1960 by Ray Solomonoff [199]. An extended version in two parts followed in 1964 [200, 201]. It would, therefore, be adequate to speak of *Solomonoff complexity*, but only a few authors employ this term. Occasionally, algorithmic complexity is called *descriptive complexity* or *algorithmic information*.

- By *program*, we mean a program written in a specific programming language. Which language we use is irrelevant to our considerations. We only require the programs to start with no input, perform a series of calculations, and, if they terminate, output a binary string s. In more formal terms, our program model is given by Turing machines starting on a blank tape. If such a machine terminates, we interpret the written tape content as the string s.

- By a program's *length*, we mean the number of bits of its encoding. If we take the set of all Java programs as a basis, we mean the length of the program file in bits. If we measure algorithmic complexity using Turing machines, the program length corresponds to the number of zeros and ones needed to represent its Gödel number in binary.

Always bear in mind that algorithmic complexity is not an absolute measure as the value $\kappa(s)$ depends not only on the binary sequence s but also on the agreed programming language and the chosen Gödelization. If we change one of these parameters, the algorithmic complexity $\kappa(s)$ changes, too. In the following considerations, it is vital to understand Gödel numbers not as natural numbers but as bit sequences. In particular, $0, 00, 000, \ldots$ are different Gödel numbers for us, although they denote the same natural number.

We are now well-positioned to answer the first of the two questions raised at the beginning of this section. If we equate the *information content* of an object with its algorithmic complexity, the fractal image in Figure 6.1 contains comparatively little information despite its richness in detail. A manageable program can generate the bit sequence of the image in arbitrary resolution. Overall, the discussion so far has demonstrated the following:

- The algorithmic complexity $\kappa(s)$ is a measure we cannot grasp with the naked eye. The binary sequences discussed at the beginning emphasize that similar-looking objects may possess entirely different complexities.

- $\kappa(s)$ sometimes diverges significantly from our intuitive notion of complexity. Highly complex structures, such as the fractal from Figure 6.1, can exhibit comparably low algorithmic complexity.

Andrej Nikolajewitsch Kolmogorov
(1903 – 1987)

Figure 6.5: The Russian mathematician Andrej Kolmogorov is the namesake of *Kolmogorov complexity*, a frequently used synonym for algorithmic complexity.

At this juncture, a natural question arises: Can we devise a procedure capable of systematically calculating the algorithmic complexity for a given binary sequence s? As you might have already guessed, the answer is negative.

6.1 Algorithmic Complexity

Theorem 6.1

> There is no systematic procedure for calculating $\kappa(s)$.

Proof: We prove the theorem by contradiction and assume the existence of a program that correctly calculates $\kappa(s)$ for any binary sequence s. As shown in Figure 6.6, we can embed this program in a larger program P. The inner working of P is straightforward: Variable n is initialized with 0 and then incremented one by one in a loop. In each iteration, P calculates the n-th binary sequence (we call it s) and determines its algorithmic complexity. As soon as $\kappa(s)$ exceeds a predefined constant N, the program outputs s and stops. How does the program P behave for large values of N?

- We observe that P will eventually terminate with an output for any given N. The reason is straightforward: since there are only finitely many binary sequences with algorithmic complexity less than N, and we iterate through an infinite number of sequences, the condition $\kappa(s) > N$ must eventually hold, causing the program to terminate.

- The program length $|P|$ is crucial for our analysis. P essentially consists of a constant number of bits, with the actual choice of N having a minimal impact on its overall length. In order to encode N, $\lceil \log_2 N \rceil + 1$ bits suffice, which leads to the following estimate:

$$|P| \leq c + \lceil \log_2 N \rceil$$

The exact value of the constant c is not relevant in our analysis. In fact, it would be cumbersome to quantify as it depends on the selected programming language and the chosen Gödelization.

- By choosing 2^c for N, we get $|P| \leq c + \lceil \log_2 2^c \rceil = 2 \cdot c$, meaning that N far exceeds the program length. After starting P, the search for a binary sequence s with $\kappa(s) > N$ begins. According to what was said above, P will find such a sequence, output it, and terminate. However, this contradicts the fact that s was output by a program shorter than N. Consequently, $\kappa(s) < N$ and $\kappa(s) > N$ must both apply. We must therefore reject the assumption that $\kappa(s)$ is computable for any binary sequence s, which concludes the proof. □

Remember that Theorem 6.1 does not imply the impossibility of calculating the algorithmic complexity in all cases; a calculation may cer-

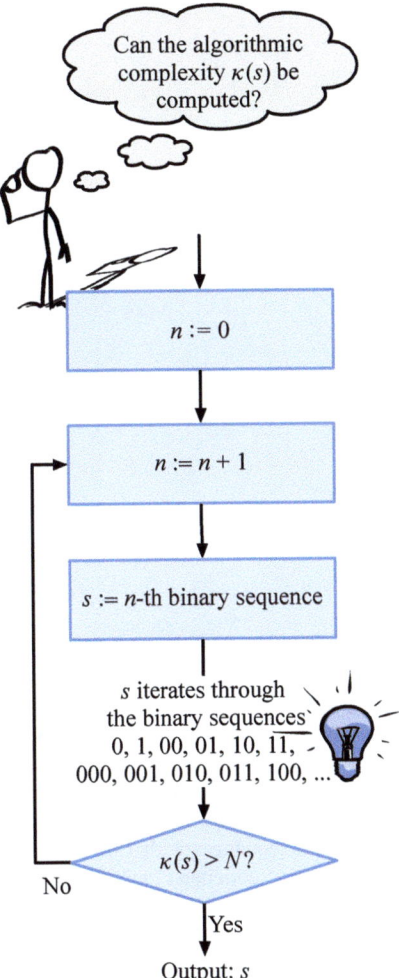

Figure 6.6: For each constant value N, the program will eventually output a binary sequence s with a complexity exceeding N. However, by choosing N large enough, s is output by a program shorter than N. The resulting contradiction leaves no doubt: no systematic method can always correctly calculate the algorithmic complexity for any given binary sequence.

tainly succeed for individual sequences. However, it excludes the existence of a general procedure that always correctly calculates the algorithmic complexity for *every* given binary sequence.

Let us revisit the initial question: What defines a binary sequence as random? The concept of algorithmic complexity serves as a powerful tool in our quest to differentiate between random and non-random sequences:

Definition 6.2 (Finite Random Sequence)

- A finite binary sequence s is *random* if its algorithmic complexity $\kappa(s)$ is approximately equal to $|s|$.

- A finite binary sequence s is *regular* or *compressible* if its algorithmic complexity $\kappa(s)$ is significantly smaller than $|s|$.

The definition reflects the idea of considering a sequence random if it is not reconstructible from a significantly smaller sequence. In other words, random sequences are not compressible to any significant degree. Of course, this definition is somewhat imprecise, as it remains entirely open to what is meant by *"significantly smaller"* or *"compressible to any significant degree."* Luckily, this problem will vanish in thin air when we transition from finite to infinite binary sequences below.

Before, let us tackle the problem of systematically deciding whether a given string is random. For this purpose, we need to decide whether the algorithmic complexity of a binary sequence s is about the length of s, so Theorem 6.1 does not exclude the existence of such a procedure from the outset. However, a similar argument to the one utilized in the proof of Theorem 6.1 dashes this hope, too.

Theorem 6.2

There is no procedure that always correctly decides for a given binary sequence s whether s is random.

A randomly selected binary sequence is rarely regular. To understand the reason for this phenomenon, let us consider all sequences with 1,000,000 binary digits. A sequence s shall be considered random if its algorithmic complexity is at least $|s| - 10$. In our example scenario, there are $2^{1,000,000}$ binary sequences of length $|s|$, but only $2^{999,990}$ of length $|s| - 10$. Hence, less than $2^{999,990}$ programs are available for generating regular binary sequences. The percentage of regular sequences can then be estimated as follows:

$$\frac{2^{999,990}}{2^{1,000,000}} = \frac{1}{2^{10}} < 0,1\%$$

Consequently, among 1,000 binary sequences, at most one sequence is regular on average. The calculation leaves no doubt: Almost all binary sequences are random!

Proof: Suppose there is a program that accepts any binary sequence as input and always correctly decides whether it is random or not. We could then set up a program P that implements the flowchart from Figure 6.7. It initializes the variable n with N and continuously increases the value in a loop. In each iteration, the program checks whether the *n*-th binary sequence is random. As soon as *P* finds such a sequence, it outputs it and halts. Once again, we are interested in how *P* behaves for large *N*.

- First, we note that *P* terminates for every chosen *N*. We have already mentioned the reason in the proof of Theorem 6.1. Since there are infinitely many random sequences, there is, for any *N*, a number *n* with $n > N$ that corresponds to a random binary sequence *s*.

- Once again, the program length $|P|$ is decisive for our analysis. Essentially, *P* is made up of a constant number of bits, and the specific choice of *N* only slightly influences its size. Since $\lceil \log_2 N \rceil + 1$ bits suffice to encode *N*, we can estimate the program length by

$$|P| \leq c + \lceil \log_2 N \rceil.$$

Again, the exact value of the constant *c* plays no role.

- If we choose 2^c for *N*, then $|P| \leq c + \lceil \log_2 2^c \rceil = 2 \cdot c$, which means the constant *N* significantly exceeds the program length $|P|$. Given what we said before, the program will eventually output a random binary sequence *s* with a length greater than *N* and terminate. However, this contradicts the fact that we have just output *s* with a program significantly shorter than *N*.   □

Again, be aware that Theorem 6.2 only excludes the existence of a systematic procedure that correctly decides for *every* given binary sequence, whether it is random or not. For particular sequences, a proof may still succeed. For example, we successfully proved that the first of the two binary sequences presented at the beginning of this chapter was regular.

Let us move on and extend the concept of algorithmic complexity to infinitely long binary sequences. The following definition seems like an obvious choice: An infinitely long binary sequence *s* is *regular* if some program outputs all digits of *s* one after the other in an infinite loop. Instead of generally speaking of a program, we could also claim the existence of a Turing machine that writes all digits of *s* one after the other on a blank tape. Again, what machines we have in mind is apparent: Machines that work just like the *computing machines* from Turing's original work.

Although this definition is appealing initially, we will refrain from using it. Consider the binary sequence in Figure 6.8 to understand the reason. It was created by interleaving the bits of a random sequence with the bits

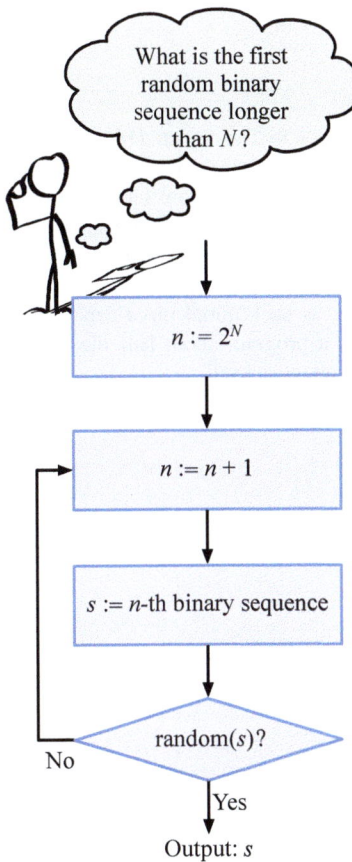

Figure 6.7: The depicted program proves that no procedure can always correctly decide whether or not a binary sequence is random.

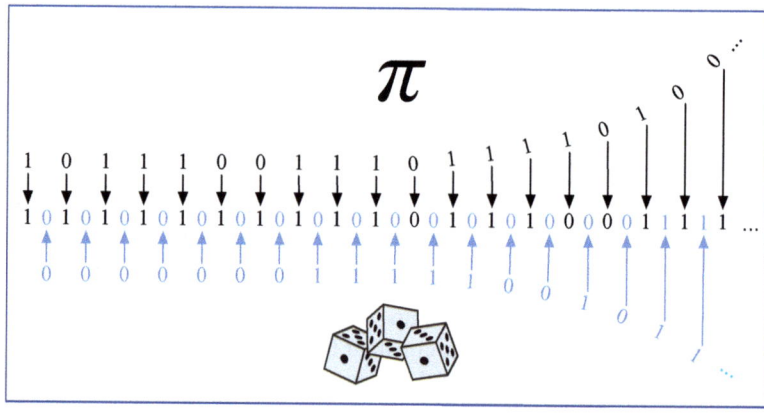

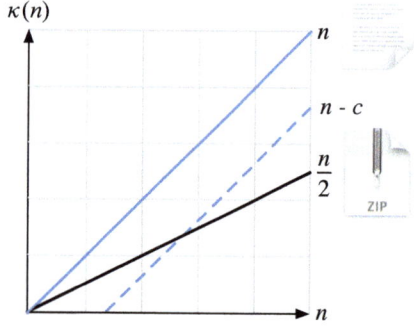

Figure 6.8: The depicted binary sequence results from interleaving a regular bit sequence with a random bit sequence. As it contains an unpredictable subsequence, no program can generate it. Nevertheless, we would not consider the sequence random, as each initial piece can be output with a program about half the length of the sequence itself.

Figure 6.9: Every initial piece of the binary sequence from Figure 6.8 can be output with a program about half the length of the sequence itself. Therefore, the complexity of the initial segment will eventually be less than $n-c$, regardless of the actual choice of the constant c. Consequently, the bit sequence is not random in the sense of Definition 6.3.

of a regular sequence. The bits at the odd positions are generated from the digits of π and are easy to calculate. The bits at the odd positions originate from our random sequence in Figure 6.2. According to the proposed definition, the constructed bit sequence would be random; it contains an unpredictable subsequence, thus being unproducible by any program. Nevertheless, we would hardly call it random; after all, we could output each initial piece with a program significantly shorter than the sequence itself (Figure 6.9). For this reason, we will take a different approach and trace the concept of an infinitely long random sequence back to the finite case.

 Definition 6.3 (Infinite Random Sequence)

An infinitely long binary sequence $s = s_1, s_2, s_3, \ldots$ with $s_i \in \{0, 1\}$ is called *random* if there exists a constant $c \in \mathbb{N}$ such that

$$\kappa(s[1\ldots n]) > (n-c) \quad \text{for all } n$$

$s[1\ldots n] := s_1, \ldots, s_n$ denotes the first n elements of s.

A binary sequence that is not random is called *regular* or *compressible*.

According to this definition, the sequence from Figure 6.8 is no longer random. Since any initial piece of length n can be output with a program that is only about half the size of n, we can choose c as we like: Even for gigantic values for c, the complexity of the initial segment is guaranteed to be smaller than $n-c$ beyond a certain n. Note that the fuzziness attached to Definition 6.2 has vanished completely after transitioning to infinitely long sequences.

6.2 Chaitin's Constant

> *"Omega [...] embodies an enormous amount of wisdom in a very small space [...] inasmuch as its first few thousand digits, which could be written on a small piece of paper, contain the answers to more mathematical questions than could be written down in the entire universe."*
>
> Charles Bennett, Martin Gardner [68]

In the previous section, we laid the foundation for formally capturing the concept of an infinitely long random sequence. In doing so, we realized that a binary sequence can be compressible without being computable. A prominent example is the *halting sequence* $H := h_1, h_2, h_3 \ldots$ given by

$$h_i := \begin{cases} 1 & \text{if the } i\text{-th program terminates} \\ 0 & \text{otherwise} \end{cases} \tag{6.1}$$

Again, we consider programs written in a predefined programming language. Since these programs do not take any input, they either run indefinitely or eventually output a binary sequence and terminate. We think of all programs as being ordered according to their Gödel numbers and refer to the program with the Gödel number i as the i-th program (Figure 6.10).

Making a statement about the exact succession of zeros and ones in H requires the following (cf. [169]):

- We need to know which program corresponds to the i-th bit in H. In short, we need to know the applied Gödelization. Since different Gödelizations lead to different bit distributions, the halting sequence is not an absolute but a relative entity.

- We must be able to decide whether the i-th program terminates. However, this can only succeed in individual cases due to the halting problem's undecidability. There can be no procedure that generates all bits of H one after the other. In short, the halting sequence H is undecidable.

We will now attempt to compress the halting sequence. To do so, we take advantage of the fact that the first m bits of H tell us

- how many of the first m programs terminate   (☞ amount of ones)

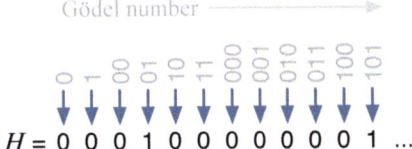

The program with the Gödel number 01 terminates.

The program with the Gödel number 101 terminates.

Figure 6.10: The i-th bit in the halting sequence H equals 1 precisely when the i-th program terminates. If we knew the entire bit sequence, we could solve the halting problem. Conversely, we can immediately conclude from the halting problem's undecidability that no systematic procedure can calculate H.

- HelloWorld.pas

```
PROGRAM HelloWorld;

BEGIN
    WRITELN('Hello World');
END.
```

```
50 52 4f 47 52 41 4d 20 48 65
6c 6c 6f 57 6f 72 6c 64 3b 0a
0a 42 45 47 49 4e 0a 20 20 20
57 52 49 54 45 4c 4e 28 27 48
65 6c 6c 6f 20 57 6f 72 6c 64
27 29 3b 0a 45 4e 44 2e
```

Figure 6.11: Pascal programs are prefix-free by design. All programs end with the keyword 'END', followed by a dot.

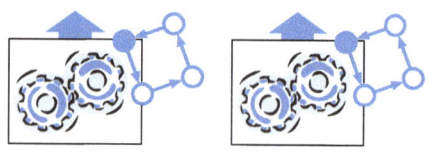

(q_1, S_0, S_1, R, q_2) (q_1, S_0, S_1, R, q_2)
(q_2, S_0, S_0, R, q_3) (q_2, S_0, S_0, R, q_3)
(q_3, S_0, S_2, R, q_4) (q_3, S_0, S_2, R, q_4)
(q_4, S_0, S_0, R, q_1) (q_4, S_0, S_0, R, q_1)
                              (q_1, S_1, S_1, R, q_2)

```
73 13 32 53 11 73 11 33 53 11
17 31 11 33 22 53 11 11 73 11
11 33 53 17 31 32 32 53 11
```

Figure 6.12: Turing's original coding from 1936 is not prefix-free. The Gödel number of the left-hand machine begins with the Gödel number of the right-hand machine.

- and which programs these are.   (☞ position of ones)

The second piece of information is indeed redundant. Once we know how many of the first m programs terminate, we can determine them individually. All we need to do is simulate the execution of the first m programs in parallel and note the stopping ones. Since we know exactly how many programs terminate, we also know when to end the simulation.

Next, let us try to specify the number of terminating programs as a probability, guided by the following idea: We put the binary sequences of the first m programs into a bucket and take out a sequence at random. Sometimes, we will take out a binary sequence, which is the Gödel number of a terminating program. We call the probability of extracting a bit sequence that corresponds to the Gödel number of a terminating program the *halting probability*. Knowing the probability lets us quickly recover the number of terminating programs by multiplying with m.

However, this idea becomes troublesome once we put all finite binary sequences into the container rather than the first m. Does the idea of randomly extracting a binary sequence then still make sense? What, for example, would be the expected average length of the sequences drawn? We immediately arrive at a contradiction for each particular value of l since there are only finitely many sequences smaller than l but an infinite number of larger ones. We will see in a moment that we can avoid contradictions of this kind if we constrain the allowed Gödelizations. Specifically, we will only allow Gödelizations that are *prefix-free*. The following definition clarifies the details.

 Definition 6.4 (Prefix-Free Program Encoding)

> A program encoding is called *prefix-free* if the Gödel number of a program never begins with the Gödel number of another program.

Prefix-free encodings ensure that each binary sequence begins with the Gödel number of at most one program; no program is the beginning of another. Some programming languages, such as Pascal, are inherently prefix-free. Here, each program ends with the unique keyword 'END.' (Figure 6.11). Many Gödelizations of Turing machines do not meet the criterion, though. Turing's original encoding, for example, allows us to generate a different Gödel number from any other Gödel number by appending additional characters (Figure 6.12). With a slight modification, however, we can also encode Turing machines such that no Gödel number is the beginning of another, so we can safely limit ourselves to prefix-free encodings. All the prerequisites are now in place for a solid definition of the halting probability (Figure 6.13):

Definition 6.5 (Halting Probability Ω_n)

Let s be a sequence randomly selected from all 2^n binary sequences of length n. The *halting probability* Ω_n is the probability that s begins with the Gödel number of a terminating program.

Let us give the halting probability an even more intuitive meaning. Imagine we are trying to generate a program's Gödel number through a random experiment by repeatedly tossing a coin and recording a 1 for *heads* and a 0 for *tails*. This way, we will eventually obtain some program's Gödel number with some luck. In this case, we finish our experiment, as an extension of the generated binary sequence can no longer lead to success due to the prefix-free encoding. Of course, it can also happen that we never arrive at a Gödel number this way. In this case, we will continue to toss the coin forever.

Figure 6.14 shows the decision tree of our experiment. Each node has two successors, each representing one of the two alternatives of adding another bit to the sequence. The result is a tree whose paths are either infinitely long or lead to leaves marked with the Gödel number of a program. In our random experiment, the halting probability has an intuitive meaning: Ω_n is the probability of generating the Gödel number of a terminating program with a maximum of n coin tosses.

The experiment also tells us how to calculate the halting probability. Our example has five terminating programs, given by the Gödel numbers 01, 101, 0010, 0011, and 1110. Our experiment generates the program 01 with probability $\frac{1}{4}$ and the program 101 with probability $\frac{1}{8}$. The programs 0010, 0011, and 1110 each have probability $\frac{1}{16}$. Because all events are pairwise disjoint, the total probability of generating one of the five machines is the sum of the individual probabilities. With

$$\Omega_1 = 0$$
$$\Omega_2 = \frac{1}{4}$$
$$\Omega_3 = \frac{1}{4} + \frac{1}{8} = \frac{3}{8}$$
$$\Omega_4 = \frac{1}{4} + \frac{1}{8} + \frac{1}{16} + \frac{1}{16} + \frac{1}{16} = \frac{9}{16}$$

■ Example

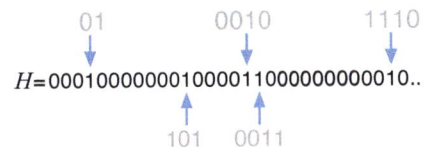

⇒ The programs with the Gödel numbers 01, 101, 0010, 0011, 1110 terminate.

■ Halting probabilities

0	00	000	0000
1	01	001	0001
	10	010	0010
	11	011	0011
		100	0100
		101	0101
		110	0110
		111	0111
			1000
			1001
			1010
			1011
			1100
			1101
			1110
			1111

$\Omega_1 = 0$ $\Omega_2 = \frac{1}{4}$ $\Omega_3 = \frac{3}{8}$ $\Omega_4 = \frac{9}{16}$

Figure 6.13: In this example, 5 programs terminate. We can determine the halting probability Ω_n by listing all bit sequences of length n and counting how many start with the Gödel number of a terminating program.

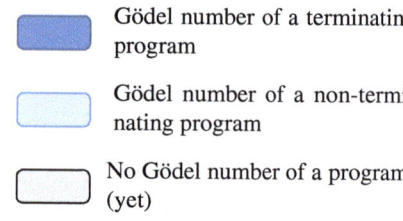
- Gödel number of a terminating program
- Gödel number of a non-terminating program
- No Gödel number of a program (yet)

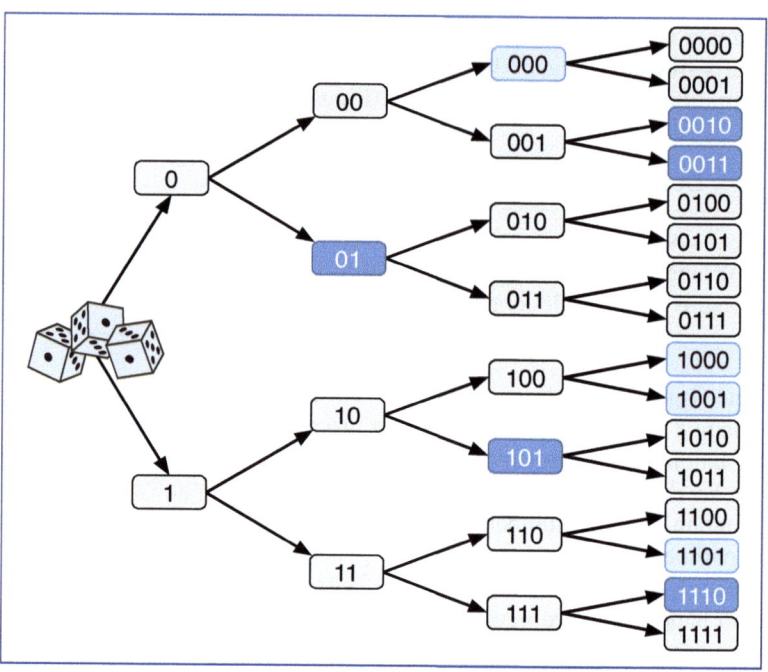

Figure 6.14: By tossing a coin, new bits are generated until the Gödel number of a program appears. The halting probability Ω_n has an intuitive meaning in this coin-tossing experiment. It is the probability of generating a terminating program with a maximum of n tosses.

we obtain the same probabilities we determined by counting in Figure 6.13. We can straightforwardly generalize our observation. If P denotes a terminating program and $|P|$ the length of its Gödel number, then P is included in the calculation of the total probability with the weight $\frac{1}{2^{|P|}}$. Thus, we have:

$$\Omega_n = \sum_{\substack{P \text{ halts,} \\ |P| \leq n}} \frac{1}{2^{|P|}} \qquad (6.2)$$

The halting probabilities Ω_n are mind-boggling objects of mathematical logic. They contain the necessary knowledge to decide all refutable questions by specifying a computable counterexample. We can translate any such question into a program P that searches for a counterexample and stops as soon as it finds one (Figure 6.15). If P has the length n, we can employ Ω_n to determine the number of binary sequences that begin with the Gödel number of a terminating program. By simulating all programs in question step by step, we can filter out those that terminate and thus determine whether P is terminating or running indefinitely.

The number of mathematical problems we could prove or disprove this way is tremendous. They include many famous yet unsolved mathematics problems, such as Goldbach's conjecture or Riemann's hypothesis.

The answers to these questions hide in the halting probability, encoded in a sequence of bits that would easily fit on a few book pages.

However, can we really hope to decide Goldbach's conjecture or Riemann's hypothesis using this method one day? To do so, we would have to find a way to calculate the halting probability Ω_n. For small values of n, this is indeed doable. Since we only need to consider a few Gödel numbers in this case, we can hope to identify all terminating programs through individual analysis. However, it is easy to see that the calculation of Ω_n is only possible for a finite number of n. If we could calculate infinitely many Ω_n, we would possess enough knowledge to decide the halting problem. To determine whether a program P of length n stops, we only had to determine the smallest number m greater than n for which Ω_m is computable. The halting probability Ω_m would then tell us how many programs of maximum length m will terminate. Thus, we could decide whether P is among them by simultaneously simulating all possible programs. It follows directly from the undecidability of the halting problem that Ω_n must be uncomputable as soon as n exceeds some limit.

Next, let us analyze how the halting probabilities Ω_n evolve for larger values of n. We know that the sequence $\Omega_1, \Omega_2, \Omega_3, \ldots$

- is monotonically increasing and ☞ $\Omega_n \leq \Omega_{n+1}$)
- has an upper bound. ☞ $\Omega_n \leq 1$)

Consequently, the sequence must approach a limit. It converges against *Chaitin's constant* Ω, one of the most fascinating objects ever encountered in mathematics.

 Definition 6.6 (Chaitin's Constant)

Chaitin's constant Ω is defined as the limit

$$\Omega := \lim_{n \to \infty} \Omega_n = \sum_{P \text{ halts}} \frac{1}{2^{|P|}}$$

Is it possible to calculate Ω systematically? Because Ω_n is uncomputable beyond a certain n, it is clear that Ω cannot be computed straight from its definition. Therefore, we will attempt to obtain Ω by other means and commence with the following definition:

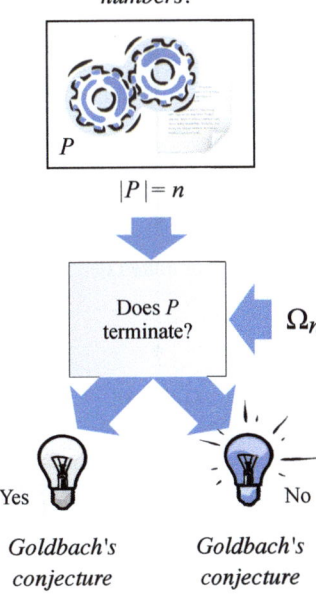

"Find the first even natural number > 2 that cannot be written as the sum of two prime numbers."

Figure 6.15: The halting probability Ω_n holds the answers to all mathematical questions decidable by the termination property of a program of maximum length n.

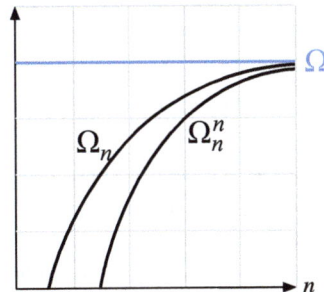

Figure 6.16: The monotonically increasing sequences Ω_n and Ω_n^n converge towards the same limit. Both strive towards Chaitin's constant Ω.

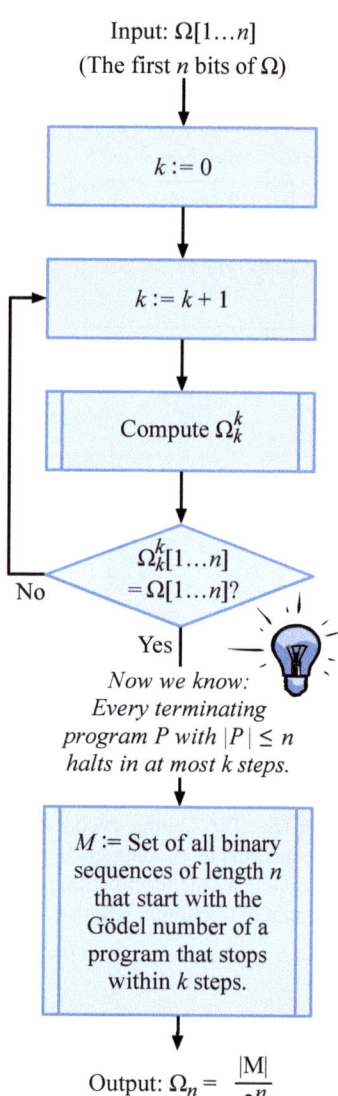

Figure 6.17: Ω unites all halting probabilities Ω_n in a single number. The first n bits of Chaitin's constant ($\Omega[1\ldots n]$) suffice to reconstruct the halting probability Ω_n.

 Definition 6.7 (Halting Probability Ω_n^k)

> Let s be a sequence randomly chosen from all 2^n binary sequences of length n. The *halting probability* Ω_n^k is the probability that s begins with the Gödel number of a program that terminates in at most k steps.

Ω_n and Ω_n^k differ in that the calculation of Ω_n^k no longer includes all terminating programs of maximum length n, but only those that terminate within k steps. The following properties are easy to recognize (Figure 6.16):

$$\Omega_n^n \leq \Omega_n \leq \Omega \qquad (6.3)$$

$$\lim_{n\to\infty} \Omega_n^n = \lim_{n\to\infty} \Omega_n = \Omega \qquad (6.4)$$

Although the sequences $\Omega_1, \Omega_2, \Omega_3, \ldots$ and $\Omega_1^1, \Omega_2^2, \Omega_3^3, \ldots$ approach the same limit, they are fundamentally different. In contrast to the first sequence, all elements of the second sequence are computable. We can determine the value Ω_n^n by simulating all programs P with $|P| \leq n$ for a maximum of n steps and counting how many terminate. With larger values for n, we get closer and closer to the value of Ω. Numbers that can be approximated arbitrarily close, as here, by a sequence of computable digits, are called *recursively enumerable*.

Does this mean that Chaitin's constant Ω is computable? At first glance, it is. The convergence property ensures that, over time, the decimal bits of Ω_n^n stabilize from left to right.

However, we are facing a substantial problem: We can never say with certainty whether a bit has reached its final value. Even if Ω_n^n is so close to Ω that the transition from Ω_n^n to Ω_{n+1}^{n+1} only affects bits on the far right, these changes may propagate to the left through the generation of carry bits and thus also change bits at the front. Thus, there could well be recursively enumerable numbers that are not computable. The fact that Chaitin's constant Ω is such a number is a consequence of the following theorem:

 Theorem 6.3

> Ω_n can be reconstructed from the first n bits of Ω.

Proof: To reconstruct the halting probability Ω_n from the first n bits of Ω, we proceed according to the flow chart in Figure 6.17:

6.2 Chaitin's Constant

- We start computing the sequence elements $\Omega_1^1, \Omega_2^2, \Omega_3^3, \ldots$ one after the other. In this way, we approach Ω from below, and at some point, Ω_k^k and Ω will match in the first n bits. As soon as this happens, we record the value of k and denote it by k_0.

- Suppose we increase k beyond k_0. Can the first n bits of Ω_k^k change? The answer is no! If only one of the first n bits were to take on a different value, we would have $\Omega_k^k > \Omega$, contradicting (6.3). Since every program P contributes $\frac{1}{2^{|P|}}$ to the halting probability Ω_k^k and the first n bits can no longer change, every program P that stops after more than k_0 steps must be longer than n bits.

- We now know that no program P with $|P| \leq n$ terminates after more than k_0 steps. This explains how we can determine the halting probability Ω_n. Simulating all programs P with $|P| \leq n$ for k_0 steps is sufficient. If a program terminates, we include it in the halting probability Ω_k^k with weight $\frac{1}{2^{|P|}}$. When it has not yet terminated after k_0 steps, we know it will never do. □

The impact of Theorem 6.3 is much more profound than it might seem at first glance. As Chaitin's constant combines the knowledge of all halting probabilities, it contains the answer to an unimaginable plethora of mathematical questions. Among others, Ω contains the knowledge to decide the halting problem for arbitrary programs and must be uncomputable for that reason alone. Theorem 6.3 unmistakably emphasizes that most of this spectacular bit sequence will be hidden from our eyes forever.

At the same time, Theorem 6.3 paves the way towards compressing the halting sequence H, introduced at the beginning of this section. Since we can recover the first 2^n bits of H from the halting probability Ω_n and reconstruct Ω_n from the first n bits of Ω, we can generate each initial piece of H with a program whose length only grows logarithmically. As a result, H is not a random number, and the individual bits are not independent.

Chaitin's constant Ω is a truly miraculous object of modern mathematics. Rarely has its nature been described more aptly than in an article by Charles Bennett and Martin Gardner published in 1979. We do not want to ignore a now-famous quote from this article at this point:

> "Throughout history mystics and philosophers have sought a compact key to universal wisdom, a finite formula or text which, when known and understood, would provide the answer to every

The proof of Theorem 6.3 has introduced a constructive procedure for extracting the halting probability Ω_n from the first n bits of Ω, which raises the question of whether we could use this algorithm in practice. The answer is no! To understand the reason, recall the core element of the algorithm: the calculation of the sequence

$$\Omega_1^1, \Omega_2^2, \Omega_3^3, \Omega_4^4, \ldots$$

We must estimate new sequence elements until the first n bits of Ω_k^k match the first n bits of Ω. For what values of k will this approximately be the case? Let us denote the time at which the first n bits of Ω_k^k match the first n bits of Ω by $f(n)$:

$$f(n) := \min\{k \mid \Omega_k^k[1\ldots n] = \Omega[1\ldots n]\}$$

It is easy to show that the function f must strive towards infinity faster than any computable function. If it did not, $f(n)$ would have a computable function $g(n)$ as an upper bound. Then, the first n bits of $\Omega_{g(n)}^{g(n)}$ were identical to the first n bits of Ω, paving the way to calculate all of Chaitin's constant.

The uncomputability of Ω implies that the function $g(n)$ and thus the effort to extract the halting probabilities Ω_n from Ω must grow faster than any computable function. Chaitin's constant conceals more knowledge than we can dream of while simultaneously encrypting the information in the best possible way. Ω reveals itself as the perfect guardian of the Grail, who will never disclose his entire knowledge. What a genuinely depressing result.

> Always remember that the bit sequence of Ω depends on both the chosen programming language and the agreed Gödelization. Altering just one of them lets the decimal bits of Ω change, too. However, all results we have worked out above remain intact. For example, Ω is always a recursively enumerable random number, regardless of the chosen programming language.
>
> In 2001, Antonín Kučera and Theodore Slaman achieved the astonishing result that the inverse also holds [123]: For any recursively enumerable random number x between 0 and 1, there is a programming language and an encoding with $x = \Omega$. Therefore, the halting probabilities are anything but lonely whimsicalities in the infinite space of the real numbers. Quite the opposite: they are everywhere!

question. The use of the Bible, the Koran and the I Ching for divination and the tradition of the secret books of Hermes Trismegistus, and the medieval Jewish Cabala exemplify this belief or hope. Such sources of universal wisdom are traditionally protected from casual use by being hard to find, hard to understand when found, and dangerous to use, tending to answer more questions and deeper ones than the searcher wishes to ask. The esoteric book is, like God, simple yet undescribable. It is omniscient, and transforms all who know it. Omega is in many senses a cabalistic number. It can be known of, but not known, through human reason. To know it in detail, one would have to accept its uncomputable digit sequence on faith, like words of a sacred text."

Charles Bennett, Martin Gardner [68]

The uncomputability of Chaitin's constant unequivocally clarifies that we will never be able to capture it as a whole. Nevertheless, since the discovery of Ω, there had been hope of getting to know at least the initial piece of this graceful string of digits. This wish came true to a reasonable extent in 2002. In that year, Cristian Calude, Michael Dinneen, and Chi-Kou Shu succeeded in calculating the first 64 digits of Ω. To reveal their value, the three researchers examined all binary sequences with a length of 84 bits. In the first step, they sorted out the sequences that did not represent the Gödel number of a program. In the next step, they further reduced the dataset by identifying functionally identical programs, and for the remaining, the authors individually examined whether it halts. This way, they compiled the knowledge encoded in the first bits of Ω piece by piece. After identifying the terminating programs, it remained to prove that the identified bit sequence was stable. The decisive discovery was that every terminating program with a length greater than 84 bits must begin with a particular bit sequence, from which Calude, Dinneen, and Shu concluded that at least 64 of the 84 calculated bits are correct. Here is the result:

$$\Omega = 0,0000001000000100000110001000010\ldots$$
$$\ldots 100011111100101110111010000010000\ldots$$

At first glance, the sequence of zeros and ones appears purely random. Is Chaitin's constant Ω perhaps a proper random number in the sense of Definition 6.3? The answer is yes, and the flowchart in Figure 6.18 reveals why. The depicted program – let us call it P – utilizes the first n bits of Ω to determine all terminating programs of maximum length n and remembers their output. Then, the variable i is initialized with 1 and incremented step by step in a loop. In each iteration, the program assigns the variable s the i-th binary sequence and compares it with

6.2 Chaitin's Constant

the previously determined outputs. If s is not the output of one of the previously simulated programs, P outputs s and stops.

What is the algorithmic complexity of the output s? Obviously, $\kappa(s)$ cannot be greater than $|P|$ because s is the output of P. On the other hand, s is not the output of any program with a length $\leq n$, and therefore $\kappa(s)$ must exceed n. Thus, we have:

$$n < \kappa(s) \leq |P| \tag{6.5}$$

We can also make a statement about $|P|$. To implement the algorithm as outlined, we need a constant number of bits for the program logic. In addition, we need space to store the first n bits of Ω, that is, at least $\kappa(\Omega[1\ldots n])$ bits. Thus, we can quantify the length of P as follows:

$$|P| = c + \kappa(\Omega[1\ldots n])$$

Now we can rewrite (6.5) as

$$n < \kappa(s) \leq c + \kappa(\Omega[1\ldots n]).$$

Solving the inequality for $\kappa(\Omega[1\ldots n])$ gives us

$$\kappa(\Omega[1\ldots n]) > n - c,$$

which means that the algorithmic complexity of any initial segment $\Omega[1\ldots n]$ is only a constant number of bits smaller than n. However, this is precisely the distinctive property we used to establish the concept of randomness in Definition 6.3. Thus, we have:

Theorem 6.4

> The bit sequence of Chaitin's constant Ω is random.

It follows directly from Theorem 6.4 that Chaitin's constant combines the knowledge about all halting probabilities Ω_n in perfectly condensed form. The information is encoded so clevery that no further compression is possible. Once again, Ω exposes itself as a truly marvelous object of modern mathematics!

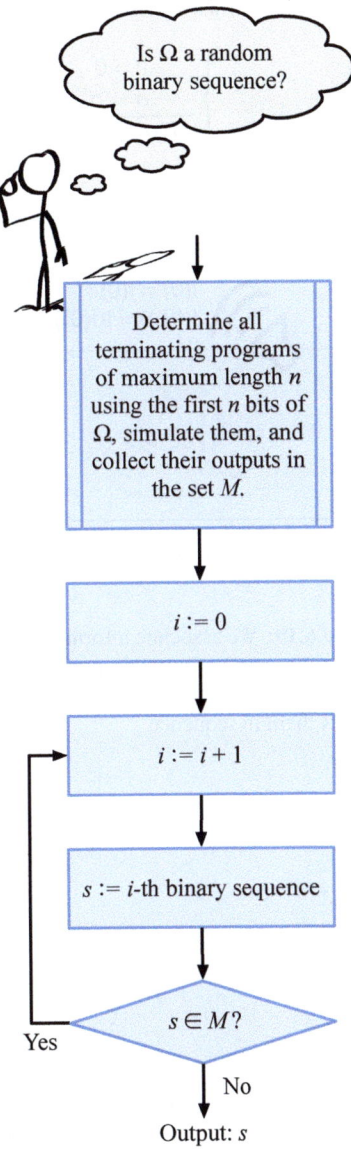

Figure 6.18: Analysing this program reveals that the bit sequence of Chaitin's constants cannot follow any regularity: Ω is a random number.

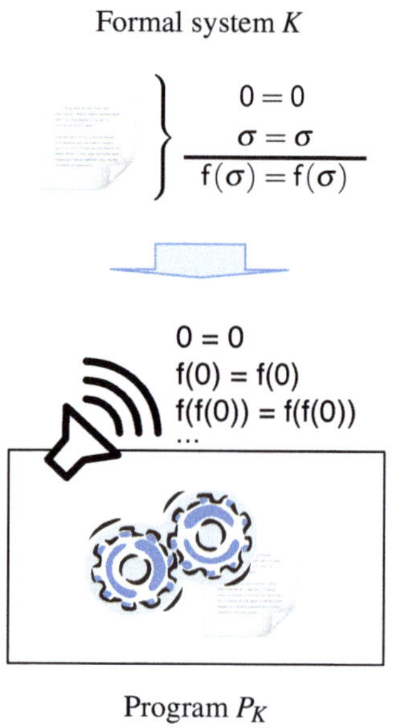

6.3 Incompleteness of Formal Systems

This section will establish a connection between algorithmic information theory and proof theory. Our effort will be well-rewarded with Chaitin's incompleteness theorem, a splendid result of Gödelian nature. This theorem will illuminate why, in any sufficiently expressive formal system, there must be true propositions that escape formal provability.

Bridging the gap between algorithmic information theory and proof theory is surprisingly simple if we relate a formal system K to the smallest possible program P_K that enumerates the theorems of K one after the other (Figures 6.19 and 6.20). The fact that such a program exists for every formal system results from what we have achieved in Section 2.2. There, we learned that the theorems of a formal system K are enumerable even if K is not finitely axiomatizable. Note that the choice of P_K is not guaranteed to be unique since there may be more than one smallest possible program. In this case, any of them will do.

We commence by approximating the program length $|P_K|$. First, we realize that P_K must contain the axioms and inference rules and denote the required number of bits by $|K|$. In addition, we need a constant number of bits to implement the enumeration routine, which allows us to quantify the length of P_K as follows:

$$|P_K| = |K| + c$$

The exact value of the constant c is irrelevant to our considerations.

When we speak of a formal system of length n or simply an n-bit system below, we mean that the considered formal system K satisfies $|P_K| = n$. In other words, some program of length n enumerates the theorems of K in sequence.

Suppose K is a formal system with the expressive power to formalize formulas $\varphi_n(s)$ with the following meaning:

"The algorithmic complexity of s is greater than n." (6.6)

For a fixed value of n, let us ask ourselves whether a formula $\varphi_n(s)$ is provable within K. For this purpose, we could start the program P_K and observe the printed theorems. With some luck, one of the formulas $\varphi_n(s)$ will eventually appear, thus demonstrating its provability within K. For small values of n, this could actually happen.

We will now convert the program P_K into a program P'_K that no longer outputs all theorems but silently searches for the proof of one of the

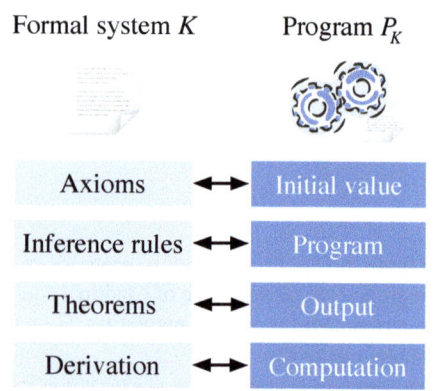

Figure 6.19: We associate a formal system K with the smallest possible program, P_K, which enumerates the theorems of K and outputs them in sequence.

Figure 6.20: Relationship between formal systems (left) and programs (right).

6.3 Incompleteness of Formal Systems

formulas $\varphi_n(s)$ (Figure 6.21). Once it finds such a proof, P'_K prints the binary sequence s and stops. If no suitable proof comes up, P'_K continues to run indefinitely without producing any output.

Estimating the size of the program in the same way as above, we get

$$|P'_K| \leq |K| + \lceil \log_2 n \rceil + c'$$

for a constant $c' \in \mathbb{N}$. By choosing n sufficiently large, we can establish

$$|P'_K| < n.$$

How will the program P'_K behave for those values of n? By construction, it will output a string s if and only if a formula $\varphi_n(s)$ is provable within K. Substantively, $\varphi_n(s)$ claims that the algorithmic complexity of the binary string s exceeds n. However, this is impossible as we have managed to output s with a program shorter than n. If n is large enough, P'_K can no longer produce any output if the formal system K is correct. Consequently, in a correct formal system, any statement of the form (6.6) must be unprovable above a particular value of n. At this point, we have arrived at Chaitin's incompleteness theorem:

Theorem 6.5 (Chaitin's Incompleteness Theorem)

In a correct formal system, all statements of the form $\kappa(s) > n$ are unprovable beyond a certain n.

We already know that for any value of n, some s satisfies $\kappa(s) > n$. Thus, it follows from Theorem 6.5 that true but unprovable sentences exist in every correct formal system expressive enough to formalize sentences of the form (6.6).

Next, let us tackle the question of when a formal system is subject to Chaitin's incompleteness theorem. How expressive does it have to be? To find an answer, we first reformulate the colloquial formulation (6.6). The statement is equivalent to the assertion that no program P with $|P| \leq n$ exists that outputs s. A corresponding formula thus has the following form:

$$\varphi_n(s) = \neg \exists x \, (\text{program}(x) \wedge |x| \leq n \wedge \text{output}(x) = s) \quad (6.7)$$

The subexpression $|x| \leq n$ is an arithmetic statement, and the subformulas $\text{program}(x)$ and $\text{output}(x)$ express that x is the Gödel number of a program that produces the output s. In Section 5.4.2, we used the

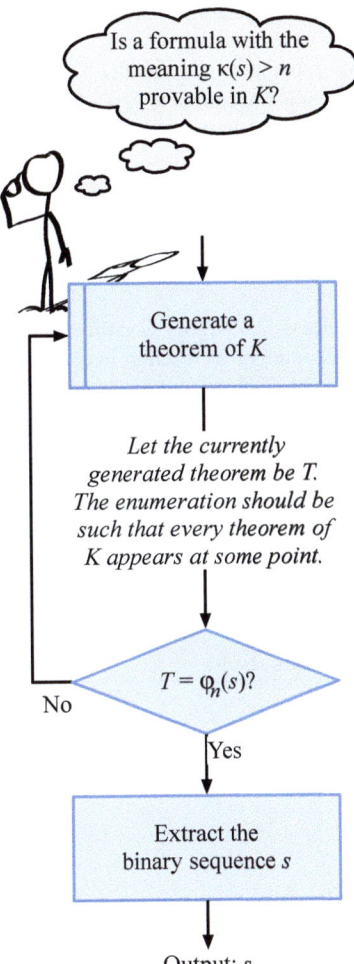

Figure 6.21: For a given constant n, the depicted program searches for a theorem substantively expressing $\kappa(s) > n$. If the search succeeds, the binary sequence s is output, and the program stops. When the formal system K is correct, the program will not produce any output for large values of n. Otherwise, we had $\kappa(s) < n$. At the same time, K would prove a theorem with the substantive meaning $\kappa(s) > n$.

example of the Turing machine to demonstrate in detail that statements of this type can be arithmetized. Even though many technical details remain open, the result is already foreseeable: Even systems as simple as Peano arithmetic are expressive enough to formalize statements of the form (6.7). Consequently, every formal system with the necessary means to talk about the additive and multiplicative properties of the natural numbers fulfills the requirements of Chaitin's incompleteness theorem. Thus, we can draw the following conclusion from Theorem 6.5:

Corallary 6.1

> Any correct formal system expressive enough to formalize Peano arithmetic is incomplete.

This theorem precisely matches the formulation of the semantic variant of Gödel's first incompleteness theorem presented on Page 194. Unlike before, however, we have reached the incompleteness theorem on a surprisingly comfortable path. All we had to do was to resort to elementary properties of algorithmic complexity and establish a suitable connection between formal systems and programs. Our considerations explain why algorithmic information theory is an established branch of mathematical logic today. It provides a fast track to some of the most pivotal theorems of mathematical logic, with Gödel's first incompleteness theorem among them.

The uncomputability of Ω leads to yet another remarkable incompleteness result. Since it is impossible to systematically determine all bits of Ω, no correct formal system can prove statements of the form

$$\text{"The } i\text{-th bit of } \Omega \text{ is } 0\text{"} \quad \text{or} \quad (6.8)$$

$$\text{"The } i\text{-th bit of } \Omega \text{ is } 1\text{"} \quad (6.9)$$

for all $i \in \mathbb{N}$. Gregory Chaitin established an astounding connection between statements of this kind and the axioms and inference rules of a formal system. He succeeded in showing that a formal system K, whose theorems can be enumerated with a program P_K and output in sequence, can prove at most

$$|P_K| + 15328$$

statements of the form (6.8) or (6.9) [28]. A fantastic result!

6.4 Exercises

Exercise 6.1 Let r_i denote the i-th digit of a random binary sequence ($i \geq 0$). In which of the following cases is the sequence $s_0, s_1, s_2, \ldots$ also random?

a) $s_i = \begin{cases} 1 & \text{if } i < 10 \\ r_i & \text{otherwise} \end{cases}$

b) $s_i = \begin{cases} r_{r_i} & \text{if } i \text{ is even} \\ r_i & \text{otherwise} \end{cases}$

c) $s_i = \begin{cases} r_i & \text{if } i < 10 \\ 1 & \text{otherwise} \end{cases}$

d) $s_i = \begin{cases} r_i & \text{if } i \text{ is a prime number} \\ 1 & \text{otherwise} \end{cases}$

e) $s_i = \begin{cases} r_i & \text{if } i \text{ is even} \\ 1 & \text{otherwise} \end{cases}$

f) $s_i = \begin{cases} r_i & \text{if } i \text{ is a square number} \\ 1 & \text{otherwise} \end{cases}$

Exercise 6.2 Name, if possible, an example of a *finitely* long binary sequence that is

- computable and compressible:

- computable but uncompressible:

- uncomputable but compressible:

- uncomputable and uncompressible:

Name, if possible, an example of an *infinitely* long binary sequence that is

- computable and compressible:

- computable but uncompressible:

- uncomputable but compressible:

- uncomputable and uncompressible:

Exercise 6.3   This exercise puts the content of Theorem 6.2 to the test. The theorem states that no systematic procedure can always correctly decide whether a given binary sequence s is random. Nevertheless, the following program seems to do the trick:

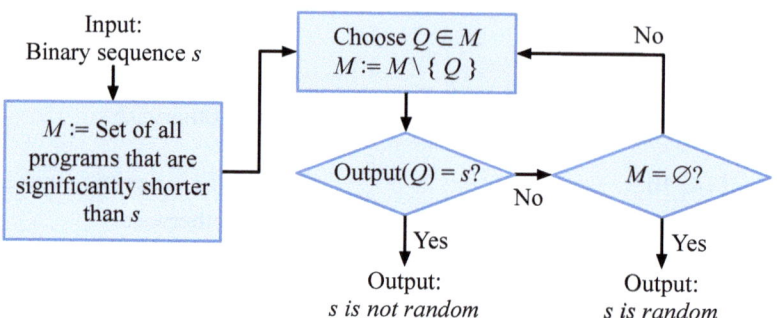

The design of the algorithm is motivated by the following observations: Only a finite number of programs are shorter than the given binary sequence s. If s were not random, it would be the output of one of them. Therefore, it suffices to iterate over the (finitely many) programs significantly shorter than s and compare their output against s. If both match, s is not a random sequence. However, if s is not output by these programs, it must be random. Needless to say, the result contradicts Theorem 6.2, but where exactly is the reasoning flawed?

Exercise 6.4   Chapter 1 discussed the twin prime conjecture:

> "There are infinitely many numbers n with the property that n and $n+2$ are prime numbers."

This exercise assumes the halting probability Ω_n is known for any given n. On page 336, we explained how to exploit this knowledge to prove or disprove statements such as Goldbach's conjecture. Could we use the same method to decide the conjecture about the existence of infinitely many prime twins?

Exercise 6.5   Section 6.2 revealed that the bit sequence of Chaitin's constants is random. Therefore, whether 0 or 1 occurs at a specific bit position of Ω is independent of the bits at other positions. Can we nevertheless make a statement about *how many* ones and zeros an initial piece $\Omega[1\ldots n]$ contains for larger values of n?

6.4 Exercises

Exercise 6.6

In the investigations on algorithmic complexity, we have assumed that the applied Gödelizations must be prefix-free; that is, the Gödel number of a program never starts with the Gödel number of another program. This exercise will elucidate why we need this assumption.

Consider the following two decision trees:

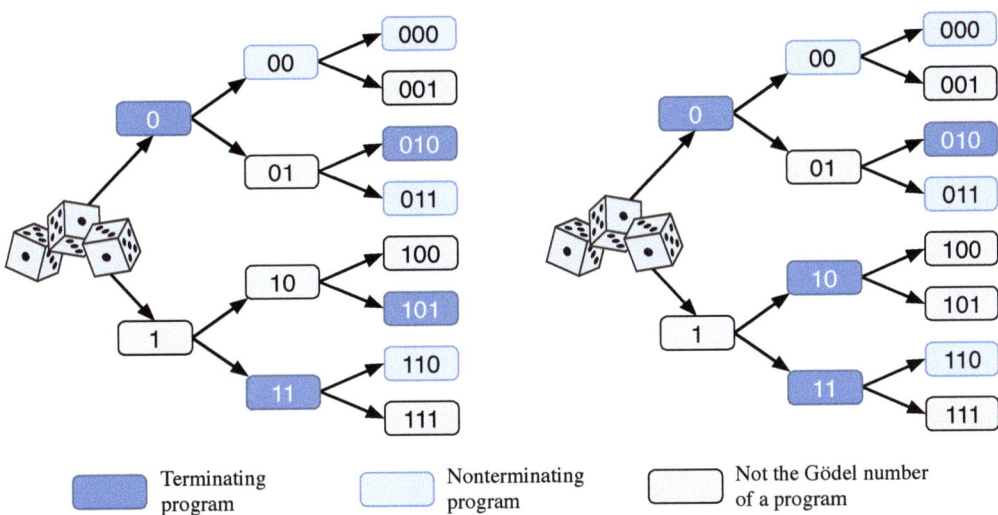

- Explain why the respective Gödelizations are not prefix-free.
- For each decision tree, calculate the halting probability Ω_3.
- Repeat the calculation with the following formula from Section 6.2:

$$\Omega_n = \sum_{\substack{P \text{ halts,} \\ |P| \leq n}} \frac{1}{2^{|P|}}$$

Explain your findings.

7 Model Theory

> *"Thus all human knowledge begins with intuitions, proceeds from thence to concepts, and ends with ideas."*
>
> Immanuel Kant [194]

The previous chapters repeatedly emphasized the two levels of a formal system. The first one is the syntactic level. Here, formulas are no more than sequences of symbols, translatable into other formulas by applying formally defined inference rules. The second one is the semantic level, also called the *model level*. It assigns substantive meaning to symbols and formulas, allowing us to interpret them in contentual terms.

Model theory is the branch of mathematical logic that deals with the semantic properties of a formal system. The following problem areas are of primary interest:

■ **Model construction**

Model construction aims to create a model for a given set of formulas. Due to a simple reason, however, proving that a model exists is often more relevant than creating the model itself: If a set of formulas has a model, it must be free of contradictions; in that case, it cannot contain a formula φ together with its negation $\neg\varphi$. We have already come across an application along these lines in Chapter 1 in connection with Hilbert's proof of the consistency of Euclidean geometry.

■ **Model analysis**

Model analysis attempts to classify sets of formulas based on the number and structure of their models. Typical questions are, for instance, whether the models of a set of formulas all have a finite or an infinite domain. Of particular interest are sets of formulas being *categorical*; that is, sets with the property that they have a unique model up to isomorphism. A prominent result in this regard is Dedekind's isomorphism theorem, stating that the Peano axioms, when written down in second-order predicate logic, uniquely characterize the natural numbers.

- **Axiomatizability**

  Many investigations in model theory are concerned with the axiomatizability of mathematical concepts, such as countability. The question here is whether a formula exists that is substantively true under precisely those interpretations with a countable domain. Some investigations deal with the general axiomatizability of concepts, i.e., with the question of whether a concept is definable within any formal logic at all. Others try to clarify whether axiomatization is possible in specific formal systems. We have already carried out investigations of this kind ourselves. In Sections 2.5 and 2.6, we learned that neither the concept of equality nor the concept of finiteness is definable within first-order predicate logic.

7.1 Meta-Results on Predicate Logic

Model theory centers around four core theorems discovered in the first half of the twentieth century (Figure 7.1). All of them are meta-results of first-order predicate logic.

- **Completeness Theorem**

  The *completeness theorem* asserts that all universally valid PL1 formulas are derivable from the axioms of first-order predicate logic. Kurt Gödel proved the theorem in 1929 as part of his dissertation, which is why many refer to it as *Gödel's completeness theorem*. We have already dealt extensively with it in Section 2.4.2 and will only touch on the theorem in passing in the rest of this chapter.

- **Model existence theorem**

  The model existence theorem establishes a connection between a formal system's consistency and the existence of a model. Alongside the completeness theorem, it is the second core theorem that interconnects the syntactic and semantic levels of a formal system. We will discuss the model existence theorem in detail in Section 7.1.1.

- **Compactness theorem**

  The *compactness theorem* is the subject of Section 7.1.2. It states that an infinite set of first-order predicate-logic formulas is satisfiable if and only if each finite subset is. We will soon realize that the compactness theorem is a powerful tool for deriving many negative results about first-order predicate logic with surprising ease.

1915 — Leopold Löwenheim:

Löwenheim proves the precursor of the theorem we now refer to as the Löwenheim-Skolem theorem. [127]

1920 — Thoralf Skolem:

Skolem advances Löwenheim's work. He formulates and proves the Löwenheim-Skolem theorem. [188]

1929 — Kurt Gödel:

In his dissertation, Gödel formulates and proves the completeness theorem and the model existence theorem. [71]

1930 — Kurt Gödel:

Gödel restructures the results of his dissertation. Among others, he formulates and proves the compactness theorem. [72]

Figure 7.1: Milestones of model theory

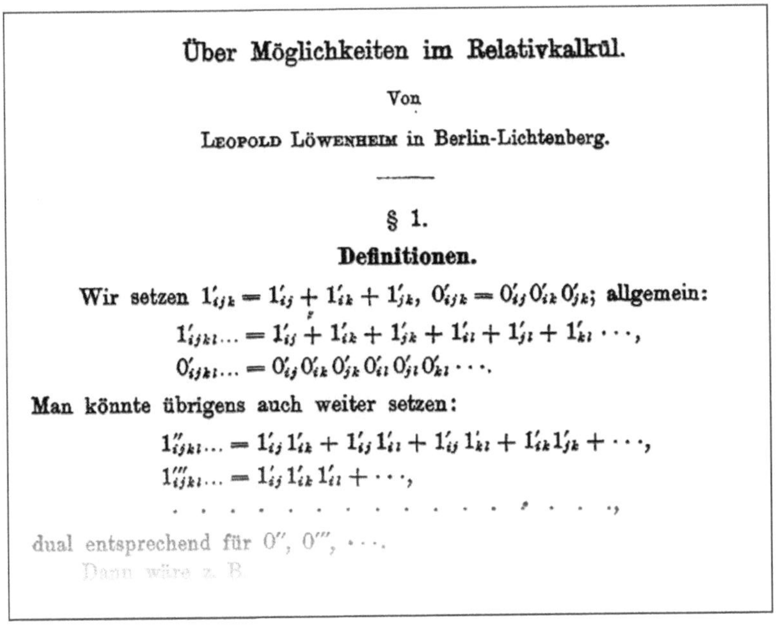

Über Möglichkeiten im Relativkalkül.

Von

Leopold Löwenheim in Berlin-Lichtenberg.

§ 1.

Definitionen.

Wir setzen $1'_{ijk} = 1'_{ij} + 1'_{ik} + 1'_{jk}$, $0'_{ijk} = 0'_{ij}0'_{ik}0'_{jk}$; allgemein:

$$1'_{ijkl\ldots} = 1'_{ij} + 1'_{ik} + 1'_{jk} + 1'_{il} + 1'_{jl} + 1'_{kl} \cdots,$$
$$0'_{ijkl\ldots} = 0'_{ij}0'_{ik}0'_{jk}0'_{il}0'_{jl}0'_{kl} \cdots.$$

Man könnte übrigens auch weiter setzen:

$$1''_{ijkl\ldots} = 1'_{ij}1'_{ik} + 1'_{ij}1'_{il} + 1'_{ij}1'_{kl} + 1'_{ik}1'_{jk} + \cdots,$$
$$1'''_{ijkl\ldots} = 1'_{ij}1'_{ik}1'_{il} + \cdots,$$
$$\ldots\ldots\ldots\ldots,$$

dual entsprechend für $0''$, $0'''$, $\cdots$.

Dann wäre z. B.

Figure 7.2: The beginnings of model theory trace back to 1915, the year the German mathematician Leopold Löwenheim published the precursor of what we call the Löwenheim-Skolem theorem today. Even though his theorem was correct in content, Löwenheim's proof was not. It was not until 1920 that the Norwegian mathematician Thoralf Skolem not only delivered an accurate proof but also extended the content of Löwenheim's theorem to formula sets. Skolem's variant from 1920 is the theorem we call the Löwenheim-Skolem theorem today. Later, the theorem was generalized even further, giving rise to the Löwenheim-Skolem-Tarski theorem, which we will examine in more detail in Section 7.1.3.

■ Löwenheim-Skolem and Löwenheim-Skolem-Tarski theorems

If a formula of first-order predicate logic has an infinite model, the *Löwenheim-Skolem-Tarski theorem* stipulates that models of any other transfinite cardinality exist. The theorem answers key questions about cardinality and categoricity in first-order predicate logic and is also the source of alleged contradictions. In Section 7.1.3, we will deal with the substantive meaning of the Löwenheim-Skolem theorems and look closer at the consequences and misunderstandings that arise from them.

This chapter deliberately presents the model-theoretic theorems in an order that diverges from the chronological order of their discovery. The historically oldest is the theorem mentioned last, the Löwenheim-Skolem theorem. The German mathematician Leopold Löwenheim discovered its original version as early as 1915 (Figure 7.2) [18, 127]. Although his proof was partially flawed, his work is of such great relevance that we can call 1915 the birth year of model theory. In 1920, the Norwegian mathematician Thoralf Skolem proved the theorem now known as the Löwenheim-Skolem theorem. Skolem successfully generalized Löwenheim's original formulation and supplied a precise proof. As part of this work, he introduced the *Skolem normal form*, which is part of almost all logic lectures taught today (e.g. [183]). In the follow-

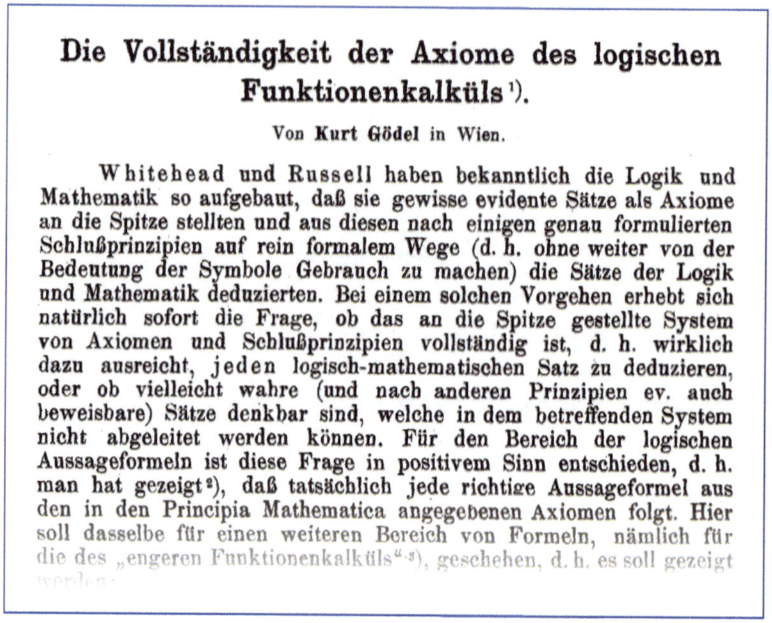

Figure 7.3: Kurt Gödel succinctly formulated three of the four core theorems of model theory in a paper published in 1930 [72]. In his publication, these are theorems I, IX, and X. Theorem I is the main result of his 1929 dissertation, the completeness of first-order predicate logic. The model existence theorem is Theorem IX, and the compactness theorem is Theorem X. Even if the order suggests otherwise, Gödel proved the compactness theorem first and obtained the model existence theorem as a corollary [49].

A small side note: You will not encounter the expression *compactness* anywhere in Gödel's work. Alfred Tarsk coined the term in the 1950s.

> **Die Vollständigkeit der Axiome des logischen Funktionenkalküls**[1]**.**
>
> Von **Kurt Gödel** in Wien.
>
> Whitehead und Russell haben bekanntlich die Logik und Mathematik so aufgebaut, daß sie gewisse evidente Sätze als Axiome an die Spitze stellten und aus diesen nach einigen genau formulierten Schlußprinzipien auf rein formalem Wege (d. h. ohne weiter von der Bedeutung der Symbole Gebrauch zu machen) die Sätze der Logik und Mathematik deduzierten. Bei einem solchen Vorgehen erhebt sich natürlich sofort die Frage, ob das an die Spitze gestellte System von Axiomen und Schlußprinzipien vollständig ist, d. h. wirklich dazu ausreicht, jeden logisch-mathematischen Satz zu deduzieren, oder ob vielleicht wahre (und nach anderen Prinzipien ev. auch beweisbare) Sätze denkbar sind, welche in dem betreffenden System nicht abgeleitet werden können. Für den Bereich der logischen Aussageformeln ist diese Frage in positivem Sinn entschieden, d. h. man hat gezeigt[2]), daß tatsächlich jede richtige Aussageformel aus den in den Principia Mathematica angegebenen Axiomen folgt. Hier soll dasselbe für einen weiteren Bereich von Formeln, nämlich für die des „engeren Funktionenkalküls"[3]), geschehen, d. h. es soll gezeigt

ing years, Skolem released a series of publications containing additional simplifications of his proof [189, 190].

The other three core theorems were explicitly formulated for the first time by Kurt Gödel. In his dissertation from 1929, Gödel proved the completeness theorem and the model existence theorem for consistent sets of formulas. In 1930, he published his results in the *Monatsheft für Mathematik und Physik* [72] (Figure 7.3). In his revised work, he had considerably restructured the proofs and introduced several new theorems. Among these is the compactness theorem, giving rise to the model existence theorem as a corollary. He proved the model existence theorem from scratch in his earlier published dissertation.

There is a good reason why we deviate from the historical order in this chapter and only present Löwenheim-Skolem's theorem in Section 7.1.3. We will see that crucial aspects of the Löwenheim-Skolem theorem follow with little effort from the compactness theorem, which, in turn, can be derived elegantly from the model existence theorem. The chosen order will emphasize how the core theorems relate to each other substantively.

Let us add a brief note on the objective of this chapter. Unlike in classic textbooks, we will not provide detailed proofs of the core theorems. If at all, we will merely sketch the general line of argument. For this

7.1 Meta-Results on Predicate Logic

reason alone, this chapter cannot – and does not intend to – replace the classical literature on model theory. This chapter is not about technical proofs; it is about the content of the core theorems of model theory, their manyfold connections and the astonishing consequences for the entire field of mathematics.

7.1.1 Model Existence Theorem

By the end of this section, we will understand that a formal system's syntactic and semantic levels are not mutually independent. However, to avoid unnecessary hurdles, let us commence with a few linguistic agreements:

 Definition 7.1 (Models of a Formal System)

> Let K be a formal system and (U, I) be an interpretation.
>
> ■ (U, I) is a *model* of K if it is a model for every theorem of K.
>
> ■ K is called *satisfiable* if K has at least one model.

Suppose the interpretation (U, I) is a model of a formal system K. Then, we know that K must be consistent, as any formula φ that is provable alongside its negation $\neg\varphi$ would immediately contradict the definition of the model relation. Let us recall the following line from Definition 2.16:

$$(U, I) \models (\neg\varphi) \Leftrightarrow (U, I) \not\models \varphi \tag{7.1}$$

Now, let us ask the converse and suppose that K is a first-order theory, i.e., a formal system with a set of theory axioms and the logical reasoning apparatus of PL1. Can we infer from the consistency of K that the theory has a model? In fact, the answer is yes!

 Theorem 7.1 (Model Existence, Gödel 1930)

> Let K be a first-order theory. Then:
>
> $$K \text{ has a model} \Leftrightarrow K \text{ is consistent}$$
>
> Or, equivalently:
>
> $$K \text{ is satisfiable} \Leftrightarrow K \text{ is consistent}$$

Are three of the four core theorems, i.e., the theorems on completeness, compactness, and model existence, the sole work of Kurt Gödel? Neither a resounding yes nor a resounding no would do justice to the historical development. Undoubtedly, Gödel was the first to formulate the three core theorems in precise form. However, the lines of reasoning he employed in his proofs were by no means new; they trace back to a work by Skolem from 1923 [189], in which the Norwegian proved both the completeness theorem and the model existence theorem without being fully aware. Unlike Gödel, Skolem distinguished less rigidly between the syntactic and semantic levels in his investigations, which is for historians the reason why Skolem failed to recognize the full scope of his results [49]. In hindsight, we can say that Skolem paved the way, but Gödel was the first to walk it to the end. For this reason, the model existence theorem is sometimes, and quite aptly, referred to as the *Skolem-Gödel theorem* in the literature.

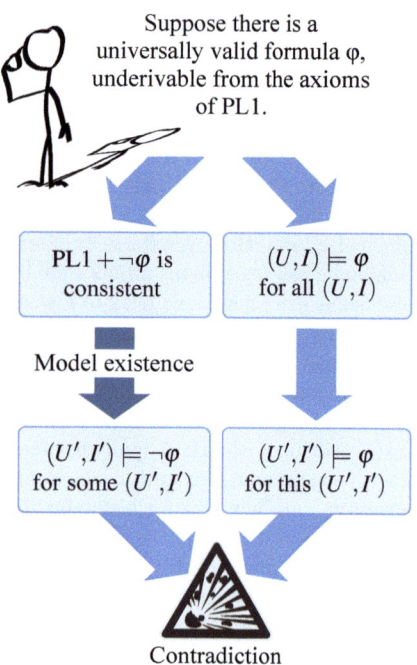

Figure 7.4: Gödel's completeness theorem is a direct consequence of the model existence theorem of first-order predicate logic.

The significance of the model existence theorem is far greater than its inconspicuous wording might suggest. For one thing, it is a valuable tool for conducting proofs. For another, it has profound philosophical significance. At the beginning of the twentieth century, many of the formalists' opponents believed that the syntactic consistency of a theory was insufficient to qualify as a meaningful construct [158]. However, the model existence theorem stipulates precisely that. It is possible to give a consistent meaning to any first-order theory incapable of deriving a formula φ simultaneously with its negation $\neg\varphi$.

We have already mentioned above that the model existence theorem is also called the *Skolem-Gödel-Theorem* in honor of its discoverers. The American mathematician Abraham Robinson coined a different term in a publication from 1951 [171]. He calls it the *extended completeness theorem* in analogy to Gödel's completeness theorem of first-order predicate logic. On the surface, the substantive meaning of both theorems appears to be completely different. At a second glance, however, it becomes apparent that the two theorems are closely intertwined. So, let us take a closer look!

We assume that φ is a universally valid first-order formula underivable from the axioms. Then, we obtain a new theory, $PL1^{\neg\varphi}$, by adding the formula $\neg\varphi$ to PL1 as an extra axiom (Figure 7.4). Since, by assumption, we cannot prove φ within PL1, $PL1^{\neg\varphi}$ is consistent. According to Theorem 7.1, $PL1^{\neg\varphi}$ has a model; that is, there exists an interpretation (U', I') under which all theorems of $PL1^{\neg\varphi}$ are substantively true. In particular, $\neg\varphi$ is true in (U', I'). On the other hand, as φ is universally valid, φ is true in all possible interpretations. Thus, it is a fortiori true in (U', I'). Then, $(U', I') \models \neg\varphi$ and $(U', I') \models \varphi$ would apply simultaneously, thus contradicting (7.1). Hence, we must drop the assumption of the unprovability of φ within first-order predicate logic and obtain Gödel's completeness theorem as a corollary to Theorem 7.1:

 Corallary 7.1 (Gödel's completeness theorem, 1929)

> First-order predicate logic is complete; every universally valid PL1 formula is derivable from the axioms.

Does the model existence theorem apply to higher-order theories, too? The answer is negative: The guaranteed existence of models is an exclusive property of first-order theories. The reason for this is straightforward to recognize. If every consistent higher-order theory had a model, we could prove the completeness of higher-order predicate logic by applying the exact same chain of reasoning. However, we already know that second-order predicate logic is incomplete; not all universally valid formulas are derivable from the axioms. The fact that the model existence theorem fails to apply is thus a direct consequence of the incompleteness of higher-order predicate logic.

7.1.2 Compactness Theorem

Let M be a set of countable first-order predicate-logic formulas. We assume that M is satisfiable; that is, all formulas from M are substantively true statements for some interpretation (U,I). We will now establish a connection between the set M and its finite subsets. An obvious result is this: If (U,I) is a model of M, then (U,I) is also a model for every finite subset. The converse question is much more interesting, though: From knowing that every finite subset of M has a model, can we conclude that M has a model, too? The *compactness theorem* of first-order predicate logic affirms this question (Figure 7.5):

Theorem 7.2 (Compactness Theorem, Gödel 1930)

For any countable set M of PL1 formulas, the following holds:

M has a model $\Leftrightarrow$ Every finite subset of M has a model

Or, equivalently:

M is satisfiable $\Leftrightarrow$ Every finite subset of M is satisfiable

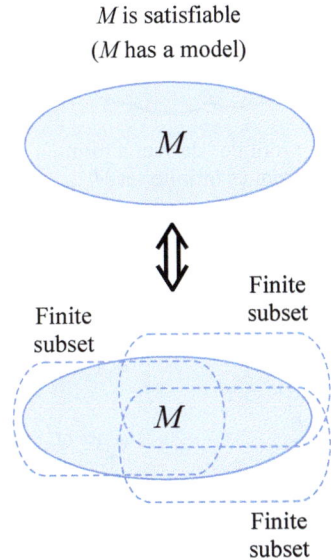

Figure 7.5: According to the compactness theorem, a countable set of PL1 formulas is satisfiable if and only if all finite subsets are.

Proof: The direction from left to right is trivial. To show the direction from right to left, let us assume the existence of an unsatisfiable set M of PL1 formulas whose finite subsets are all satisfiable. Since M is unsatisfiable, this set has no model. Consequently, every PL1 formula is a logical consequence of M, for example, this one:

$$\varphi := \exists x\,(P(x) \wedge \neg P(x)) \qquad (7.2)$$

For this formula we have:

- φ is unsatisfiable because the formula cannot become a true statement under any interpretation (U,I). So φ has no model.

- φ is derivable from the set M using the logical inference apparatus of PL1. This is a consequence of Gödel's completeness theorem.

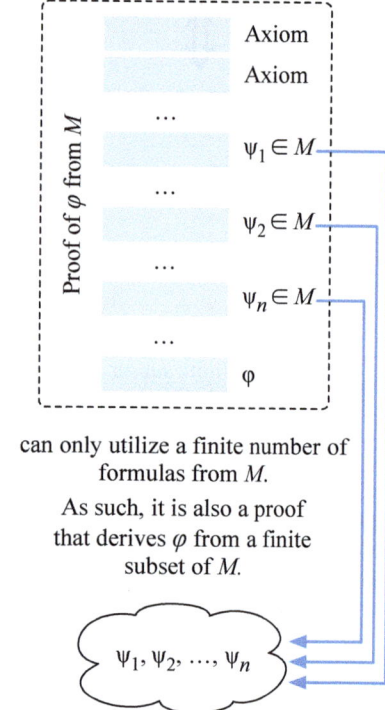

Figure 7.6: The core argument in the proof of the compactness theorem

We know that the derivation of φ can only comprise a finite number of proof steps and, therefore, only include a finite number of formulas from M. If we collect all the formulas used in the proof, as shown in Figure 7.6, we obtain a finite subset of M rich enough to derive a contradiction. Contrary to our assumption, a finite subset with no model exists. □

The compactness theorem is a precious result, as it helps us firmly grasp the elusive concept of infinity. However, the conclusions it allows us to draw are even more spectacular.

Among other consequences, the compactness theorem implies that the concept of finiteness is undefinable within PL1. We had already anticipated this result in Section 2.6.1 and are now in a position to provide solid justification. Let us begin by assuming the existence of a PL1 formula $\varphi_{<\mathbb{N}}$ with the following property:

$$(U, I) \models \varphi_{<\mathbb{N}} \Leftrightarrow U \text{ is finite}$$

Without loss of generality, we assume that the symbols P_1, P_2, P_3 etc. do not occur in $\varphi_{<\mathbb{N}}$. We utilize them to define the following set of formulas:

$$\varphi_1 := \exists x_1\, P_1(x_1)$$
$$\varphi_2 := \exists x_2\, (P_2(x_2) \wedge \neg P_1(x_2))$$
$$\varphi_3 := \exists x_3\, (P_3(x_3) \wedge \neg P_2(x_3) \wedge \neg P_1(x_3))$$
$$\varphi_4 := \exists x_4\, (P_4(x_4) \wedge \neg P_3(x_4) \wedge \neg P_2(x_4) \wedge \neg P_1(x_4))$$
$$\ldots$$

Combining $\varphi_1, \ldots, \varphi_n$ conjunctively leads to the following formula:

$$\varphi_{\geq n} := \varphi_1 \wedge \varphi_2 \wedge \varphi_3 \wedge \varphi_4 \wedge \ldots \wedge \varphi_n$$

This formula can only be true if the domain contains at least n elements:

$$(U, I) \models \varphi_{\geq n} \Rightarrow U \text{ contains at least } n \text{ elements} \quad (7.3)$$

Next, consider the set

$$M := \{\varphi_{<\mathbb{N}}, \varphi_{\geq 1}, \varphi_{\geq 2}, \varphi_{\geq 3}, \ldots\} \quad (7.4)$$

Every finite subset is satisfiable. To construct a model (U, I), we determine the formula $\varphi_{\geq n}$ with the largest index n and let the domain U consist of n elements $v_1, \ldots, v_n$. By interpreting the predicate sign P_i as the relation $\{v_i\}$, (U, I) becomes a model. According to the compactness theorem, M would also have a model, which is impossible. Every model of M would need to have a finite domain because of $\varphi_{<\mathbb{N}} \in M$ and simultaneously satisfy $|U| \geq n$ for any natural number n. The contradiction shows that $\varphi_{<\mathbb{N}}$ cannot exist.

 Corollary 7.2

> The concept of finiteness is not definable within first-order predicate logic.

In addition, the discussed example lets us conclude that the compactness theorem cannot apply in higher-order predicate logic. That is because as soon as we are allowed to quantify over functions, we can pen down $\varphi_{<\mathbb{N}}$ with ease:

$$\varphi_{<\mathbb{N}} := \forall f\,(\varphi_I \to \varphi_S)$$
$$= \forall f\,(\forall x\,\forall y\,(f(x) \doteq f(y) \to x \doteq y) \to \forall y\,\exists x\,(y \doteq f(x)))$$

The above formula is familiar to us. It has already been discussed in Section 2.6.1 and matches the formula (2.7) one-to-one.

Malcev's Contribution

The Russian mathematician Anatolij Ivanovič Malcev gave the compactness theorem an inconspicuous but powerful boost. In 1936, he proved that the assumption that M is countable can be dropped [88, 129, 130]. Leon Henkin's 1947 dissertation contains a comprehensible derivation of this result [84].

 Theorem 7.3 (Compactness Theorem, Malcev 1936)

> Let M be a set of PL1 formulas. Then:
>
> M is satisfiable $\Leftrightarrow$ Every finite subset of M is satisfiable

Malcev's theorem guarantees the compactness property even if M is uncountable. In fact, he demonstrated even more. Unlike us, he had never demanded a finite formula alphabet. Malcev's result even applies to formulas drawing from an uncountable symbol set. From a logician's perspective, there are good reasons why we have restricted ourselves to finite alphabets. Nobody will ever feel the need for infinitely many symbols when writing down a formula. So why did Malcev care about

Clearly, the natural numbers with ordinary addition and multiplication form a model for Peano arithmetic. Furthermore, the discussion in this chapter has shown that the existence of a model enforces the underlying formal system to be free of contradictions. So why did Hilbert spend so long seeking a consistency proof for arithmetic?
The reason is that we have justified the consistency of PA with a semantic argument rather than a formal proof. The fact that the PA axioms are true under their standard interpretation is evident to us for intuitive reasons. After all, we have learned to deal with natural numbers since childhood. However, if we tried to formalize our intuitive arguments, we would have to consciously or unconsciously resort to the knowledge and reasoning of set theory. We would then have proved consistency in a system that contains PA as a subset – and gained nothing. No one who seriously questions the consistency of PA would trust a proof carried out in ZF, which is a potentially more uncertain system.

> In his original paper from 1915, Löwenheim formulated his famous theorem as follows:
>
> Satz 2: *"Jede Fluchtzählgleichung ist bereits in einem abzählbaren Denkbereich nicht mehr für beliebige Werte der Relativkoeffizienten erfüllt."*
> ([127], page 450)
>
> This phrase is a closed book, even for native German speakers. Löwenheim's "Denkbereich" is what we call an interpretation today, and the noun "Fluchtzählgleichung" refers to a first-order logic formula that is not necessarily universally valid, but true under all interpretations with a finite domain. In order to see how to derive Theorem 7.4 from Löwenheim's original theorem, we will first put the original formulation into a form that is easier to read for us:
>
> *"Every 'Fluchtzählgleichung' is false under at least one denumerable interpretation."*
>
> Now let φ be a formula that fulfills all preconditions of Theorem 7.4 (φ is a satisfiable formula of first-order predicate logic). If φ has a finite model, that is, a model (U,I) with $|U| < |\mathbb{N}|$, then the statement of Theorem 7.4 trivially holds. If φ has no finite model, then $\neg\varphi$ must be true under all finite interpretations. In short: $\neg\varphi$ is what Löwenheim refers to as "Fluchtzählgleichung". Then, it follows from Löwenheim's original theorem that $\neg\varphi$ is false under at least one denumerable interpretation. This interpretation is then a model for φ, and this is precisely the statement of Theorem 7.4.

this generalization at all? In the current light of argument, it seems like a trivial matter, but in the next section, we will recognize the true strength of Malcev's result. We will show that it delivers an integral part of the Löwenheim-Skolem-Tarski theorem virtually for free.

7.1.3 Löwenheim-Skolem Theorem

The Löwenheim-Skolem theorem is the fourth primary result of model theory alongside the model existence theorem, the completeness theorem, and the compactness theorem. Its historically eldest variant was formulated in 1915 by the German mathematician Leopold Löwenheim [127]. In modern terminology, it reads as follows:

Theorem 7.4 (Löwenheim, 1915)

> Let φ be a formula of first-order predicate logic. Then:
>
> ■ φ has a model $\Rightarrow$ φ has a model (U,I) with $|U| \leq |\mathbb{N}|$

Löwenheim's theorem exposes the limits of PL1 in striking clarity. It rules out the possibility of restricting a formula's models to those with uncountable domains. If a PL1 formula has a model at all, it also has a model with a countable domain.

Thoralf Skolem generalized Löwenheim's theorem in 1920 when he realized that the theorem applies not only to individual formulas but also to countable sets of formulas. His work led to the theorem we now regard as the classical formulation of the Löwenheim-Skolem theorem:

Theorem 7.5 (Löwenheim-Skolem Theorem, Skolem 1920)

> Let M be a set of PL1 formulas with $|M| \leq |\mathbb{N}|$. Then:
>
> ■ M has a model $\Rightarrow$ M has a model (U,I) with $|U| \leq |\mathbb{N}|$

In fact, Skolem proved an even stronger result. He realized that the model (U,I) can be chosen as a *submodel*, which means that U is a subset of the original domain and the interpretation of the predicate and function symbols remains unchanged for all elements remaining in U.

Unlike Löwenheim, Skolem published a correct proof. However, his work from 1920 was not without drawbacks, as he achieved his result

7.1 Meta-Results on Predicate Logic

through the axiom of choice. Nevertheless, he believed in the possibility of proving the theorem without this controversial construct of set theory. His efforts soon came to fruition, and as early as 1922, he published a variant that no longer required the axiom of choice [189].

In the following years, the assertion of the Löwenheim-Skolem theorem was strengthened even further. The most notable result in this direction is the Löwenheim-Skolem-Tarski theorem or *LST theorem* for short. According to Skolem, the Polish-American mathematician Alfred Tarski proved the theorem in a seminar in 1928; unfortunately, no written record exists [49, 191, 214]. The first published proof came from Malcev in 1936.

Referring to Theorem 7.6 as the theorem of Löwenheim-Skolem-Tarski reflects its namesake only partially. It is historically accurate to associate the content of the LST theorem with Alfred Tarski, as he is the originator of the ascending part, i.e., the part that allows us to conclude the existence of models with higher cardinalities. In comparison, the mention of Thoralf Skolem is debatable. The fact that we blatantly associate the theorem with his name today belies that Skolem could never identify himself with it. In fact, he rejected the existence of uncountable sets throughout his life and never accepted the ascending variant of the LST theorem as a meaningful theorem of mathematics. Bruno Poizat comments on the problem in [159] as follows:

"Legend has it that Thoralf Skolem, up until the end of his life, was scandalized by the association of his name to a result of this type, which he considered an absurdity, nondenumerable sets being, for him, fictions without real existence."

Theorem 7.6 (Löwenheim-Skolem-Tarski Theorem)

Let M be a set of PL1 formulas, and λ, κ be two transfinite cardinal numbers. Then:

- M has a model with $|U| = \lambda \Rightarrow M$ has a model with $|U| = \kappa$

Once again, we are facing a startling result. Suppose a set M of PL1 formulas has an infinite model. In that case, we can choose any transfinite cardinal number κ and conclude from the Löwenheim-Skolem-Tarski theorem that M also has a model of cardinality κ. In Sections 7.2 and 7.3, we will work out the far-reaching consequences of this result.

Sometimes, the Löwenheim-Skolem-Tarski theorem is split into two clauses. In this case, we speak of a *descending* and an *ascending* variant (Figure 7.7):

Theorem 7.7 (Descending LST Theorem)

Let M be a set of PL1 formulas, and λ, κ be two transfinite cardinal numbers with $\lambda > \kappa$. Then:

- M has a model with $|U| = \lambda \Rightarrow M$ has a model with $|U| = \kappa$

Put simply, the descending variant states that the existence of an infinite model implies the existence of a model for all *smaller* infinite cardinalities.

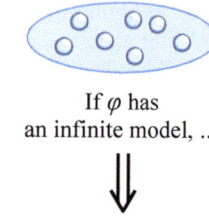

 Theorem 7.8 (Ascending LST Theorem)

Let M be a set of PL1 formulas, and λ, κ be two transfinite cardinal numbers with $\lambda < \kappa$. Then:

- M has a model with $|U| = \lambda$ $\Rightarrow$ M has a model with $|U| = \kappa$

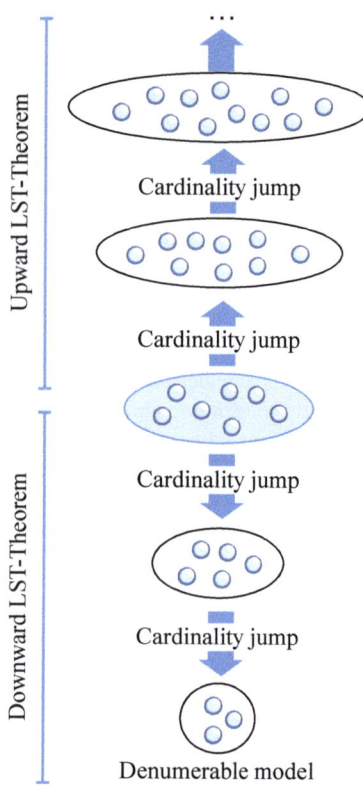

Figure 7.7: In first-order predicate logic, the Löwenheim-Skolem-Tarski theorem lets us infer the existence of models of arbitrary transfinite cardinality from the existence of an infinite model.

The ascending variant stipulates that the existence of an infinite model always implies the existence of models for all *greater* cardinalities. In particular, the ascending variant implies that every PL1 formula with a denumerable model also has an uncountable model.

We can elegantly derive the ascending variant of the LST theorem from the descending variant and the compactness theorem. As required in Theorem 7.8, let λ and κ be two transfinite cardinals with $\lambda < \kappa$. Furthermore, let M be a set of PL1 formulas with a model (U, I) satisfying $|U| = \lambda$. Then, we can obtain a model (U', I') with $|U'| = \kappa$ in a few simple steps. First, we extend our logic's alphabet with κ new constant symbols. Afterward, we express with several PL1 formulas that the new symbols are pairwise different. $\kappa \cdot \kappa$ formulas do suffice, which we combine with φ to a set M'. Every finite subset of M' is clearly satisfiable, so we can conclude from the compactness theorem that M' must also have a model.

Note that we cannot legitimize this proof step with the classical variant of the compactness theorem since M' may contain uncountably many formulas. However, the extension proved by Malcev in 1936 will do (Theorem 7.3). It guarantees that the compactness theorem still applies if we consider an uncountable number of formulas over an uncountable set of symbols.

From the compactness theorem, we know that M' has a model, but what conclusions can we draw about its cardinality? First, we note that M' is constructed such that every model must be of cardinality $\geq \kappa$. It follows from the descending LST theorem that M' also has a model of cardinality κ, which is precisely the statement of the ascending LST theorem.

Finally, it is worth noting that the Löwenheim-Skolem-Tarski theorem can be strengthened in the same sense as Theorem 7.5; the postulated models can be obtained either as submodels or as model extensions. We do not require this strengthening here and have deliberately presented the theorem in a slightly weaker but more comprehensible formulation.

7.2 Nonstandard Models of PA

As usual, we assume the consistency of Peano arithmetic; that is, no formula shall be derivable from the axioms simultaneously with its negation. Under this assumption, the natural numbers with the usual addition and multiplication are a (denumerable) model of PA, abbreviated as $(\mathbb{N}, \{s, +, \times\})$. We have called it the *standard model* of Peano arithmetic since it assigns the arithmetic formulas their intended meaning.

A *nonstandard model* of Peano arithmetic is an interpretation that satisfies the PA axioms but is not isomorphic to the natural numbers. Despite our best efforts to keep these unwelcome guests away, our acquired knowledge lets us conclude that this is doomed to fail for PA and ZF. The existence of nonstandard models is an inevitable consequence of the Löwenheim-Skolem-Tarski theorem and the fact that both Peano arithmetic and Zermelo-Fraenkel set theory are first-order theories.

Hence, it was by no means without consequences that we formulated the principle of induction as a first-order axiom schema in Section 3.1.3. Peano arithmetic thereby loses the property of being categorical. Initially, this news is difficult to digest. By abandoning the second-order induction axiom, we have created ghost models whose existence is guaranteed by the Löwenheim-Skolem-Tarski theorem beyond doubt. But how are these models structured? How can they deviate from the structure of natural numbers and still be compatible with all Peano axioms?

In the next section, we will try to lure two of these mysterious structures out of the dark. However, too much euphoria is not advisable at this point. We will soon realize that we are dealing with highly timid creatures here.

7.2.1 Denumerable Nonstandard Models

We can derive the existence of a denumerable nonstandard model with little effort from the compactness theorem and the Löwenheim-Skolem-Tarski theorem. To do so, we start by defining a series of formulas φ_n of the following kind:

$$\varphi_n := (c > \bar{n})$$

Combining these formulas with the PA axioms into the set

$$M := \text{PA} \cup \{\varphi_0, \varphi_1, \varphi_2, \ldots\},$$

■ Observation 1

"There is an element λ greater than all natural numbers."

■ Observation 2

"The number line extends to infinity on both sides of λ. Thus, λ is located on a copy of ℤ."

■ Observation 3

"The elements $2\lambda, 3\lambda, \ldots$ must be located on separate copies of ℤ."

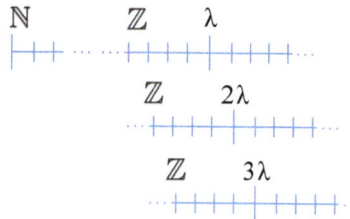

Figure 7.8: Necessary properties of a denumerable nonstandard model of Peano arithmetic

every finite subset of M is clearly satisfiable, and we can employ the compactness theorem to infer that M has a model. Given how we have constructed the formulas φ_n, the range of individuals must comprise infinitely many elements. Thus, the Löwenheim-Skolem-Tarski theorem guarantees the existence of a denumerable model. Since all formulas φ_n hold simultaneously in this model, an element that exceeds any natural number must exist. Since the set ℕ has no such element, we apparently face a model that is not isomorphic to the standard interpretation $(\mathbb{N}, \{\mathsf{s}, +, \times\})$.

Let us take a closer look at its structure. First, it must contain a zero, followed by a one, a two, and so on. In other words, the nonstandard model's structure begins with the sequence of natural numbers. In addition, there must be an element, let us call it λ, which exceeds all natural numbers (Figure 7.8 above):

$$\lambda > x \quad \text{for all } x \in \mathbb{N} \tag{7.5}$$

The Peano axioms establish that zero is the only element without a predecessor. Consequently, infinitely many elements must be left of λ that are also greater than all natural numbers. Likewise, the number line right of λ must extend to infinity because the Peano axioms demand every number to have a successor. Our nonstandard model thus begins with the structure of the natural numbers ℕ, followed by a copy of ℤ. Somewhere on this copy resides the element λ (Figure 7.8 center).

Yet, the structure is still not fully described. The existence of λ implies the existence of 2λ which cannot lie on the number line of λ, as otherwise, it would be reachable from λ by a finite number of successor steps. In particular, there would be a natural number $x \in \mathbb{N}$ with $\lambda + x = 2\lambda$. From this, $x = \lambda$ would follow, contradicting inequality (7.5). Consequently, the initial piece ℕ must be followed by an infinite number of copies of ℤ, each containing one of the elements $\lambda, 2\lambda, 3\lambda, \ldots$ (Figure 7.8 below).

However, this is not the end of the story either. Let us consider the element 3λ and suppose it is an even number; if odd, we could perform a similar analysis for the successor number $3\lambda + 1$. Even numbers can be written as twice the sum of another number, resulting in $3\lambda = 2\kappa$. The element κ satisfies $\lambda < \kappa < 2\lambda$, thus lying somewhere between λ and 2λ. Let us try to find out more about its exact location:

■ κ cannot lie on the same number line as λ. If it did, we had $\kappa = \lambda + x$ for some $x \in \mathbb{N}$, resulting in $2\kappa = 2\lambda + 2x = 3\lambda$. Thus $2x = \lambda$, contradicting inequality (7.5).

7.2 Nonstandard Models of PA

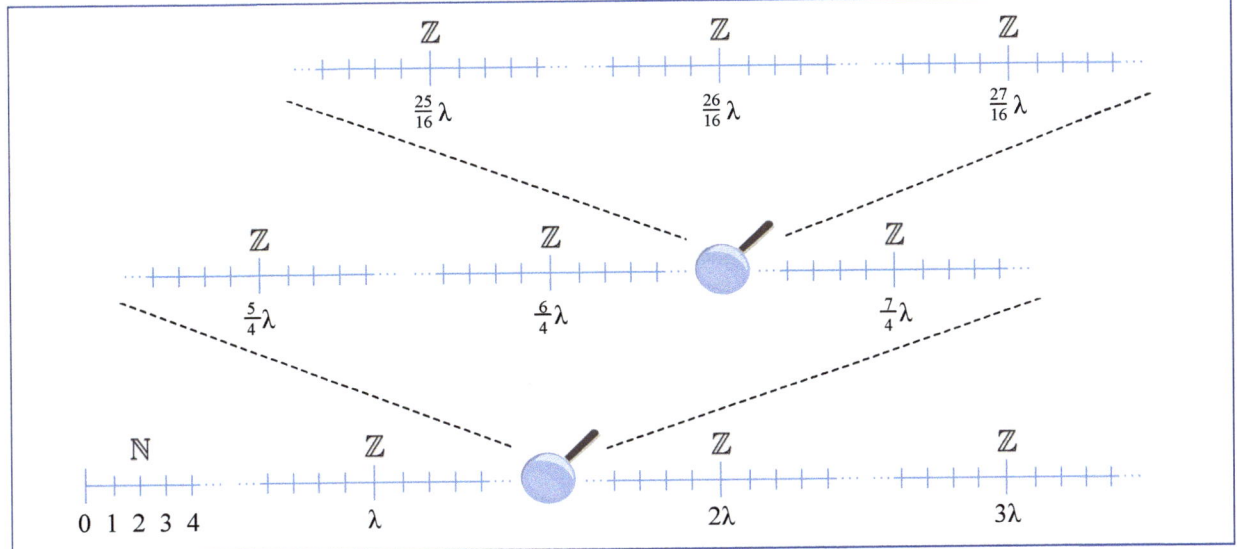

Figure 7.9: Ordering structure of the denumerable nonstandard models of Peano arithmetic

■ κ cannot lie on the same number line as 2λ. If it did, we had $\kappa + x = 2\lambda$ for some $x \in \mathbb{N}$, resulting in $2\kappa + 2x = 3\lambda + 2x = 4\lambda$. Again, we arrive at the contradiction $2x = \lambda$.

Our analysis has uncovered an additional number line between the number lines of λ and 2λ. Generalize the argument, we can always expose another one between two copies of $\mathbb{Z}$. Together, these copies form a dense, open-ended interval, thus having the same order as the rational numbers. Figure 7.9 graphically summarizes the result. The structure of the nonstandard model begins with the natural numbers $\mathbb{N}$, followed by $\mathbb{Q}$ copies of $\mathbb{Z}$.

The described properties are necessary properties of all denumerable nonstandard models with an element λ larger than all natural numbers. It can be shown that such an element exists in every nonstandard model of Peano arithmetic. Consequently, every model that is not isomorphic to the natural numbers exhibits the structure in Figure 7.9 [114].

All that's left is to define addition and multiplication consistently, that is, consistent with the Peano axioms. In fact, no one has ever come up with a suitable definition. Today, we know that no constructive proof exists. This is the astonishing result by Stanley Tennenbaum from 1959 [114, 210]:

Theorem 7.9 (Tennenbaum, 1959)

> No nonstandard model of Peano arithmetic is computable.

Specifically, Tennenbaum's theorem states that there is no denumerable nonstandard model with a computable addition and a computable multiplication. This result leaves us with a dilemma. On the one hand, we know that a countable standard model exists; we are aware of its structure, and we know there must be an interpretation of the symbols '+' and '×' compatible with the Peano axioms. On the other hand, we will never be able to precisely define the arithmetical operators; we will never succeed in always correctly calculating the sum and the product of two given elements.

Although we have gotten close to the denumerable nonstandard models of PA, we cannot walk the final step. Tennenbaum's theorem makes it unmistakably clear that the nonstandard models are surrounded by a veil we can never fully penetrate. What a fascinating result.

7.2.2 Uncountable Nonstandard Models

In Section 7.2.1, we utilized the compactness theorem and the theorem of Löwenheim-Skolem-Tarski to prove the existence of a denumerable nonstandard model. However, the Löwenheim-Skolem-Tarski theorem tells us even more. It affirms that Peano arithmetic has uncountable models, too. This section will outline their construction.

The basic building blocks of our construction are sequences of natural numbers; that is, elements of $\mathbb{N}^\mathbb{N}$ with the following form:

$$(x_0, x_1, x_2, \ldots) \text{ with } x_i \in \mathbb{N}.$$

The successor operation, addition, and multiplication are defined componentwise, as shown in Figure 7.10.

The number sequences include the natural numbers as the set

$$N := \{(x, x, x, \ldots) \mid x \in \mathbb{N}\}$$

is isomorphic to the standard interpretation $(\mathbb{N}, \{\mathsf{s}, +, \times\})$ according to the arithmetic rules we have agreed upon above (Figure 7.11).

It is almost trivial to recognize that the agreed arithmetic operations comply with the Peano axioms. Yet, does this imply that the set of

■ Successor operation

$$\begin{array}{rl} X = (& x_0, \quad x_1, \quad x_2, \ldots) \\ \hline s(X) = (& s(x_0), \ s(x_1), \ s(x_2), \ldots) \end{array}$$

■ Addition

$$\begin{array}{rl} X = (& x_0, \quad x_1, \quad x_2, \ldots) \\ Y = (& y_0, \quad y_1, \quad y_2, \ldots) \\ \hline X+Y = (x_0+y_0, x_1+y_1, x_2+y_2, \ldots) \end{array}$$

■ Multiplication

$$\begin{array}{rl} X = (& x_0, \quad x_1, \quad x_2, \ldots) \\ Y = (& y_0, \quad y_1, \quad y_2, \ldots) \\ \hline X \times Y = (x_0 \times y_0, x_1 \times y_1, x_2 \times y_2, \ldots) \end{array}$$

Figure 7.10: Componentwise addition and multiplication of number sequences

number sequences, together with componentwise addition and multiplication, is already an uncountable model of PA? The answer is no! Every model must fulfill all order properties derivable from the Peano axioms, including trichotomy, which we have already encountered in connection with ordinal numbers:

"For all X and Y, it is either $X < Y$ or $X = Y$ or $X > Y$."

As a result, for the two different number sequences

$$X := (1,0,0,0,0,0,0,0,\ldots) \tag{7.6}$$
$$Y := (0,1,0,0,0,0,0,0,\ldots) \tag{7.7}$$

either $X < Y$ or $X > Y$ has to apply. In the first case, there would be a sequence Z with $X + Z = Y$; in the latter, a sequence Z with $X = Y + Z$. However, since our number sequences must not contain negative values, such a sequence Z cannot exist.

We will circumvent the issue by permitting different number sequences to represent the same individual element. More precisely, the domain of our uncountable nonstandard model (U, I) consists of equivalence classes rather than individual number sequences (Figure 7.12):

$$U := \{[X] \mid X \in \mathbb{N}^{\mathbb{N}}\}$$

For the construction to work, we must translate the arithmetic operations, as agreed to in Figure 7.10, to the set of equivalence classes. This translation happens as usual:

$$s([X]) := [s(X)]$$
$$[X] + [Y] := [X + Y]$$
$$[X] \times [Y] := [X \times Y]$$

Next, we need to establish the rules governing the formation of equivalence classes. We do this indirectly by establishing an order on $\mathbb{N}^{\mathbb{N}}$. As soon as for two number sequences, X and Y, either $X < Y$ or $X > Y$ applies, we assign them to different equivalence classes. If, to the contrary, neither $X < Y$ nor $X > Y$ applies, we consider them as representatives of the same class.

Let us examine some examples to see how an adequate order is structured. First, consider the following sequence:

$$(1,2,3,4,5,6,\ldots)$$

Compared with a sequence of the form

$$(x,x,x,\ldots),$$

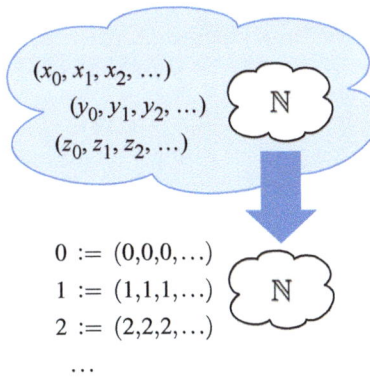

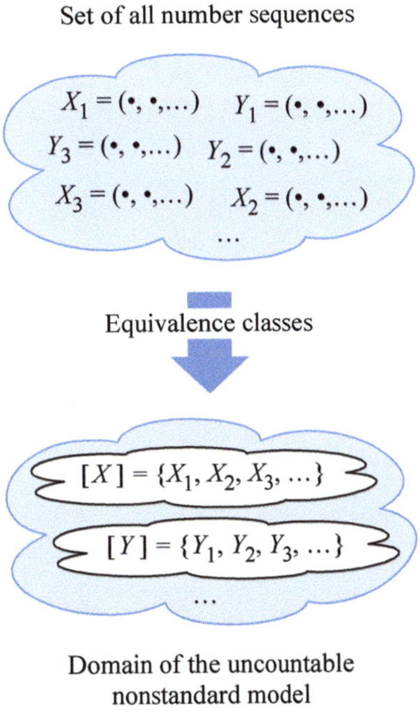

Figure 7.11: Embedding of $\mathbb{N}$ in the set of all number sequences

Figure 7.12: The domain of PA's uncountable nonstandard model is obtained by forming equivalence classes.

it is smaller at only finitely many positions but larger at infinitely many. Intuitively, we can consider it the representative of a number greater than any natural number $x \in \mathbb{N}$. It has the same property as the element λ encountered in constructing denumerable nonstandard models.

By adding a natural number to the sequence $(1,2,3,4,5,6,\ldots)$ we obtain further numbers:

$$1 + (1,2,3,4,\ldots) = (1+1, 1+2, 1+3, 1+4, \ldots) = (2,3,4,5,\ldots)$$
$$2 + (1,2,3,4,\ldots) = (2+1, 2+2, 2+3, 2+4, \ldots) = (3,4,5,6,\ldots)$$
$$3 + (1,2,3,4,\ldots) = (3+1, 3+2, 3+3, 3+4, \ldots) = (4,5,6,7,\ldots)$$

As the obtained sequences exceed their summands at all index positions, we intuitively regard them as larger numbers.

By multiplying the sequence $(1,2,3,4,5,6,\ldots)$ with 2, the resulting sequence

$$2 \times (1,2,3,4,\ldots) = (2\times 1, 2\times 2, 2\times 3, 2\times 4, \ldots) = (2,4,6,8,\ldots)$$

is again larger than anything obtainable by adding a natural number. Its ordering property is similar to the element 2λ encountered in the previous section. This element was larger than all elements of the form $\lambda + x$ with $x \in \mathbb{N}$. Putatively, we can establish an order on sequences of natural numbers reminiscent of the countable nonstandard models.

Let us step back and recap what we have accomplished so far. In the considered examples, two number sequences, X and Y, satisfied the relationship $X > Y$ precisely when the elements of X exceeded the elements of Y beyond some index position. Accordingly, $X < Y$ was true if the elements of X were eventually smaller than those of Y. Unfortunately, this definition does not suffice to define a suitable order, as two issues are yet to be solved:

- Suppose X and Y are sequences that only differ in a finite initial segment. In other words, they match beyond a certain index. We face two different number sequences in this case, but neither $X < Y$ nor $Y > X$ applies. We solve this problem by considering two sequences as representants of the same equivalence class if they only differ at a finite number of index positions. Accordingly, the sequences (7.6) and (7.7) represent the same number, zero, in this case.

- So far, we have only included constant and monotonic sequences in our analysis, where the order relation was intuitively obvious. The

7.2 Nonstandard Models of PA

The uncountable domain of the nonstandard model embeds the natural numbers in the same way the complex numbers embed the real numbers. We can imagine each number $x + yi \in \mathbb{C}$ as a point in the *Gaussian number plane*, whose coordinates are given by the real part x and the imaginary part y.

In the construction of the uncountable nonstandard model, we proceeded analogously. We have extended the set of natural numbers $\mathbb{N}$ in the same way as the set of real numbers $\mathbb{R}$ extends to the set of complex numbers $\mathbb{C}$. We find the natural numbers in the uncountable individual set of our nonstandard model in the form of the sequences $(x, x, x, \ldots)$, and just as in the case of the complex numbers, these are closed under addition and multiplication.

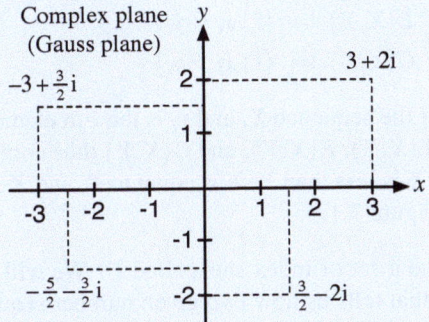

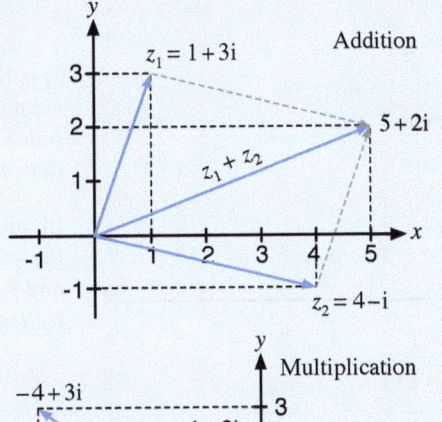

By considering the complex numbers as vectors, addition and multiplication take on a geometric interpretation. We can calculate the sum of two complex numbers by adding the vectors of the two summands. Multiplication takes on a geometric interpretation as soon as we represent the complex numbers in *polar coordinates* (r, α), where r denotes the vector length, and α is the angle to the x-axis. In this case, the product of two complex numbers (r_1, α_1) and (r_2, α_2) is the vector $(r_1 \cdot r_2, \alpha_1 + \alpha_2)$.

The complex numbers along the x-axis are closed under addition and multiplication. They behave one-to-one like the real numbers, and it is safe to say that the complex numbers on the x-axis *are* the real numbers.

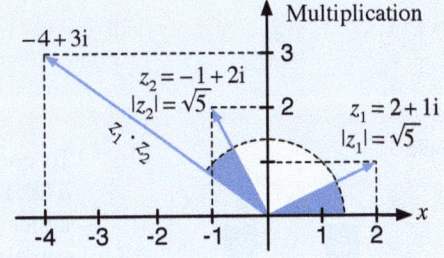

following example illustrates that the situation is generally more involved:

$$X := (1, 0, 1, 0, 1, 0, 1, 0, \ldots), \quad Y := (0, 1, 0, 1, 0, 1, 0, 1, \ldots)$$

Due to our current understanding, neither $X < Y$ nor $X > Y$ applies. Since X and Y differ at any index position, we certainly do not want to put them in the same equivalence class.

The previous example has pushed us to the boundary of what is intuitively recognizable. We can no longer provide a watertight argument as to why X or Y should be considered the larger number sequence.

- **Example 1**

	1	2	3	4	5	
X = (	0	1	2	3	4	, ...)
Y = (	2	2	2	2	2	, ...)

$$L = \{\ 1\ ,\ 2\ ,\qquad\qquad\ \dots\}$$
$$E = \{\qquad\qquad 3\ ,\qquad\ \dots\}$$
$$G = \{\qquad\qquad\qquad 4\ ,\ 5\ ,\dots\}$$

- **Example 2**

	1	2	3	4	5	
X = (	1	2	3	4	5	, ...)
Y = (	1	3	5	7	9	, ...)

$$L = \{\qquad 2\ ,\ 3\ ,\ 4\ ,\ 5\ ,\dots\}$$
$$E = \{\ 1\ ,\qquad\qquad\qquad\ \dots\}$$
$$G = \{\qquad\qquad\qquad\qquad\ \dots\}$$

- **Example 3**

	1	2	3	4	5	
X = (	0	1	0	1	0	, ...)
Y = (	1	0	1	0	1	, ...)

$$L = \{\ 1\ ,\qquad 3\ ,\qquad 5\ ,\dots\}$$
$$E = \{\qquad\qquad\qquad\qquad\ \dots\}$$
$$G = \{\qquad 2\ ,\qquad 4\ ,\qquad\dots\}$$

Figure 7.13: Calculation of the index sets $L(X,Y)$, $E(X,Y)$, and $G(X,Y)$

Ultra Filter Construction

From now on, we take a more systematic approach and define three index sets, L, E, and G, supplying us with information about the componentwise proportions of two number sequences (cf. [169]):

$$L(X,Y) := \{i \mid x_i < y_i\}$$
$$E(X,Y) := \{i \mid x_i = y_i\}$$
$$G(X,Y) := \{i \mid x_i > y_i\}$$

x_i is the i-th element of the sequence X, and y_i is the i-th element of the sequence Y. The sets $L(X,Y)$, $E(X,Y)$, and $G(X,Y)$ thus contain those index positions where X is less than Y, X is equal to Y, and X is greater than Y, respectively (Figure 7.13).

In addition, we imagine a set of index sets called F. We will use F in the sense of an oracle that tells us how two given number sequences, X and Y, compare in magnitude. More specifically, questioning the oracle follows these three simple rules (Figure 7.14):

$$L(X,Y) \in F \quad\Rightarrow\quad X < Y$$
$$E(X,Y) \in F \quad\Rightarrow\quad X = Y$$
$$G(X,Y) \in F \quad\Rightarrow\quad X > Y$$

To consult our oracle, we must first calculate the index sets $L(X,Y)$, $E(X,Y)$, and $G(X,Y)$. A simple inclusion test then determines the larger number sequence among X and Y. Intuitively, the oracle set F contains patterns of index positions. The patterns determine which positions matter in the comparison of two number sequences.

Naturally, not every set is eligible as an oracle, and it is by no means self-evident that a suitable oracle set exists at all. For this reason, let us first consider some criteria F must necessarily fulfill.

- If X and Y are identical sequences, then $E(X,Y) = \mathbb{N}$. For our oracle set to recognize the equality of X and Y, $\mathbb{N}$ must be an element of F. If X and Y match nowhere, then $E(X,Y) = \emptyset$. Since we do not want the oracle to recognize X and Y as equal, F must not contain the empty set.

  ☞ $\emptyset \notin F,\ \mathbb{N} \in F$

- Let X, Y, and Z be three number sequences. Furthermore, let $M := E(X,Y)$ and $N := E(Y,Z)$. If M and N are both contained in F, then $X = Y$ and $Y = Z$. Due to the transitivity of equality, we must require F to classify X and Z as equal, too. However, all we can say about the sequences X and Z is that they match at least at the index positions $M \cap N$. Therefore, we require that for every two sets $M, N \in F$, F must also contain the intersection $M \cap N$.

☞ $M, N \in F \Rightarrow M \cap N \in F$

- We again consider two sequences, X and Y, which F classifies as equal, that is, $E(X,Y) \in F$. Now, suppose another sequence Z matches X even better, that is, $E(X,Z) \supset E(X,Y)$. Then, the oracle set F should also confirm equality between X and Z. As a result, for every set $M \in F$, all supersets must also be part of F.

☞ $M \in F, N \supset M \Rightarrow N \in F$

- Our oracle set F can only fulfill its purpose if it provides an answer for any given number sequences X and Y. Let us assume that X and Y do not match at any index position, so $E(X,Y) = \emptyset$ applies. Since $\emptyset$ must not belong to F, either $L(X,Y)$ or $G(X,Y)$ is an element of F. In this case, however, $G(X,Y)$ is the complement of $L(X,Y)$. Consequently, for each set M, either M itself or its complement $\overline{M}$ must belong to F.

☞ $M \subseteq \mathbb{N} \Rightarrow (M \in F \Leftrightarrow \overline{M} \notin F)$

- We want to prevent two sequences, X and Y, from being recognized as equal, smaller, or larger if they satisfy the relationships $x_i = y_i$, $x_i < y_i$, or $x_i > y_i$ only for finitely many indices i, respectively. As a result, F may only contain infinite index sets.

☞ $M \in F \Rightarrow |M| = |\mathbb{N}|$

All conditions mentioned above are conditions an oracle set F must necessarily fulfill. However, this by no means answers the question of whether such a set exists. At this point, however, we are fortunate! Our requirement catalog describes a well-known and well-studied mathematical structure.

In the terminology of ordinary mathematics, a set fulfilling the first three conditions is called a *filter*. The set is a so-called *ultrafilter* if the complement property applies on top. Such filters are not refinable; adding another element will violate at least one filter property. Because of this, an ultrafilter is also called a *maximal filter*.

When an ultrafilter is made up exclusively of infinite sets, we speak of a *free ultrafilter* (Figure 7.15). The promising news is that free ultrafilters do exist. The postulated oracle set F is reality, and it can be

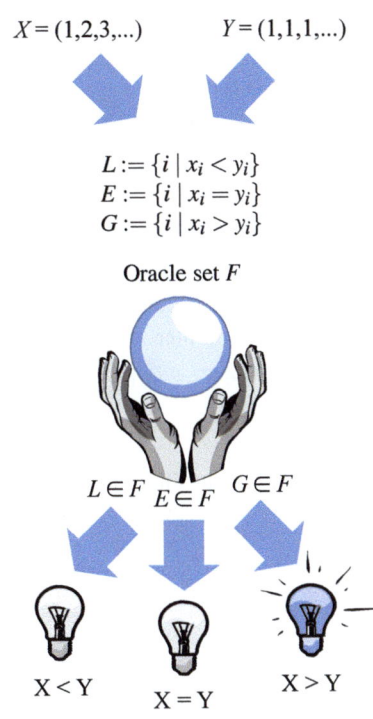

Figure 7.14: To compare two elements, X and Y, we first determine the index sets L, E, and G. Afterward, we query the oracle set F to determine which of the three index sets it contains. The result of the inclusion test determines whether X and Y are equal or one of the sequences is greater than the other.

Figure 7.15: Various filters in comparison

shown that the resulting equivalence class structure does indeed lead to an uncountable nonstandard model of Peano arithmetic [1, 114].

On the downside, the existence of free ultrafilters is only provable non-constructively using the axiom of choice. The consequences are sobering. On the one hand, we know that an ultrafilter F exists that divides the set $\mathbb{N}^{\mathbb{N}}$ into equivalence classes such that an uncountable model of Peano arithmetic emerges. On the other hand, we will never be able to generate F explicitly. The non-constructiveness of the free ultrafilters ensures that we can approach the nonstandard model only to a certain degree. Just as with denumerable nonstandard models, we are denied the final step.

7.3 Skolem's Paradox

In 1923, the Norwegian mathematician Thoralf Skolem published an article whose English translation is titled *Some remarks on axiomatized set theory* [192]. Skolem presents the reader with an alleged contradiction that sparked controversy in the scientific community at the beginning of the twentieth century. The bone of contention was the third of eight remarks on the axiomatic system of **Zermelo**'s set theory. Skolem writes:

> "If Zermelo's axiom system, when made precise, is consistent, it mus be possible to introduce an infinite sequence of symbols 1,2,3,... in such a way that they form a domain B in which all of Zermelo's axioms hold [...]. So far as I know, no one has called attention to this peculiar and apparently paradoxical state of affairs. *By virtue of the axioms we can prove the existence of higher cardinalities of higher number classes, and so forth. How can it be, then, that the entire domain B can already be enumerated by means of the finite positive integers?*"
>
> Thoralf Skolem [192]

The *Skolem paradox* is a seemingly toxic combination of two facts that appear harmless on their own:

- In first-order predicate logic, we can pen down a formula φ_{Skolem} with the substantive meaning that the domain comprises an uncountable number of elements. φ_{Skolem} has an uncountable model that is easy to construct.

7.3 Skolem's Paradox

Thoralf Albert Skolem was born on May 23, 1887, in Sandsvaer, located in southern Norway. In 1905, he completed school in Kristiania, the later Oslo, 70km from his home. In the same year, he enrolled as a mathematics student at Oslo university and graduated in 1913.

Skolem's scientific interests went far beyond mathematics, and he published his first research papers in physics. After several assistantships and a research semester at the University of Göttingen, he accepted a lecturer position in Oslo in 1918. Skolem did not originally intend to obtain a doctorate but made up for this in 1926. Between 1930 and 1938, he held a research position in Bergen in western Norway. In 1938, Skolem was finally appointed professor by the University of Oslo at the age of 51.

Skolem significantly contributed to the field of mathematical logic and set theory. His name is inextricably linked with the Löwenheim-Skolem theorem, one of the four core theorems of model theory. His proof introduced a normal form representation for predicate-logic formulas, which we now call *Skolem normal form*. Today, it is part of the curriculum of almost all logic lectures. Gödel's line of reasoning in the proofs of the compactness theorem and the theorem on the existence of models also roots back into Skolem's work. Skolem also made seminal contributions to the field of set theory. For instance, he was the first to develop a precise predicate-logical formulation of Zermelo-Fraenkel set theory [190]. Also inextricably linked to his name is the *Skolem paradox*, a putative contradiction that provides valuable insights into the nature of first-order predicate logic [189].

Skolem officially retired in 1957. Nevertheless, he continued to visit numerous universities and retained most of his official duties. As in his entire life, Skolem was an active man, even in old age, and his scientific and creative urge was unbroken. His death came suddenly and unexpectedly. Thoralf Albert Skolem died on March 23, 1963, aged 75.

■ The Löwenheim-Skolem-Tarski theorem assures that φ_{Skolem} has models of arbitrary transfinite cardinalities. Consequently, there must also be a model with a denumerable domain.

Hold on! How can a true formula assert that the domain of a countable model is uncountable? If this is not a profound paradoxical situation, then what else is?

As disturbing as the Skolem paradox may sound, we are by no means left without defense. In the previous chapters, we have sharpened our mathematical tools far enough to resolve the alleged misconception. Nevertheless, let us first convince ourselves of the existence of a formula φ_{Skolem} with the intended meaning. We will then disarm Skolem's paradox by resolving its putative contradiction.

To construct the formula φ_{Skolem}, we follow the idea from [16]. First, we establish an order '<' on the set of number pairs (x, y) with $x, y \in \mathbb{N}$, as follows:

$$(x_1, y_1) < (x_2, y_2) :\Leftrightarrow \begin{cases} x_1 + y_1 < x_2 + y_2 \text{ or} \\ x_1 + y_1 = x_2 + y_2 \text{ and } y_1 < y_2 \end{cases}$$

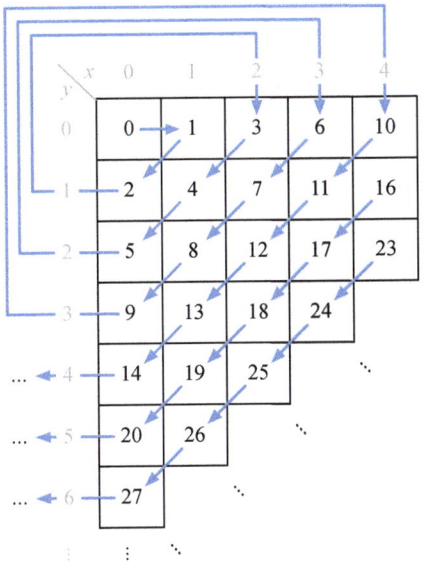

Figure 7.16: One of infinitely many enumerations of the set $\mathbb{N}^2$

The ordering compares two pairs of numbers, (x_1, y_1) and (x_2, y_2), by their component sums. If the values $x_1 + y_1$ and $x_2 + y_2$ are different,

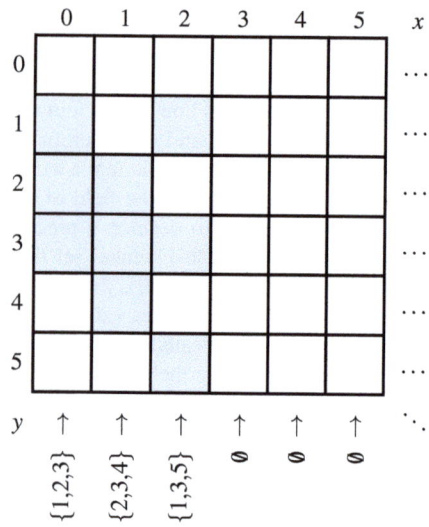

Figure 7.17: Matrix representation of the set $\{\emptyset, \{1,2,3\}, \{2,3,4\}, \{1,3,5\}\}$

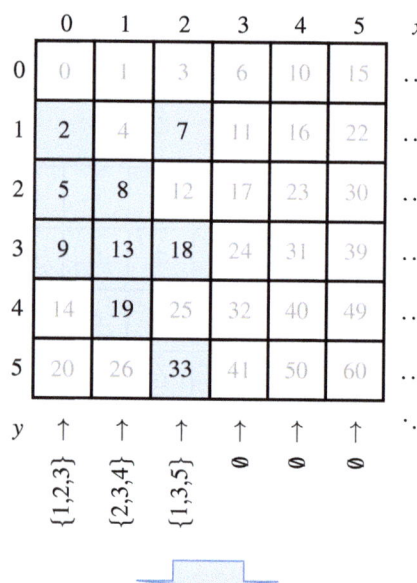

$E_M = \{2, 5, 7, 8, 9, 13, 18, 19, 33\}$

Figure 7.18: An enumerator for the set $\{\emptyset, \{1,2,3\}, \{2,3,4\}, \{1,3,5\}\}$

the pair with the smaller sum is also the smaller element. If the sums match, the elements y_1 and y_2 decide the comparison. This way, we obtain a total order on the set $\mathbb{N}^2$, plotted in Figure 7.16.

The established ordering is a familiar one. We had already used it in Chapter 1 to prove the equality of the sets $\mathbb{N}$ and $\mathbb{N}^2$ (revisit Figure 1.16). We had also shown how to calculate the position of an element (x,y) with a closed formula. Our reflections resulted in *Cantor's pairing function*, given by:

$$\pi_{\mathbb{N}}(x,y) = y + \sum_{i=0}^{x+y} i = y + \frac{(x+y)(x+y+1)}{2}$$

Next, we consider a set $M \subseteq \mathcal{P}(\mathbb{N})$, a set whose elements are subsets of the natural numbers. For instance, M may be chosen as follows:

$$M := \{\emptyset, \{1,2,3\}, \{2,3,4\}, \{1,3,5\}\}. \tag{7.8}$$

For the moment, we assume M is countable; that is, the set is either finite or has the cardinality of the natural numbers. We can then represent M as a matrix by assigning each element $\{x_1, x_2, x_3, \ldots\} \in M$ a separate column and coloring the cells in rows $x_1, x_2, x_3, \ldots$ Figure 7.17 illustrates this kind of matrix representation for our example set.

Figure 7.18 shows the matrix again, with the value of Cantor's pairing function $\pi_{\mathbb{N}}(x,y)$ denoted in each field (x,y). These numbers are the key to the next step: Representing M by a subset of the natural numbers. We can quickly obtain such a set E_M by including $\pi(x,y)$ in E_M for each marked field (x,y). For our example set M, we get:

$$E_M = \{2, 5, 7, 8, 9, 13, 18, 19, 33\} \tag{7.9}$$

We call the set E_M an *enumerator* for M.

We additionally agree that every enumerator of a set M enumerates all subsets of M. In particular, we do not require that the elements described by E_M correspond precisely to the elements of M, but only that every set contained in M occurs somewhere in the matrix representation of E_M. For example, besides E_M, the set

$$E'_M = \{2, 5, 6, 7, 8, 9, 13, 18, 19, 22, 24, 32, 33, 39, 49\}$$

also enumerates M (Figure 7.19). Conversely, E_M and E'_M are enumerators for the sets

$$\{\{1,2,3\}, \{2,3,4\}, \{1,3,5\}\}, \{\{1,2,3\}, \{2,3,4\}\}, \{\{1,2,3\}\}, \ldots$$

Enumerators and countable sets are closely related. On the one hand, we can find an enumerator for every countable set $M \subset \mathcal{P}(\mathbb{N})$. All we need to do is to translate M into its matrix representation, where we can read off the enumerator directly. On the other hand, enumerators are subsets of natural numbers and thus countable. Summing up:

 Theorem 7.10

> A set $M \subseteq \mathcal{P}(\mathbb{N})$ has an enumerator $\Leftrightarrow |M| \leq |\mathbb{N}|$

Next, we will formalize the enumerator concept within first-order predicate logic. The resulting formula will assert that the domain has no enumerator. According to Theorem 7.10, it is the formula we seek: it claims that the individual domain is uncountable.

Our formalization is based on the following colloquial characterization of enumerators:

> "E is an enumerator for the domain U if, for every set of natural numbers $z \in U$, there exists a natural number x with the property that a natural number y is in z if and only if $\pi(x,y)$ is in E."

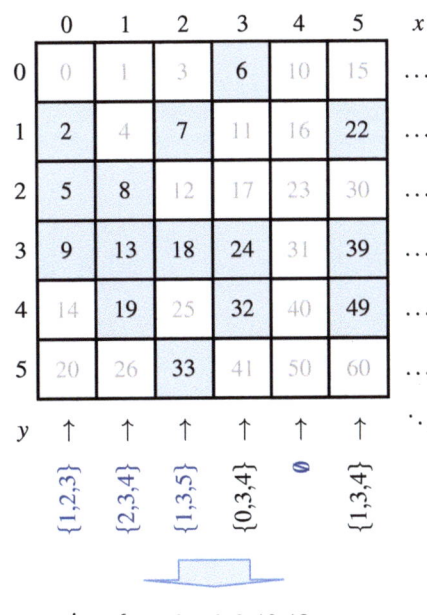

$E'_M = \{2, 5, 6, 7, 8, 9, 13, 18,$
$19, 22, 24, 32, 33, 39, 49\}$

Figure 7.19: One of many enumerators for the set
$M = \{\emptyset, \{1,2,3\}, \{2,3,4\}, \{1,3,5\}\}$

Now, let (U, I) be an interpretation with

$$U = \mathbb{N} \cup \mathcal{P}(\mathbb{N})$$

In this case, the domain comprises all natural numbers and all subsets of natural numbers. Furthermore, suppose I assigns the symbols '$\in$', S, and N the following meaning:

$(U,I) \models \mathsf{N}(\xi) :\Leftrightarrow \xi$ is a natural number
$(U,I) \models \mathsf{S}(\xi) :\Leftrightarrow \xi$ is a set of natural numbers
$(U,I) \models \xi \in \nu :\Leftrightarrow \xi$ is an element of ν

Furthermore, let π be a binary function symbol that (U,I) interprets as Cantor's pairing function $\pi_\mathbb{N}(x,y)$.

We call (U,I) the standard interpretation since it assigns the symbols their intended meaning. Now, we can easily translate the colloquial characterization of the enumerator into a formula of first-order predicate logic:

"*E is an enumerator for the domain U,*

if, for every set of natural numbers $z \in U$,
☞ $\forall z\, (S(z) \rightarrow ($

there exits a natural number x with the property
☞ $\exists x\, (N(x) \wedge ($

that a natural number y
☞ $\forall y\, (N(y) \rightarrow$

is in z if and only if $\pi(x,y)$ is in E."
☞ $(y \in z \leftrightarrow \pi(x,y) \in E))))))$

By claiming that no enumerator exists for the domain, we are done. We obtain the formula φ_{Skolem} we were looking for:

$\varphi_{\text{Skolem}} := \neg \exists w\, (S(w) \wedge$
$\forall z\, (S(z) \rightarrow (\exists x\, (N(x) \wedge (\forall y\, (N(y) \rightarrow (y \in z \leftrightarrow \pi(x,y) \in w))))))$

Clearly, φ_{Skolem} is a true statement under the interpretation (U, I) constructed above. It is true because the range of individuals $U = \mathbb{N} \cup \mathcal{P}(\mathbb{N})$ is uncountable, which is precisely what φ_{Skolem} claims.

Since φ_{Skolem} has an uncountable model, it must also have a denumerable one. This is a direct consequence of the Löwenheim-Skolem theorem and is the origin of the Skolem paradox.

The alleged contradiction disappears as soon as we analyze the content of φ_{Skolem} in detail. Above, we have associated the formula with the following meaning:

$$\varphi_{\text{Skolem}} \;\widehat{=}\; \text{"There exists no enumerator for the domain"} \qquad (7.10)$$

However, does φ_{Skolem} truly have this meaning? The answer depends on the chosen interpretation (U, I). If we write down the formula's contents accurately, it reads as follows:

$$\varphi_{\text{Skolem}} \;\widehat{=}\; \begin{array}{l}\textit{„Within the domain,}\\ \textit{there exists no enumerator for the domain."}\end{array} \qquad (7.11)$$

Under the standard interpretation $(\mathbb{N} \cup \mathcal{P}(\mathbb{N}), I)$, the statements (7.10) and (7.11) are equivalent, as the domain includes all subsets of the natural numbers, and thus, a fortiori includes all enumerators. Therefore, under the standard interpretation, the formula φ_{Skolem} indeed claims the uncountability of U.

The situation changes if we consider a model whose individual domain only comprises countably many enumerators. In such an interpretation,

7.3 Skolem's Paradox

(7.10) and (7.11) are no longer equivalent since an enumerator could exist for the individual domain that is not an element of the individual domain itself. The model would be countable, and the formula φ_{Skolem} would still be true.

Such a model can be obtained directly from the standard interpretation by replacing the set of individuals U with the set

$$U' := \mathbb{N} \cup \{\{0\}, \{0,1\}, \{0,1,2\}, \{0,1,2,3\}, \ldots\}$$

The new domain U' is denumerable, thus being representable in a matrix, as shown in Figure 7.20.

As a countable set, the new domain has an enumerator that we can easily read off from the matrix representation:

$$E_{U'} = \{0, 1, 3, 4, 6, 7, 10, 11, 12, 15, 16, 17, \ldots, \}$$

As U' contains neither the enumerator $E_{U'}$ nor any of its supersets, the formula φ_{Skolem} is a substantively true statement.

The discussion has clarified that Skolem's paradox is no contradiction at all. It dissolves as soon as we correctly formulate the content of the formula φ_{Skolem}.

In his 1923 paper, Skolem had not only formulated his alleged paradox but also provided a mathematically sound explanation. Therefore, there was never any reason to fear that the paradox would damage the edifice of mathematics like Russell's antinomy did years before. It bluntly demonstrated that some formulas' meaning is relative to the domain over which we interpret their symbols; these formulas have no absolute substantive meaning. Skolem, who refered to set theory in the context of his discussion, calls this phenomenon *"a relativity of set-theoretic notions,"* which is *"inseparably bound up with every thoroughgoing axiomatization."* [192]

Back then, Skolem's result led to controversial discussions about the meaningfulness of the formal method. The discovered interplay between the syntactic and semantic levels conflicts with our intuition, and several critics argued against building mathematics on a foundation that was mathematically consistent but lacking the strength to define the concept of set in an absolute sense. One of the critics was Skolem himself. He considered his discovery severe enough to call the entire axiomatic set theory, although still being developed, into question. His paper concludes with the following words:

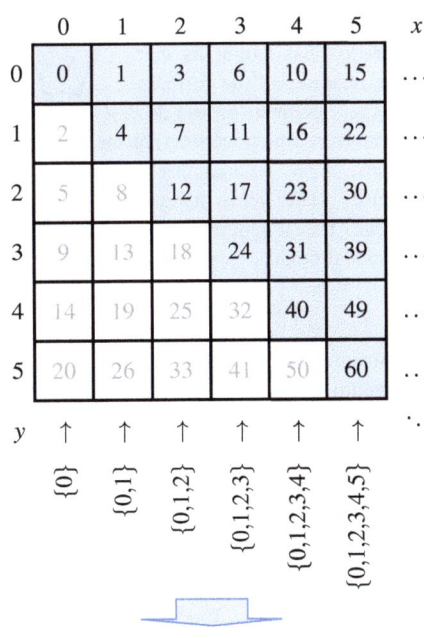

Figure 7.20: The domain U' is countable and has an enumerator according to Theorem 7.10. However, this enumerator is not part of the domain, and the formula φ_{Skolem} is thus a true statement.

> The discussion of Skolem's paradox has taught us that the meaning of the formula φ_{Skolem} depends on the domain over which we interpret its symbols. Thus, it has no meaning in an absolute sense. The following example demonstrates that this phenomenon is not ubiquitous though:
>
> $$\varphi := \exists x \, \exists y \, (x \neq y)$$
>
> Substantively, the formula states that the domain comprises at least two elements. In fact, φ has this connotation in an absolute sense; its substantive meaning is always the same under all interpretations. Therefore, we must be cautious not to draw the wrong conclusions from the paradox. It would be wrong to interpret Skolem's result as if a formula's meaning were always relative. All we learn from it is that not every formula is meaningful in an absolute sense.

"The most important result above is that set-theoretic notions are relative. I had already communicated it orally to F. Bernstein in Göttingen in the winter of 1915–1916. There are two reasons why I have not published anything about it until now: first, I have in the meantime been occupied with other problems; second, I believed that it was so clear that axiomatization in terms of sets was not a satisfactory ultimate foundation of mathematics that mathematicians would, for the most part, not be very much concerned with it. But in recent times I have seen to my surprise that so many mathematicians think that these axioms of set theory provide the ideal foundation for mathematics; therefore it seemed to me that the time had come to publish a critique."

Thoralf Skolem [192]

How do we deal with Skolem's legacy today? At present, there is little discussion on the philosophical significance of the paradox. Most mathematicians have learned to live with it; they see it as a phenomenon rather than an issue, so almost all modern treatises confine themselves to dealing with the mathematical component and cleanly resolving the supposed contradiction. From this perspective, the reactions Skolem's work provoked in the first half of the twentieth century seem exaggerated. However, were they really? Do we merely share this opinion because philosophical considerations hardly play a role in modern mathematics? It is not easy to give an objective answer because we must never forget one thing: We are all children of our time.

7.4 Boolean Models

So far, we have imagined an interpretation as a tuple (U,I) with U being a non-empty set of individuals and I a mapping that assigns a relation or a function to each predicate and function symbol. Given an interpretation (U,I), a formula φ is either true or false under this interpretation. In the first case, we write $(U,I) \models \varphi$; in the second case, $(U,I) \not\models \varphi$. It is thus possible to imagine an interpretation as a mapping of formulas to the truth values 1 (true) and 0 (false). Denoting the function value by $[\![\varphi]\!]_{(U,I)}$, this relationship reads as follows:

$$[\![\varphi]\!]_{(U,I)} := \begin{cases} 1 & \text{if } (U,I) \models \varphi \\ 0 & \text{if } (U,I) \not\models \varphi \end{cases} \qquad (7.12)$$

7.4 Boolean Models

We can generalize this principle by extending the range of $[\![\varphi]\!]_{(U,I)}$ to the elements of a Boolean algebra. In this way, we arrive at novel interpretations referred to in the literature as *Boolean-valued models*. They were pioneered at the end of the 1970s by Dana Scott, Robert Solovay, and Petr Vopěnka [185, 215]. The original idea came from Solovay in 1965 and was brought into its modern form by Scott two years later [185]. Vopěnka developed his theory independently of the others but arrived at similar conclusions.

Boolean-valued model theory was developed to facilitate access to Paul Cohen's forcing technique, created in 1963. A forcing proof involves the construction of a specific Boolean model from which several ordinary models are then derived. Suppose the construction succeeds in such a way that the theorems of a formal system are true in all models, and another formula, φ, is true in one model and false in another. Then, we know that neither φ nor $\neg\varphi$ is logically deducible from the axioms.

We want to expose the idea behind the forcing technique in broad strokes by outlining the construction of a model of ZFC in which the continuum hypothesis (CH) is false. If we assume ZFC to be correct, the existence of such a model implies the unprovability of the continuum hypothesis within ZFC.

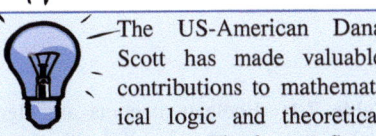

The US-American Dana Scott has made valuable contributions to mathematical logic and theoretical computer science. Thanks to Scott, we call our own a comparatively intuitive approach to Paul Cohen's proof of the independence of the continuum hypothesis. In 1972, he was awarded the Leroy P. Steele Prize for his work on Boolean-valued models.
Scott achieved his most notable success in theoretical computer science by inventing the concept of non-deterministic automata in 1959, together with Michael O. Rabin [165]. It gave rise to a new school of thought that influences computability and complexity theory to this day. For their pioneering work, Rabin and Scott were honored with the Turing Award in 1976, the most prestigious award in the field of theoretical computer science.

We already know from Chapter 1 that the forcing technique relies on the principle of model extension; that is, the newly constructed models are created by adding elements to an existing model. Furthermore, we know from Chapter 4 that we cannot guarantee ZFC to have a model due to Gödel's second incompleteness theorem. Therefore, a forcing proof is always conducted under the assumption that a model exists at all. As a matter of fact, we have to strengthen this assumption even further by requiring a *countable transitive standard model*. A standard model is a model that interprets the formula symbol '$\in$' as the element relation.

Studying standard models entails some simplifications that affect, among others, the notation we use to describe the model property. Unlike in plain first-order predicate logic, all function symbols are removed in the language of ZFC, and so are all predicates other than '$=$' and '$\in$'. To classify a ZFC formula as true or false, we no longer have to worry about the meaning of freely definable functions or predicates but only about the domain of an interpretation and the individuals assigned to the free variables. For a closed formula φ, only the domain is relevant. Thus, instead of

$$(U,I) \models \varphi, \tag{7.13}$$

Table 7.1: Boolean algebras are algebraic structures developed in the mid-nineteenth century by the British mathematician and philosopher George Boole (Figure 7.21) [12, 13].
In 1904, the American mathematician Edward Vermilye Huntington established that Boolean algebras are uniquely characterized by four laws, today known as Huntington's axioms.

- **Commutative laws**

$$a \wedge b = b \wedge a$$
$$a \vee b = b \vee a$$

- **Neutral elements**

There exist 1, 0 with

$$a \wedge 1 = a$$
$$a \vee 0 = a$$

- **Distributive laws**

$$a \wedge (b \vee c) = (a \wedge b) \vee (a \wedge c)$$
$$a \vee (b \wedge c) = (a \vee b) \wedge (a \vee c)$$

- **Inverse elements**

For all a, a $\neg a$ exists with

$$a \wedge \neg a = 0$$
$$a \vee \neg a = 1$$

A forcing proof presupposes the existence of a countable transitive standard model of ZFC, which raises the question of whether such a model necessarily exists. The honest answer to this question is no, even if we assume the consistency of ZFC. If ZFC is consistent, the model existence theorem and the Löwenheim-Skolem-Tarski theorem guarantee the existence of countable models. However, they do not guarantee that one of these models is a transitive standard model. Nevertheless, there are reasons why most mathematicians consider the strengthened precondition as reasonable. It partly stems from discoveries in the theory of large cardinal number axioms, which we cannot discuss in this book.
You may have wondered why we demanded that the assumed model be transitive. At a second glance, this requirement turns out to be natural. If a model is not transitive, at least one set needs to contain an element that is not part of the set universe. Although this property is compatible with Zermelo's original idea that sets are hierarchical aggregations of separately existing objects, it is not compatible with the modern conception that all elements of sets are themselves sets.

we will employ the short-hand notation

$$U \models \varphi. \qquad (7.14)$$

We will also adapt this notation for open formulas. In this case, (7.14) shall express that the relationship (7.13) holds for all possible assignments of the free variables.

If a set U meets (7.14) for every formula provable in ZFC, we call it a model of ZFC. The postulated model we will use to create new models via forcing is referred to as $\mathcal{M}$ below. As mentioned above, we assume that $\mathcal{M}$ comprises countably many elements, $\mathcal{M}$ is transitive in the sense of Definition 3.7, and the symbol '$\in$' is interpreted as the ordinary membership relation. In short, $\mathcal{M}$ is a countable transitive standard model of ZFC.

Our first intermediate goal is to extend $\mathcal{M}$ to a so-called *Boolean-valued model* $\mathcal{M}^{(B)}$. To do so, we need to do a little groundwork first. Specifically, we must clarify the exact meaning of a *Boolean algebra* and a *Boolean set*.

7.4.1 Boolean Algebras

Let '$\wedge$' and '$\vee$' be binary operators on a non-empty set B. We call the triple $(B, \wedge, \vee)$ a *Boolean algebra* if the two operations satisfy the four *Huntington axioms* from Table 7.1. Be careful not to misinterpret the term *axiom* at this point, as we have only used it in the context of formal systems so far. Here, Huntington's axioms are to be understood as ordinary algebraic relations and not as the axioms of a formal system. The same applies to the symbols '$\wedge$' and '$\vee$', which carry a double

7.4 Boolean Models

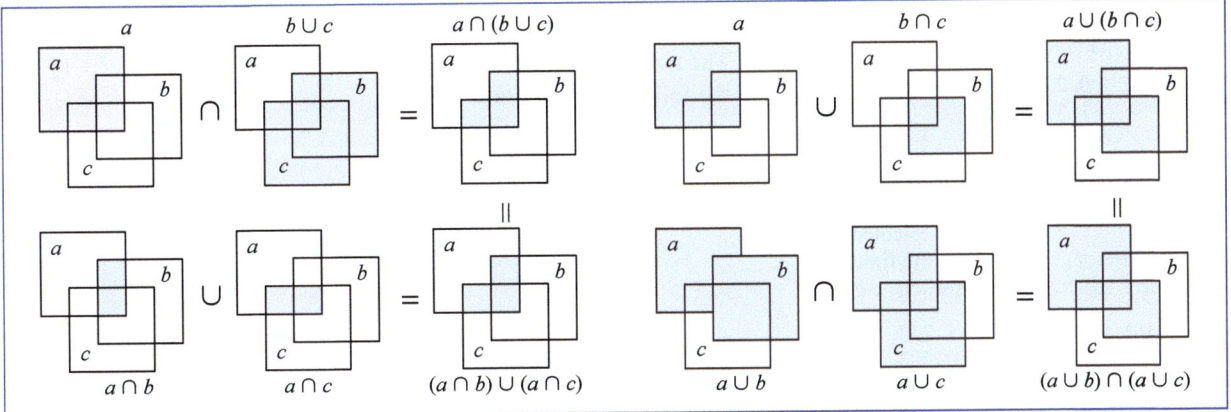

Figure 7.22: Graphical demonstration of the distributive laws using Venn diagrams

meaning now. In the context of Boolean algebras, they stand for the operations defined on B and yet have nothing to do with the two symbols we frequently use in logic formulas.

Let us bring the concept of a Boolean algebra to life by examining two prominent examples:

■ **Power-set Algebra**

For a non-empty but otherwise arbitrary set T, the triple $(B, \wedge, \vee)$ given by

$$B := \mathcal{P}(T)$$
$$a \wedge b := a \cap b$$
$$a \vee b := a \cup b$$

is a Boolean algebra. We can easily verify that $(B, \wedge, \vee)$ satisfies Huntington's axioms. The validity of the commutative laws is almost self-evident, as is the existence of neutral and inverse elements: The neutral elements 0 and 1 are the empty set $\emptyset$ and the element T, respectively, and the inverse element of a set a is the complement set $T \backslash a$. The validity of the distributive laws becomes evident by representing the sets in the form of *Venn diagrams*, as depicted in Figure 7.22.

Figure 7.24 illustrates the structure of the power set algebras for $T_1 = \{1\}$, $T_2 = \{1,2\}$, and $T_3 = \{1,2,3\}$.

George Boole
(1815 – 1864)

Figure 7.21: The British mathematician and philosopher George Boole was one of the most influential logicians of the nineteenth century. His work, *The Laws of Thought*, ranks among the most significant publications of early mathematical logic. [12].

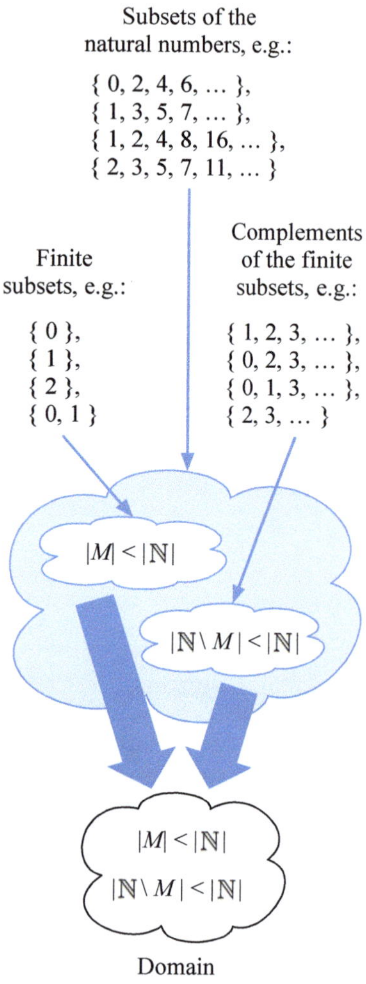

Figure 7.23: Constructing the domain of an incomplete Boolean algebra

■ **Two-element Boolean Algebra**

The two-element Boolean algebra is given by the set $B = \{0, 1\}$ and the following operator definition:

$$a \wedge b := \begin{cases} 1 & \text{if } a = 1 \text{ and } b = 1 \\ 0 & \text{otherwise} \end{cases}$$

$$a \vee b := \begin{cases} 1 & \text{if } a = 1 \text{ or } b = 1 \\ 0 & \text{otherwise} \end{cases}$$

From a practical point of view, the two-element algebra is the most important among all Boolean algebras. It is the mathematical foundation of digital circuit design and, in this sense, nothing less than the theoretical underpinning of computer engineering [99]. For this reason, the literature on computer science often refers to the two-element Boolean algebra as *switching algebra*. Paul Halmos coined another name by simply calling it the Boolean algebra **2**.

From a mathematical point of view, the switching algebra is a particular case of a power-set algebra. By setting $0 := \emptyset$ and $1 := \{u\}$ for an arbitrary element u, the two-element algebra becomes isomorphic to the set algebra $(\mathcal{P}(\{u\}), \cap, \cup)$.

A necessary property all Boolean algebras must fulfill is the closure of '$\wedge$' and '$\vee$'; that is, we must obtain another element from B when combining two elements $a, b \in B$ to either $a \wedge b$ or $a \vee b$. The power-set algebra meets this property even for an infinite number of elements, permitting us to safely extend the operator definition to arbitrary subsets $A \subseteq B$:

$$\bigwedge_{a \in A} a := \bigcap_{a \in A} a \quad \text{(Infimum of } A\text{)}$$

$$\bigvee_{a \in A} a := \bigcup_{a \in A} a \quad \text{(Supremum of } A\text{)}$$

The structure of the power set ensures that for any subset $A \subseteq B$, both the supremum and the infimum are elements of B. Boolean algebras with this property are called *complete*.

All Boolean algebras over a finite set B are complete, just like all power-set algebras, which inevitably raises the question of whether incomplete Boolean algebras exist at all. The answer is yes!

We can obtain such a structure from the power-set algebra. To do so, we aim to remove some elements from B such that Huntington's axioms are

7.4 Boolean Models

still satisfied, but at the same time, not every subset continues to possess an infimum or a supremum. The following example demonstrates the construction. Starting with the natural numbers, we define B as follows:

$$B := \{M \in \mathcal{P}(\mathbb{N}) \mid |M| < |\mathbb{N}|\} \cup \{M \in \mathcal{P}(\mathbb{N}) \mid |\mathbb{N} \setminus M| < |\mathbb{N}|\}$$

Simply put, a subset M of the natural numbers belongs to B precisely if M or its complement $\mathbb{N} \setminus M$ is finite (Figure 7.23). It is easy to check that $(B, \cap, \cup)$ remains a Boolean algebra. The elements $0 = \emptyset$ and $1 = \mathbb{N}$ are present in B, and for each $a \in B$, the inverse element $\neg a$ also belongs to B. Furthermore, the set operations '$\cap$' and '$\cup$' never lead outside of B, thus assuring that Huntington's axioms continue to hold.

However, unlike the abovementioned algebras, $(B, \cap, \cup)$ is no longer complete. For example,

$$A := \{\{0\}, \{2\}, \{4\}, \{6\}, \ldots\}$$

is a subset of B whose supremum

$$\bigcup_{a \in A} a = \{0, 2, 4, 6, \ldots\}$$

does not belong to B. Neither the set $\{0, 2, 4, 6, \ldots\}$ nor its complement $\{1, 3, 5, 7, \ldots\}$ is finite.

Only complete Boolean algebras are relevant for the considerations in this section. The reasons for this will become almost self-evident later in this chapter.

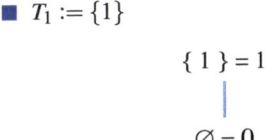

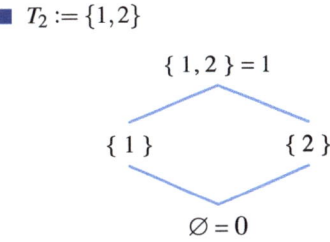

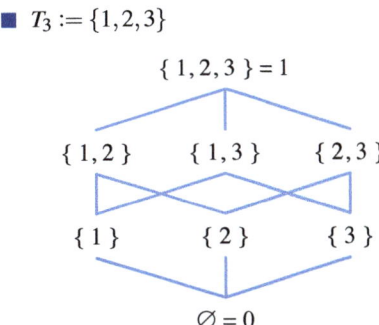

Figure 7.24: Order structures of the first three power-set algebras

At this juncture, let us take another look at the illustrations in Figure 7.24. We can easily spot two peculiarities. First, the sub-set relation '$\subseteq$' is a partial order bounded by a largest and a smallest element. Second, the largest and smallest elements are represented by two prominent members of the Boolean algebra, the two neutral elements 1 and 0. Interestingly, such an ordering structure is not only present in power-set algebras but in every Boolean algebra. By defining

$$a \leq b :\Leftrightarrow a \wedge b = a,$$

every Boolean algebra $(B, \wedge, \vee)$ becomes a partial ordering $(B, \leq)$. Note that the order is not total in general since two elements $a, b \in B$ not necessarily satisfy $a \leq b$ or $b \leq a$.

Defining an order on the elements of B allows us to translate important order-theoretic concepts to Boolean algebras. Of particular importance for us is the idea of a filter, as given by the following definition:

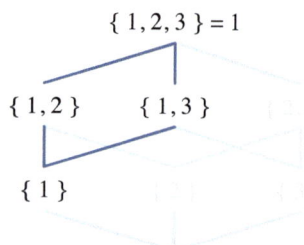

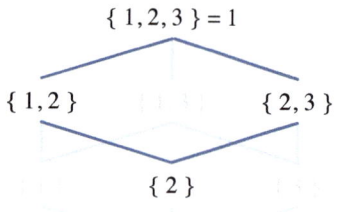

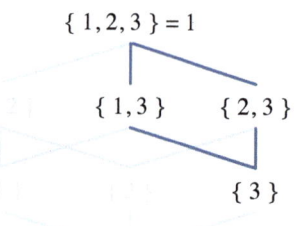

Figure 7.25: Ultrafilters of the power-set algebra $(\mathcal{P}(\{1,2,3\}), \cap, \cup)$.

 Definition 7.2 (Filter of a Boolean Algebra)

Let $(B, \wedge, \vee)$ be a Boolean algebra. A *filter* is a subset $F \subset B$ satisfying the following properties:

$$1 \in F, \, 0 \notin F \tag{7.15}$$
$$a \in F \text{ and } b \in B \text{ and } b \geq a \;\Rightarrow\; b \in F \tag{7.16}$$
$$a \in F \text{ and } b \in F \;\Rightarrow\; a \wedge b \in F \tag{7.17}$$

F is an *ultrafilter*, if additionally:

$$a \in F \;\Leftrightarrow\; \neg a \notin F \tag{7.18}$$

We are already familiar with the ultrafilter concept from Section 7.2.2, where we used it in a more specific context.

Figure 7.25 depicts the ultrafilters of the power set algebra

$$(\mathcal{P}(\{1,2,3\}), \cap, \cup).$$

Only three subsets of B, namely

$$F_1 := \{\{1,2,3\}, \{1,2\}, \{1,3\}, \{1\}\}$$
$$F_2 := \{\{1,2,3\}, \{1,2\}, \{2,3\}, \{2\}\}$$
$$F_3 := \{\{1,2,3\}, \{1,3\}, \{2,3\}, \{3\}\}$$

qualify as ultrafilters.

We will come across the term *generic filter* several times below. To explain its exact meaning, we need to introduce the notion of a *dense subset* of a Boolean algebra. The term refers to a subset $D \subseteq B\setminus\{0\}$ that contains for each element of $B\setminus\{0\}$ either the element itself or a smaller one:

$$\text{For all } b \in B\setminus\{0\}, \text{ there exists a } d \in D \text{ such that } d \leq b. \tag{7.19}$$

We can translate the concept of a dense subset to any partial order $(P, \leq)$ by simply replacing the set $B\setminus\{0\}$ in (7.19) with P. In this case, no special treatment of the zero element is required.

The introduced notions put the following definition on a solid footing:

7.4 Boolean Models

 Definition 7.3

Let $(B, \wedge, \vee)$ be a Boolean algebra. A filter $G \subset B$ is called *generic* if it satisfies the following intersection property:

$$G \cap D \neq \emptyset \text{ for all dense subsets } D \subseteq B \backslash \{0\} \quad (7.20)$$

It can be shown that every generic filter of a Boolean algebra is also an ultrafilter, but not vice versa. Consequently, we can consider a generic filter as a special ultrafilter that shares at least one element with each dense subset of B.

It is only a small step from here to another concept we need below: the $\mathcal{M}$-generic filter. This expression refers to a filter that restricts the intersection property to sets contained in $\mathcal{M}$. Therefore, every generic filter is also $\mathcal{M}$-generic, but not vice versa.

7.4.2 Boolean Sets

For all ordinary sets, a set u may or may not be an element of another set x. In the first case, we write $u \in x$. In the second case, we write $u \notin x$. In the Boolean world, we will interpret formulas using Boolean sets, which we can think of as a variant of fuzzy sets. They build upon the idea that an element does not have to belong to a set entirely; instead, it can be an element of another set to only a certain degree.

Among the best-known fuzzy sets are those that express the membership property by a probability out of the interval $[0; 1]$. Boolean sets follow the same idea. They differ from fuzzy sets only in describing the degree of set membership by an element of a Boolean algebra rather than a number.

Mathematically, a fuzzy set is represented by an ordinary set; in the case of a Boolean set, by a function that assigns the elements of its domain, which consists of Boolean sets itself, to the elements of a Boolean algebra.

By examining two example sets,

$$x := \{(\emptyset, \{2,3\}), (\{(\emptyset, \{1,3\})\}, \{1,3\})\},$$
$$y := \{(\emptyset, \{1,3\}), (\{(\emptyset, \{1,2,3\})\}, \{1,2,3\})\},$$

The model construction in this section relies on an $\mathcal{M}$-generic filter, which inevitably raises the question of its existence. A lemma named after the Polish mathematicians Helena Rasiowa and Roman Sikorski tells us the conditions under which we can assume the existence of such a filter. In the formulation for Boolean algebras, it reads as follows:

> Let $(B, \wedge, \vee)$ be a Boolean algebra, $b \in B \backslash \{0\}$ and D the set of all dense subsets of $B \backslash \{0\}$. If D contains countably many elements, then there exists a generic filter $G \subset B$ with $b \in G$.

The prerequisite that D contains a countable number of elements is essential, as the lemma becomes false otherwise. Fortunately, we are safe in this respect, as we only need an $\mathcal{M}$-generic filter and, by assumption, $\mathcal{M}$ only contains countably many elements. Consequently, the set of all dense subsets in $\mathcal{M}$ is also countable, which guarantees the existence of an $\mathcal{M}$-generic filter. In fact, the Rasiowa-Sikorski lemma tells us even more. It stipulates that for every element $b \in B \backslash \{0\}$, there exists a generic filter containing b.

$x = \{(\emptyset, \{2,3\}), (\{(\emptyset, \{1,3\})\}, \{1,3\})\}$
$y = \{(\emptyset, \{1,3\}), (\{(\emptyset, \{1,2,3\})\}, \{1,2,3\})\}$

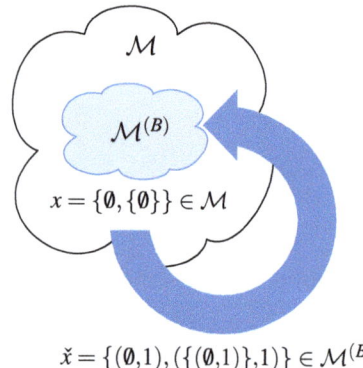

World 1
World 2
World 3

$x = \{\emptyset, \{\emptyset\}\}$
$y = \{\emptyset, \{\emptyset\}\}$
$x = y$

$x = \{\emptyset\}$
$y = \{\{\emptyset\}\}$
$x \in y$

$x = \{\{\emptyset\}\}$
$y = \{\emptyset, \{\emptyset\}\}$
$x \subseteq y$

Figure 7.26: Multi-world interpretation of Boolean sets

$\mathcal{M}$
$\mathcal{M}^{(B)}$
$x = \{\emptyset, \{\emptyset\}\} \in \mathcal{M}$

$\check{x} = \{(\emptyset, 1), (\{(\emptyset, 1)\}, 1)\} \in \mathcal{M}^{(B)}$

Figure 7.27: Embedding of $\mathcal{M}$ in $\mathcal{M}^{(B)}$ by forming standard representatives

let us elaborate on what this means. Both are Boolean sets over the power-set algebra

$$(\mathcal{P}(\{1,2,3\}), \cap, \cup).$$

Although the elements of a Boolean algebra could also be interpreted as probabilities similar to the fuzzy sets mentioned above, thinking in terms of different worlds is more intuitive here. For example, we can imagine the three elements of the set $\{1,2,3\}$ as three possible worlds and interpret a Boolean set in each world as an ordinary set. Figure 7.26 illustrates this idea for our example sets.

The Boolean truth value $[\![\varphi]\!]$ of a ZFC formula φ can be interpreted similarly: $[\![\varphi]\!]$ indicates where the formula φ is a substantively true statement. A formula with $[\![\varphi]\!] = 1$ is true in all worlds, a formula with $[\![\varphi]\!] = 0$ is false in all worlds, and a formula with $0 < [\![\varphi]\!] < 1$ is true in some worlds and false in others. For the sets x and y defined above, the following holds:

$$[\![x \subseteq y]\!] = \{1, 3\} \tag{7.21}$$
$$[\![x \in y]\!] = \{2\} \tag{7.22}$$
$$[\![x = y]\!] = \{3\} \tag{7.23}$$

At this stage, we want to introduce a distinguished set that will accompany us until the end of this chapter: set $\mathcal{M}^{(B)}$. For a complete Boolean algebra $(B, \wedge, \vee)$, this set contains precisely those elements from the postulated ground model $\mathcal{M}$ that exhibit the structure of a Boolean set. In short, $\mathcal{M}^{(B)}$ is the set of all Boolean sets from $\mathcal{M}$.

Although $\mathcal{M}$ is a proper superset of $\mathcal{M}^{(B)}$, it is entirely embedded in $\mathcal{M}^{(B)}$. With

$$\check{x} := \{(\check{u}, 1) \mid u \in x\}, \tag{7.24}$$

every element $x \in \mathcal{M}$ has a counterpart in $\mathcal{M}^{(B)}$. In the terminology of Boolean models, $\check{x}$ is called the *standard representation* or the *standard representative* of x (Figure 7.27).

7.4.3 Boolean Semantics

At this point, we have introduced all the necessary means to define function $[\![\cdot]\!]$ formally. However, to write the definition down as clearly as possible, let us agree on several notational simplifications. First, we recall that the Boolean truth value of a formula φ depends on the domain $\mathcal{M}^{(B)}$ and the assignment of the free variables in φ to elements from $\mathcal{M}^{(B)}$. The notation $[\![\varphi]\!]_{\mathcal{M}^{(B)}, I}$ emphasizes this dependency. We will utilize an intuitive notation that we had already anticipated in (7.21) to (7.23) without explicitly pointing it out. There, we took the liberty of replacing the free variables in a formula with the sets assigned by I. Below, we will continue this way. In particular, by writing

$$[\![x \in y]\!]_{\mathcal{M}^{(B)}} = b \tag{7.25}$$

we mean to express the following relationship:

$$[\![\xi \in \zeta]\!]_{\mathcal{M}^{(B)}, I} = b \text{ with } I(\xi) = x \text{ and } I(\zeta) = y$$

Mentioning I in expressions of the form (7.25) is no longer necessary, as the assignment of the free variables becomes apparent.

When writing $[\![\varphi]\!]_{\mathcal{M}^{(B)}} = b$, without explicitly mentioning the assignment of the free variables of φ, we express that the truth value equals b for all possible assignments of the free variables. The agreed simplifications allow us to define function $[\![\cdot]\!]_{\mathcal{M}^{(B)}}$ rather clearly:

Definition 7.4 (Boolean Truth Function)

Let $\mathcal{M}$ be a countable transitive standard model of ZFC and $(B, \wedge, \vee)$ a complete Boolean algebra. The *truth function* $[\![\cdot]\!]$ is defined inductively:

$$[\![1]\!]_{\mathcal{M}^{(B)}} := 1$$
$$[\![0]\!]_{\mathcal{M}^{(B)}} := 0$$
$$[\![x \in y]\!]_{\mathcal{M}^{(B)}} := \bigvee_{v \in \mathrm{dom}(y)} \left(y(v) \wedge [\![x = v]\!]_{\mathcal{M}^{(B)}} \right) \tag{7.26}$$
$$[\![x = y]\!]_{\mathcal{M}^{(B)}} := [\![x \subseteq y]\!]_{\mathcal{M}^{(B)}} \wedge [\![y \subseteq x]\!]_{\mathcal{M}^{(B)}}$$
$$[\![x \subseteq y]\!]_{\mathcal{M}^{(B)}} := \bigwedge_{u \in \mathrm{dom}(x)} \left(x(u) \to [\![u \in y]\!]_{\mathcal{M}^{(B)}} \right) \tag{7.27}$$
$$[\![\neg \varphi]\!]_{\mathcal{M}^{(B)}} := \neg [\![\varphi]\!]_{\mathcal{M}^{(B)}}$$
$$[\![\varphi \wedge \psi]\!]_{\mathcal{M}^{(B)}} := [\![\varphi]\!]_{\mathcal{M}^{(B)}} \wedge [\![\psi]\!]_{\mathcal{M}^{(B)}}$$
$$[\![\varphi \vee \psi]\!]_{\mathcal{M}^{(B)}} := [\![\varphi]\!]_{\mathcal{M}^{(B)}} \vee [\![\psi]\!]_{\mathcal{M}^{(B)}}$$
$$[\![\forall \xi\, \varphi(\xi)]\!]_{\mathcal{M}^{(B)}} := \bigwedge_{x \in \mathcal{M}^{(B)}} [\![\varphi(x)]\!]_{\mathcal{M}^{(B)}} \tag{7.28}$$
$$[\![\exists \xi\, \varphi(\xi)]\!]_{\mathcal{M}^{(B)}} := \bigvee_{x \in \mathcal{M}^{(B)}} [\![\varphi(x)]\!]_{\mathcal{M}^{(B)}} \tag{7.29}$$

The unmentioned connectives '$\to$', '$\leftrightarrow$', and '$\nleftrightarrow$' are broken down to the connectives '$\neg$', '$\wedge$', and '$\vee$' in the usual way.

It is worth taking a closer look at how Definition 7.4 defines the semantics of the following two expressions:

$$[\![x \in y]\!]_{\mathcal{M}^{(B)}}$$
$$[\![x \subseteq y]\!]_{\mathcal{M}^{(B)}}$$

In order to intuitively grasp the seemingly complex right-hand sides of (7.26) and (7.27), let us recall the multiverse idea used above to illustrate Boolean sets. To decide in which worlds the relationship $[\![x \in y]\!]$ holds, we can check in each world separately whether y contains an element v that matches x. The set of all worlds where a set v is contained in y is $y(v)$, and the set of all worlds where x and v match is $[\![x = v]\!]$.

The definition of the subset relation follows a similar line of reasoning: The relation $x \subseteq y$ is true in a world if all elements u that are contained in x in this world also occur in y in the same world. Putting the above considerations into symbolic form results directly in the expressions defined in Definition 7.4.

At first glance, the definitions are subject to prohibited self-references, as the element relationship traces back to the subset relationship and the subset relationship back to the element relationship. At second glance, however, it becomes evident that we are dealing with a conjoint inductive definition. The foundation of the $\in$ relation ensures that the recursion eventually reaches one of the base cases, $[\![x \in \emptyset]\!]$ or $[\![\emptyset \subseteq y]\!]$.

The definition makes it apparent why we solely include complete Boolean algebras in our analysis. Only these algebras ensure that the expressions (7.28) and (7.29) take on a definite value.

We have already elucidated above why we cannot speak of true and false formulas in the Boolean case without further ado. However, there is an exception if the truth function yields either 0 or 1. It is legitimate to regard 1 as true and 0 as false, which we also want to express linguistically. We say a formula φ is true in $\mathcal{M}^{(B)}$ if it satisfies $[\![\varphi]\!]_{\mathcal{M}^{(B)}} = 1$. If $[\![\varphi]\!]_{\mathcal{M}^{(B)}} = 0$, we speak of a false formula.

We are now ready to present the central theorem of this section. It states that all theorems of ZFC are true in $\mathcal{M}^{(B)}$:

 Theorem 7.11 (Boolean Model Property)

> Let $\mathcal{M}$ be a countable transitive standard model of ZFC and $(B, \wedge, \vee)$ a complete Boolean algebra. Then, each ZFC formula φ satisfies:
>
> $$\vdash \varphi \;\Rightarrow\; [\![\varphi]\!]_{\mathcal{M}^{(B)}} = 1 \qquad (7.30)$$

A detailed proof of this heavyweight theorem is presented, e.g., in [7]. In the following, when we call the set $\mathcal{M}^{(B)}$ a *boolean model*, we verbally paraphrase the symbolical relationship (7.30) in Theorem 7.11.

7.5 Forcing

We know from Section 7.4.3 that $\mathcal{M}^{(B)}$ is a Boolean model of ZFC. Despite being a remarkable fact on its own, its real power only unfolds in combination with another key result: the ability to translate Boolean models back into ordinary models.

The model generated from $\mathcal{M}^{(B)}$ is called $\mathcal{M}[G]$ and guarantees that the true and false formulas of $\mathcal{M}^{(B)}$ are also true or false in $\mathcal{M}[G]$. Expressed symbolically, the following applies:

$$[\![\varphi]\!]_{\mathcal{M}^{(B)}} = 1 \;\Rightarrow\; \mathcal{M}[G] \models \varphi \qquad (7.31)$$
$$[\![\varphi]\!]_{\mathcal{M}^{(B)}} = 0 \;\Rightarrow\; \mathcal{M}[G] \not\models \varphi \qquad (7.32)$$

In addition, $\mathcal{M}^{(B)}$ contains formulas with truth values other than 0 or 1. These formulas are neither true nor false in $\mathcal{M}^{(B)}$. The set G, which is a subset of B, controls whether these formulas become true or false in $\mathcal{M}[G]$. In particular, G determines the truth content as following:

$$\mathcal{M}[G] \models \varphi \;\Leftrightarrow\; [\![\varphi]\!]_{\mathcal{M}^{(B)}} \in G \qquad (7.33)$$

Thus, φ becomes a true statement in $\mathcal{M}[G]$ precisely when its Boolean truth value is an element of G (Figure 7.28). However, the described construction will not succeed for arbitrary subsets of B; it will only succeed for those that meet certain criteria. For example, in order to bring the relationship (7.33) in line with the two previously formulated properties (7.31) and (7.32), we must impose the following two constraints on G:

$$1 \in G,\; 0 \notin G \qquad (7.34)$$

Do these relationships sound familiar? A closer look at Definition 7.2 shows that we have also imposed these requirements on those subsets of B forming a filter. The nature of G is thus being disclosed. The above transformation requires G to be an $\mathcal{M}$-generic filter to succeed.

Overall, the following procedure seems plausible: To identify a statement such as the continuum hypothesis as undecidable, we mentally switch from the postulated ground model $\mathcal{M}$ to the Boolean model $\mathcal{M}^{(B)}$ and determine the Boolean truth value of the continuum hypothesis. Let us denote this truth value with b and assume b is different from 0 and 1, as only then can an independence proof succeed. Next, we create an $\mathcal{M}$-generic filter G_1 with $b \in G_1$ and an $\mathcal{M}$-generic filter G_2 with $b \notin G_2$. Transforming the Boolean model $\mathcal{M}^{(B)}$ back into ordinary models yields a model $\mathcal{M}[G_1]$ in which the continuum hypothesis

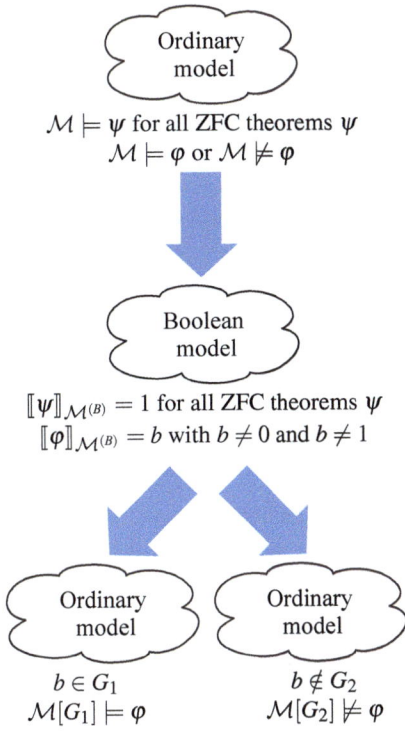

Figure 7.28: To prove the independence of a formula φ, the postulated model $\mathcal{M}$ is translated into a Boolean model $\mathcal{M}^{(B)}$. In $\mathcal{M}^{(B)}$, every formula provable in ZFC is true. The formula φ, however, is neither true nor false. In the next step, $\mathcal{M}^{(B)}$ is translated back into two ordinary models such that φ is true in one model and false in the other. A generic filter G controls the construction. The formula φ is true in $\mathcal{M}[G]$ precisely if the Boolean truth value of φ belongs to G.

In technical terms, the reverse transformation from $\mathcal{M}^{(B)}$ to $\mathcal{M}[G]$ takes two steps. In the first step, $\mathcal{M}^{(B)}$ is converted into a so-called *quotient model* $\mathcal{M}^{(B)}/G$, consolidating the elements of $\mathcal{M}^{(B)}$ into equivalence classes. The classes are formed by the following equivalence relation on the elements of $\mathcal{M}^{(B)}$:

$$x \sim_G y \;:\Leftrightarrow\; [\![x = y]\!]_{\mathcal{M}^{(B)}} \in G$$

The set of all equivalence classes is the quotient model $\mathcal{M}^{(B)}/G$:

$$\mathcal{M}^{(B)}/G := \{\, [x]_G \mid x \in \mathcal{M}^{(B)} \,\}$$

The quotient model is an ordinary model of ZFC in which the fuzziness of $\mathcal{M}^{(B)}$ is gone. However, in contrast to $\mathcal{M}$, where the construction started, the quotient model is a nonstandard model because the relation '$\in$' is no longer interpreted as the ordinary element relation. Instead, the symbol '$\in$' bears the following meaning:

$$\mathcal{M}[G] \models [x]_G \in [y]_G \;:\Leftrightarrow\; [\![x \in y]\!]_{\mathcal{M}^{(B)}} \in G$$

In the second step, the quotient model $\mathcal{M}^{(B)}/G$ undergoes a transformation discovered by the Polish mathematician Andrzej Mostowski in 1949 [141]. The outcome of this transformation is called the *Mostowski collapse*. It is the transitive standard model $\mathcal{M}[G]$ we sought.

is true and a model $\mathcal{M}[G_2]$ in which it is false. The existence of these models then implies the undecidability of the continuum hypothesis in ZFC.

As plausible as this plan may sound, it is challenging to implement in the described form. First and foremost, we are confronted with the problem of hardly knowing anything about the postulated model $\mathcal{M}$. In fact, we cannot even be sure whether this model exists at all. Even less do we know about the Boolean truth value of CH in $\mathcal{M}^{(B)}$, let alone which Boolean algebra $(B, \wedge, \vee)$ we should use to form the model $\mathcal{M}^{(B)}$. Therefore, we choose a different starting point, granted by the following theorem:

Theorem 7.12 (Model Theorem for Boolean Algebras)

> Let $\mathcal{M}$ be a countable transitive standard model of ZFC, $(B, \wedge, \vee)$ a complete Boolean algebra from $\mathcal{M}$, and $G \subset B$ an $\mathcal{M}$-generic filter. Then ZFC has a countable transitive standard model $\mathcal{M}_G$ with the following properties:
>
> $$\mathcal{M} \subseteq \mathcal{M}[G] \qquad (7.35)$$
> $$G \in \mathcal{M}[G] \qquad (7.36)$$

The model theorem discloses that the generated model $\mathcal{M}[G]$ contains the generic filter G as an element. Unlike the Boolean algebra $(B, \wedge, \vee)$, G is not required to belong to $\mathcal{M}$. In this case, $\mathcal{M}[G]$ is a proper model extension with G as a new element. This property is our window of opportunity, opening a gateway to sneak in the set G into $\mathcal{M}$.

Before implementing this plan, we will first generalize the model theorem into a form that is easier to handle. It particular, it does no longer rely on the concept of a Boolean algebra:

Theorem 7.13 (Model Theorem for Partial Orders)

> Let $\mathcal{M}$ be a countable transitive standard model of ZFC, $(P, \leq)$ a partial order from $\mathcal{M}$, and $G \subseteq P$ an $\mathcal{M}$-generic filter. Then ZFC has a countable transitive standard model $\mathcal{M}[G]$ with the following properties:
>
> $$\mathcal{M} \subseteq \mathcal{M}[G] \qquad (7.37)$$
> $$G \in \mathcal{M}[G] \qquad (7.38)$$

7.5 Forcing

In essence, the generalization involves replacing the Boolean algebra $(B, \wedge, \vee)$ with a partial order $(P, \leq)$ and the filter $G \subset B$ with a corresponding filter $G \subset P$. The following definition clarifies what such a filter looks like:

Definition 7.5 (Filter of a Partial Order)

Let $(P, \leq)$ be a partial order. A *filter* is a non-empty subset $F \subseteq P$ with the following properties:

$$a \in F \text{ and } b \in P \text{ and } b \geq a \;\Rightarrow\; b \in F \tag{7.39}$$
$$a \in F \text{ and } b \in F \;\Rightarrow\; c \leq a \text{ and } c \leq b \text{ for some } c \in F \tag{7.40}$$

The following also applies to filters defined on orderings: If such a filter shares at least one element with all dense subsets of P, it is called *generic*, and we talk of an $\mathcal{M}$-generic filter if it fulfills the intersection property for all dense subsets contained in $\mathcal{M}$.

With the help of Theorem 7.13, we will attempt to construct a model of ZFC in which the continuum hypothesis is a false statement. For this purpose, we will adjoin a bijective mapping between the sets $\mathbb{R}$ and $\aleph_2$. Such a mapping implies that no bijection can exist between $\mathbb{R}$ and $\aleph_1$, contradicting the continuum hypothesis. Filters are no mappings, which eliminates the possibility of directly adding the bijection as G. Instead, we take advantage of the fact that G is not the only set added in the course of a model extension. In order to retain the model property, $\mathcal{M}[G]$ must necessarily include all those sets that are constructible from G. Our plan is, therefore, to find a partial order $(P, \leq)$ with an $\mathcal{M}$-generic filter that allows the construction of a bijective mapping between $\mathbb{R}$ and $\aleph_2$.

The partial order $(P, \supseteq)$ defined by

$$P := \{ f : \aleph_2^{\mathcal{M}} \times \mathbb{N} \to \{0,1\} \mid f \in \mathcal{M} \text{ and } \mathrm{dom}(f) \text{ is finite} \}$$

is appropriate for our purposes. Simply put, P includes all partial functions from $\mathcal{M}$ that map a finite subset of $\aleph_2^{\mathcal{M}} \times \mathbb{N}$ into the set $\{0,1\}$. The superset relation orders the elements of P: A function $f \in P$ is smaller than a function $g \in P$ precisely if its domain is a superset of the domain of g, and the function values match wherever both functions are defined.

As illustrated in Figures 7.29 and 7.30, we can imagine every function from P as an infinite table that extends horizontally over all elements

In a forcing proof, we need to be aware of a phenomenon already discussed in the section about Skolem's paradox: The phenomenon that the meaning of a formula may depend on the domain over which we interpret its symbols. To understand the implications in the context of model construction, recall that all theorems of ZFC are substantively true statements about the sets contained in $\mathcal{M}$. All mathematical objects whose existence is provable within ZFC also exist in $\mathcal{M}$. In addition to trivial objects, such as the empty set, the cardinal numbers $\aleph_1$ and $\aleph_2$ exist in every model, too. Viewed from the outside, however, these sets may be different. For example, the sets that take on the role of the cardinals $\aleph_1$ and $\aleph_2$ within $\mathcal{M}$ are countable from the outside and thus different from the uncountable sets labeled $\aleph_1$ and $\aleph_2$ outside of $\mathcal{M}$.

If a symbol can signify different sets, it is advisable to supplement the model to which it refers. For instance, the symbols $\aleph_1^{\mathcal{M}}$ and $\aleph_2^{\mathcal{M}}$ denote the sets that take on the role of the second or third transfinite cardinal number within $\mathcal{M}$. The same applies to the real numbers. $\mathbb{R}^{\mathcal{M}}$ refers to the (externally countable) set that takes on the role of the real numbers within $\mathcal{M}$. Not all sets are relative, though! For example, the set of natural numbers $\mathbb{N}$ is the same in all models. For sets that exist in such an *absolute* sense, we can do without supplementing the set symbol with a model name.

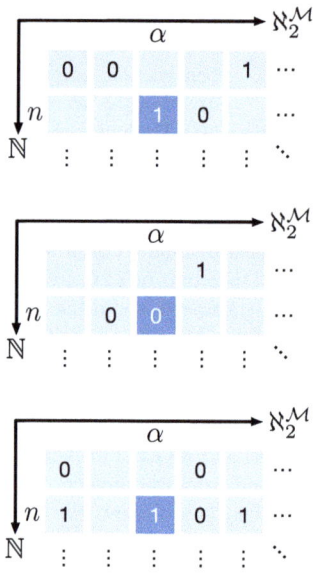

Figure 7.29: Elements of the partial order $(P, \supseteq)$

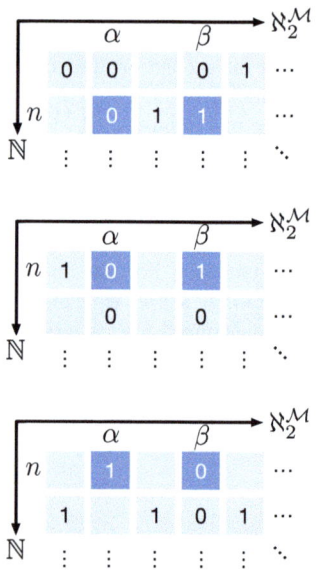

Figure 7.30: Further elements of the partial order $(P, \supseteq)$

from $\aleph_2^{\mathcal{M}}$ and vertically over all elements from $\mathbb{N}$. This way, the functions in P correspond precisely to those tables with a finite number of entries marked with zeros or ones.

Like P, a generic filter $G \subseteq P$ is also a set of functions, and we can derive several elementary relationships from the filter property of G. First of all, it follows from (7.40) that for two functions $p, q \in G$, there must exist another function $r \in G$, satisfying both $r \leq p$ and $r \leq q$. Consequently, r is a function whose domain includes the domains of p and q, and takes on the same function value as p and q wherever p or q are defined. Hence, the function values of p and q must coincide in the intersection of their domains, and the union $p \cup q$ is, in turn, a function. The same applies when combining all elements of G: The filter property of G implies that the union $\bigcup G$ must be a partial function f of the following form:

$$f : \aleph_2^{\mathcal{M}} \times \mathbb{N} \to \{0, 1\}$$

As a matter of fact, we can say even more about f. For this purpose, let us revisit Figure 7.29. We deliberately chose the depicted examples to have their domains contain the element (α, n). Let us join all functions with this property in set $D_{(\alpha,n)}$. The examples in Figure 7.30 are not random either. They obey the property that the function values in all columns marked with α and β differ in at least one row. Let us join all those functions in set $D_{\alpha,\beta}$.

The sets $D_{(\alpha,n)}$ and $D_{\alpha,\beta}$ are dense in $(P, \supseteq)$. Thus, the generic filter G has at least one element in common with each set. From this, we can draw two important conclusions:

■ From the relationship

$$G \cap D_{(\alpha,n)} \neq \emptyset \text{ for all } \alpha \in \aleph_2^{\mathcal{M}} \text{ and all } n \in \mathbb{N}$$

it follows that the domain of function f contains every element from the set $\aleph_2^{\mathcal{M}} \times \mathbb{N}$. In other words: The function f is total.

■ From the relationship

$$G \cap D_{\alpha,\beta} \neq \emptyset \text{ for all } \alpha, \beta \in \aleph_2^{\mathcal{M}}$$

it follows that the partial functions $f_\alpha(n) := f(\alpha, n)$ are pairwise distinct.

Now, we are ready for the grand finale. Since the generic filter G is an element of $\mathcal{M}[G]$, we also find function $f = \bigcup G$ in $\mathcal{M}[G]$, as well as

7.5 Forcing

the functions f_α. Each of these functions describes a real number in $\mathcal{M}[G]$. Since these functions are pairwise different, $\mathcal{M}[G]$ must contain at least $\aleph_2^\mathcal{M}$ different real numbers, symbolically written as follows:

$$\mathcal{M}[G] \models |\mathbb{R}^\mathcal{M}| \geq \aleph_2^\mathcal{M} \qquad (7.41)$$

In order to cross the finish line, one final step is necessary. We already know that the sets playing the roles of $\mathbb{R}$ and $\aleph_2$ in $\mathcal{M}[G]$ satisfy inequality (7.41) in $\mathcal{M}[G]$. However, we must prove that the inequality applies to the sets $\mathbb{R}^{\mathcal{M}[G]}$ and $\aleph_2^{\mathcal{M}[G]}$.

Since $\mathcal{M}[G]$ is a proper model extension of $\mathcal{M}$, $\mathcal{M}[G]$ contains at least as many real numbers as $\mathcal{M}$. Consequently, we can reformulate (7.41) as follows:

$$\mathcal{M}[G] \models |\mathbb{R}| \geq \aleph_2^\mathcal{M} \qquad (7.42)$$

However, our mathematical tools are not sufficiently developed to take the last step. At this juncture, we need a property known as *countable chain condition*, or CCC for short. If a Boolean algebra exhibits this property, it can be shown that the model $\mathcal{M}[G]$, when generated through Theorem 7.12, contains the same cardinal numbers as the model $\mathcal{M}$, thus satisfying $\aleph_2^\mathcal{M} = \aleph_2^{\mathcal{M}[G]}$. An analogous relationship applies if we form $\mathcal{M}[G]$ using Theorem 7.13 with a corresponding half-order. From this it follows that we can rewrite (7.42) into

$$\mathcal{M}[G] \models |\mathbb{R}| \geq \aleph_2 \qquad (7.43)$$

which is the same as

$$\mathcal{M}[G] \models 2^{\aleph_0} \geq \aleph_2.$$

At this point, we are done. With $\mathcal{M}[G]$, we have created a model in which the continuum hypothesis is a substantively false statement.

Readers with prior knowledge of the forcing technique will undoubtedly miss a particular symbol: the *forcing relation* '$\Vdash$'. Introduced by Paul Cohen in his original work, this sign has become the symbolic figurehead of this proof technique over time. The fact that this symbol appears nowhere in our presentation is due to our decision to introduce the forcing technique via Boolean-valued models. Cohen was unaware of this elegant approach when he developed his theory, which is also why his original derivation differs significantly from the one outlined here. In reality, however, the difference is not as big as it appears at first glance. The terms and concepts of the Boolean world map straight onto the terms and concepts devised by Cohen, including the forcing

If a and b are two elements of a Boolean algebra, we write $a \perp b$ if a and b are two *incompatible* elements, that is, two elements whose conjunction results in the zero element:

$$a \perp b :\Leftrightarrow a \wedge b = 0$$

An *antichain* of a Boolean algebra $(B, \wedge, \vee)$ is a subset $A \subseteq B \backslash \{0\}$ with the property that all elements of A are pairwise incompatible. Since two elements $a, b \in B \backslash \{0\}$ are always compatible if $a \leq b$, all elements of an antichain turn out to be incomparable with each other. This property makes an antichain the opposite of a *chain*, defined as a linearly ordered subset of B. For two elements a and b of a chain, $a \leq b$ or $b \leq a$ applies. The so-called *countable chain condition* demands that each antichain of a Boolean algebra has a countable number of elements. The fact that this property, as indicated in the text, affects the nature of the cardinal numbers within the Boolean model $\mathcal{M}^{(B)}$ is astonishing since no direct intuitive connection becomes apparent. In order to derive this result mathematically, numerous details from the theory of Boolean models are necessary, which would go far beyond the scope of this short introduction.

relation, which relates to the conceptual framework presented here as follows:

$$p \Vdash \varphi \Leftrightarrow p \leq [\![\varphi]\!]_{\mathcal{M}^{(B)}} \quad (7.44)$$

This also sheds some light on why Cohen called his proof technique forcing and the relation '$\Vdash$' the forcing relation. If the generic filter G includes the element p, it also includes all elements q with $p \leq q$. Consequently, every formula φ with $p \leq [\![\varphi]\!]_{\mathcal{M}^{(B)}}$ becomes a true statement in $\mathcal{M}[G]$. In other words, the choice $p \in G$ enforces the truth of φ in $\mathcal{M}[G]$. The left-hand side of (7.44) reads similarly. $p \Vdash \varphi$ means

"*p forces φ.*"

As powerful as the forcing technique may be, its details are incredibly intricate, and it takes a book of its own to go into them in mathematical rigor. Hence, the explanations on the previous pages are merely a tentative foray into this exciting area of set theory. Readers who wish to delve deeper into the theory of Boolean models and the forcing technique find the omitted details in the overview articles [33] and [52] as well as the books [7], [176], and [108]. Readers proficient in German may also enjoy reading my book on this account [102]. It bears the title *"Forcing: Eine Einführung in die Mathematik der Unabhängigkeitsbeweise"* and presents the mathematical details of the forcing method this section could only sketch.

7.6 Exercises

Exercise 7.1

In this exercise, we utilize first-order predicate logic to describe directed graphs. We select the set of individuals U of an interpretation (U,I) as the set of graph nodes and let the binary predicate symbol E determine whether there is an edge between two nodes, x and y. The following example illustrates the relationship between interpretations and graphs:

- Interpretation (U,I)
- Graph G

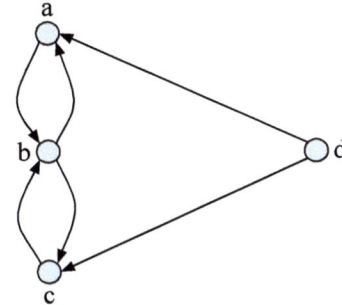

$U := \{a, b, c, d\}$

$I(E) := \{(a,b),$
$\quad\quad\quad (b,a),$
$\quad\quad\quad (b,c),$
$\quad\quad\quad (c,b),$
$\quad\quad\quad (d,a),$
$\quad\quad\quad (d,c)\}$

Furthermore, the following list of first-order predicate-logic formulas is given:

$\varphi_1 := \neg E(a,b)$
$\varphi_2 := \neg \exists z_1 \, (E(a,z_1) \wedge E(z_1,b))$
$\varphi_3 := \neg \exists z_1 \, \exists z_2 \, (E(a,z_1) \wedge E(z_1,z_2) \wedge E(z_2,b))$
$\ldots := \ldots$

a and b are constant symbols.

a) What is the intuitive meaning of the formula φ_n?

b) The property of being a connected graph is not definable within first-order predicate logic. Prove this assertion by showing that no PL1 formula φ with the following property exists:

$$(U,I) \models \varphi \Leftrightarrow (U,I) \text{ describes a connected graph}$$

Remember: A graph is connected if there is a path between two arbitrary nodes, x and y. A path from x to y is a finite sequence of edges connecting x and y in the desired direction.

Exercise 7.2

Let K be a formal system capable of deriving arithmetical statements. Suppose K fulfills the following property:

$$\vdash \varphi \Leftrightarrow \not\models \varphi$$

- Theorems of K are, for example, $1+1=3$, $\exists x \forall y\, x > y$, $\neg \exists y\, 1+y = 2, \ldots$
- No theorems of K are, for example, $1+1=2$, $\forall x \exists y\, x \leq y$, $\exists y\, 1+y = 2, \ldots$

K is the perfect liar as its axioms allow us to derive precisely those arithmetic formulas being false in the standard model of Peano arithmetic.

a) Can the symbols 's', '=', '+', and '×' be reinterpreted such that K has a model?

b) Can a formal system with the postulated property exist at all?

Exercise 7.3

Let φ be a formula of first-order predicate logic with equality. Are the following statements true or false? Justify your answers.

a) If φ has a finite model, it also has an infinite model.

b) If φ has an infinite model, it also has a finite model.

c) Can a) or b) be answered with the help of the Löwenheim-Skolem-Tarski theorem?

Exercise 7.4

Euclid's theorem postulates the existence of infinitely many prime numbers, which implies that no natural number x is dividable by every prime number. Peano arithmetic can express this fact as follows:

$$\varphi := \neg \exists x \forall y\, (\text{prime}(y) \to y \mid x)$$

The formula φ is apparently a true statement in the standard model of Peano arithmetic. As usual, we assume Peano arithmetic is free of contradictions.

- Show that φ is unprovable within Peano arithmetic.
- Is the unprovability of φ a consequence of Gödel's incompleteness theorem?

7.6 Exercises

Exercise 7.5

In our wording, the model existence theorem also applies to predicate logic with equality. If we restrict ourselves to PL1 without equality, we can even tighten it a bit:

Theorem 7.14 (Model Existence, PL1 Without Equality)

Let K be a first-order theory. Then:

$$K \text{ has a denumerable model} \Leftrightarrow K \text{ is consistent}$$

The difference, though subtle, is significant: In predicate logic without equality, consistency does not only imply the existence of a model but the existence of a *denumerable* model

a) What are the consequences regarding the Löwenheim-Skolem-Tarski theorem?

b) Show that the generalization does not apply to predicate logic with equality.

Exercise 7.6

Let $M := \{N \subseteq \mathbb{N} \mid 1 \in N\}$

a) Which of the following statements are true? Which are false?

	True	False
■ M is a filter.	○	○
■ M is a maximal filter.	○	○
■ M is an ultrafilter.	○	○
■ M is a free ultrafilter.	○	○

b) Let F be an ultrafilter. What can we say about its elements if it contains the set $\{1\}$?

Exercise 7.7 In this exercise, let E_1 and E_2 be two enumerators with:

$$E_1 := \{5,8,9,18,19,25,33\}$$
$$E_2 := \{8,12,19,23,24,25,31,32,39,49,60\}$$

a) Create a matrix representation of E_1 and E_2:

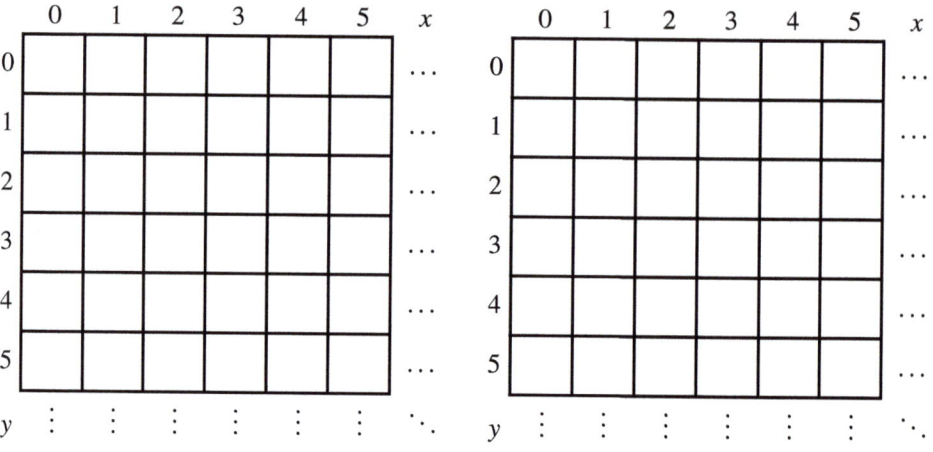

b) For which of the following sets is E_1 or E_2 an enumerator?

$M_1 := \{\emptyset, \{2,3\}, \{2,4\}\}$ $\qquad$ $M_3 := \{\{2,3\}, \{2,4\}\}$

$M_2 := \{\emptyset, \{2,3\}, \{2,4\}, \{3,4\}, \{3,4,5\}\}$ $\qquad$ $M_4 := \{\{2,3\}, \{2,4\}, \{3,4\}, \{3,4,5\}\}$

Exercise 7.8 From Section 7.3, we know that an enumerator of a set M is, by definition, also an enumerator for all subsets of M.

Suppose for the moment that we had dispensed with the subset rule. Then, a set $E \in \mathcal{P}(\mathbb{N})$ would be an enumerator for a set $M \subset \mathcal{P}(\mathbb{N})$ if M contains precisely those sets that occur in the matrix representation of E. The set

$$E = \{0,4,12,24,40,60\}$$

would, therefore, only be an enumerator for the set

$$M = \{\{0\}, \{1\}, \{2\}, \{3\}, \{4\}, \{5\}\}.$$

7.6 Exercises

	0	1	2	3	4	5	x
0	0	1	3	6	10	15	...
1	2	4	7	11	16	22	...
2	5	8	12	17	23	30	...
3	9	13	18	24	31	39	...
4	14	19	25	32	40	49	...
5	20	26	33	41	50	60	...
y	↑	↑	↑	↑	↑	↑	⋱
	$\{0\}$	$\{1\}$	$\{2\}$	$\{3\}$	$\{4\}$	$\{5\}$	

At first glance, this definition better captures our intuitive idea of an enumerator than the original, yet we have deliberately refrained from using it. What might be the reasons?

Exercise 7.9

a) Does the triple $(\mathcal{P}(\emptyset), \cap, \cup)$ constitute a Boolean algebra?

b) Prove the existence of a finite Boolean algebra with an odd number of elements.

Exercise 7.10

The following list includes several laws that hold in every Boolean algebra. Derive each law from Huntington's four axioms.

- **Associative laws**

  $a \vee (b \vee c) = (a \vee b) \vee c$
  $a \wedge (b \wedge c) = (a \wedge b) \wedge c$

- **Idempotence laws**

  $a \vee a = a$
  $a \wedge a = a$

- **Absorption laws**

  $a \vee (a \wedge b) = a$
  $a \wedge (a \vee b) = a$

- **De Morgan's laws**

  $\neg(a \vee b) = \neg a \wedge \neg b$
  $\neg(a \wedge b) = \neg a \vee \neg b$

- **Annulment laws**

  $a \vee 1 = 1$
  $a \wedge 0 = 0$

- **Double negation law**

  $\neg\neg a = a$

Exercise 7.11   In Section 7.4.1, we introduced the '≤' relation as an order on the elements of a Boolean algebra. Show the following equivalences:

$$a \leq b \Leftrightarrow a \vee b = b \Leftrightarrow a \to b = 1$$

Exercise 7.12   In Section 7.5, we have utilized the forcing technique to extend the ground model $\mathcal{M}$ to a model $\mathcal{M}[G]$ that contains a bijection between $\mathbb{R}$ and $\aleph_2$. Applying the forcing machinery to $\mathcal{M}[G]$ a second time generates another model with a bijection between $\mathbb{R}$ and $\aleph_3$, for instance. Consequently, this model would also contain a bijection between $\aleph_2$ and $\aleph_3$, which is impossible.

Where is this argument flawed?

Exercise 7.13   In this exercise, let us try to apply Theorem 7.13 to the half-order $(P, \supseteq)$ given by:

$$P := \{ f : \aleph_1^{\mathcal{M}} \to \mathbb{R}^{\mathcal{M}} \mid f \in \mathcal{M} \text{ and } \mathrm{dom}(f) \text{ is countable in } \mathcal{M} \}$$

Expressed in plain words, P comprises all partial functions from $\mathcal{M}$ that are defined on a countable subset of $\aleph_1^{\mathcal{M}}$ and map to sets that belong to the real numbers in $\mathcal{M}$. The superset relation orders the elements of P.

Next, let us define the following sets:

$$D_\alpha := \{ p \in P \mid \alpha \in \mathrm{dom}(p) \} \qquad (7.45)$$
$$D_x := \{ p \in P \mid x \in \mathrm{ran}(p) \} \qquad (7.46)$$

Verbally put, set D_α contains those functions from P defined at $\alpha \in \aleph_1^{\mathcal{M}}$, and set D_x contains all functions from P that map at least one element to the real number $x \in \mathcal{M}$.

a) Show that the two sets are dense in P for all $\alpha \in \aleph_1^{\mathcal{M}}$ and all $x \in \mathbb{R}^{\mathcal{M}}$.

b) Let G be a generic filter on P. Show that the set $\bigcup G$ is a total and surjective function from $\aleph_1^{\mathcal{M}}$ to $\mathbb{R}^{\mathcal{M}}$.

c) Part b) implies $\mathcal{M}[G] \models |\mathbb{R}^{\mathcal{M}}| \leq |\aleph_1^{\mathcal{M}}|$. Does this mean that the continuum hypothesis is true in $\mathcal{M}[G]$?

Bibliography

[1] Abian, A.: Nonstandard models for arithmetic and analysis. In: *Studia Logica* 33 (1974), Nr. 1, S. 11–22

[2] Ackermann, W.: Zum Hilbert'schen Aufbau der reellen Zahlen. In: *Mathematische Annalen* 99 (1928), S. 118–133

[3] Ackermann, W.: Zur Axiomatik der Mengenlehre. In: *Mathematische Annalen* 131 (1956), August, Nr. 4, S. 336–345

[4] Aliprand, J.: *The Unicode Standard Version 5.0.* Boston, MA: Addison-Wesley, 2007

[5] Amos, M.: *Theoretical and Experimental DNA Computation.* Berlin, Heidelberg, New York: Springer-Verlag, 2005

[6] Baez, J.: *This Week's Finds in Mathematical Physics.* http://math.ucr.edu/home/baez/week236.html

[7] Bell, J. L.: *Set Theory: Boolean-Valued Models and Independence Proofs in Set Theory.* Oxford: Oxford University Press, 1985 (Oxford Logic Guides)

[8] Bergman, G. M.: *Portraitphoto von Alfred Tarski.* 1968. – GNU-Lizenz für freie Dokumentation, Version 1.2

[9] Bernays, P.: Axiomatische Untersuchung des Aussagen-Kalküls der „Principia Mathematica". In: *Mathematische Zeitschrift* 25 (1926), S. 305–320

[10] Bernays, P.: A system of axiomatic set theory. II. In: *Journal of Symbolic Logic* 6 (1941)

[11] Boole, G.: *The Mathematical Analysis of Logic, Being an Essay Toward a Calculus of Deductive Reasoning.* Cambridge: Macmillan, 1847

[12] Boole, G.: *An Investigation of the Laws of Thought.* London: Walton and Maberley, 1854. – Nachgedruckt in [13]

[13] Boole, G.; Corcoran, J.: *The Laws of Thought (Reprint).* New York: Prometheus Books, 2003

[14] Boolos, G.: A New Proof of the Gödel Incompleteness Theorem. In: *Notices of the American Mathematical Society* (1989), Nr. 36, S. 388–390

[15] Boolos, G. S.; Burgess, J. P.; Jeffrey, R. C.: *Computability and Logic.* Cambridge: Cambridge University Press, 2007

[16] Boolos, G. S.; Jeffrey, R. C.: *Computability and Logic.* Cambridge: Cambridge University Press, 1989

[17] The Busy Beaver Game and the Meaning of Life. In: Brady, A. H.: *The Universal Turing Machine: A Half Century Survey.* Oxford: Oxford University Press, 1991, S. 259–277

[18] Brady, G.: *From Peirce to Skolem: A Neglected Chapter in the History of Logic.* Amsterdam: Elsevier Scienceg, 2000 (Studies in the History and Philosophy of Mathematics)

[19] Burger, E. B.; Tubbs, R.: *Making Transcendence Transparent.* Berlin, Heidelberg, New York: Springer-Verlag, 2004

[20] Cantor, G.: Über eine Eigenschaft des Inbegriffes aller reellen algebraischen Zahlen. In: *Crelles Journal für Mathematik* 77 (1874), S. 258–262

[21] Cantor, G.: Über unendliche, lineare Punktmannigfaltigkeiten. V. In: *Mathematische Annalen* 21 (1883), S. 545–591

[22] Cantor, G.: Beiträge zur Begründung der transfiniten Mengenlehre (Zweiter Artikel). In: *Mathematische Annalen* 49 (1897), S. 207–246

[23] Cantor, G.; Jourdain, P. E. B. (Hrsg.): *Contributions to the Founding of the Theory of Transfinite Numbers.* New York: Dover Publications, 1955 (Dover Books on Mathematics). – 137–201 S.

[24] Cantor, G.; Jourdain, P. E. B. (Hrsg.): *Contributions to the Founding of the Theory of Transfinite Numbers.* New York: Dover Publications, 1955 (Dover Books on Mathematics). – 85–136 S.

[25] Cantor, G.: From Kant to Hilbert: A Source Book in the Foundations of Mathematics. II (2005), S. 840–843

[26] Carnap, R.: *Logical Syntax of Language*. London: Routledge, 2001 (Philosophy of Mind and Language 4)

[27] Carnap, R.; Reichenbach, H.: *Bericht über die 2. Tagung für Erkenntnislehre der exakten Wissenschaften Königsberg 1930*. Leipzig: Felix Meiner Verlag, 1931

[28] Chaitin, G.: *The Limits of Mathematics*. Berlin, Heidelberg, New York: Springer-Verlag, 1998

[29] Chaitin, G.: *The Unknowable*. Berlin, Heidelberg, New York: Springer-Verlag, 1999

[30] Chaitin, G.: *Conversations with a Mathematician*. Berlin, Heidelberg, New York: Springer-Verlag, 2002

[31] Chaitin, G.: *Meta Maths: The Quest for Omega*. London: Atlantic Books, 2007

[32] Chaitin, G. J.: On the Simplicity and Speed of Programs for Computing Infinite Sets of Natural Numbers. In: *Journal of the ACM* 16 (1969), Nr. 3, S. 407–422

[33] Chow, T. Y.: A Beginner's Guide to Forcing. In: *Communicating Mathematics, Contemporary Mathematics* 479 (2009), S. 25–40

[34] Church, A.: A Note on the Entscheidungsproblem. In: *Journal of Symbolic Logic* 1 (1936), Nr. 1, S. 40–41

[35] Church, A.: An Unsolvable Problem of Elementary Number Theory. In: *American Journal of Mathematics* 58 (1936), Nr. 2, S. 345–363

[36] Church, A.; Rosser, J. B.: Some properties of conversion. In: *Transactions of the American Mathematical Society* 39 (1936), S. 472–482

[37] Cohen, P.: The Independence of the Continuum Hypothesis. In: *Proceedings of the National Academy of Sciences of the United States of America* Bd. 50. Washington, DC: National Academy of Sciences, 1963, S. 1143–1148

[38] Cohen, P.: The Independence of the Continuum Hypothesis II. In: *Proceedings of the National Academy of Sciences of the United States of America* Bd. 51. Washington, DC: National Academy of Sciences, 1964, S. 105–110

[39] Cohen, P.: *Set Theory and the Continuum Hypothesis*. New York: Benjamin, 1966

[40] Corporation, Rand: *A million random digits with 100,000 normal deviates*. New York: Free Press, 1955

[41] Dauben, J. W.: Georg Cantor and the Origins of Transfinite Set Theory. In: *Scientific American* 248 (1983), Nr. 6, S. 122–131

[42] Dauben, J. W.: Georg Cantor and the Battle for Transfinite Set Theory. In: *Proceedings of the 9th ACMS Conference*, 1993, S. 1–22

[43] Corrections to Turing's Universal Computing Machine. In: Davies, D. W.: *The Essential Turing*. Oxford University Press, 2004, S. 97–101

[44] Davis, M.: Arithmetical Problems and Recursively Enumerable Predicates. In: *Journal of Symbolic Logic* 18 (1953), S. 33–41

[45] Davis, M.: *Computability and Unsolvability*. New York: McGraw-Hill, 1958

[46] Davis, M.: *The Undecidable*. Mineola, NY: Dover Publications, 1965

[47] Davis, M.: Mathematical logic and the origin of modern computers. (1987), S. 137–167

[48] Davis, M.; Putnam, H.; Robinson, J.: The Decision Problem for Exponential Diophantine Equations. In: *Annals of Mathematics* 74 (1961), Nr. 3, S. 425–436

[49] Dawson, J. W.: The Compactness of First-Order Logic: From Gödel to Lindström. In: *History and Philosophy of Logic* 14 (1993), S. 15–37

[50] Dawson, J. W.: *Kurt Gödel. Leben und Werk*. Berlin, Heidelberg, New York: Springer-Verlag, 1999

[51] Deiser, O.: *Einführung in die Mengenlehre*. Berlin, Heidelberg, New York: Springer-Verlag, 2009

[52] Easwaran, K.: A Cheerful Introduction to Forcing and the Continuum Hypothesis. In: *ArXiv e-prints* (2007), December

[53] Ebbinghaus, H.-D.; Flum, J.; Thomas, W.: *Einführung in die mathematische Logik*. Spektrum Akademischer Verlag, 1996

[54] Feferman, S.: *In the Light of Logic*. Oxford: Oxford University Press, 1999 (Logic and Computation in Philosophy)

[55] Fraenkel, A.: Zu den Grundlagen der Cantor-Zermeloschen Mengenlehre. In: *Mathematische Annalen* 86 (1922), S. 230–237

[56] Franzen, T.: *Gödel's Theorem: An Incomplete Guide to Its Use and Abuse*. Wellesley, MA: Transatlantic Publishers, 2005

[57] Frege, G.: *Begriffsschrift. Eine der arithmetischen nachgebildete Formelsprache des reinen Denkens.* Ditzingen: Verlag Louis Nebert, 1879

[58] Frege, G.: *Die Grundlagen der Arithmetik – Eine logisch mathematische Untersuchung über den Begiff der Zahl.* Breslau: Wilhelm Koebner Verlag, 1884

[59] Frege, G.: Rezension von Georg Cantor: Zur Lehre vom Transfiniten. In: *Zeitschrift für Philosophie und Philosophische Kritik* 100 (1892), S. 269–272

[60] Frege, G.: *Grundgesetze der Arithmetik.* Bd. 1. Hildesheim: Verlag Olms, 1962

[61] Frege, G.: *Grundgesetze der Arithmetik.* Bd. 2. Hildesheim: Verlag Olms, 1962

[62] Frege, G.; Ebert, P. A. (Hrsg.); Rossberg, M. (Hrsg.): *Basic Laws of Arithmetic.* Oxford, UK: Oxford University Press, 1964

[63] Frege, G.: *Die Grundlagen der Arithmetik: Eine logisch-mathematische Untersuchung über den Begriff der Zahl.* Halle: Reclam, 1986

[64] Frege, G.: *Begriffsschrift und andere Aufsätze.* Hildesheim: Verlag Olms, 2007

[65] Frege, G.; P. A. Evert, C. W. M. Rossberg R. M. Rossberg (Hrsg.): *Basic Laws of Arithmetic. Volumes I and II.* Oxford: Oxford University Press, 2013

[66] Frege, G.; Austin, J. L.: *The Foundations of Arithmetic: A Logico-Mathematical Enquiry Into the Concept of Number.* Evanston, Illinois: Northwestern University Press, 1980

[67] Galilei, Galileo: *Two New Sciences.* Madison, WI: University of Wisconsin Press, 1974

[68] Gardner, M.: The random number omega bids fair to hold the mysteries of the universe. In: *Scientific American* 241 (1979), December, S. 22–31

[69] Gauß, C. F.: *Brief an H. C. Schumacher vom 12. Juli. 1831.* – Abgedruckt in [137]

[70] Gentzen, G.: Die Widerspruchsfreiheit der reinen Zahlentheorie. In: *Mathematische Annalen* 112 (1936), S. 493–565

[71] Gödel, K.: *Über die Vollständigkeit des Logikkalküls,* Universität Wien, Diss., 1929

[72] Gödel, K.: Die Vollständigkeit der Axiome des logischen Funktionenkalküls. In: *Monatshefte für Mathematik* 37 (1930), S. 349–360

[73] Gödel, K.: Über formal unentscheidbare Sätze der Principia Mathematica und verwandter Systeme I. In: *Monatshefte für Mathematik und Physik* 38 (1931), S. 173–198

[74] Gödel, K.: *On undecidable propositions of formal mathematical systems.* Princeton, NJ: Institute for Advanced Study, 1934

[75] Gödel, K.: The Consistency of the Axiom of Choice and of the Generalized Continuum-Hypothesis. In: *Proceedings of the U.S. National Academy of Sciences* Bd. 24, 1938, S. 556–557

[76] Gödel, K.: What is Cantor's Continuum Problem? In: *American Mathematical Monthly* 54 (1947), S. 515–525

[77] Gödel, K.: On Formally Undecidable Propositions of Principia Mathematica and Related Systems I. In: Heijenoort, J. van (Hrsg.): *From Frege to Gödel : A Source Book in Mathematical Logic, 1879-1931 (Source Books in the History of the Sciences).* Cambridge, MA: Harvard University Press, 1977, S. 596–616

[78] Gödel, K.: *Collected Works V. Correspondence, H-Z.* New York: Oxford University Press, 2003

[79] Gómez-Torrente, M.; Zalta, Edward N. (Hrsg.): *Alfred Tarski.* Spring 2011. http://plato.stanford.edu/archives/spr2011/entries/tarski/, 2011

[80] Goodstein, R.: On the restricted ordinal theorem. In: *Journal of Symbolic Logic* 9 (1944), S. 33–41

[81] Gostanian, R.: Constructible Models of Subsystems of ZF. In: *Journal of Symbolic Logic* 45 (1980), Nr. 2, S. 237–250

[82] Hardy, G. H.: *Bertrand Russell and Trinity.* Cambridge: Cambridge University Press, 1970

[83] Hasenjäger, G.: Eine Bemerkung zu Henkins Beweis für die Vollstandigkeit des Prädikatenkalküls der Ersten Stufe. In: *Journal of Symbolic Logic* 18 (1953), Nr. 1, S. 42–48

[84] Henkin, L. A.: The Completeness of Formal Systems. (1947)

[85] Henkin, L. A.: The Completeness of the First-Order Functional Calculus. In: *Journal of Symbolic Logic* 14 (1949), S. 159–166

[86] Henkin, L. A.: Completeness in the Theory of Types. In: *Journal of Symbolic Logic* 15 (1950), S. 81–91

[87] Henkin, L. A.: A Problem Concerning Provability. In: *Journal of Symbolic Logic* 17 (1952), S. 160

[88] Henkin, L. A.; Mostowski, A.: Review of Maltsev 1941 and 1956. In: *Journal of Symbolic Logic* 24 (1959), S. 55–57

[89] Herken, R.: *The Universal Turing Machine. A Half-Century Survey: A Half-century Survey*. Berlin, Heidelberg, New York: Springer-Verlag, 1995

[90] Hermes, H.: *Einführung in die mathematische Logik*. Stuttgart: Teubner Verlag, 1963

[91] Hilbert, D.: *Grundlagen der Geometrie*. Leipzig: Teubner Verlag, 1899

[92] Hilbert, D.: Über das Unendliche. In: *Mathematische Annalen* 95 (1926), Nr. 1, S. 161–190

[93] Hilbert, D.: Die Grundlagen der Mathematik. In: *Abhandlungen aus dem mathematischen Seminar* VI (1928), S. 80

[94] On the Infinite. In: Hilbert, D.: *Philosophy of Mathematics: Selected Readings*. Cambridge University Press, 1984

[95] Hilbert, D.: The foundations of mathematics. In: Heijenoort, J. van (Hrsg.): *From Frege to Gödel : A Source Book in Mathematical Logic, 1879-1931 (Source Books in the History of the Sciences)*. Cambridge, MA: Harvard University Press, 2002, S. 464–479

[96] Hilbert, D.; Ackermann, W.: *Grundzüge der theoretischen Logik*. 1. Auflage. Berlin, Heidelberg: Springer-Verlag, 1928

[97] Hilbert, D.; Bernays, P.: *Die Grundlehren der mathematischen Wissenschaften in Einzeldarstellungen*. Bd. 50: *Grundlagen der Mathematik – Band II*. Berlin, Heidelberg: Springer-Verlag, 1939

[98] Hilbert, David: Mathematical problems. In: *Bulletin of the American Mathematical Society* 8 (1902), Nr. 10, S. 437 – 479

[99] Hoffmann, D. W.: *Grundlagen der Technischen Informatik*. München: Hanser-Verlag, 2009. – 2. Auflage

[100] Hoffmann, D. W.: *Theoretische Informatik*. München: Hanser-Verlag, 2009

[101] Hoffmann, D. W.: *Die Gödel'schen Unvollständigkeitssätze. Eine geführte Reise durch Kurt Gödels historischen Beweis*. 2. Auflage. Heidelberg: Springer Spektrum, 2017

[102] Hoffmann, D. W.: *Forcing: Eine Einführung in die Mathematik der Unabhängigkeitsbeweise*. Norderstedt: BoD – Books on Demand, 2018

[103] Hoffmann, D. W.: *Grenzen der Mathematik. Eine Reise durch die Kerngebiete der mathematischen Logik*. 3. Auflage. Heidelberg: Springer, 2018

[104] Hoffmann, D. W.: *Gödel's Incompleteness Theorems. A Guided Tour Through Kurt Gödel's Historic Proof*. Heidelberg: Springer, 2024

[105] Hofstadter, D. R.: *Gödel, Escher, Bach: An Eternal Golden Braid*. New York: Basic Books, 1999

[106] Hofstadter, D. R.: *Gödel, Escher, Bach: Ein endloses geflochtenes Band*. Stuttgart: Klett-Cotta, 2006

[107] Jacobs, K.: *Portraitphoto von Ernst Zermelo*. http://creativecommons.org/licenses/by-sa/2.0. Version: 1953. – Creative Commons License 2.0, Typ: Attribution-ShareAlike

[108] Jech, T.: *Set Theory: The Third Millennium Edition*. Berlin, Heidelberg, New York: Springer-Verlag, 2011

[109] Jones, J. P.; Matijasevič, Y. V.: Register Machine Proof of the Theorem on Exponential Diophantine Representation of Enumerable Sets. In: *Journal of Symbolic Logic* 49 (1984), Nr. 3, S. 818–829

[110] Jones, J. P.; Matijasevič, Y. V.: Proof of Recursive Unsolvability of Hilbert's Tenth Problem. In: *The American Mathematical Monthly* 98 (1991), Nr. 8, S. 689–709

[111] Jones, J. P.; Sato, D.; Wada, H.; Wiens, D.: Diophantine Representation of the Set of Prime Numbers. In: *The American Mathematical Monthly* 83 (1976), S. 449–464

[112] Kalmár, L.: Über die Axiomatisierbarkeit des Aussagenkalküls. In: *Acta Scientiarum Mathematicarum* 7 (1935), S. 222–243

[113] Katz, V. J.: *A History of Mathematics: An Introduction*. Boston: Pearson, 2017 (Pearson Modern Classics for Advanced Mathematics Series)

[114] Kaye, R.: *Models of Peano arithmetic*. Oxford: Oxford University Press, 1991 (Oxford Logic Guides)

[115] Kelley, J. L.: *General Topology*. New York: Van

Nostrand Reinhold, 1955

[116] Kirby, L.; Paris, J.: Accessible independence results for Peano arithmetic. In: *Bulletin of the London Mathematical Society* 14 (1982), S. 725–731

[117] Kleene, S. C.: General recursive functions of natural numbers. In: *Mathematische Annalen* 112 (1936), S. 727–742

[118] Kleene, S. C.: *Introduction to Metamathematics*. Amsterdam: North Holland, 1952 (Bibliotheca Mathematica)

[119] Kleene, S. C.: *Mathematical Logic*. Mineola, NY: Dover Publications, 2002

[120] Knuth, D. E.: Mathematics and Computer Science: Coping with Finiteness. In: *Science Magazine* 194 (1976), Nr. 4271, S. 1235–1242

[121] Kolmogorov, A. N.: Three approaches to the quantitative definition of information. In: *Problems in Information Transmission* 1 (1965), S. 1–7

[122] Kuratowski, K.: Sur la notion d'ordre dans la théorie des ensembles. In: *Fundamenta Mathematicae* 2 (1921), S. 161–171

[123] Kučera, A.; Slaman, T.: Randomness and Recursive Enumerability. In: *SIAM Journal on Computing* 31 (2001), Nr. 1, S. 199–211

[124] Leibniz, G. W.; Latta, R. (Translator): *The Monadology and Other Philosophical Writings*. Oxford University Press, 1965 (Philosophy of Leibniz)

[125] Löb, M. H.: Solution of a problem of Leon Henkin. In: *Journal of Symbolic Logic* 20 (1955), S. 115–118

[126] Löb, M. H.; Wainer, S. S.: Hierarchies of number theoretic functions I, II: A Correction. In: *Archive for Mathematical Logic* 14 (1970), S. 198–199

[127] Löwenheim, L.: Über Möglichkeiten im Relativkalkül. In: *Mathematische Annalen* 76 (1915), S. 447–470

[128] Investigations into the Sentential Calculus. In: Łukasiewicz, L.; Tarski, A.: *Logic, Semantics, Metamathematics: Papers from 1923 to 1938*. Hackett Publishing Company, 1989

[129] Maltsev, A.: Untersuchungen aus dem Gebiete der mathematischen Logik. In: *Matematicheskii Sbornik* 43 (1936), S. 323–336

[130] Maltsev, A.: Ob odnom obscem metode polucenia lokal'nyh teorem grupp. In: *Ivanovskii Gosudarstvennyi Pedagogiceskii Institut, Ucenye zapiski, Fiziko-matematicheskie nauki* 1 (1941), Nr. 1, S. 3–9

[131] Matijasevič, Y. V.: *Portraitphoto von Yuri Matiyasevich*. http://creativecommons.org/licenses/by/3.0. Version: 1969. – Creative Commons License 3.0, Typ: Attribution

[132] Matijasevič, Y. V.: Enumerable Sets are Diophantine. In: *Soviet Mathematics Doklady* 11 (1970), S. 354–358

[133] On Formally Undecidable Propositions of Principia Mathematica and Related Systems I. In: Mendelson, E.: *The Undecidable*. Mineola, NY: Dover Publications, 1965. – Translation of the German original by Kurt Gödel

[134] Mendelson, E.: *Introduction to Mathematical Logic*. 4th edition. Boca Raton, FL: Chapman & Hall, CRC Press, 1997

[135] Menzel, W.; Schmitt, P. H.: *Formale Systeme. Vorlesungsskript Wintersemester 94/95*. Universität Karlsruhe, 1994

[136] Meschkowski, H.: *Problemgeschichte der neueren Mathematik (1800–1950)*. Mannheim: B. I. Wissenschaftsverlag, 1978

[137] Meschkowski, H.; Nilson, W.: *Georg Cantor. Briefe*. Berlin, Heidelberg, New York: Springer-Verlag, 1991

[138] Minsky, M. L.: Recursive Unsolvability of Post's Problem of 'Tag' and Other Topics in the Theory of Turing Machines. In: *Mathematische Annalen* 74 (1961), Nr. 2, S. 437–455

[139] Minsky, M. L.: *Computation: Finite and Infinite Machines*. Englewood Cliffs, N. J.: Prentice Hall, 1967 (Prentice Hall Series in Automatic Computation)

[140] Morse, A. P.: *A Theory of Sets*. New York: Academic Press, 1965

[141] Mostowski, A.: An undecidable arithmetical statement. In: *Fundamenta Mathematicae* 36 (1949), Nr. 1, S. 143–164

[142] Nagel, E.; Newman, J. R.: *Der Gödel'sche Beweis*. München: Oldenbourg Wissenschaftsverlag, 2006

[143] Neumann, J. von: Zur Einführung der transfiniten Zahlen. In: *Acta Litterarum ac Scientiarum Regiae Universitatis Hungaricae* (1923)

[144] Neumann, J. von: Various techniques used in connection with random digits. In: *Applied Mathematics Series* (1951), Nr. 12, S. 36–38

[145] Neumann, J. von; Burks, A. W. (Hrsg.): *Theory of self-reproducing automata*. Urbana: University of Illinois Press, 1966

[146] Brief an Ernst Zermelo vom 15. August 1923. In: Neumann, J. von: *Problemgeschichte der neueren Mathematik (1800–1950)*. Mannheim: B. I. Wissenschaftsverlag, 1978

[147] Neumann, J. von: First Draft of a Report on the EDVAC. In: *IEEE Annals of the History of Computing* 15 (1993), Nr. 4, S. 27–75

[148] Nielsen, M. A.; Chuang, I. L.: *Quantum Computation and Quantum Information*. Cambridge: Cambridge University Press, 2000 (Cambridge Series on Information & the Natural Sciences)

[149] Noether, E. (Hrsg.); Cavailles, J. (Hrsg.): *Actualités scientifiques et industrielles*. Bd. 518: *Briefwechsel Cantor-Dedekind*. Paris: Hermann, 1937

[150] Oberschelp, A.: On Pairs and Tuples. In: *Zeitschrift für Mathematische Logik und Grundlagen der Mathematik* 37 (1991), S. 55–56

[151] Pappas, T.: *The Joy of Mathematics: Discovering Mathematics All Around You*. San Carlos, CA: Wide World Publishing, Tetra, 1989

[152] Peano, G.: *Arithmetices principia, nova methodo exposita*. Torino: Fratelli Bocca, 1889

[153] Peano, G.: The principles of arithmetic, presented by a new method. In: Heijenoort, J. van (Hrsg.): *From Frege to Gödel : A Source Book in Mathematical Logic, 1879-1931 (Source Books in the History of the Sciences)*. Cambridge, MA: Harvard University Press, 1977, S. 83–97

[154] Péter, R.: Über den Zusammenhang der verschiedenen Begriffe der rekursiven Funktionen. In: *Mathematische Annalen* 110 (1934), S. 612–632

[155] Péter, R.: Konstruktion nichtrekursiver Funktionen. In: *Mathematische Annalen* 111 (1935), S. 42–60

[156] Péter, R.: *Rekursive Funktionen*. Berlin: Akademie-Verlag, 1957

[157] Péter, R.: *Deutsche Taschenbücher*. Bd. 37: *Das Spiel mit dem Unendlichen*. Frankfurt: Verlag Harri Deutsch, 1984

[158] Podnieks, K.: *About Model Theory*. http://www.ltn.lv/~podnieks/gt.html

[159] Poizat, B.: *A Course in Model Theory: An Introduction to Contemporary Mathematical Logic*. Berlin, Heidelberg, New York: Springer-Verlag, 2000

[160] Pöppe, C.: Mandelbrot Dreidimensional. In: *Spektrum der Wissenschaft* 4 (2010), April

[161] Post, E. L.: Recursively enumerable sets of positive integers and their decision problems. In: *Bulletin of the American Mathematical Society* 50 (1944), S. 284–316

[162] Post, E. L.: Recursive Unsolvability of a Problem of Thue. In: *Journal of Symbolic Logic* 12 (1947), Nr. 1, S. 1–11

[163] Quine, W. V. O.: New Foundations for Mathematical Logic. In: *American Mathematical Monthly* 44 (1937), S. 70–80

[164] Quine, W. V. O.: *Mathematical Logic, Revised Edition*. Harvard: Harvard University Press, 1981

[165] Rabin, M.; Scott, D.: Finite automata and their decision problems. In: *IBM Journal of Research and Development* 3 (1959), S. 114–125

[166] Rado, T.: On Non-Computable Functions. In: *The Bell System Technical Journal* 41 (1962), Nr. 3, S. 877–884

[167] Ramsey, F. P.: On a Problem in Formal Logic. In: *Proceedings of the London Mathematical Society* 30 (1930), S. 264–286

[168] Rautenberg, W.: *Einführung in die Mathematische Logik*. 3. Auflage. Wiesbaden: Vieweg+Teubner Verlag, 2008

[169] Resag, J.: *Die Grenzen der Berechenbarkeit. Unvollständigkeit und Zufall in der Mathematik*. http://www.joerg-resag.de

[170] Riesel, H.: *Prime Numbers and Computer Methods for Factorization*. Boston, MA: Birkhäuser-Verlag, 1994

[171] Robinson, A.: *On the Metamathematics of Algebra*. Amsterdam: North-Holland, 1951

[172] Rosser, J. B.: A mathematical logic without variables. In: *Annals of Mathematics* 36 (1935), S. 127–150

[173] Rosser, J. B.: Extensions of Some Theorems of Gödel and Church. In: *Journal of Symbolic Logic* 1 (1936),

S. 87–91

[174] Rosser, J. B.: *Many-valued logic.* Amsterdam: North-Holland Publishing, 1952

[175] Rosser, J. B.: *Logic for Mathematicians.* New York: McGraw-Hill, 1953

[176] Rosser, J. B.: *Simplified Independence Proofs: Boolean Valued Models of Set Theory.* New York: Academic Press, 1969

[177] Rucker, R.: *Infinity and the Mind. The Science and Philosophy of the Infinite.* Basel: Birkhäuser-Verlag, 1982

[178] Russell, B.: *Basic Writings of Bertrand Russell.* London: Routledge, 1992

[179] The Philosophy of Logical Atomism. In: Russell, B.: *The Collected Papers of Bertrand Russell.* London: Routledge, 1993

[180] Russell, B.: *Autobiography.* London: Routledge, 1998

[181] Salerno, Joe: *Portraitphoto von Leon Henkin.* http://creativecommons.org/licenses/by-sa/3.0. Version: 2007. – Creative Commons License 3.0, Typ: Attribution

[182] Schiemer, Georg: Fraenkel's Axiom of Restriction: axiom choice, intended models, and categoricity. In: Löwe, Benedikt (Hrsg.); Müller, Thomas (Hrsg.): *Philosophy of Mathematics: Sociological Aspects and Mathematical Practice* Bd. 11. London: College Publications, 2010, S. 307–340

[183] Schöning, U.: *Logik für Informatiker.* Heidelberg: Spektrum Akademischer Verlag, 2000

[184] Schöning, U.: *Theoretische Informatik – kurzgefasst.* Heidelberg: Spektrum Akademischer Verlag, 2001

[185] Scott, D.: A proof of the independence of the continuum hypothesis. In: *Mathematical Systems Theory* 1 (1967), S. 89–111

[186] Sierpinski, W.: Sur une courbe dont tout point est un point de ramification. In: *Comptes Rendus hebdomadaires des séances de l'Académie des Sciences* (1915), S. 302–305

[187] Sigmund, K.; Dawson, J.; Mühlberger, K.: *Kurt Gödel: Das Album.* Wiesbaden: Vieweg+Teubner Verlag, 2006

[188] Skolem, T.: Logisch-kombinatorische Untersuchungen über die Erfüllbarkeit und Beweisbarkeit mathematischer Sätze nebst einem Theorem über dichte Mengen. In: *Det Norske Videnskaps-Akademi* 4 (1920), S. 1–36

[189] Skolem, T.: Einige Bemerkungen zur axiomatischen Begründung der Mengenlehre. In: *Matematikerkongressen i Helsingfors den 4-7 Juli 1922.* Helsinki: Akademiska Bokhandeln, 1923, S. 217–232

[190] Skolem, T.: Über einige Grundlagenfragen der Mathematik. In: *Skrifter utgitt av Det Norske Videnskaps-Akademi i Oslo, I. Matematisk-naturvidenskapelig klasse* 7 (1929), S. 1–49

[191] Skolem, T.: Über die Nichtcharakterisierbarkeit der Zahlenreihe mittels endlich oder abzählbar unendlich vieler Aussagen mit ausschließlich Zahlvariablen. In: *Fundamenta Mathematicae* 23 (1934), S. 150–161

[192] Skolem, T.: Some remarks on axiomatized set theory. In: Heijenoort, J. van (Hrsg.): *From Frege to Gödel: A Source Book in Mathematical Logic, 1879-1931 (Source Books in the History of the Sciences).* Cambridge, MA: Harvard University Press, 1977, S. 290, Äì301

[193] Smith, J. T.: David Hilbert's Radio Address – English Translation. In: *Convergence: Where Mathematics, History, and Teaching Meet* (2014), February

[194] Smith, N. K.: *Immanuel Kant's Critique of Pure Reason.* La Vergne, Tennessee: Lightning Source Incorporated, 2008

[195] Smith, P.: *An Introduction to Gödel's Theorems.* Cambridge: Cambridge University Press, 2007 (Cambridge Introductions to Philosophy)

[196] Smoryński, C.: The incompleteness theorems. In: *Handbook of Mathematical Logic.* Amsterdam: North-Holland Publishing, 1993, S. 821–866

[197] Smullyan, R. M.: *Gödel's Incompleteness Theorems.* Oxford: Oxford University Press, 1992 (Oxford Logic Guides)

[198] Socher, R.: *Theoretische Grundlagen der Informatik.* München: Hanser-Verlag, 2008

[199] Solomonoff, R. J.: A Preliminary Report on a General Theory of Inductive Inference. (1960), Februar, Nr. V-131

[200] Solomonoff, R. J.: A Formal Theory of Inductive

Inference. Part I. In: *Information and Control* 7 (1964), March, Nr. 1, S. 1–22

[201] Solomonoff, R. J.: A Formal Theory of Inductive Inference. Part II. In: *Information and Control* 7 (1964), June, Nr. 2, S. 224–254

[202] St-Louis, Jerome: *The Mandelbulb at 5 iterations.* http://creativecommons.org/licenses/by-sa/3.0. Version: 2009. – Creative Commons License 3.0, Typ: Attribution-Share Alike

[203] Tarski, A.: Sur les ensembles définissables de nombres réels. In: *Fundamenta Mathematicae* 17 (1931), S. 210–239

[204] Tarski, A.: Pojęcie prawdy w językach nauk dedukcyjnych. In: *Nakładem Towarzystwa Naukowego Warszawskiego* (1933)

[205] Tarski, A: Der Wahrheitsbegriff in den formalisierten Sprachen. In: *Studia Philosophica* (1933), Nr. 1, S. 261–405

[206] Tarski, A.: The Concept of Truth in Formalized Languages. In: Corcoran, John (Hrsg.): *Logic, semantics, metamathematics*. Indianapolis, IN: Hackett Publishing Co., 1983, S. 152–278

[207] The Concept of Truth in Formalized Languages. In: Tarski, A: *Logic, Semantics, Metamathematics: Papers from 1923 to 1938*. Indianapolis: Hackett Publishing Company, 1983, S. 152–278

[208] Tarski, A.: On Definable Sets of Real Numbers. In: Corcoran, John (Hrsg.): *Logic, semantics, metamathematics*. Indianapolis, IN: Hackett Publishing Co., 1983, S. 110–142

[209] Tarski, A.; Givant, S. R.: *A Formalization of Set Theory Without Variables*. American Mathematical Society, 1987 (Colloquium Publications 41)

[210] Tennenbaum, S.: Non-archimedian models for arithmetic. In: *Notices of the American Mathematical Society* 6 (1959), S. 270

[211] Toenniessen, F.: *Das Geheimnis der transzendenten Zahlen: Eine etwas andere Einführung in die Mathematik*. Berlin, Heidelberg, New York: Springer-Verlag, 2009

[212] Turing, A. M.: On computable numbers with an application to the Entscheidungsproblem. In: *Proceedings of the London Mathematical Society* 2 (1936), Juli – September, Nr. 42, S. 230–265

[213] Turing, A. M.: On computable numbers with an application to the Entscheidungsproblem. A correction. In: *Proceedings of the London Mathematical Society* 2 (1937), Nr. 43, S. 544–546

[214] Vaught, R.: Model theory before 1945. In: Henkin, L. (Hrsg.): *Proceedings of Symposia in Pure Mathematics* Bd. 25, American Mathematical Society, 1974, S. 153–172

[215] Vopěnka, P.: General theory of ∇-models. In: *Commentationes Mathematicae Universitatis Carolinae* 8 (1967), S. 145–170

[216] Wainer, S. S.: A classification of the ordinal recursive functions. In: *Archive for Mathematical Logic* 13 (1970), S. 136–153

[217] Wang, L.: The date of the Sunzi suanjing and the Chinese remainder theorem. In: *Proceedings of the tenth international conference on history of science*, 1962, S. 489–492

[218] Weber, H.: Leopold Kronecker. In: *Jahresbericht der Deutschen Mathematiker-Vereinigung* 2 (1893), S. 19

[219] Weisstein, E. W.: *Prime Diophantine Equations.* From MathWorld – A Wolfram Web Ressource. http://mathworld.wolfram.com/PrimeDiophantineEquations.html

[220] Whitehead, A. N.; Russell, B.: *Principia Mathematica. Volume I*. London: Merchant Books, 1910

[221] Wiener, N.: A Simplification of the Logic of Relations. In: *Proceedings of the Cambridge Philosophical Society* 17 (1914), S. 387–390

[222] Wiener, N.: A Simplification of the Logic of Relations. In: Heijenoort, J. van (Hrsg.): *From Frege to Gödel: A Source Book in Mathematical Logic, 1879-1931 (Source Books in the History of the Sciences)*. Cambridge, MA: Harvard University Press, 1977, S. 224–227

[223] Wikipedia: *Ordinal number*. Public domain image. http://en.wikipedia.org/wiki/Ordinal_number

[224] Wolf, R. S.: *The Carus Mathematical Mongraphs*. Bd. 30: *A Tour through Mathematical Logic*. Washington, DC: Mathematical Association of America, 2005

[225] Wolfram, S.: *A New Kind of Science*. Champaign, IL:

Wolfram Media, Inc., 2002

[226] Wußing, H.: *6000 Jahre Mathematik: Eine kulturgeschichtliche Zeitreise*. Bd. Band 2: Von Euler bis zur Gegenwart. Berlin, Heidelberg, New York: Springer-Verlag, 2008

[227] Yourgrau, P.: *Gödel, Einstein und die Folgen: Vermächtnis einer ungewöhnlichen Freundschaft*. München: C. H. Beck, 2005

[228] Zermelo, E.: Beweis, dass jede Menge wohlgeordnet werden kann. In: *Mathematische Annalen* 59 (1904), S. 514–516

[229] Zermelo, E.: Untersuchungen über die Grundlagen der Mengenlehre I. In: *Mathematische Annalen* 65 (1908), S. 261–281

[230] Zermelo, E.: Über Grenzzahlen und Mengenbereiche. In: *Fundamenta Mathematicae* 16 (1930), S. 29–47

[231] Zermelo, E.: Über Stufen der Quantifikation und die Logik des Unendlichen. In: *Jahresbericht der Deutschen Mathematiker-Vereinigung* 41 (1932), S. 85–88

[232] Zermelo, E.: Investigations in the foundations of set theory I. In: Heijenoort, J. van (Hrsg.): *From Frege to Gödel : A Source Book in Mathematical Logic, 1879-1931 (Source Books in the History of the Sciences)*. Cambridge, MA: Harvard University Press, 1977, S. 199–215

Image Credits

Page 1   Gottfried Wilhelm Leibniz
commons.wikimedia.org/wiki/File:Gottfried_Wilhelm_Leibniz_c1700.jpg   PD

Page 3   Goldbach's conjecture
commons.wikimedia.org/wiki/File:Goldbach-1000000.png   CC

Page 5   Leonhard Euler (Jakob Emanuel Handmann – Kunstmuseum Basel)
commons.wikimedia.org/wiki/File:Leonhard_Euler.jpg   PD

Page 5   Pierre de Fermat
commons.wikimedia.org/wiki/File:Pierre_de_Fermat.jpg   PD

Page 6   Arithmetica
commons.wikimedia.org/wiki/File:Diophantus-cover.jpg   PD

Page 10   Joseph Liouville
commons.wikimedia.org/wiki/File:Joseph_liouville.jpeg   PD

Page 13   Georg Cantor
commons.wikimedia.org/wiki/File:Georg_Cantor2.jpg   PD

Page 24   Carl Friedrich Gauß
commons.wikimedia.org/wiki/File:Carl_Friedrich_Gauss.jpg   PD

Page 25   Gottlob Frege
commons.wikimedia.org/wiki/File:Young_frege.jpg   PD

Page 29   David Hilbert
commons.wikimedia.org/wiki/File:Hilbert.jpg   PD

Page 32   Giuseppe Peano
commons.wikimedia.org/wiki/File:Giuseppe_Peano.jpg   PD

Page 38   Ernst Zermelo
commons.wikimedia.org/wiki/File:Ernst_Zermelo.jpeg   CC

Page 42   Kurt Gödel
commons.wikimedia.org/wiki/File:Kurt_godel_tomb_2004.jpg   PD

Page 45   John von Neumann
commons.wikimedia.org/wiki/File:JohnvonNeumann-LosAlamos.gif   PD

Page 47   ENIAC
commons.wikimedia.org/wiki/File:Classic_shot_of_the_ENIAC.jpg   PD

Page 51   Alan Turing
commons.wikimedia.org/wiki/File:Turing_statue_Surrey.jpg   PD

Page 55   Emil Leon Post
commons.wikimedia.org/wiki/File:Emil_Leon_Post.jpg   PD

Page 55   Yuri Matiyasevich
commons.wikimedia.org/wiki/File:Yuri-Matiyasevich-1969.png   CC

Page 76   Wilhelm Ackermann
commons.wikimedia.org/wiki/File:Ackermann_Wilhelm.jpg   PD

Page 114   Leon Henkin
commons.wikimedia.org/wiki/File:Leon.Henkin.jpg   CC

Page 121   Peter Gustav Lejeune Dirichlet
commons.wikimedia.org/wiki/File:Peter_Gustav_Lejeune_Dirichlet.jpg   PD

Page 132   Euclid
commons.wikimedia.org/wiki/File:Euklid-von-Alexandria_1.jpg   PD

Page 154   Georg Cantor
commons.wikimedia.org/wiki/File:Georgcantor01.png   PD

Image Credits

Page 200 Rózsa Péter
commons.wikimedia.org/wiki/File:RozsaPeter.jpg (PD)

Page 223 Alfred Tarski
commons.wikimedia.org/wiki/File:AlfredTarski1968.jpeg (GNU)

Page 302 Joseph-Louis Lagrange
commons.wikimedia.org/wiki/File:Joseph-Louis_Lagrange.jpeg (PD)

Page 313 Wacław Sierpiński
de.wikipedia.org/w/index.php?title=Datei:Sierpinski.jpg (PD)

Page 326 Three-dimensional Mandelbrot set
commons.wikimedia.org/wiki/File:Power_8_mandelbulb_fractal_overview.jpg (CC)

Page 328 Andrej Kolmogorov
commons.wikimedia.org/wiki/File:Kolmogorov-m.jpg (PD)

Page 379 George Boole
commons.wikimedia.org/wiki/File:PSM_V17_D740_George_Boole.jpg (PD)

- (PD) Public Domain
- (CC) Creative Commons License
- (GNU) GNU Licence

All clipart images are from **www.openclipart.org**.

Name Index

A

Abbe, Ernst, 26
Ackermann, Wilhelm F., 42, 76, 142, 201
Aristotle, 24

B

Bennett, Charles, 333, 339
Bernays, Paul, 95, 123, 142, 170, 234
Bernstein, Felix, 21, 376
Boole, George, 25, 378, 379
Brouwer, Luitzen E. J., 41, 42
Burali-Forti, Cesare, 35
Burgess, John P., 267

C

Calude, Cristian, 340
Cantor, Georg, **13**, 17, 21, 35, 154, 167, 170, 177, 179, 184
Carnap, Rudolf, 42, 192
Cauchy, Augustin L., 12
Chaitin, Gregory, 325, 328, 344
Church, Alonzo, 33, 51, 53, 216, **274**, 276
Cohen, Paul J., **57**, 58, 377

D

Dauben, Joseph, 17
Davies, Donald W., 268, 269
Davis, Martin, 54, 55, 300, 301
Dedekind, J. W. Richard, 12, 24, 134, 349
Dinneen, Michael, 340
Diophantos of Alexandria, 6
Dirichlet, Peter Gustav, 121
Dummett, Michael, 41

E

Eckert, J. Presper, 47
Einstein, Albert, 44
Euclid, 132, 243, 394
Euler, Leonhard, 4, 10

F

Fermat, Pierre de, 5, 8
Fraenkel, Abraham A. H., 38, 149
Frege, Gottlob, 24, 25, **26**, 30, 33
Furtwängler, Philipp, 44

G

Galilei, Galileo, 24
Gardner, Martin, 333, 339
Gauss, J. Carl Friedrich, 12, 24, 367
Gentzen, Gerhard, 244
Goldbach, Christian, 4
Goodstein, Reuben L., 48, **246**
Gödel, Kurt, 42, **44**, 59, 142, 170, 191, 204, 211, 267, 350, 352

H

Harrington, Leo Anthony, 48
Henkin, Leon Albert, 114, **115**, 274, 357
Hermes, Hans, 80
Hermite, Charles, 10
Heyting, Arend, 41
Hilbert, David, **29**, 43, 76, 201, 234, 300
Hofstadter, Douglas R., 192, 276
Huntington, Edward Vermilye, 378

J

Jones, James P., 55, 272, 300, 301

K

Kant, Immanuel, 349
Kelley, John L., 142
Kirby, Laurie, 48, 246, 251
Kleene, Stephen C., 41, 200, 216, 274
Knuth, Donald E., 251
Kolmogorov, Andrej, 325, 328
Kreisel, Georg, 250
Kripke, Saul Aaron, 141
Kronecker, Leopold, 12, 13
Kučera, Antonín, 340
Kummer, Ernst Eduard, 13
Kuratowski, Kazimierz, 159

L

Lagrange, Joseph-Louis, 131, 302
von Lindemann, C. L. Ferdinand, 10
Liouville, Joseph, 10, 18
Löb, Martin Hugo, 234, 250
Löwenheim, Leopold, 350, 351, 358
Łukasiewicz, Jan, 95, 224

M

Malcev, Anatolij I., 357, 359

Matiyasevich, Yuri W., 55, 272, 300, 301
Mauchly, John W., 47
Mendelson, Elliott, 207
Minsky, Marvin L., 269
Morse, Anthony P., 142
Mostowski, Andrzej, 388

N

Neumann, John von, 42, 44, 142, 165, 325
Nylander, Paul, 326

P

Paris, Jeffrey B., 48, 246, 251
Peano, Giuseppe, 31, 134
Péter, Rózsa, 200
Platek, Richard A., 141
Poizat, Bruno, 359
Post, Emil Leon, 54, 268, 269
Presburger, Mojżesz, 241
Putnam, Hilary W., 55, 300, 301
Péter, Rózsa, 200

Q

Quine, Willard Van Orman, 141

R

Rabin, Michael Oser, 274, 377
Radó, Tibor, 252
Ramsey, Frank Plumpton, 48
Rasiowa, Helena, 383
Rice, Henry G., 284
Riemann, G. F. Bernhard, 12
Robinson, Abraham, 354
Robinson, Julia, 55, 300, 301
Rosser, J. Barkley, 196, **216**, 274
Russell, Bertrand A. W., 1, 33, **34**, 47, 48, 141, 234
Russell, Lord John, 34

S

Schlick, Moritz, 44
Schröder, Ernst, 21
Scott, Dana S., 58, 274, 377
Shu, Chi-Kou, 340
Sierpiński, Wacław, 224, 311
Sikorski, Roman, 383
Skolem, Thoralf, 38, 350, 351, 353, 358, 359, **371**, 375
Slaman, Theodore A., 340
Smith, Alex, 270
Smith, Peter, 207
Smullyan, Raymond M., 253, 274
Solomonoff, Ray, 325, 328

Solovay, Robert M., 58, 377
Sunzi, 205

T

Tarski, Alfred, **224**, 352, 359
Tennenbaum, Stanley, 363
Thue, Axel, 275
Turing, Alan M., 33, 49, 51, **53**, 258, 267, 274

V

Vopěnka, Petr, 58, 377

W

Wainer, Stanley Scott, 250
Weierstraß, Karl T. W., 12, 13
White, Daniel, 326
Whitehead, Alfred N., 34, 36, 141, 234
Wiener, Norbert, 187
Wolfram, Stephen, 263, 270

Z

Zermelo, Ernst F. F., 38, 149, 156, 165, 192
Zorn, Max August, 157

Timeline

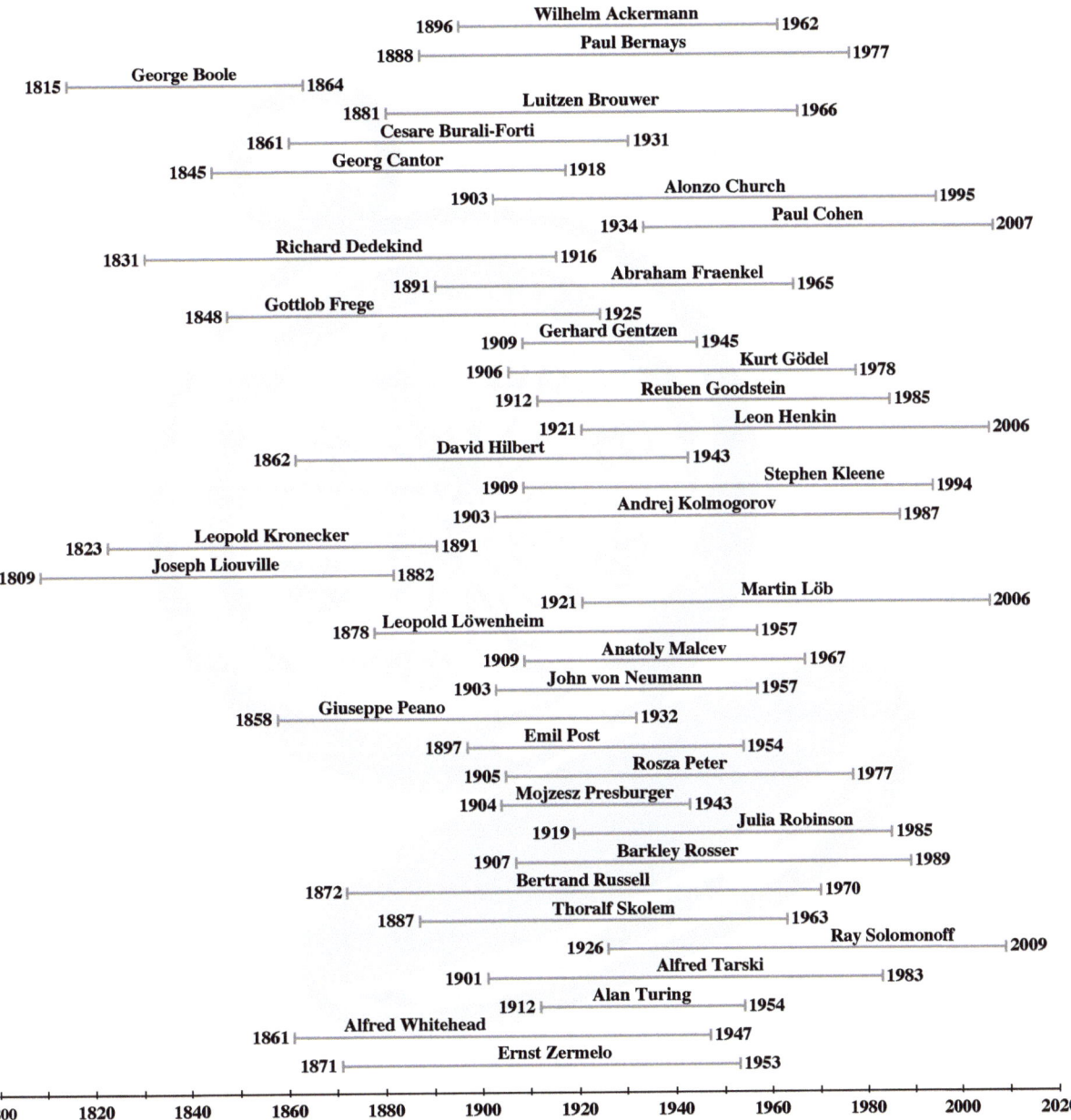

Index

A

absolute consistency proof, 32
absorption law, 85, 397
acceptor, 317
Ackermann function, 76, 201, 251
    diagonalized, 251
actual infinity, 23, 24
addition
    of ordinal numbers, 174
algebra
    Boolean, 378
        complete, 380
        power-set, 379
algebraic
    equation
        height, 16
        number, 10
algorithm, 33, 54, 257
algorithmic
    complexity, 326
    information, 328
    theory, 325
alphabet, 66
    of a Turing-machine, 258
annulment law, 85, 397
antichain, 391
antinomy
    Russell's, 35
antisymmetric
    relation, 153
architecture
    von Neumann, 45, 47
arithmetic
    formula, 128
    mean, 9
    Presburger, 241, 242
    representable
        function, 202
        relation, 202
    Robinson, 242
    term, 128
Arithmetica, 5, 6
Arithmetices Principia, 31, 134
arithmetization
    of syntax, 195
arrow notation, 251
assignment
    of variables, 81
associative
    law, 85, 397
asymmetric
    relation, 153
atomic
    formula, 80
    statement, 80
automaton
    cellular, 263, 318
    linear, 263
axiom, 65
    of choice, 39, 40, 151, **151**
    of constructibility, 56
    of definiteness, *see* axiom of extensionality
    of elementary sets, 39
    of empty set, 144
    of extensionality, 39, 144
    of foundation, 38, 39, 150
    Huntington, 378
    of induction, 133
    of infinity, 39, 59, **147**
    non-logical, 127
    of pairing, 39, 145
    Peano, 134
    of power set, 39, 148
    proper, 127
    of regularity, *see* axiom of foundation
    of replacement, 38, 39, 149
    of restriction, 150
    schema, 73
    of separation, 39, 147
    theory, 127
    of union, 39, 146
    ZF, 143
axiomatic
    set theory, 38, **141**
axiomatizability, 350
    finite, 142
axiomatized theory, 127

B

b-adic representation, 246
    expanded, 246
barber paradox, 35, 194
base, **246**, 305
    bumping, 247
beaver, 252
    function, 252
Begriffsschrift, 25, 61
β function, **204**
binary
    coding, 261
binomial
    coefficient, 311
    theorem, 312
blank-tape halting problem, 282
bombe (decryption device), 53
Boolean
    algebra, 378
        complete, 380
        two-element, 380
    function, 82
    semantics, 384

set, 383
  -valued model, 58, **376**, 377, 386
bottom-up
  decision procedure, 77
bounded
  quantifier theorem, 301
  Turing machine, 261
  variable, 96
Burali-Forti paradox, 35, 171
busy beaver, 252
  function, 252

C

calculus, *see* formal system
  lambda, 53, 275
Calculus ratiocinator, **2**
Cantor
  normal form, 177
  pairing function, **15**, 16, 372
  -Schröder-Bernstein theorem, 15, **21**
Cantor's
  theorem, 20
capturable
  function, 207
  relation, 207
cardinal number, 21, **184**
cardinality, 13, **185**
cartesian product, 188
categorical model, 349
CCC, *see* countable chain condition
cellular automaton, 263, 318
  cell, 264
chain, 391
Chaitin's
  constant, 171, 333, **337**
  incompleteness theorem, 325, 342
characteristic function, 277
  partial, 277
Characteristica universalis, 2
Chinese remainder theorem, 205
choice
  axiom, 39, 40, 151, **151**
Church's thesis, **273**, 276
Church-Rosser theorem, 216
class, 141

sign, 209
cleanst formula, 99
closed formula, 98
coding
  binary, 261
  unary, 261
collision-free substitution, 103
commutative
  law, 85, 378
compactness theorem, 113, 350, **355**
complement law, 85
complete
  Boolean algebra, 380
  formal system, 40, 70
  operator set, 121
completeness
  semantic, 70
  syntactic, 70
  theorem, 106, 350, 354
complexity
  algorithmic, 326
  descriptive, 328
  Kolmogorov, 328
  theory, 257
comprehension
  schema
    general, 34, 115
computability, 5, 48
  limits of, 280
  theory, 33, **257**
computing machine, 51, 53
conditional quantifier
  existential, 131, 143
  universal, 131, 143
configuration
  of a Turing machine, 260
conjecture
  Erdős-Straus, 63
  Goldbach's, **4**, 62, 186
  twin-prime, 4, 186, 346
conjunction, 80
consequence
  logical, 84
consistency, 41, 70
  proof
    absolute, 32
    relative, 32
constant, 98

Chaitin's, 171, 333, **337**
constructibility
  axiom, 56
constructible set, 56
continuity, 96
continuum, 6, 9
  hypothesis, 23, 389, 398
    generalized, 23
contraposition, 86
control-flow matrix, 305
coordinate
  polar, 367
correct formal system, 70
correspondence problem
  Post's, 55
countable chain condition, 391
CSB theorem, *see* Cantor-Schröder-Bernstein theorem

D

data-flow matrix, 305
De Morgan's law, 85, 397
decidability, 41, 276
decision
  problem
    semantic, 79
    syntactic, 77
  procedure, 54, **76**, 77, 292
    bottom-up, 77
    top-down, 78
deduction theorem
  in predicate logic, 104
  in propositional logic, 87
dense set, 382
derivability conditions, 234
description number
  of a Turing machine, 267
descriptive complexity, 328
diagonalization, 13, 18, **209**, 251
  lemma, 218
diagonalized Ackermann function, 251
diophantine
  representable
    relation, 303
Diophantine
  equation, 7
    exponential, 8

Index

representability, 303
Dirichlet's drawer principle, 121
disjunction, 80
distributive law, 85, 378
DNA computing, 276
domain, 96, 99
double negation law, 85, 397
drawer principle, 121

E

elementary sets
    axiom, 39
elliptical geometry, 243
empty set
    axiom, 144
engerer Funktionenkalkül, 42
enumerability, 278
enumerable
    recursively, 338
enumerator, 372
epsilon-delta criterion, 96
equation
    Diophantine, 7
        exponential, 8, **300**
equipotence, 13
equivalence
    operator, 80
    in propositional logic, 84
Erdős-Straus conjecture, 63
Euclid's
    parallel postulate, 243
    theorem, 394
Euclidean geometry, 30, 243, 349
Euler's number, 10
exclusive or, 80
existential
    quantifier, 27
        conditional, 131, 143
b-adic representation, 246
exponential diophantine equation, **300**
expressible
    function, 207
    relation, 207
extended completeness theorem, 354
extensional equality, *see* extensionality
extensionality, 144

axiom, 39, 144
principle of, 202

F

filter, 369, 382, 389
    generic, 382, 389
    $\mathcal{M}$-generic, 383
    maximal, 369
    ultra, 368, 369, 382
        free, 369
finite
    axiomatizability, 142
    means, 41
finiteness
    in second-order logic, 112
first incompleteness theorem, 43, **192**, 344
first-order theory, 114
fixpoint
    of an ordinal number, 176
    theorem, 218
forcing, 57, **387**
    relation, 391
formal system, 65
    complete, 40, 70
    consistent, 41, 70
    correct, 70
    decidable, 41, 276
    model, 353
    negation-complete, 70
    Peano
        arithmetic, 133
    satisfiable, 353
formula
    arithmetic, 128
    atomic, 80
    cleanst, 99
    closed, 98
    Henkin, 237
    instance, 66
    open, 98
    of PA, 128
    predicate logic, 98
    propositional, 80
    satisfiable
        predicate logic, 83

        propositional logic, 102
    schema, 66
    universally valid
        predicate logic, 102, 243
        propositional logic, 83
    unsatisfiable
        predicate logic, 102
        propositional logic, 83
foundation
    axiom, 38, 39, 150
foundational crisis, 33
four-square theorem, 131, 302
free
    ultrafilter, 369
    variable, 96
function
    Ackermann, 76, 201, 251
        diagonalized, 251
    arithmetic representable, 202
    Boolean, 82
    busy beaver, 252
    continuous, 96
    Goodstein, 249
    injective, 112, 188
    μ-recursive, 200, 274
    partial, 159
    primitive-recursive, **200**, 274
    representable
        semantically, 203
        syntacially, 208
    surjective, 112, 188
    symbol, 97
    total, 159, 188, 250
    truth, 385
    Turing-computable, 260

G

Gaussian number plane, 367
general comprehension schema, 34, 115
generic filter, 382, 389
geometry
    elliptical, 243
    euclidean, 30, 243, 349
    hyperbolic, 243
Gödel
    -Rosser theorem, 218

number, 197
    of a Turing machine, 267
Gödel's
    completeness theorem, 106, 350, 354
        extended, 354
    incompleteness theorem
        first, 43, **192**, 344
        second, 45, 231
    incompleteness theorems, **191**
Gödelization, 197
Goldbach's conjecture, **4**, 62, 186
Goodstein
    function, 249
    sequence, 245
Goodstein's theorem, 48, **245**
Goto
    language, 271
    program, 271
grammar
    phrase structure, 275

H

halting
    probability, 325, **335**, 338
    problem, 54, 280, **280**
        blank-tape, 282
    sequence, 333
head cell, 265
height
    of an algebraic equation, 16
Henkin
    formula, 237
    interpretation, 114
    semantics, 114
heptation, 246
hexation, 246
higher-order predicate logic, 110
Hilbert's
    program, 40
    tenths problem, 6, 32, 54, **300**
Hilbert-Bernays-Löb
    provability criteria, 233
Huntington axiom, 378
hyperbolic geometry, 243
hypothesis

continuum, 23, 389, 398
    generalized, 23
Riemann, 57, 336

I

idempotence law, 85, 397
identity law, 85
imitation game, *see* Turing test
implication operator, 80
incompleteness
    theorem, **191**
        Chaitin's, 325, 342
        common misconceptions, 240
        first, 43, **192**, 344
        second, 45, 231
indeterministic
    Turing machine, 259
individual set, *see* domain
induction
    axiom, 133
    transfinite, 181
inference rule, 65
infimum, 380
infinity, 12
    actual, 23, 24
    axiom, 39, 59, **147**
    potential, 23
information
    algorithmic, 328
    content, 325, 328
    theory
        algorithmic, 325
initial state
    of a Turing machine, 258
injective function, 112, 188
instance
    of a formula, 66
instantiation
    universal, 34
instruction set
    of a register machine, 271
    of a Turing-machine, 258
interpretation
    Henkin, 114
    predicate logic, 99
    propositional logic, 81

    standard, 69, 130
intuitionism, 41
inverse element, 378
irreflexive relation, 153
isomorphism
    order, 179

K

Kleene-Rosser paradox, 216
Kolmogorov complexity, 328
Kripke-Platek set theory, 141

L

Lagrange's
    four-square theorem, 131, 302
lambda calculus, 53, 275
language, 66
    Goto, 271
    While, 274
law
    absorption, 85, 397
    annulment, 85, 397
    associative, 85, 397
    commutative, 85, 378
    complement, 85
    De Morgan's, 85, 397
    distributive, 85, 378
    double negation, 85, 397
    of excluded middle, 41
    idempotence, 85, 397
    identity, 85
left-comparative relation, 124
lemma
    Rasiowa–Sikorski, 383
    Zorn's, 157
limit ordinal, **173**
linear
    automaton, 263
    order, *see* total order
Liouville number, 10
Löb's theorem, 237
logic
    calculus
        predicate logic, 103
        propositional logic, 86

predicate, 96
    with equality, 106
    higher-order, 110
    term, 98
    third-order, 111
propositional, 80
symbolic, 25
logic calculus, *see* formal system
logical consequence, 84
logicism, 26, 27
Löwenheim theorem, 358
Löwenheim-Skolem theorem, 351, 358, 371
Löwenheim-Skolem-Tarski theorem, 113, 351
LST theorem, *see* Löwenheim-Skolem-Tarski theorem

M

$\mathcal{M}$-generic filter, 383
manifold, 13
masking relation, 304
matrix
    control-flow, 305
    data-flow, 305
maximal filter, 369
mean
    arithmetic, 9
meta
    level, 66
    theory
        Tarski's, 224
metamathematics, 2
minimality principle, 181, 182
modal operator, 234
model, 57, 68, 82
    analysis, 349
    Boolean-valued, 58, **376**, 377, 386
    categorical, 349
    of computation, 257, **258**
    construction, 349
    existence theorem, 350, 353
    of a formal system, 353
    level, 349
    nonstandard
        of PA, 361
        of ZFC, 388
    quotient, 388
    relation, 68, 129
        predicate logic, 100
        propositional logic, 82
    standard
        of PA, 361
    theorem, 388
    theory, **349**
modus
    barbara, 90
    ponens, 74, 86
Morse-Kelley set theory, 142
Mostowski collapse, 388
μ operator, 274
μ-recursive function, 200, 274
multi-tape Turing machine, 262
multi-track Turing machine, 262
multiplication
    of ordinal numbers, 174

N

natural
    independence phenomenon, 246
    numbers, 133
        embedding in ZF, 165
NBG set theory, 142
negation, 80
negation-complete formal system, 70
neutral element, 378
New foundations, 141
non-logical axiom, 127
nonstandard model
    of PA, 361
    of ZFC, 388
normal form
    Cantor, 177
    Skolem, 351, 371
null set, 39, 144
number
    algebraic, 10
    cardinal, 21, **184**
    Euler's, 10
    Liouville's, 10
    natural, 133
    ordinal, 167, **169**, 248
    plane
        Gaussian, 367
    rational, 8
    real, 9
    series
        Zermelo's, 165
    transcendental, 10

O

object level, 66
octation, 246
open formula, 98
operator
    conjunction, 80
    disjunction, 80
    equivalence, 80
    exclusive or, 80
    implication, 80
    μ, 274
    negation, 80
    set
        complete, 121
order, 152
    partial, 153
    total, 153
    type, 178
    well-, **153**, 178
order isomorphism, 179
ordered pair, 159
ordinal
    limit, **173**
    transfinite, 173
ordinal number, 167, **169**, 248
    addition, 174
    fixpoint, 176
    multiplication, 174
    successor, 172

P

pair
    ordered, 159
pairing
    axiom, 39, 145
    function
        Cantor's, **15**, 16, 372
paradox

barber, 35, 194
Burali-Forti, 35, 171
Kleene-Rosser, 216
Skolem's, 370, 371
parallel postulate, 243
partial
    characteristic function, 277
    function, 159
    order, 153
partially-bounded Turing machine, 261
Pascal's triangle, 311
Peano
    arithmetic, 128
        formal system, 133
        formula, 128
        semantics, 129
        syntax, 128
    axiom, 134
        trichotomy, 365
pentation, 246
phrase structure grammar, 275
pigeonhole principle, 121
polar coordinate, 367
Polish notation, 95
Post correspondence problem, 55
potential infinity, 23
power-set
    algebra, 379
    axiom, 39, 148
predicate
    logic, 96
        calculus, 103
        with equality, 106
        formula, 98
        higher-order, 110
        semantics, 97
        syntax, 97
        term, 98
        third-order, 111
        truth, 223
        variable, 110
prefix-free, 334
Presburger arithmetic, 241, 242
primitive
    recursion, 200
    recursive
        function, **200**, 274
        relation, 202

Principia Mathematica, 36
principle
    of extensionality, 202
    of minimality, 181, 182
probability
    halting, 325, **335**, 338
product
    Cartesian, 188
production
    of a rewriting system, 275
program, 325
    Goto, 271
    While, 274
proof, 67
    consistency
        absolute, 32
        relative, 32
    by reduction, 282
proper axiom, 127
propositional
    formula, 80
    logic, 80
        calculus, 86
        semantics, 80
        syntax, 80
provability, 1
    criteria
        Hilbert-Bernays-Löb, 233
    relation, 68
Pythagorean
    theorem, 8
    triple, 8

Q

quantifier, 96
    existential, 27
        conditional, 131, 143
    universal, 27
        conditional, 131, 143
quantum computing, 276
quotient model, 388

R

ramified type theory, 141
Ramsey's theorem, 48

Rasiowa–Sikorski lemma, 383
rational number, 8
real number, 9
recursion
    primitive, 200
recursively enumarable, 338
reduction proof, 282
reflexive relation, 124, 153
register
    equation, 308
    machine, 55, 271, **271**, 286, 301
        encoding, 305
        instruction set, 271
relation, 159
    antisymmetric, 153
    arithmetic representable, 202
    asymmetric, 153
    forcing, 391
    irreflexive, 153
    left-comparative, 124
    masking, 304
    primitive-recursive, 202
    reflexive, 124, 153
    representable
        diophantine, 303
        semantically, 203
        syntactically, 207
    right-unique, 159
    symmetric, 124
    transitive, 153
relative consistency proof, 32
replacement axiom, 38, 39, 149
representable
    function
        semantically, 203
        syntactically, 208
    relation
        diophantine, 303
        semantically, 203
        syntactically, 207
restriction axiom, 150
rewriting system, 275
    production, 275
Rice's theorem, 282, 320
Riemann hypothesis, 57, 336
right-unique relation, 159
Robinson arithmetic, 242
Rossers trick, 214

rule
    of inference, 65
Russell's antinomy, 35

S

satisfiable
    formal system, 353
    formula
        predicate logic, 102
        propositional logic, 83
saturated subtraction, 271
second incompleteness theorem, 45, 231
second-order
    logic
        standard semantics, 111
semantic
    completeness, 70
    decision problem, 79
semantically
    representable
        function, 203
        relation, 203
semantics, 68
    Boolean, 384
    Henkin, 114
    predicate logic, 97
    propositional logic, 80
semi-Thue system, 275
separation axiom, 39, 147
sequence
    Goodstein, 245
set
    constructible, 56
    theory
        axiomatic, 38, **141**
        Kripke-Platek, 141
        Morse-Kelley, 142
        NBG, 142
        New Foundations, 141
        Zermelo, 38
        Zermelo-Fraenkel, 38, 141, 142
        transitive, 167
Skolem
    -Gödel theorem, 353, 354
    normal form, 351, 371
    paradox, 370, 371

sound formal system, *see* correct formal system
squaring the circle, 11
standard
    description
        of a Turing machine, 266
    interpretation, 69, 130
    model
        of PA, 361
    representation
        of a set, 384
    semantics
        of second-order logic, 111
starting state
    of a Turing machine, 258
statement
    atomic, 80
subformula, 80
submodel, 358
substitution
    collision-free, 103
subtraction
    saturated, 271
successor
    operation, 23
    ordinal, 172
Sunzi Suanjing, 205
supremum, 380
surjective function, 112, 188
symbolic logic, 25
symmetric relation, 124
syntactic
    completeness, 70
    decision problem, 77
syntactically
    representable
        function, 208
        relation, 207
syntax, 65
    arithmetization, 195
    predicate logic, 97
    propositional logic, 80
    ZF, 143

T

tape alphabet, 258

tautology, 83
tenths problem
    Hilbert's, 6, 32, 54, **300**
term
    arithmetic, 128
    predicate logic, 98
    rewriting system, 275
        production, 275
tertium non datur, *see* law of excluded middle
tetration, 246
theorem, 66, 67
    binomial, 312
    bounded
        quantifier, 301
    Cantor's, 20
    Cantor-Schröder-Bernstein, 15, **21**
    Church-Rosser, 216
    compactness, 113, 350, **355**
    Euclid's, 394
    four-square, 131, 302
    Gödel-Rosser, 218
    Goodstein's, 48, **245**
    Löwenheim, 358
    Löwenheim-Skolem, 351, 358, 371
    Löwenheim-Skolem-Tarski, 113, 351
    Löb's, 237
    model existence, 350, 353
    Pythagorean, 8
    Ramsey's, 48
    Rice's, 282, 320
    Skolem-Gödel, 353, 354
theory, 127
    axiom, 127
    axiomatized, 127
    first-order, 114
thesis
    Church's, **273**, 276
third-order
    predicate logic, 111
top-down
    decision procedure, 78
total
    function, 159, 188, 250
    order, 153
transcendental number, 10

transducer, 272, 317
transfinite
    induction, 181
    ordinal, 173
transitive
    relation, 153
    set, 167
triangle
    Pascal's, 311
trichotomy
    of natural numbers, 365
    of ordinal numbers, 170
triple
    Pythagorean, 8
truth, 1
    function, 385
    predicate, 223
    table, 82
Turing
    bombe, 53
    -computable function, 260
    machine, 49, **258**
        alphabet, 258
        configuration, 260
        extensions, 261
        Gödel number, 267
        indeterministic, 259
        initial state, 258
        instruction set, 258
        multi-tape, 262
        multi-track, 262
        partially bounded, 261
        standard description, 266
        starting state, 258
        universal, 266
    test, 53
twin prime conjecture, 4, 186, 346
two-element Boolean algebra, 380
type, 141
type theory
    ramified, 141

U

ultrafilter, 368, 369, 382
    free, 369
unary coding, 261
union axiom, 39, 146
universal
    instantiation, 34
    quantifier, 27
        conditional, 131, 143
    Turing machine, 266
universally valid formula
    predicate logic, 102, 243
    propositional logic, 83
unsatisfiable formula
    predicate logic, 102
    propositional logic, 83
Urelement, 142

V

variable
    assignment, 81
    bounded, 96
    free, 96
    predicate, 110
    propositional logic, 97
    unbound, 96
Venn diagram, 379
Vienna Circle, 44
von Neumann
    architecture, 45, 47

W

weakening rule, 86
well-ordering, **153**, 178
While
    language, 274
    program, 274

Z

Zermelo
    number series, 165
    set theory, 38
Zermelo-Fraenkel
    set theory, 38, 141, 142
        axiom, 143
        syntax, 143
        with AC, 40, 141, 151
Zorn's lemma, 157

MIX
Papier aus verantwortungsvollen Quellen
Paper from responsible sources
FSC® C105338

If you have any concerns about our products,
you can contact us on
ProductSafety@springernature.com

In case Publisher is established outside the EU,
the EU authorized representative is:
Springer Nature Customer Service Center GmbH
Europaplatz 3, 69115 Heidelberg, Germany

Printed by Libri Plureos GmbH
in Hamburg, Germany